City of the Future

Sushobhan Majumdar
Moharana Choudhury
Editors

City of the Future

Navigating Smart Technologies and Circular Economies

palgrave
macmillan

Editors
Sushobhan Majumdar
Department of Geography
Pandit Raghunath Murmu Smriti Mahavidyalaya
Bankura, West Bengal, India

Moharana Choudhury
Environmental Research and Management Division
Voice of Environment (VoE)
Guwahati, Assam, India

ISBN 978-3-032-18025-4 ISBN 978-3-032-18026-1 (eBook)
https://doi.org/10.1007/978-3-032-18026-1

This Palgrave Macmillan imprint is published by the registered company Springer Nature Switzerland AG.
The registered company address is: Gewerbestrasse 11, 6330 Cham, Switzerland

The editors of this book dedicate the book to the Great Almighty and their family members

Foreword

The cities of tomorrow are no longer distant dreams; they are rapidly taking shape around us. As we stand at the intersection of accelerating technological innovation and escalating environmental urgency, urban spaces are being reimagined from the ground up. *City of the Future: Navigating Smart Technologies and Circular Economies* offers a compelling exploration of how this transformation is unfolding and, more importantly, how we can steer it toward a sustainable and inclusive future.

For centuries, cities have been the engines of human progress, centres of culture, commerce, and creativity. Yet, the same concentration of activity that drives growth also magnifies our most significant challenges: congestion, waste, inequality, and the strain on natural resources. The twenty-first century demands a new urban paradigm, one that integrates intelligent systems, data-driven governance, and circular design principles to create cities that not only function efficiently but also regenerate and thrive.

This book brings together the vision of innovators, planners, and policymakers who recognise that the "smart city" is not merely a technological ambition but a social contract. It explores how artificial intelligence, the Internet of Things, and renewable infrastructure can intersect with circular economic models where waste becomes a resource and growth is decoupled from consumption. Through case studies, critical analyses, and forward-looking insights, it presents a roadmap for building cities that are adaptive, equitable, and resilient.

As readers journey through these pages, they are invited to rethink what a city can be: not a static network of roads and buildings, but a living, learning ecosystem designed for both people and planet. The future of our cities will not be defined solely by the technologies we adopt, but by the values we embed within them.

It is hoped that this book will inspire leaders, scholars, and citizens alike to engage in the vital work of shaping urban environments that embody intelligence, sustainability, and shared prosperity. The cities of the future are ours to create, and the time to begin is now.

School of Chemical and Environmental Engineering Technical University of Crete, Greece

Alexandros Stefanakis

Preface

The twenty-first century has ushered in an era of unprecedented urban transformation. Cities, once shaped primarily by geography, industry, and human migration, are now being redefined by data, digital infrastructures, and a growing global consciousness of environmental limits. *City of the Future: Navigating Smart Technologies and Circular Economies* emerges from this crossroads—where rapid technological innovation meets the urgent need for sustainable, inclusive urban development.

This book has been written to bridge a widening gap between two accelerating movements. On one side are smart technologies: sensors embedded in streets and buildings, artificial intelligence powering mobility and energy systems, and digital platforms reshaping public services and civic engagement. On the other hand, the circular economy is an approach that challenges the traditional "take–make–waste" model and seeks to design out waste, regenerate natural systems, and keep materials cycling at their highest value for as long as possible.

While each movement is powerful on its own, its true potential lies in its convergence. Innovative technologies can enable cities to monitor consumption, predict demand, and optimise resource flows. At the same time, circular principles provide the ecological and economic foundation to ensure these innovations lead to long-term prosperity rather than short-lived efficiency gains. Together, they offer a framework for cities that are not only technologically advanced but also resilient, regenerative, and equitable.

This book invites policymakers, urban planners, technologists, business leaders, students, and curious citizens to explore how this convergence can be practically achieved. Drawing on global projects, emerging research, and real-world challenges, *City of the Future* highlights both the opportunities and the complexities of designing cities that are smarter—not just in how they collect data, but in how they use it to build sustainable futures for all.

Bankura, India — Sushobhan Majumdar
Guwahati, India — Moharana Choudhury

Acknowledgement

Writing a book is more complex than we thought and more rewarding than we imagined. The editors thank Springer International Publishing for their spontaneous support and cooperation for this edited book volume. Support from Editorial Director **Rachael Ballard** and Production Editor **TK Santhosh** of Palgrave/Springer is highly appreciated from the start of the final book submission process. This endeavour would not have been possible without the active participation of esteemed contributing authors and their quality writings, and it is a matter of pride for all the editors of this book. The contributed chapters highlighted the broader dissemination of the methods and outcomes of smart technologies: sensors embedded in streets and buildings, artificial intelligence powering mobility and energy systems, and digital platforms reshaping public services and civic engagement. On the other hand, the circular economy is an approach that challenges the traditional "take–make–waste" model and seeks to design out waste, regenerate natural systems, and keep materials cycling at their highest value for as long as possible.

The editors express their special thanks to the parenting organizations for providing support and encouragement as and when needed. Indirect yet quality vital support has been obtained from the "Voice of Environment" (A Scientific and Environmental Research Organization) to complete the task.

We, the editors, extend our thanks to all the contributing authors from around the world in this book, our colleagues, research supervisors, collaborators, and scholars for their direct or indirect support, and to the team at Springer for their enthusiasm and commitment to publishing this book in various ways.

Finally, we would like to acknowledge the vital contributions of all the esteemed reviewers and researchers worldwide who contributed to this volume. This includes those whose written words may not appear in this book but whose ideas, experiences, and inputs into discussion parts are fully reflected in the printed work.

Finally, the editorial team would like to express gratitude to their parents, family members, and friends for their continuous support, motivation, and valuable suggestions throughout the book editing period.

This book is dedicated to all striving to create a more sustainable future through innovative technologies, the **City of the Future, and the circular economy**. The book reflects our vision and collaborates with many brilliant minds and kind hearts who believe in this volume.

Contents

About the Editors

Sushobhan Majumdar is an assistant professor in the Department of Geography at Pandit Raghunath Murmu Smriti Mahavidyalaya, affiliated with Bankura University, India. He is an applied geographer who holds a doctorate in Geography from the Department of Geography, Jadavpur University, Kolkata, and a postgraduate degree from Presidency University, Kolkata. He has completed a postgraduate diploma in Remote Sensing and GIS from Jadavpur University. He has over 13 years of experience in both academic research and professional practice. His main areas of specialization include urban geography, social geography, regional development and planning, remote sensing and GIS, spatial intelligence and the geography of transport. He is currently serving as a reviewing editor and book reviewer in Springer. He is now editor and reviewer of various renowned publishing houses like Springer, Elsevier, Royal Society of Chemistry, CRC Press, AAP Press, Wiley Scrivener, Cambridge Scholars Publishing, and Frontiers. He is presently the Co-Principal Investigator of a major research project supported by the Indian Council of Social Science Research (ICSSR). Dr. Majumdar has edited more than 18 books with internationally reputed publishing houses. He has more than 50 research papers published in various reputable journals and more than 95 book chapters in the edited volumes. He has presented papers at over 35 national and international seminars and conferences across India and abroad. Additionally, he has delivered invited lectures and special talks in a range of national and regional academic programs. Dr. Majumdar is an active member of several academic organizations like Geographical Society of India, Regional Science Association of India (RSAI), Institute of Indian Geographers (IIG), and National Association of Geographers of India (NAGI).

Moharana Choudhury is an Indian researcher and environmentalist. He is currently a research coordinator in the Environmental Research and Management Division of "Voice of Environment". Choudhury has over 14 years of experience in environmental research and services, spanning both research and professional work. His research interests encompass environmental research, waste management, wastewater treatment, environmental risk assessment, pollution control, plastic pollution, and climate change mitigation. He has authored or edited more than 20 books with internationally renowned publishing houses and published more than 50 research articles, review articles, and more than 70 book chapters in internationally recognized journals and publishers. He has been working as an editor and reviewer for various prestigious international publishing houses worldwide, such as Elsevier, Royal Society of Chemistry (RSC), Springer, Frontiers, Wiley, CRC Press/Routledge group, Apple Academic Press, Cambridge Scholar Publishing, etc. He is also well known in India for his efforts to implement environmental education and safeguard the privileges of environmental degree holders in India. He holds an MSc, PGDGIS, PGDEM, and PGCGARD from Gurukul Kangri Deemed to be University, Haridwar, Uttrakhand; North Eastern Hill University, Shillong, Meghalaya; Tezpur University, Assam; and National Institute of Rural Development Hyderabad, India, respectively.

Abbreviations

3D	Three Dimensions
5G	Fifth-Generation Mobile Network
AI	Artificial Intelligence
AIA	Artificial Intelligence Applications
AMP	Autonomous Mobile Platforms/AMP Robotics (company)
API	Application Programming Interface
BDA	Big Data Analytics
BIM	Building Information Modeling
CAGR	Compound Annual Growth Rate
CCTV	Closed-Circuit Television
CE	Circular Economy
CPS	Cyber-Physical Systems
DC	Data Center
DOI	Digital Object Identifier
DPP	Digital Product Passport
DT	Digital Twin
EoL	End-of-Life
ESP32-S3	Espressif Systems Microcontroller Series
EU	European Union
GIS	Geographic Information System
GPS	Global Positioning System
ICT	Information and Communication Technology
IIoT	Industrial Internet of Things
IoT	Internet of Things
ISO	International Organization for Standardization
LED	Light-Emitting Diode
LTE	Long-Term Evolution (cellular network standard)

MaaS	Mobility-as-a-Service
MDPI	Multidisciplinary Digital Publishing Institute
ML	Machine Learning
MOOCs	Massive Open Online Courses
OEM	Original Equipment Manufacturer
OS	Operating System
QoS	Quality of Service
PaaS	Product-as-a-Service
PPP	Public–Private Partnership
QoS	Quality of Service
R&D	Research and Development
RFID	Radio-Frequency Identification
ROI	Return-on-Investment
SHM	Structural Health Monitoring
SME	Small and Medium-sized Enterprise
STI	Science, Technology, and Innovation
SVM	Support Vector Machine
ToF	Time of Flight (sensing technology)
UOS	Urban Operating System
UNCTAD	United Nations Conference on Trade and Development
USD	United States Dollar
VPN	Virtual Private Network
VR	Virtual Reality
WCAG	Web Content Accessibility Guidelines
WLAN	Wireless Local Area Network

List of Figures

List of Images

List of Tables

PART I

Smart City Foundations: Technology, Economy, and Sustainability

CHAPTER 1

Technological Foundations of Smart Urbanism: The Tools Shaping Our Cities

Sushobhan Majumdar and Moharana Choudhury

1.1 Introduction

Smart urbanism refers to the strategic integration of advanced technologies, data-driven governance, and participatory planning practices to create cities that are more efficient, sustainable, resilient, and inclusive (Ababneh, 2023). As an evolving paradigm, it encompasses a wide range of digital tools and urban innovations, including sensor networks, artificial intelligence (AI), big data analytics, Internet of Things (IoT) systems, intelligent mobility platforms, and digitally enabled public services that collectively transform how cities are planned, managed, and experienced (Piosik & Karmańska, 2023). At its core, smart urbanism extends beyond the deployment of technology; it emphasizes the intersection of digital infrastructure with social equity, environmental stewardship, and economic development (Okafor, Aigbavboa, & Thwala, 2022). The scope of

S. Majumdar
Department of Geography, Pandit Raghunath Murmu Smriti Mahavidyalaya, Bankura, West Bengal, India

M. Choudhury (✉)
Environmental Research and Management Division, Voice of Environment (VoE) Guwahati, Guwahati, Assam, India

S. Majumdar, M. Choudhury (eds.), *City of the Future*,
https://doi.org/10.1007/978-3-032-18026-1_1

smart urbanism, therefore, includes not only the optimization of urban systems such as transportation, energy, water, waste management, and public safety, but also the enhancement of citizen engagement, collaborative governance, and data ethics frameworks that ensure the responsible use of emerging technologies (Deep & Verma, 2023; Narayan et al., 2025). By aligning technological capabilities with inclusive urban strategies, smart urbanism seeks to address complex contemporary challenges, ranging from climate change and resource scarcity to rapid urbanization and socioeconomic inequality, while shaping adaptable, future-ready cities (Cao, 2023). In this context, this chapter explores the digital architectures, analytical methodologies, and innovation ecosystems that underpin smart urban development, providing a comprehensive understanding of how technology is reshaping urban life and governance.

Since 2009, research on smart cities and smart urbanism has experienced a significant rise in both publications and citations, reflecting the growing academic and practical interest in the field (Fig. 1.1). The early years saw relatively few publications, but by the early 2010s, output and citation rates increased rapidly, signalling the emergence of smart

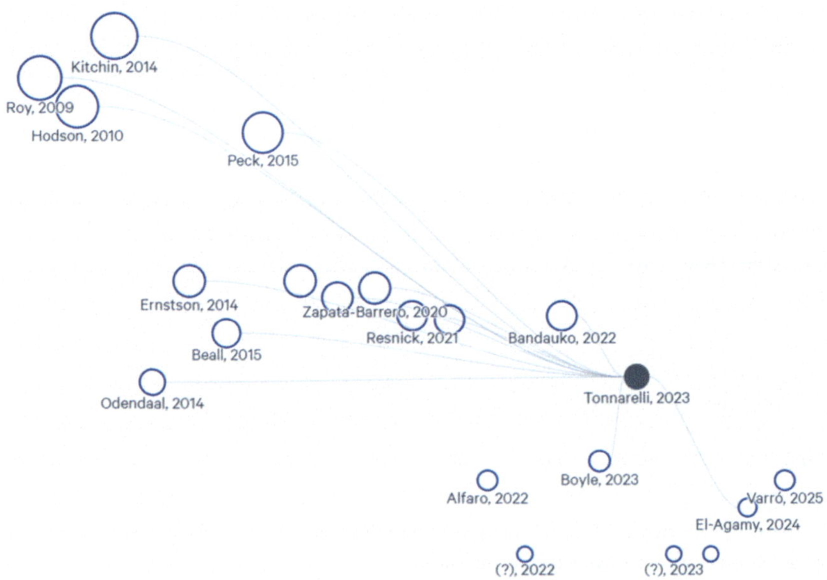

Fig. 1.1 Citation analysis of smart urbanism

urbanism as a distinct interdisciplinary domain. Citation patterns over the years indicate the development of key thematic clusters, including information and communication technologies, urban governance, sustainability, mobility, and environmental management. In recent years, from 2020 to 2025, citations have continued to grow, particularly for studies focusing on advanced technologies such as digital twins, artificial intelligence, the Internet of Things, and data-driven urban planning, as well as integrated approaches linking sustainability and social inclusion. Overall, the trajectory of citations demonstrates an initial phase of rapid expansion, followed by the consolidation of research into specialized and highly interconnected clusters, highlighting the maturation and increasing sophistication of the smart urbanism research landscape.

1.1.1 Evolution from Traditional to Smart Cities

The evolution from traditional to smart cities represents a fundamental shift in how urban areas are designed, governed, and experienced, mainly driven by advances in digital technology and the growing complexity of contemporary urban challenges (Ghosh, Majumdar, & Cheshmehzangi, 2024; Majumdar, Kandpal, & Anthopoulos, 2024; Neves, Aparício, & Neto, 2024). Traditional cities relied on centralized planning, physical infrastructure, and reactive management approaches that often struggled to keep pace with rapid population growth, resource pressures, and environmental degradation (Majumdar & Chatterjee, 2020; Majumdar & Sivaramakrishnan, 2017; Porse, 2013). As cities expanded, limitations in conventional systems such as inefficient transportation networks, fragmented service delivery, and limited data for decision-making became increasingly apparent (Piera-Jiménez et al., 2024). The emergence of information and communication technologies (ICTs), sensor networks, and real-time data analytics laid the foundation for smart city development, enabling more proactive, integrated, and adaptive urban management (Folorunsho, Adenekan, Ezeigweneme, Somadina, & Okeleke, 2024). Smart cities leverage interconnected digital systems, automation, and predictive analytics to optimize energy consumption, enhance mobility, improve public safety, strengthen environmental monitoring, and streamline public services (Apanavičienė & Shahrabani, 2023). At the same time, the shift towards smart cities reflects a broader reorientation towards citizen-centred governance, participatory planning, and sustainable development, positioning technology as a tool not only for efficiency

Table 1.1 Evolution from traditional cities to smart cities

Feature/ dimension	*Traditional cities*	*Smart cities*	*Key differences/impacts*
Governance	Centralized, manual	Data-driven, participatory	Transparency, citizen engagement
Infrastructure	Static, reactive	Adaptive, sensor-enabled	Real-time monitoring, predictive maintenance
Services	Siloed	Integrated	Efficiency, cross-sector coordination
Mobility	Fixed, manual systems	Intelligent, multimodal	Reduced congestion, optimized routes

Source: Compiled by the Authors

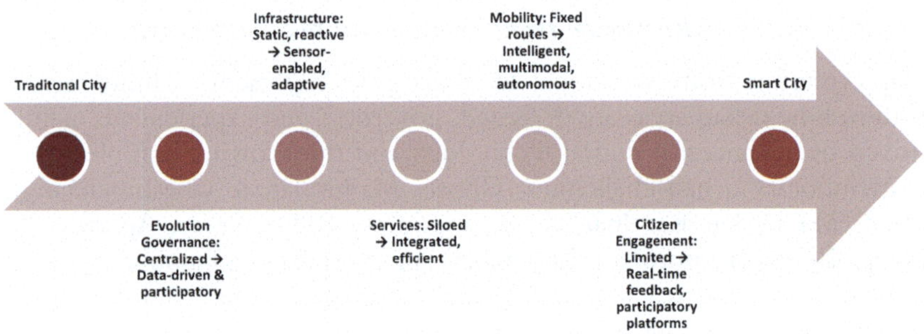

Fig. 1.2 Evolution from traditional cities to smart cities

but also for transparency, resilience, and social inclusion (Sameer, Alalouch, Al-Saadi, & Saleh, 2023). This transformation signifies an ongoing shift from cities that merely accommodate growth to those that intelligently anticipate and respond to their residents' needs. Table 1.1 and Fig. 1.2 show the evolution of cities from traditional to smart cities.

1.1.2 Interdisciplinary Nature: Urban Planning, ICT, Governance

The interdisciplinary nature of smart urbanism arises from the convergence of urban planning, information and communication technologies (ICT), and modern governance frameworks, each contributing essential perspectives and capabilities to the design and management of intelligent

Table 1.2 Key ICT infrastructure components for smart cities

Component	*Function*	*Benefits*
Broadband and 5G	High-speed connectivity	Real-time data transmission, IoT support
Edge computing	Local data processing	Low latency, faster decision-making
IoT and sensors	Data collection	Continuous monitoring, predictive analytics
Cloud and data centres	Storage and processing	Scalable storage, centralized analysis

Source: Compiled by the Authors

cities (Mozūriūnaitė, 2018). Urban planning provides the spatial, social, and environmental foundations upon which smart city strategies are built, ensuring that technological solutions align with human needs, public spaces, mobility patterns, and long-term urban development goals (Pancewicz, 2022). ICT introduces the digital infrastructure, such as sensors, data platforms, communication networks, and automation systems, which enable real-time monitoring, efficient resource management, and data-driven decision-making across urban systems (D'Amico, L'Abbate, Liao, Yiğitcanlar, & Ioppolo, 2020). Governance, meanwhile, establishes the policies, regulatory frameworks, institutional structures, and participatory mechanisms necessary to ensure that smart city initiatives are transparent, inclusive, and accountable (Adenekan, Ezeigweneme, & Chukwurah, 2024). Together, these fields create a holistic approach in which physical design, technological innovation, and democratic oversight reinforce one another, enabling cities to respond more effectively to complex challenges such as climate change, urbanization, and social inequality (Chatterjee, Majumdar, & Bhunia, 2021; Ijiga et al., 2024; Majumdar, 2020; Majumdar & Sivaramakrishnan, 2025). This interdisciplinary interplay not only drives the functionality of smart cities but also shapes their ethical, social, and environmental impacts, ensuring that technological advancement is integrated with responsible and equitable urban development. Table 1.2 shows key ICT infrastructure for smart cities.

1.1.3 The Role of Technology in Urban Transformation

Technology plays a pivotal role in urban transformation by reshaping how cities function, grow, and interact with their residents, driving a shift from traditional, reactive urban management to dynamic, data-informed systems capable of anticipating and addressing complex challenges (Yan & Lv, 2024). Through digital tools such as sensor networks, artificial intelligence, big data analytics, Internet of Things (IoT) platforms, and geospatial technologies, cities can monitor environmental conditions, optimize infrastructure performance, enhance mobility, and improve the delivery of public services in real time (Olimat, Liu, & Abudayyeh, 2023). These technologies support smarter energy grids, intelligent transportation systems, predictive maintenance of infrastructure, and more responsive emergency and public health services, contributing to greater efficiency, sustainability, and resilience (Baburaj, Jayadevan, Aliyana, & Stylios, 2025). Beyond operational improvements, technology also transforms citizen engagement by enabling participatory governance through digital platforms that facilitate feedback, transparency, and collaboration (Hafel, 2023). As urban environments face pressures from rapid population growth, climate change, and resource scarcity, technology serves as a critical enabler of innovative solutions that help cities adapt, reduce their ecological footprint, and promote social inclusion. In this way, technology is not merely an add-on to urban systems but a foundational driver of how contemporary cities evolve and how future cities will be imagined.

1.1.4 Overview of Key Technological Drivers

The transformation of modern cities into smart urban environments is driven by a set of key technological innovations that enable greater efficiency, connectivity, and responsiveness across urban systems (Mohamed, 2023). Central among these drivers are the Internet of Things (IoT) and sensor networks, which provide real-time monitoring of energy, transportation, water, and waste systems, facilitating data-driven decision-making and predictive management. Big data analytics and artificial intelligence (AI) further enhance urban intelligence by processing vast amounts of information to optimize operations, identify patterns, forecast demand, and support evidence-based policy development (Yiğitcanlar et al., 2022). Cloud computing and edge computing enable scalable, decentralized, and accessible data storage and processing, while Geographic Information Systems (GIS) and spatial analytics enable precise urban planning and resource allocation. Additionally, advanced communication networks such as 5G ensure seamless connectivity among devices, systems, and stakeholders, supporting smart mobility, emergency response, and citizen engagement platforms. Emerging technologies, including blockchain for secure data sharing, autonomous vehicles, and digital twins for virtual city modelling, further expand the potential for integrated, sustainable, and resilient urban management (Delavari, Pflanzner, & Kertész, 2025). Together, these technological drivers form the backbone of smart urbanism, shaping how cities operate, interact with their residents, and address the complex challenges of the twenty-first century.

This chapter aims to provide a comprehensive understanding of the technological foundations of smart urbanism by exploring the key tools, systems, and innovations shaping modern cities. It seeks to examine the evolution from traditional to smart cities, highlight the interdisciplinary nature of urban planning, ICT, and governance, and analyse the role of technology in urban transformation. Additionally, the chapter outlines the primary technological drivers enabling smarter, more sustainable, and resilient urban environments, offering insights into how these innovations support efficient resource management, citizen engagement, and inclusive urban development. To provide a consolidated view of these themes before examining them in detail, the key technological pillars of smart urbanism can be organized into a layered framework. Table 1.3 synthesizes the core functional domains, the main technologies that underpin them, and indicative supporting literature that has shaped contemporary debates on smart urban development.

As outlined in Table 1.3, the technological foundations of smart urbanism range from ICT infrastructure and urban sensing to analytics, platform governance, applied urban systems, and cross-cutting ethical and emerging concerns. The following sections unpack these layers sequentially, beginning with the digital backbone of ICT infrastructure in Sect. 1.2.

1.1.5 The Digital Backbone: ICT Infrastructure

1.1.5.1 Broadband Networks and 5G

Broadband networks and 5G technologies form the critical backbone of smart urban infrastructure, enabling high-speed, reliable, and low-latency connectivity essential for the seamless operation of modern cities (Folorunsho et al., 2024). Broadband networks provide the foundational capacity for data transmission across homes, businesses, and public services, supporting cloud computing, IoT devices, and digital platforms that

Table 1.3 Synthesis of technological foundations of smart urbanism and indicative supporting literature

Pillar of smart urbanism	*Key function in cities*	*Main technologies/systems*	*Indicative supporting literature*
ICT infrastructure backbone	Provides high-speed, reliable connectivity for all smart city services and data flows	Broadband, 5G, edge computing, cloud platforms, data centres	Trigka and Dritsas (2025)
Sensing and data generation	Captures real-time information on environment, mobility, infrastructure, and citizen activity	IoT sensor networks, mobile and crowdsourced data, surveillance and computer vision systems	Liu and Chen (2024)
Urban intelligence: analytics and AI	Transforms raw data into actionable insight for predictive and adaptive urban management	Big data analytics, ML/AI models, digital twins, simulation and scenario modelling	Rane (2023)
Platforms, operating systems, and governance	Integrates sectoral data and coordinates decisions across agencies while enabling citizen participation	Urban operating systems, integrated urban platforms, e-governance portals, open data and participation tools	Ishrat, Khan, Faisal, & Ansari (2025)
Application domains: mobility, infrastructure, and environment	Delivers tangible benefits in transport, infrastructure reliability, and environmental sustainability	Smart transportation and traffic systems, infrastructure monitoring, green infrastructure, smart energy, water and waste systems	Annapareddy, Botlagunta, Kommaragiri, Gadi, & Kalisetty (2022)
Cross-cutting risks, ethics, and inclusion	Addresses privacy, security, digital divide, lock-in, and equity to ensure responsible deployment	Cybersecurity frameworks, data governance policies, inclusion and accessibility programmes, open standards	Belsian, Tamhane, Kumar, & Nicholas (2025)
Emerging and human-centric directions	Explores next-generation tools while keeping human well-being and participation at the centre	Quantum computing, edge AI, blockchain, smart-citizen and ageing-in-place solutions	Nguyen, Tran, Zeng, & Tran (2024)

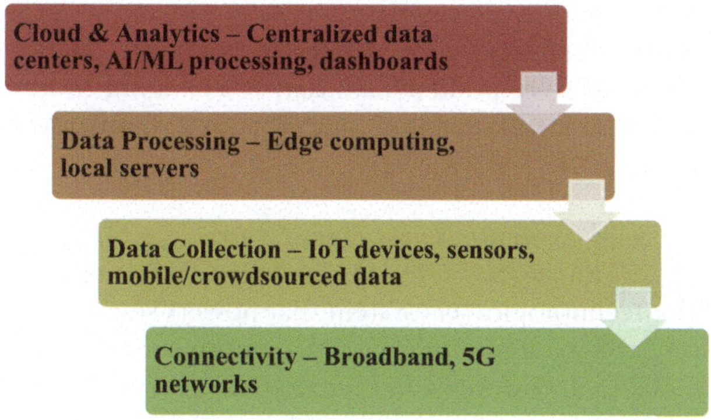

Fig. 1.3 Smart city ICT infrastructure architecture

underpin urban intelligence (Singh & Angurala, 2022). Building on this, 5G networks offer significantly faster speeds, reduced latency, and the ability to connect a massive number of devices simultaneously, facilitating real-time monitoring, autonomous mobility, smart energy management, and responsive public services (Pradhan et al., 2023). By ensuring ubiquitous and resilient connectivity, broadband and 5G networks empower cities to harness data-driven insights, support advanced analytics, and enhance citizen engagement, making them indispensable enablers of smart urban transformation. Figure 1.3 shows the framework of smart city ICT infrastructure architecture.

1.1.5.2 High-Speed Connectivity and Edge Computing

High-speed connectivity and edge computing are essential components of the digital infrastructure that enable responsive, data-driven urban systems (Obioha-Val, Olaniyi, Selesi-Aina, Gbadebo, & Kolade, 2024). High-speed networks ensure rapid and reliable transmission of large volumes of data generated by IoT devices, sensors, and urban applications, supporting real-time monitoring, smart traffic management, and efficient service delivery (Sathishbabu, Dhanalakshmi, Bharathiraja, & Thirunavukkarasu, 2025). Edge computing complements this by processing data closer to the source, reducing latency, minimizing bandwidth usage, and enabling faster decision-making for critical applications such as autonomous vehicles, emergency response, and energy grid management (Sodiya et al.,

2024). Together, high-speed connectivity and edge computing enhance the performance, scalability, and resilience of smart city systems, allowing urban environments to respond dynamically to changing conditions while improving efficiency, sustainability, and the overall quality of life for residents.

1.1.5.3 IoT, Cloud Computing and Data Centres

The Internet of Things (IoT), cloud computing, and data centres form the core of the technological ecosystem that powers smart cities, enabling the collection, storage, and analysis of vast amounts of urban data (Desai & Kulkarni, 2024). IoT devices and sensor networks continuously generate real-time information on mobility, energy usage, environmental conditions, and public services, providing the raw data needed for informed decision-making (García, Morales, & Kraak, 2019). Cloud computing offers scalable, flexible, and remote processing capabilities, enabling cities to store and analyse data efficiently, deploy applications rapidly, and support collaborative platforms for governance and citizen engagement (Coteț, Deac, Deac, & Popa, 2020). Data centres serve as the physical and virtual infrastructure that underpins both IoT and cloud systems, ensuring secure, reliable, and high-capacity data management. Together, these technologies enable cities to leverage actionable insights, optimize resource use, enhance service delivery, and build adaptive, resilient, and sustainable urban environments. Table 1.4 shows urban intelligence tools and their various applications.

Table 1.4 shows urban intelligence tools and its various applications

Table 1.4 Urban intelligence tools and applications

Tool/technology	*Purpose*	*Urban application*
Big data analytics	Data processing and insights	Traffic management, energy use
Artificial intelligence/ML	Predictive modelling	Public safety, service delivery
Digital twins	Virtual city modelling	Infrastructure planning, disaster simulation
Simulation models	Scenario analysis	Emergency response, urban planning

Source: Compiled by the Authors

1.1.5.4 Data Storage, Access, and Processing at Scale

Efficient data storage, access, and processing at scale are fundamental to the operation of smart cities, as urban systems generate massive volumes of structured and unstructured data from IoT devices, sensors, and citizen interactions (Pioli, Macedo, Costa, & Dantas, 2024). Scalable storage solutions, including cloud platforms and distributed databases, enable cities to securely store this information while ensuring redundancy and reliability (Hui, McLernon, & Zaidi, 2018). Advanced data processing frameworks allow real-time analytics, predictive modelling, and machine learning (ML) applications that optimize urban operations such as traffic management, energy distribution, waste collection, and emergency response (Shahrour et al., 2020). Equally important is seamless data access, which ensures that stakeholders from city planners to service providers and citizenscan retrieve relevant information when needed to support informed decision-making. By integrating storage, access, and processing capabilities at scale, cities can transform raw data into actionable insights, enabling smarter, more responsive, and resilient urban environments.

1.1.5.5 Sensors, Actuators, and Real-Time Data Collection

Sensors, actuators, and real-time data collection form the frontline of smart urban infrastructure, enabling cities to monitor, respond to, and manage dynamic urban environments effectively (Radziszewska, 2023). Sensors continuously capture data on environmental conditions, traffic flows, energy consumption, air quality, water levels, and public safety metrics, providing a real-time picture of urban operations (Kumari & Kaur, 2024). Actuators act on this data, executing automated responses such as adjusting traffic signals, regulating energy distribution, controlling water flow, or activating emergency systems, creating a responsive feedback loop that enhances efficiency and reliability (Safta, Cococi, Călinoiu, & Marin, 2023). Real-time data collection allows city managers and AI systems to identify trends, predict potential issues, and make informed, timely decisions that improve service delivery, reduce waste, and enhance quality of life (Zhang, Chatterjee, & Mukherjee, 2023). Together, these technologies form the sensory and operational foundation of smart cities, enabling proactive, adaptive, and data-driven urban management. Building on this, the ICT backbone and sensing infrastructure can be viewed as an integrated urban data pipeline that channels information from the physical city to analytics, decision-making, and citizen-facing services (Fig. 1.4).

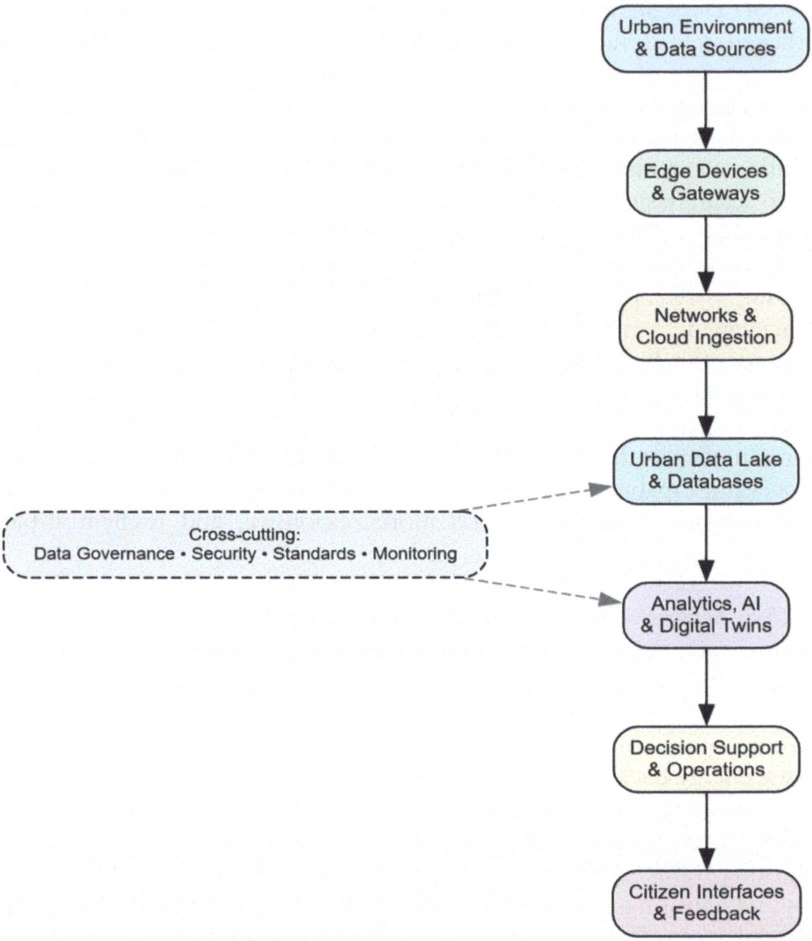

Fig. 1.4 Urban data pipeline linking environmental data sources, sensing and network layers, data platforms, analytics, and decision-support systems to citizen-facing interfaces under cross-cutting data governance

As shown in Fig. 1.4, smart urbanism relies on an end-to-end data pipeline that links sensing, connectivity, storage, and advanced analytics to practical decision-making. The upstream layers of sensors, actuators, and edge devices generate continuous streams of real-time information, while communication networks and cloud data centre infrastructure consolidate these streams into scalable urban data platforms. On top of this foundation, AI, big data analytics, and digital twins produce the "urban intelligence" that supports operational control centres, policy dashboards, and citizen interfaces. The following sections unpack this pipeline in more detail, beginning with data generation and collection (Sect. 1.3) and then examining the analytics and AI layer that converts data into actionable insight (Sect. 1.4).

1.1.6 Sensing the City: Data Generation and Collection

1.1.6.1 Sensor Networks

Sensor networks are fundamental to the generation and collection of urban data, forming the backbone of smart city monitoring and management systems (Radziszewska, 2023). These networks consist of interconnected sensors deployed across transportation systems, energy grids, water networks, environmental monitoring stations, and public spaces to capture real-time information on traffic flows, air quality, noise levels, energy usage, and infrastructure performance (Álías & Alsina-Pagès, 2019). By continuously transmitting data to centralized platforms or cloud systems, sensor networks enable cities to detect anomalies, respond to emergencies, and optimize resource allocation efficiently (Papaioannou et al., 2025). They also provide the granular, high-resolution data required for advanced analytics, predictive modelling, and automated decision-making. Through comprehensive and continuous monitoring, sensor networks empower cities to operate more intelligently, enhance sustainability, and improve the quality of life for residents.

1.1.6.2 Mobile and Crowdsourced Data

Mobile and crowdsourced data have emerged as vital sources of real-time information in smart cities, complementing traditional sensor networks by capturing dynamic human and environmental activity (Strigaro, Cannata, & Antonovic, 2019). Mobile devices, GPS-enabled applications, and wearable technologies provide continuous streams of location, mobility, and behavioural data, while crowdsourcing platforms allow citizens to report issues, share observations, and contribute to urban decision-making (Miranda et al., 2024). This participatory approach enhances situational awareness, supports traffic management, emergency response, environmental monitoring, and service optimization, and fosters greater civic engagement (Cvetković, 2024). By integrating mobile and crowdsourced data with other urban datasets, cities can gain richer, more granular insights into patterns of human activity and infrastructure use, enabling responsive, data-driven, and citizen-centred urban management.

1.1.6.3 Surveillance and Computer Vision

Surveillance systems and computer vision technologies play a critical role in monitoring urban environments and enhancing the safety, security, and efficiency of city operations (Namana & Kumar, 2024). High-resolution cameras, drones, and sensor-equipped devices capture visual data across streets, public spaces, and transportation networks, which is then processed using computer vision algorithms to detect patterns, identify anomalies, and recognize objects or behaviours in real time (Figueiredo et al., 2020). These technologies support applications such as traffic flow analysis, crime prevention, crowd management, infrastructure inspection, and environmental monitoring. Enabling automated, scalable, and continuous observation, surveillance, computer vision, situational awareness, and operational decision-making, while also allowing cities to respond proactively to emerging challenges, raises essential privacy, data security, and ethical governance considerations.

1.1.7 Urban Intelligence: Data Analytics and AI

1.1.7.1 Big Data Analytics

Big data analytics plays a central role in smart urbanism by enabling cities to process, interpret, and leverage the massive volumes of data generated from sensors, IoT devices, social media, and public services (Nyangaresi et al., 2024). Through advanced analytical techniques, including machine learning, predictive modelling, and pattern recognition, urban planners and administrators can gain actionable insights into traffic flows, energy consumption, environmental conditions, public health trends, and citizen behaviour (Adenyi, Okolo, Olorunsogo, & Babawarun, 2024). These insights support evidence-based decision-making, optimize resource allocation, and enhance operational efficiency across urban systems. By transforming raw data into meaningful information, big data analytics allows cities to anticipate challenges, improve service delivery, and develop proactive strategies for sustainable, resilient, and responsive urban management.

1.1.7.2 Artificial Intelligence and Machine Learning

Artificial intelligence (AI) and machine learning (ML) are transformative technologies in smart urbanism, enabling cities to analyse complex datasets, automate processes, and make predictive, data-driven decisions. AI and ML algorithms can optimize traffic management, energy consumption, waste collection, and emergency response by identifying patterns, forecasting demand, and recommending adaptive interventions (Rane, Rane, Paramesha, & Kaya, 2024). These technologies also enhance urban planning, environmental monitoring, and public service delivery by providing actionable insights from heterogeneous data sources, including IoT sensors, social media, and geospatial data. For instance, AI-based modelling of industrial emissions and ambient pollutant levels has been shown to improve air quality assessment and forecasting in heavily industrialized urban regions, enabling more targeted mitigation and planning responses (Halaktionov et al., 2025). By integrating AI and ML into urban systems, cities can improve efficiency, resilience, and sustainability while enabling proactive governance and personalized services that respond dynamically to the needs of residents.

1.1.7.3 Digital Twins and Simulation Models

Digital twins and simulation models are powerful tools for smart urban planning and management, enabling cities to create virtual replicas of physical infrastructure, systems, and processes (Vempati, 2024). These models integrate real-time data from IoT devices, sensors, and other urban sources to simulate traffic flows, energy consumption, environmental conditions, and emergency scenarios. By providing a dynamic, data-driven representation of the city, digital twins allow planners, policymakers, and engineers to test interventions, forecast outcomes, and optimize urban operations without disrupting real-world systems. Simulation models further support scenario analysis, risk assessment, and strategic planning, enhancing decision-making under uncertainty. Together, digital twins and simulations empower cities to anticipate challenges, improve efficiency, and implement adaptive, resilient, and sustainable urban solutions. Figure 1.5 shows about the urban intelligence and data analytics ecosystem.

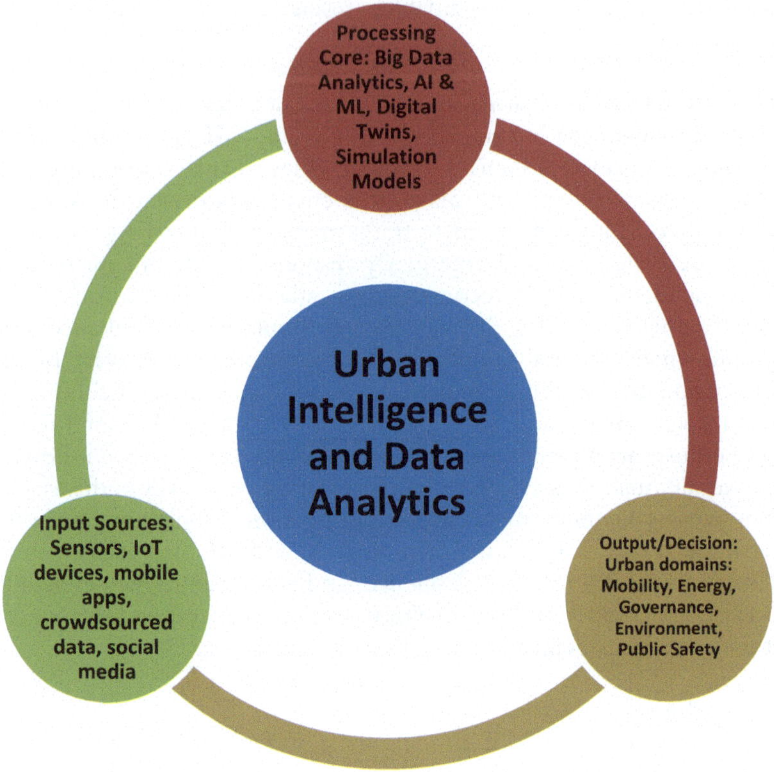

Fig. 1.5 Urban intelligence and data analytics ecosystem

1.1.8 Platform Urbanism and Smart Governance

1.1.8.1 Urban Operating Systems

Urban operating systems (UOS) serve as the foundational digital platforms that integrate, coordinate, and manage diverse urban services and infrastructure in smart cities. Acting as centralized hubs, UOS consolidate real-time data from transportation networks, energy grids, water systems, public safety, and citizen services, enabling city administrators to monitor performance, detect anomalies, and make data-driven decisions efficiently (Wilson et al., 2012). These platforms facilitate interoperability across multiple agencies and stakeholders, support predictive analytics, and enable automated or adaptive management of urban operations. By providing a cohesive digital framework, urban operating systems enhance governance transparency, optimize resource allocation, and improve responsiveness to dynamic urban challenges. Ultimately, UOS transform cities into integrated, intelligent environments where data-driven insights drive operational efficiency, policy implementation, and citizen-centric service delivery.

1.1.8.2 Smart Governance Tools

Smart governance tools are digital platforms and applications designed to enhance transparency, accountability, and citizen participation in urban management. These tools include e-governance portals, mobile applications, open data platforms, participatory decision-making systems, and real-time feedback mechanisms that enable residents to interact directly with municipal authorities (Nnenna et al., 2024). By facilitating data-driven policymaking, monitoring public services, and streamlining administrative processes, smart governance tools improve the efficiency and responsiveness of city governments. They also empower citizens to contribute to urban planning, report issues, and engage in collaborative problem-solving, fostering inclusive and participatory governance. Collectively, these tools support the creation of more transparent, accountable, and resilient urban systems, where technology bridges the gap between government operations and citizen needs.

1.1.8.3 Urban Platforms and Tech

Urban platforms and technology form the backbone of smart city operations by providing integrated digital infrastructures that connect urban services, stakeholders, and citizens. These platforms aggregate data from transportation systems, energy grids, public services, environmental sensors, and social interactions, enabling real-time monitoring, analytics, and coordinated management of city functions (Adenekan et al., 2024). They support applications such as mobility as a service (MaaS), digital twins, resource optimization tools, and citizen engagement interfaces, allowing cities to respond dynamically to changing conditions and demands. By fostering interoperability, collaboration, and scalable service delivery, urban platforms and technologies enhance operational efficiency, promote data-driven governance, and empower citizens to participate actively in decision-making. Ultimately, they transform urban ecosystems into responsive, intelligent, and resilient environments capable of addressing

Fig. 1.6 Smart city applications across sectors

complex challenges and improving quality of life. Figure 1.6 shows smart city applications across sectors.

1.1.9 *Mobility and Infrastructure Technologies*

1.1.9.1 Smart Transportation Systems

Smart transportation systems are a cornerstone of intelligent urban mobility, leveraging advanced technologies to enhance efficiency, safety, and sustainability in city transit networks (Majumdar, 2020; Majumdar & Sarkar, 2020). These systems integrate IoT-enabled sensors, real-time traffic monitoring, predictive analytics, and adaptive signal control to optimize traffic flow, reduce congestion, and minimize travel times. Intelligent public transport solutions, including connected buses, trains, and shared mobility platforms, enable dynamic route planning, demand-responsive services, and improved service reliability (Majumdar et al., 2024). Additionally, smart transportation systems support the adoption of electric and autonomous vehicles, integrated mobility-as-a-service (MaaS) platforms, and multimodal connectivity, reducing carbon emissions and promoting sustainable urban mobility. By combining technology with data-driven decision-making, these systems enhance urban accessibility, improve commuter experiences, and contribute to safer, greener, and more resilient transportation networks.

1.1.9.2 Traffic Management Technologies

Traffic management technologies are essential for optimizing urban mobility, reducing congestion, and enhancing road safety in smart cities. These technologies utilize real-time data from IoT sensors, connected vehicles, cameras, and GPS systems to monitor traffic flow, detect incidents, and predict congestion patterns. Intelligent traffic signal control, adaptive routing algorithms, and dynamic lane management enable more efficient use of road infrastructure while minimizing delays and emissions. Integration with public transport and shared mobility platforms allows cities to coordinate multimodal transportation, prioritize emergency vehicles, and support sustainable commuting options. By leveraging data-driven analytics and automated control systems, traffic management technologies improve the overall efficiency, safety, and environmental performance of urban transportation networks, contributing to smarter, more responsive, and sustainable cities.

1.1.9.3 Infrastructure Monitoring

Infrastructure monitoring is a critical component of smart urban management, enabling cities to maintain, optimize, and extend the lifespan of essential assets such as roads, bridges, tunnels, public transit systems, and utilities. Advanced monitoring technologies, including IoT sensors, drones, remote sensing, and structural health monitoring systems, provide real-time data on stress, wear, environmental conditions, and potential faults. Predictive analytics and AI-driven insights allow city managers to identify maintenance needs proactively, prevent failures, and allocate resources efficiently, reducing costs and enhancing safety. Integrating infrastructure monitoring with digital twin models and centralized urban management platforms supports coordinated planning, emergency preparedness, and long-term resilience. By enabling continuous, data-informed oversight, infrastructure monitoring ensures that urban systems remain reliable, safe, and adaptable to the demands of growing populations and changing environmental conditions.

1.1.10 *Sustainability and Environmental Technologies*

1.1.10.1 Green Infrastructure and Energy Systems

Green infrastructure and sustainable energy systems are central to the environmental objectives of smart urbanism, enabling cities to reduce their ecological footprint while enhancing resilience and liveability. Green infrastructure, including parks, green roofs, urban forests, and permeable surfaces, supports ecosystem services such as stormwater management, air quality improvement, heat mitigation, and biodiversity conservation. In coastal and peri-urban regions, blue–green infrastructure also extends to marine ecosystems, where studies on wild seaweeds and their antioxidant properties highlight how coastal biodiversity can contribute to health-oriented and bioresource-based sustainability strategies (Cokrowati et al., 2025). Complementing this, smart energy systems leverage renewable energy sources, intelligent grids, energy storage solutions, and real-time monitoring to optimize consumption, reduce greenhouse gas emissions, and enhance the reliability and efficiency of urban energy networks. Integrating green infrastructure with advanced energy technologies creates synergistic benefits, allowing cities to address climate change, improve public health, and promote sustainable urban development. Within this context, converting food and agro-industrial wastes into biofuels provides a dual benefit of low-carbon energy generation and waste reduction, and can be integrated into urban energy systems as part of circular bioeconomy strategies (Choudhary, Harne, Dhar, Alqahtani, & Patel, 2025). This combination of ecological design and energy innovation forms a cornerstone of environmentally responsible, resilient, and future-ready smart cities. In parallel, the development of bio-based materials and biopolymers derived from food and agro-industrial wastes offers additional opportunities to replace conventional, fossil-based urban materials with more sustainable, circular alternatives in green infrastructure and built-environment applications (Choudhary et al., 2025).

1.1.10.2 Water and Waste Management

Effective water and waste management is a critical component of sustainable smart urbanism, ensuring the efficient use of resources while minimizing environmental impact (Sreenath, Dhanaraj, Rath, Bhardwaj, & Choudhary, 2025). Smart water systems utilize sensors, IoT devices, and data analytics to monitor consumption, detect leaks, optimize distribution, and support real-time decision-making for water quality and supply management. Similarly, intelligent waste management leverages technologies such as sensor-enabled bins, automated collection systems, and data-driven route optimization to enhance recycling rates, reduce landfill use, and lower operational costs, thereby supporting urban resource recovery and the development of circular supply chains that link waste streams with secondary raw material markets (Narayan et al., 2025). Integrated approaches that combine water and waste management with broader urban planning and sustainability strategies help cities minimize resource waste, reduce pollution, and promote circular economy principles, supported by structured performance indicators that track progress towards circularity (Narayan et al., 2025). By adopting these technologies, urban areas can achieve more resilient, efficient, and environmentally responsible resource management, contributing to long-term sustainability and improved quality of life for residents. Empirical evidence from community-bin systems in rural settlements also shows that local socio-spatial conditions and user behaviour strongly influence the effectiveness of waste management schemes, underscoring the need to adapt smart and circular waste strategies to regional contexts (Bredun, Choudhary, & Kumar, 2024). In addition, biological transformation pathways for food and agro-industrial wastes reinforce this transition by converting organic residues into bio-based products, materials, and energy streams that support circular urban resource management (Kumar et al., 2025).

1.1.10.3 Climate Resilience and Disaster Response

Climate resilience and disaster response are essential components of smart urbanism, enabling cities to anticipate, withstand, and recover from environmental hazards and extreme events. Advanced technologies such as real-time monitoring systems, IoT sensors, remote sensing, and predictive analytics allow for early detection of floods, storms, heatwaves, and other climate-related threats. Smart emergency response platforms facilitate rapid communication, resource allocation, and coordination among government agencies, first responders, and citizens, enhancing preparedness and minimizing damage. Additionally, integrating climate adaptation strategies into urban planning such as resilient infrastructure, green spaces, and adaptive building designs strengthens long-term urban resilience. By combining technological innovation with proactive planning, cities can better protect lives, infrastructure, and ecosystems while ensuring sustainable and adaptive development in the face of increasing climate uncertainty. The utilization of biofuels derived from food and agro-industrial wastes offers important mitigation co-benefits by reducing greenhouse gas emissions from urban energy systems while simultaneously valorizing waste streams within circular economy frameworks (Choudhary et al., 2025).

1.1.11 *Challenges and Ethical Considerations*

1.1.11.1 Privacy and Data Security

Privacy and data security represent significant challenges and ethical considerations in the development of smart cities, where extensive data is collected from citizens, sensors, and urban systems. The aggregation of personal, behavioural, and operational information raises risks of unauthorized access, data breaches, surveillance misuse, and identity theft, potentially compromising individual rights and public trust. Ensuring robust cybersecurity, encryption, secure storage, and controlled access is essential, alongside transparent data governance policies that define how data is collected, shared, and used. Ethical considerations also require balancing innovation with respect for privacy, informed consent, and accountability, ensuring that the deployment of smart technologies does not disproportionately affect vulnerable populations. Addressing these challenges is crucial for fostering citizen confidence, promoting responsible data practices, and enabling the sustainable and inclusive growth of smart urban environments. Table 1.5 shows challenges and ethical considerations of urbanism process.

Table 1.5 Challenges and ethical considerations in smart urbanism

Challenge	*Impact on urban development*	*Mitigation strategies*
Privacy and data security	Loss of trust, legal issues	Encryption, governance policies
Digital divide	Exclusion of vulnerable populations	Affordable connectivity, digital literacy programmes
Technological lock-in	Reduced flexibility, higher costs	Open standards, modular architecture
Environmental concerns	Sustainability risks	Green infrastructure, smart energy systems

Source: Compiled by the Authors

1.1.11.2 Digital Divide and Accessibility

The digital divide and accessibility are critical ethical and practical challenges in smart urbanism, as unequal access to digital technologies and connectivity can exacerbate existing social and economic inequalities (Turin et al., 2022). Disparities in internet access, device availability, digital literacy, and affordable services may prevent marginalized communities from fully participating in or benefiting from smart city initiatives (Jaafreh & Allouzi, 2023). Ensuring inclusive access requires targeted policies, investment in digital infrastructure, and programmes that enhance technological literacy, affordability, and usability for all residents. Moreover, accessibility considerations extend to designing urban services, platforms, and applications that accommodate diverse needs, including those of people with disabilities, the elderly, and other vulnerable groups (Apostolidou & Fokaides, 2023). Addressing the digital divide is essential not only for equitable participation but also for fostering social cohesion, enhancing citizen engagement, and realizing the full potential of smart cities as inclusive, human-centred environments.

1.1.11.3 Technological Lock-In and Vendor Dependence

Technological lock-in and vendor dependence are significant challenges in smart urban development, arising when cities rely heavily on proprietary systems, platforms, or technologies provided by a limited number of vendors (Voorwinden, 2021). Such dependence can restrict flexibility, limit interoperability, and increase costs over time, making it difficult for cities to adapt to emerging technologies or switch providers without substantial disruption. Lock-in may also hinder innovation, as reliance on specific vendors can constrain the integration of alternative solutions or open-source platforms that could better meet evolving urban needs (Schreiber & Schreiber, 2023). To mitigate these risks, cities must prioritize open standards, modular system architectures, and multi-vendor strategies that promote interoperability, scalability, and long-term resilience. Addressing technological lock-in ensures that smart city initiatives remain adaptable, sustainable, and capable of evolving alongside technological advancements and changing urban priorities.

1.1.12 Future Directions and Emerging Technologies

1.1.12.1 Quantum Computing, Edge AI, and Blockchain

Emerging technologies such as quantum computing, edge AI, and blockchain have the potential to become transformative disruptors in urban technology, reshaping how cities process information, make decisions, and deliver services (Beshaj & Crislip, 2025). Quantum computing promises unprecedented computational power, enabling the rapid simulation and optimization of complex urban systems, from traffic networks and energy grids to climate modelling and urban planning scenarios (Celsi & Celsi, 2024). Edge AI combines artificial intelligence with edge computing, enabling data analysis and action at the source, reducing latency, improving responsiveness, and enabling real-time decision-making for critical applications such as autonomous transportation, smart infrastructure, and emergency response (Chataut, Nankya, & Akl, 2024). Blockchain technology offers secure, transparent, and decentralized mechanisms for data sharing, transactions, and governance, enhancing trust, accountability, and efficiency across urban services, including supply chains, energy trading, and citizen identity management (Douaioui & Benmoussa, 2024). Together, these technologies signal a new era of urban innovation, where cities can become more intelligent, resilient, and adaptive, while addressing challenges related to scalability, security, and sustainability in increasingly complex urban environments.

1.1.12.2 Human-Centric and Inclusive Smart Urbanism

Human-centric and inclusive smart urbanism emphasizes placing people at the core of technological and urban development, ensuring that smart city initiatives enhance quality of life, equity, and social participation. Beyond efficiency and sustainability, this approach prioritizes the needs, experiences, and well-being of diverse urban populations, including marginalized and underserved communities (Woolrych & Man-lin, 2024). By integrating participatory governance, accessible digital services, and inclusive design principles, cities can empower citizens to actively engage in decision-making, co-create solutions, and benefit from technological advancements (Buyannemekh, Gascó-Hernández, & Gil-García, 2024). Human-centric smart urbanism also addresses digital divides, ensuring equitable access to connectivity, information, and services while safeguarding privacy, security, and ethical use of data. Ultimately, this approach envisions smart cities not just as technology-driven systems, but as responsive, resilient, and socially inclusive environments that support the aspirations, safety, and well-being of all residents. Taken together, the technological, governance, and human-centric dimensions discussed in this chapter can be synthesized into an integrated framework that links digital infrastructure, urban intelligence, platforms, applications, and cross-cutting ethical considerations (Fig. 1.7).

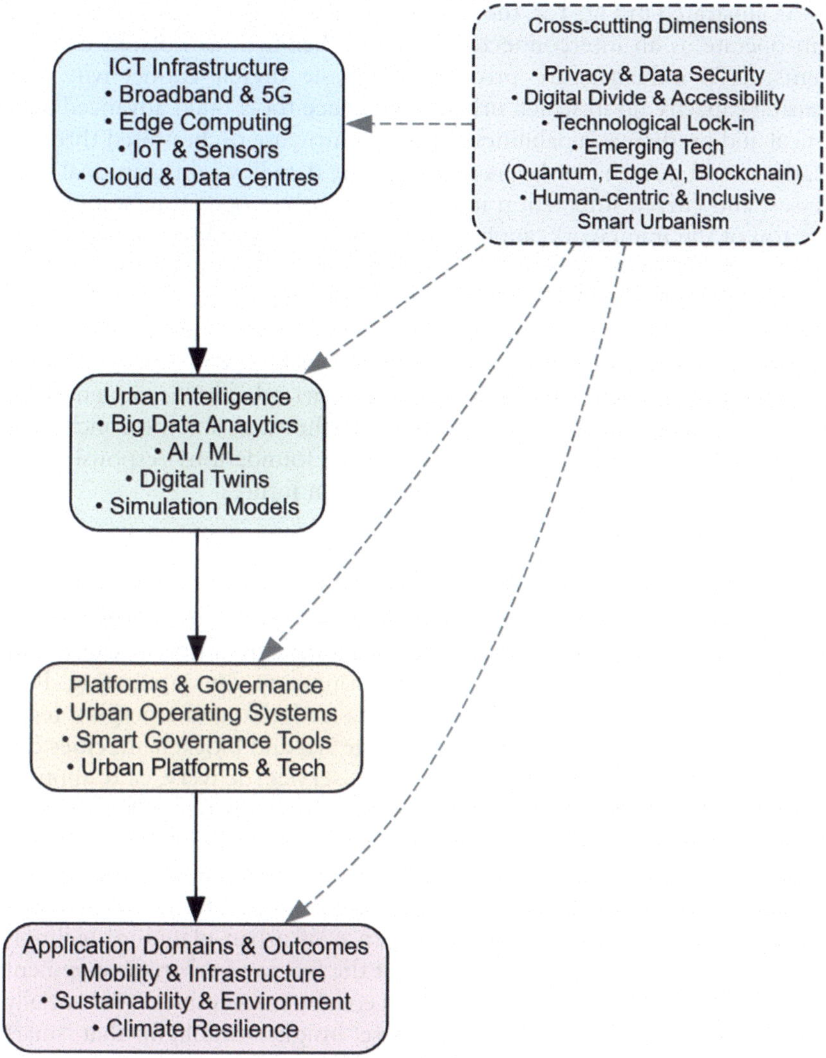

Fig. 1.7 Conceptual framework showing how ICT infrastructure, urban intelligence, and platforms support smart city applications under cross-cutting ethical and human-centric dimensions

As illustrated in Fig. 1.7, the technological foundations of smart urbanism operate as an interconnected system rather than as isolated components. ICT infrastructure provides the basic digital connectivity and sensing capacity upon which urban intelligence tools build advanced analytical and predictive capabilities. These, in turn, are orchestrated through platform and governance layers that integrate data flows, institutional processes, and citizen interaction into coherent urban operating systems. At the top of the framework, application domains in mobility, infrastructure, environmental management, and climate resilience translate this technological stack into tangible improvements in urban performance and quality of life. Throughout all layers, cross-cutting dimensions such as privacy and data security, digital inclusion, technological lock-in, emerging technologies, and human-centric design shape how smart urbanism is implemented in practice. This integrated perspective sets the stage for the concluding discussion on how cities can leverage these foundations responsibly to achieve adaptive, resilient, and equitable urban futures.

1.2 Conclusion

In conclusion, the technological foundations of smart urbanism provide the tools and frameworks necessary to transform cities into more efficient, sustainable, resilient, and inclusive environments. By integrating ICT infrastructure, IoT, cloud computing, data analytics, and emerging technologies such as quantum computing, edge AI, and blockchain, cities can optimize resource management, enhance service delivery, and support real-time, data-driven decision-making. Equally important is the human-centric and inclusive dimension, which ensures that technological advancements benefit all citizens, foster participatory governance, and address social and digital inequalities. The evolution from traditional to smart cities reflects an interdisciplinary convergence of urban planning, technology, and governance, demonstrating that the future of urban development relies not only on innovation but also on collaborative, ethical, and socially responsible approaches. Together, these insights highlight that smart urbanism is not merely a technological upgrade but a comprehensive transformation of how cities function, interact with residents, and respond to the complex challenges of the twenty-first century.

References

Ababneh, A. (2023). Smart urban management of green space. *Journal of Design for Resilience in Architecture and Planning, 4*(3), 339–353. https://doi.org/10.47818/drarch.2023.v4i3101

Adenekan, O., Ezeigweneme, C., & Chukwurah, E. (2024). The evolution of smart cities: Integrating technology, governance, and sustainable development. *International Journal of Applied Research in Social Sciences, 6*(5), 891–902. https://doi.org/10.51594/ijarss.v6i5.1131

Adenyi, A., Okolo, C., Olorunsogo, T., & Babawarun, O. (2024). Leveraging big data and analytics for enhanced public health decision-making: A global review. *GSC Advanced Research and Reviews, 18*(2), 450–456. https://doi.org/10.30574/gscarr.2024.18.2.0078

Álías, F., & Alsina-Pagès, R. (2019). Review of wireless acoustic sensor networks for environmental noise monitoring in smart cities. *Journal of Sensors, 2019*, 1–13. https://doi.org/10.1155/2019/7634860

Annapareddy, V. N., Botlagunta, P. N., Kommaragiri, V. B., Gadi, A. L., & Kalisetty, S. (2022). Emerging technologies in smart computing, sustainable energy, and next-generation mobility: Enhancing digital infrastructure, secure networks, and intelligent manufacturing. *SSRN*. https://doi.org/10.2139/ssrn.5204960

Apanavičienė, R., & Shahrabani, M. (2023). Key factors affecting smart building integration into smart city: Technological aspects. *Smart Cities, 6*(4), 1832–1857. https://doi.org/10.3390/smartcities6040085

Apostolidou, E., & Fokaides, P. (2023). Enhancing accessibility: A comprehensive study of current apps for enabling accessibility of disabled individuals in buildings. *Buildings, 13*(8), 2085. https://doi.org/10.3390/buildings13082085

Baburaj, A., Jayadevan, S., Aliyana, A., & Stylios, G. (2025). Ai-driven tengs for self-powered smart sensors and intelligent devices. *Advanced Science, 12*(20). https://doi.org/10.1002/advs.202417414

Belsian, G., Tamhane, H., Kumar, B. A., & Nicholas, G. S. (2025). Justice in IoT and AI for sustainable development. In S. Mishra, N. Gupta, & P. Goktas (Eds.), *AI and ML techniques in IoT-based communication: A path to sustainable development goals (chapter 12)*. Wiley. https://doi.org/10.1002/9781394337262.ch12

Beshaj, L., & Crislip, S. (2025). The dual role of emerging technologies in innovation and security. *Disruptive Technologies in Information Sciences, 13480*, 134800B. https://doi.org/10.1117/12.3053753

Bredun, V., Choudhary, R., & Kumar, A. (2024). Regional specifics of using community bins in waste management: A case study of rural communities in Poltava region (Ukraine). *Trends in Ecological and Indoor Environmental Engineering, 2*(4), 10–17. https://doi.org/10.62622/TEIEE.024.2.4.10-17

Buyannemekh, B., Gascó-Hernández, M., & Gil-García, J. (2024). Fostering smart citizens: The role of public libraries in smart city development. *Sustainability, 16*(5), 1750. https://doi.org/10.3390/su16051750

Cao, H. (2023). Urban resilience: Concept, influencing factors and improvement. *Frontiers in Business Economics and Management, 9*(1), 343–346. https://doi.org/10.54097/fbem.v9i1.8777

Celsi, M., & Celsi, L. (2024). Quantum computing as a game changer on the path towards a net-zero economy: A review of the main challenges in the energy domain. *Energies, 17*(5), 1039. https://doi.org/10.3390/en17051039

Chataut, R., Nankya, M., & Akl, R. (2024). 6g networks and the ai revolution—Exploring technologies, applications, and emerging challenges. *Sensors, 24*(6), 1888. https://doi.org/10.3390/s24061888

Chatterjee, U., Majumdar, S., & Bhunia, G. S. (2021). Land restoration strategies to mitigate the problems of shrinkage of wetlands in West Bengal. In G. S. Bhunia, U. Chatterjee, A. Kashyap, & P. K. Shit (Eds.), *Modern cartography series* (Vol. 10, pp. 525–538). Academic Press. https://doi.org/10.1016/B978-0-12-823895-0.00019-1

Choudhary, R., Harne, K., Dhar, S., Alqahtani, F. M., & Patel, K. (2025). Case study on biofuels derived from food and agro-industrial wastes. In M. Sillanpää, M. Choudhury, & S. Goswami (Eds.), *Food and agro-industrial wastes: Sustainable impacts, transformation and added value of by-products.* Academic Press. https://doi.org/10.1016/B978-0-443-31654-8.00030-8

Choudhary, R., Kumar, A., Kumar, V., Elmnifi, M. H., Dhar, S., Alqahtani, F. M., & Patel, K. (2025). Case study on biopolymers derived from food and agro-industrial wastes. In M. Sillanpää, M. Choudhury, & S. Goswami (Eds.), *Food and agro-industrial wastes: Sustainable impacts, transformation and added value of by-products.* Academic Press. https://doi.org/10.1016/B978-0-443-31654-8.00026-6

Choudhary, R., Tigga, S., Bhardwaj, L. K., Goswami, S., Nguyen Tan, T., & Chaudhari, S. K. (2025). Biofuels from food and agro-industrial wastes. In M. Sillanpää, M. Choudhury, & S. Goswami (Eds.), *Food and agro-industrial wastes: Sustainable impacts, transformation and added value of by-products.* Academic Press. https://doi.org/10.1016/B978-0-443-31654-8.00004-7

Cokrowati, N., Choudhary, R., Yatin, N., Apriliyanti, F. J., Amanda, R. A., Ramadhan, A. F., Kumar, A., & Hrdzelidze, S. (2025). Identification of wild seaweeds at Nipah Beach, Indonesia, and study of their antioxidant potential. *Trends in Ecological and Indoor Environmental Engineering, 3*(3), 1–7. https://doi.org/10.62622/TEIEE.025.3.3.01-07

Coteţ, C., Deac, G., Deac, C., & Popa, C. (2020). An innovative industry 4.0 cloud data transfer method for an automated waste collection system. *Sustainability, 12*(5), 1839. https://doi.org/10.3390/su12051839

Cvetković, V. (2024). In-depth analysis of disaster (risk) management system in Serbia: A critical examination of systemic strengths and weaknesses.

D'Amico, G., L'Abbate, P., Liao, W., Yiğitcanlar, T., & Ioppolo, G. (2020). Understanding sensor cities: Insights from technology giant company driven smart urbanism practices. *Sensors, 20*(16), 4391. https://doi.org/10.3390/s20164391

Deep, G., & Verma, J. (2023). Embracing the future: Ai and ml transforming urban environments in smart cities. *Journal on Artificial Intelligence, 5*, 57–73.

Delavari, Z., Pflanzner, T., & Kertész, A. (2025). A taxonomy and survey of integrating emerging technologies to intelligent transportation systems. https://doi.org/10.21203/rs.3.rs-7715812/v1.

Desai, R., & Kulkarni, R. (2024). Energy efficient reliable data transmission for optimizing iot data transmission in smart city. *Indonesian Journal of Electrical Engineering and Computer Science, 34*(3), 1978. https://doi.org/10.11591/ijeecs.v34.i3.pp1978-1988

Douaioui, K., & Benmoussa, O. (2024). Insights into industrial efficiency: An empirical study of blockchain technology. *Big Data and Cognitive Computing, 8*(6), 62. https://doi.org/10.3390/bdcc8060062

Figueiredo, D., Braun, C., Forte, K., Brito, F., Silva, A., Lins, S., & Klautau, A.. (2020). Testbed for connected artificial intelligence using unmanned aerial vehicles and convolutional pose machines (pp. 1–6). https://doi.org/10.21528/cbic2019-131

Folorunsho, S., Adenekan, O., Ezeigweneme, C., Somadina, I., & Okeleke, P. (2024). Developing smart cities with telecommunications: Building connected and sustainable urban environments. *Engineering Science & Technology Journal, 5*(8), 2492–2519. https://doi.org/10.51594/estj.v5i8.1441

García, M., Morales, J., & Kraak, M. (2019). Integration and exploitation of sensor data in smart cities through event-driven applications. *Sensors, 19*(6), 1372. https://doi.org/10.3390/s19061372

Ghosh, S., Majumdar, S., & Cheshmehzangi, A. (2024). *Cities of tomorrow: Urban resilience and climate change preparedness.* Springer. https://doi.org/10.1007/978-981-97-9658-8

Hafel, M. (2023). Digital transformation in politics and governance in Indonesia: Opportunities and challenges in the era of technological disruption. *Society, 11*(2), 742–757. https://doi.org/10.33019/society.v11i2.577

Halaktionov, M., Bredun, V., Choudhary, R., Goroneskul, M., Kumar, A., Ouiya, F., Sydorenko, V., & Markina, L. (2025). AI-enhanced air quality assessment and prediction in industrial cities: A case study of Kryvyi Rih, Ukraine. *Ecological Engineering & Environmental Technology, 26*(6), 45–56. https://doi.org/10.12912/27197050/203725

Hui, H., McLernon, D., & Zaidi, A. (2018). Design of the security mechanism for a bpo cloud computing platform (pp. 1092–1095). https://doi.org/10.1109/icsess.2018.8663713.

Ijiga, A., Peace, A., Idoko, I., Ezebuka, C., Harry, K., Ukatu, I., & Agbo, D. (2024). Technological innovations in mitigating winter health challenges in New York City, USA. *International Journal of Science and Research Archive, 11*(1), 535–551.

Ishrat, M., Khan, W., Faisal, S. M., & Ansari, M. S. H. (2025). Smart governance and citizen engagement platforms: The role of technology in smart governance. In *Revolutionizing urban development and governance with emerging technologies* (p. 44). IGI Global. https://doi.org/10.4018/979-8-3373-1375-7.ch007

Jaafreh, M., & Allouzi, M. (2023). Smart cities adoption in Saudi Arabia: A comprehensive review and future drivers. *British Journal of Multidisciplinary and Advanced Studies, 4*(5), 20–39. https://doi.org/10.37745/bjmas.2022.0300

Kumar, P., Chauhan, S., Addai, G., Choudhary, R., Sudan, P., & Kumar, P. (2025). Valorization of food and agroindustrial wastes—Biological transformation. In M. Sillanpää, M. Choudhury, & S. Goswami (Eds.), *Food and agro-industrial wastes: Sustainable impacts, transformation and added value of by-products.* Academic Press. https://doi.org/10.1016/B978-0-443-31654-8.00008-4

Kumari, R., & Kaur, K. (2024). Robust security framework for iot-enabled smart cities: Leveraging ensemble machine learning techniques in fog computing environments. https://doi.org/10.21203/rs.3.rs-5197026/v1

Liu, W., & Chen, G. (2024). Towards optimal image processing-based internet of things monitoring approaches for sustainable cities. *International Journal of Advanced Computer Science and Applications, 15*(5), 1134. https://doi.org/10.14569/IJACSA.2024.01505115

Majumdar, S. (2020). Prospects of sustainable transport of automobile vehicles in Kolkata City. In G. R. Kanagachidambaresan, R. Maheswar, V. Manikandan, & K. Ramakrishnan (Eds.), *Internet of things in smart technologies for sustainable urban development. EAI/springer innovations in communication and computing*. Springer. https://doi.org/10.1007/978-3-030-34328-6_5

Majumdar, S., & Chatterjee, U. (2020). Modelling urban growth using urban growth deterministic model in Kolkata metropolitan area: A geo-statistical approach. *Modeling Earth Systems and Environment, 7*(4), 2241–2249. https://doi.org/10.1007/s40808-020-00985-6

Majumdar, S., Kandpal, V., & Anthopoulos, L. G. (2024). *Smart cities: Innovations, challenges and future perspectives.* Springer Nature. https://doi.org/10.1007/978-3-031-59846-3

Majumdar, S., & Sarkar, B. (2020). Developing a smart and sustainable transportation plan for a large-sized city: An approach to smart city modeling (pp. 145–152). https://doi.org/10.1007/978-981-15-2780-7_18

Majumdar, S., & Sivaramakrishnan, L. (2017). Monitoring urban growth and land use change detection in the southern fringes of Kolkata metropolitan area. *Indian Geographical Journal, 92*(2), 168–183.

Majumdar, S., & Sivaramakrishnan, L. (2025). Impacts of vegetation and impervious surfaces on urban sustainability: A case study of Kolkata. In L. Sivaramakrishnan & S. Bandyopadhyay (Eds.), *Sustainable urban forms and communities: Urban geographies of eastern India*. Routledge. https://doi.org/10.4324/9781003509400-11

Miranda, R., Alves, C., Sousa, R., Chaves, A., Montenegro, L., Peixoto, H., & Machado, J. (2024). Revolutionising the quality of life: The role of real-time sensing in smart cities. *Electronics, 13*(3), 550. https://doi.org/10.3390/electronics13030550

Mohamed, A. (2023). Nurturing sustainable urban space: Integrating smart city innovations and earthship design principles for eco-friendly futures. *International Journal of Architectural Engineering and Urban Research, 6*(2), 376–400. https://doi.org/10.21608/ijaeur.2024.265873.1066

Mozūriūnaitė, S. (2018). The role of landscape design in smart cities. *Landscape Architecture and Art, 13*, 49–55. https://doi.org/10.22616/j.landarchart.2018.13.05

Namana, M., & Kumar, B. (2024). An efficient and robust night-time surveillance object detection system using yolov8 and high-performance computing. *International Journal of Safety and Security Engineering, 14*(6), 1763–1773. https://doi.org/10.18280/ijsse.140611

Narayan, S., Khatun, B., Munasinghe, S. H., Pandey, S. R., Choudhary, R., Troilo, M., & Birje, T. S. (2025). Performance metrics for circular economy: Techniques for evaluating effectiveness and measuring progress. In S. Majumdar, M. Choudhury, & A. Cheshmehzangi (Eds.), *Sustainable urban future: Embracing circular economy in waste management* (pp. 249–270). Springer. https://doi.org/10.1007/978-981-95-2976-6_11

Narayan, S., Pathak, P., Khatun, B., Choudhary, R., Troilo, M., Raj, C., & Vambol, V. (2025). Resource recovery in urban settings: Developing circular supply chains for sustainable cities. In S. Majumdar, M. Choudhury, & A. Cheshmehzangi (Eds.), *Sustainable urban future: Embracing circular economy in waste management* (pp. 225–247). Springer. https://doi.org/10.1007/978-981-95-2976-6_10

Narayan, S., Yadav, D., Jha, S., Khatun, B., Choudhary, R., Verma, U. K., & Kumar, D. (2025). From past to present: The evolution of waste management and the rise of circular economy principles. In S. Majumdar, M. Choudhury, & A. Cheshmehzangi (Eds.), *Sustainable urban future: Embracing circular economy in waste management* (pp. 23–53). Springer. https://doi.org/10.1007/978-981-95-2976-6_2

Neves, F., Aparício, M., & Neto, M. (2024). The impacts of open data and explainable ai on real estate price predictions in smart cities. *Applied Sciences, 14*(5), 2209. https://doi.org/10.3390/app14052209

Nguyen, L. H., Tran, K. D., Zeng, X., & Tran, K. P. (2024). Human-centered edge artificial intelligence for smart factory applications in industry 5.0: A review and perspective. In K. P. Tran (Ed.), *Artificial intelligence for safety and reliability engineering (springer series in reliability engineering)*. Springer. https://doi.org/10.1007/978-3-031-71495-5_5

Nnenna, U., Perpetua, U., Nchaga, A., Hadijah, T., Mabonga, E., & Nyamboga, T. (2024). Impact of digital governance on public service delivery in East Africa. *IAA Journal of Arts and Humanities, 11*(1), 18–29. https://doi.org/10.59298/iaajah/2024/11.11829.11

Nyangaresi, V., Abduljabbar, Z., Mutlaq, K., Bulbul, S., Ma, J., Aldarwish, A., & Neamah, H. (2024). Smart city energy efficient data privacy preservation protocol based on biometrics and fuzzy commitment scheme. *Scientific Reports, 14*(1). https://doi.org/10.1038/s41598-024-67064-z

Obioha-Val, O., Olaniyi, O., Selesi-Aina, O., Gbadebo, M., & Kolade, T. (2024). Machine learning-enabled smart sensors for real-time industrial monitoring: Revolutionizing predictive analytics and decision-making in diverse sector. *Asian Journal of Research in Computer Science, 17*(11), 92–113. https://doi.org/10.9734/ajrcos/2024/v17i11522

Okafor, C., Aigbavboa, C., & Thwala, W. (2022). A bibliometric evaluation and critical review of the smart city concept—Making a case for social equity. *Journal of Science and Technology Policy Management, 14*(3), 487–510. https://doi.org/10.1108/jstpm-06-2020-0098

Olimat, H., Liu, H., & Abudayyeh, O. (2023). Enabling technologies and recent advancements of smart facility management. *Buildings, 13*(6), 1488. https://doi.org/10.3390/buildings13061488

Pancewicz, A. (2022). Urban nature as an active means of adapting public spaces to climate conditions: Case studies from Copenhagen and selected polish cities. *Civil and Environmental Engineering Reports, 32*(4), 123–146. https://doi.org/10.2478/ceer-2022-0049

Papaioannou, C., Dimara, A., Papaioannou, A., Tzitzios, I., Anagnostopoulos, C., & Krinidis, S. (2025). Hierarchical resources management system for internet of things-enabled smart cities. *Sensors, 25*(3), 616. https://doi.org/10.3390/s25030616

Piera-Jiménez, J., Carot-Sans, G., Ramiro-Pareta, M., Nogueras, M., Folguera, J., Ródenas, P., & Sust, P. (2024). A 25-year retrospective of health it infrastructure building: The example of the catalonia region. *Journal of Medical Internet Research, 26*, e58933. https://doi.org/10.2196/58933

Pioli, L., Macedo, D., Costa, D., & Dantas, M. (2024). Towards an ai-driven data reduction framework for smart city applications. *Sensors, 24*(2), 358. https://doi.org/10.3390/s24020358

Piosik, A., & Karmańska, A. (2023). Impact of digitalization on budgeting functions: An investigation into contingency theory perspectives. https://doi.org/10.20944/preprints202312.0134.v1.

Porse, E. (2013). Stormwater governance and future cities. *Water, 5*(1), 29–52. https://doi.org/10.3390/w5010029

Pradhan, B., Das, S., Roy, D., Routray, S., Benedetto, F., & Jhaveri, R. (2023). An AI-assisted smart healthcare system using 5g communication. *IEEE Access, 11*, 108339–108355. https://doi.org/10.1109/access.2023.3317174

Radziszewska, A. (2023). Data-driven approach in knowledge-based smart city management. *European Conference on Knowledge Management, 24*(2), 1090–1098. https://doi.org/10.34190/eckm.24.2.1600

Rane, N. (2023). Integrating leading-edge artificial intelligence (AI), internet of things (IoT), and big data technologies for smart and sustainable architecture, engineering and construction (AEC) industry: Challenges and future directions. *SSRN*. https://doi.org/10.2139/ssrn.4616049

Rane, N., Rane, J., Paramesha, M., & Kaya, Ö. (2024). Artificial intelligence, machine learning, and deep learning for enabling smart and sustainable cities and infrastructure. In *Artificial intelligence and industry in society 5.0*. Deep Science Publishing. https://doi.org/10.70593/978-81-981271-1-2_2

Safta, C., Cococi, V., Călinoiu, C., & Marin, A. (2023). Performance of water valves required by water supply network digitalization process. *IOP Conference Series: Earth and Environmental Science, 1136*(1), 012051. https://doi.org/10.1088/1755-1315/1136/1/012051

Sameer, N., Alalouch, C., Al-Saadi, S., & Saleh, M. (2023). Toward smart sustainable cities: Assessment of stakeholders' readiness for digital participatory planning. *International Journal of Architectural Research Archnet-IJAR, 18*(4), 914–936. https://doi.org/10.1108/arch-06-2023-0159

Sathishbabu, S., Dhanalakshmi, R., Bharathiraja, R., & Thirunavukkarasu, K. (2025). Predicting the performance of internet of things communication systems using finite element interpolated neural network in smart cities. *International Journal of Communication Systems, 38*(10). https://doi.org/10.1002/dac.70101

Schreiber, A., & Schreiber, I. (2023). Integrating cybersecurity maturity: A hybrid model for both traditional environment and cloud environment. *International Research Journal of Modernization in Engineering Technology and Science, 5*, 885. https://doi.org/10.56726/irjmets47210

Shahrour, I., Bélaïd, F., Flambard, V., Siano, P., Radu, L., Stermieri, L., & Şanta, A. (2020). Disruptive technologies in smart cities: A survey on current trends and challenges. *Smart Cities, 3*(3), 1022–1038. https://doi.org/10.3390/smartcities3030051

Singh, M., & Angurala, M. (2022). 5g technology based edge computing in uav networks for resource allocation with routing using federated learning access network and trajectory routing protocol. *International Journal on Future Revolution in Computer Science & Communication Engineering, 8*(2), 51–61. https://doi.org/10.17762/ijfrcsce.v8i2.2101

Sodiya, E., Umoga, U., Obaigbena, A., Jacks, B., Ugwuanyi, E., Daraojimba, A., & Lottu, O. (2024). Current state and prospects of edge computing within the internet of things (iot) ecosystem. *International Journal of Science and Research Archive, 11*(1), 1863–1873. https://doi.org/10.30574/ijsra.2024.11.1.0287

Sreenath, A. B., Dhanaraj, K., Rath, P., Bhardwaj, L. K., & Choudhary, R. (2025). Waste management and circular economy. In S. Kulkarni & H. Akolkar (Eds.), *Sustainable chemistry and pioneering green engineering solutions.* IGI Global. https://doi.org/10.4018/979-8-3373-1409-9.ch011

Strigaro, D., Cannata, M., & Antonovic, M. (2019). Boosting a weather monitoring system in low income economies using open and non-conventional systems: Data quality analysis. *Sensors, 19*(5), 1185. https://doi.org/10.3390/s19051185

Trigka, M., & Dritsas, E. (2025). Edge and cloud computing in smart cities. *Future Internet, 17*(3), 118. https://doi.org/10.3390/fi17030118

Turin, T., Subroto, S., Raihan, M., Koch, K., Wiles, R., Ruttan, E., & Chowdhury, N. (2022). Identifying challenges, enabling practices, and reviewing existing policies regarding digital equity and digital divide toward smart and healthy cities: Protocol for an integrative review. *Jmir Research Protocols, 11*(12), e40068. https://doi.org/10.2196/40068

Vempati, S. (2024). Securing smart cities: A cybersecurity perspective on integrating iot, ai, and machine learning for digital twin creation. *Journal of Electrical Systems, 20*(3), 1420–1429. https://doi.org/10.52783/jes.3548

Voorwinden, A. (2021). The privatised city: Technology and public-private partnerships in the smart city. *Law, Innovation and Technology, 13*(2), 439–463. https://doi.org/10.1080/17579961.2021.1977213

Wilson, D., Thomas, R., Burns, K., Hewitt, J., Osei-Waree, J., & Robertson, S. (2012). Canadian rural-urban differences in end-of-life care setting transitions. Global. *Journal of Health Science, 4*(5). https://doi.org/10.5539/gjhs.v4n5p1

Woolrych, R., & Man-lin, L. (2024). Exploring the role of smart cities in supporting ageing-in-place in Chongqing, China. *Australasian Journal on Ageing, 43*(2), 264–270. https://doi.org/10.1111/ajag.13301

Yan, H., & Lv, Z. (2024). A survey of sustainable development of intelligent transportation system based on urban travel demand. *Sustainable Social Development, 2*(1), 2399. https://doi.org/10.54517/ssd.v2i1.2399

Yiğitcanlar, T., Regona, M., Kankanamge, N., Mehmood, R., D'Costa, J., Lindsay, S., & Brhane, A. (2022). Detecting natural hazard-related disaster impacts with social media analytics: The case of Australian states and territories. *Sustainability, 14*(2), 810. https://doi.org/10.3390/su14020810

Zhang, Y., Chatterjee, P., & Mukherjee, A. (2023). Trust, privacy and security for smart cities. *Sustainability, 15*(6), 5523. https://doi.org/10.3390/su15065523

CHAPTER 2

Environmental Stewardship: The Impact of Circular Economies on Urban Sustainability

Basona Khatun, Saurav Narayan, Bhanu Pratap Singh, Moharana Choudhury, Himanshu, Pallav Suman, Chandra Kant Raj, and Deepak Kumar

2.1 Introduction

Urbanization has been advancing at an extraordinary pace; currently, over half of the world's population, approximately 4.4 billion people, lives in cities, and this figure is expected to reach around 70% by 2050. Such rapid expansion places immense pressure on limited natural resources, energy systems, and urban infrastructure, intensifying global environmental challenges, including climate change, air pollution, and biodiversity loss

B. Khatun • S. Narayan (✉)
Department of Law, School of Legal Studies, Central University of Punjab, Bathinda, Punjab, India

B. P. Singh
Department of Civil Engineering, Faculty of Engineering and Technology, SRM Institute of Science and Technology, Delhi NCR Campus, Modinagar, Uttar Pradesh, India

S. Majumdar, M. Choudhury (eds.), *City of the Future*,
https://doi.org/10.1007/978-3-032-18026-1_2

(Saxena, 2025). The prevailing linear economic approach, characterized by a "take-make-dispose" cycle, has proven to be environmentally unsustainable. It generates an estimated 2.01 billion tonnes of municipal solid waste each year, a number projected to rise to 3.8 billion tonnes by 2050 (United Nations Environment Programme, 2024), while material production alone contributes to roughly 45% of global greenhouse gas emissions (Henderson, 2023). Cities, as centers of production and consumption, are both significant contributors to and victims of environmental degradation, facing growing threats such as flooding, urban heat islands, and resource scarcity (Mishra, 2025). Environmental stewardship is the moral duty to care for, protect, and restore the natural world by using methods that meet the needs of both current and future generations (Kara, Hauschild, Sutherland, & McAloone, 2022). It emphasizes a collective duty shared among governments, industries, and citizens to preserve ecological integrity and sustain the natural systems upon which human life depends. Closely linked to this idea is the concept of urban sustainability, which focuses on developing and maintaining cities that achieve a balance between economic prosperity, social well-being, and environmental protection. This approach promotes the creation of urban systems that function efficiently, minimize resource consumption, and foster social inclusion while remaining within the planet's ecological limits (Ghisellini, Cialani, & Ulgiati, 2016). In this sense, environmental stewardship serves as the

M. Choudhury
Environmental Research and Management Division, Voice of Environment (VoE), Guwahati, Assam, India

Himanshu
Department of English, University of Delhi, New Delhi, India

P. Suman
Department of Chemistry and Biotechnology, Tallinn University of Technology, Tallinn, Estonia
e-mail: pallav.suman@taltech.ee

C. K. Raj
College of International Education, Jinzhou Medical University, Jinzhou, Liaoning Province, People's Republic of China

D. Kumar
Dnipro State Medical University, Dnipro, Ukraine

foundation for urban sustainability, guiding cities toward resilient, regenerative, and adaptive development pathways.

The rapid growth of urban populations has intensified environmental degradation, as expanding cities drive higher demands for energy, water, housing, and transportation. This growth often results in unsustainable patterns of land use, loss of natural habitats, and increasing waste generation (Das et al., 2024). Without strategic intervention, such trends could undermine not only environmental quality but also urban livability and social stability. Therefore, the establishment of sustainable systems to accommodate future urban growth is essential. Cities must shift from extractive and wasteful models to regenerative and circular systems that prioritize resource efficiency, renewable energy, and closed-loop production processes. Integrating circular economy (CE) principles into urban planning is a practical way to make this change happen. It lets cities balance economic growth with environmental protection and long-term resilience (Scheel, Aguiñaga, & Bello, 2020). Circular economies (CEs) are a promising alternative to traditional models because they focus on regeneration, using resources wisely, and cutting down on waste (Ghisellini et al., 2016; Kara et al., 2022). Unlike linear systems, CE frameworks aim to decouple economic growth from the extraction of finite resources, promoting practices such as product durability, material recycling, and industrial symbiosis (Scheel et al., 2020). This change not only lowers ecological footprints, but it also makes cities more resilient and opens up new socio-economic opportunities for everyone. For instance, adopting CE principles in the construction sector through modular designs and the use of recycled materials could potentially reduce global building-related emissions by up to 38% by 2030 (UNEP, 2020). Additionally, the development of waste-to-energy initiatives and decentralized renewable energy networks aligns directly with global climate mitigation goals (Möslinger, Ulpiani, & Vetters, 2023). This chapter explores how the adoption of circular economy principles can reshape urban systems to promote environmental stewardship and sustainable development. Drawing on bibliometric analysis of research from 2015 to 2024, it demonstrates the increasing scholarly and global attention devoted to applying CE concepts in addressing urban sustainability challenges. By combining theoretical perspectives, practical case studies, and policy implications, the chapter contends that the transition to circular models is vital for tackling the dual challenges of ecological preservation and social equity in rapidly urbanizing environments. Ultimately, this analysis emphasizes the urgent need to move away from exploitative, linear systems and toward regenerative urban ecosystems that operate within planetary boundaries while supporting inclusive and sustainable growth.

2.2 Global Research Trends on Sustainability and Circular Economy

To assess the current state of research on sustainability and CE at the international level, a bibliometric analysis was performed using the Scopus database. The search query employed was TITLE-ABS-KEY (sustainability OR circular AND economy) AND PUBYEAR > 2014 AND PUBYEAR < 2025. The analysis focused on publications from 2015 to 2024, aiming to capture recent trends and emerging themes within the field. Key findings derived from the search results are presented graphically in Figs. 2.1, 2.2, and 2.3.

The search resulted in a total of 69,881 research documents published during the 10-year period from 2015 to 2024. Beginning with only 1363 research documents in 2015 and 1748 documents in 2016, interest in the field of sustainability and CE remained modest until 2017. However,

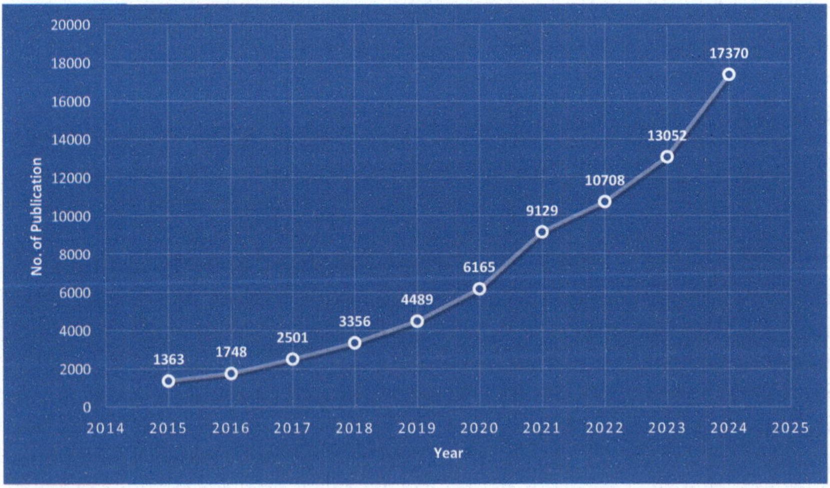

Fig. 2.1 Year-wise distribution of publications on *sustainability and circular economy* (2015–2024)

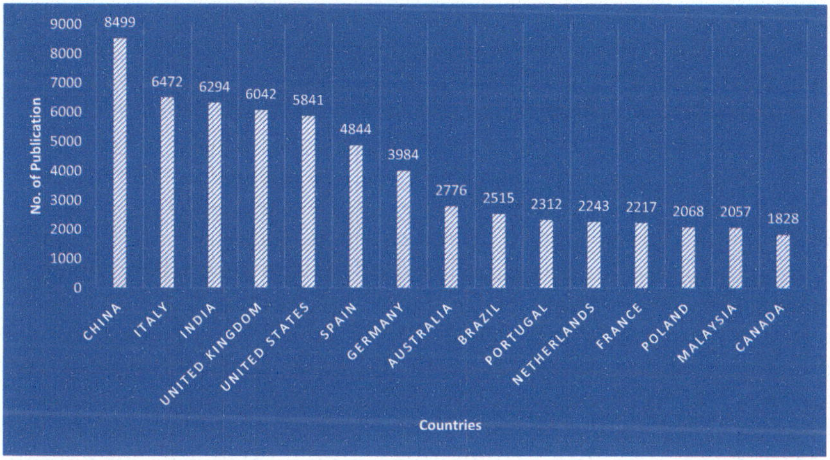

Fig. 2.2 Number of publications on *sustainability and circular economy* (2015–2024) by country

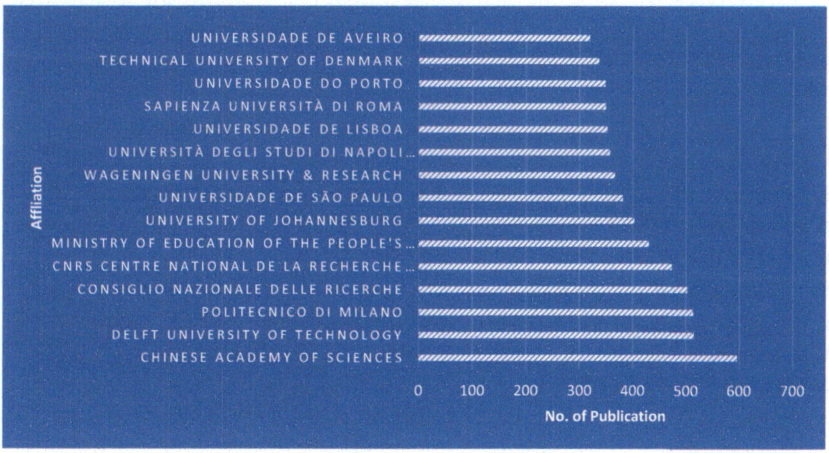

Fig. 2.3 Affiliation-wise research on *sustainability and circular economy*

research activity started gaining momentum from 2018 onward, with more than 3000 documents published in that year alone. The upward trend continued strongly, with more than 6000 publications annually since 2020. Reflecting the growing global focus on sustainable practices and CE models, 2024 witnessed a significant peak, with 17,370 research documents published in a single year (Fig. 2.1).

The country-wise distribution of research on sustainability and the CE reveals a significant concentration of studies in a handful of countries. China leads the field, producing 8499 research documents between 2015 and 2024, making it the highest contributor to the topic. Italy ranks second with 6472 documents, followed closely by India with 6294 publications. The United Kingdom produced 6042 documents, slightly trailing behind India. The United States stands fifth, contributing 5841 documents to the research on sustainability and CE. Other notable contributors include Spain (4844 documents), Germany (3984 documents), and Australia (2776 documents). Meanwhile, Brazil (2515 documents), Portugal (2312 documents), the Netherlands (2243 documents), France (2217 documents), Poland (2068 documents), Malaysia (2057 documents), and Canada (1828) also made considerable contributions. It is important to note that only the list of the top 15 countries accounts for 59,992 out of 69,881 research documents (Fig. 2.2).

From 2015 to 2024, an analysis of global research output reveals that the top 15 institutions have significantly outperformed others in terms of publication volume in the field of sustainability and CE-related topics. The Chinese Academy of Sciences leads globally, having produced 596 research documents during this period. It is followed by Delft University of Technology with 515 documents and Politecnico di Milano with 514 documents. Consiglio Nazionale delle Ricerche recorded 503 publications, while CNRS Centre National de la Recherche Scientifique contributed 474 documents. The Ministry of Education of the People's Republic of China is also prominent, with 431 documents. Other leading institutions include the University of Johannesburg (403 documents), Universidade de São Paulo (381 documents), and Wageningen University & Research (366 documents). Università degli Studi di Napoli Federico II reported 357 documents, closely followed by Universidade de Lisboa (352 documents) and Sapienza Università di Roma (349 documents). Additional significant contributors are Universidade do Porto (348 documents), Technical University of Denmark (336 documents), and Universidade de Aveiro (319 documents).

The bibliometric analysis of global research trends from 2015 to 2024 highlights a rapid and substantial growth in sustainability and CE studies. Research output surged significantly after 2017, culminating in a record number of publications in 2024. China, Italy, and India emerged as top contributors, together with strong outputs from the United Kingdom and the United States. A small group of institutions, led by the Chinese Academy of Sciences and Delft University of Technology, dominated global research production. These patterns reflect an accelerating international commitment to sustainable development and CE practices, emphasizing the field's expanding global significance.

2.3 Theoretical Foundations of Circular Economies

The evolution from linear to circular economic models represents a fundamental transformation in how societies conceptualize resource use, waste management, and environmental stewardship in urban contexts (Fratini, Georg, & Jørgensen, 2019; Fusco Girard & Nocca, 2019). This paradigm shift addresses the escalating environmental crisis generated by traditional linear systems while establishing new pathways for sustainable urban development.

2.3.1 Historical Background: Evolution from Linear to Circular Economic Models

The linear economy emerged during the Industrial Revolution as a system predicated on abundant natural resources and cheap energy (Fischer-Kowalski, Rovenskaya, Krausmann, Pallua, & Mc Neill, 2019). This "take-make-dispose" model became the dominant framework for economic development, characterized by extracting raw materials, manufacturing products, consuming them, and discarding them as waste (Rashid & Malik, 2023). The historical circular economy model actually preceded industrialization; societies traditionally operated with nothing lost or discarded, systematically recovering and reusing materials (Circular economy: Definition, importance, and benefits, 2023). Three major shifts catalyzed the decline of historical circular practices and the rise of linear systems. First, the Industrial Revolution introduced coal-fired steam engines and cheap energy, dramatically reducing raw material extraction costs, while new synthetic materials like chemical fertilizers replaced recovered materials (Stahel, 2021). Second, the hygienist movement of the late nineteenth century stigmatized waste circulation, claiming it caused epidemics and leading to the institutionalization of waste containment through landfills. Third, the 1930s marked the growth of consumer society, where modern lifestyles became defined by objects consumed or owned (Kovacic, Strand, & Völker, 2019).

The concept of circular economy re-emerged in the late twentieth century, drawing from multiple schools of thought, including industrial ecology (IE), biomimicry, and cradle-to-cradle (C2C) design principles (Sarna, Kortelainen, Plotnikova, & Saqib, 2025). A parallel evolution can be traced in the field of waste management, where policy regimes and technological systems gradually shifted from end-of-pipe disposal toward prevention, recovery, and circular value creation. Comprehensive historical analyses document how successive phases of waste governance, from uncontrolled dumping and basic sanitary landfilling to integrated waste management, have progressively incorporated circular economy logics into their core objectives. This trajectory underscores that the rise of circular economy principles is deeply intertwined with the maturation of waste management systems, particularly in urban contexts where material flows are most concentrated (Narayan et al., 2025). The foundational vision was articulated by Walter Stahel and Genevieve Reday in their 1976 research report, "*The Potential for Substituting Manpower for Energy*," which sketched an economy in loops and its impact on job creation, economic competitiveness, resource savings, and waste prevention (Stahel, 2020). China integrated circular economy principles into its industrial and environmental policies in the early 2000s, while the Ellen MacArthur Foundation played an instrumental role in diffusing the concept throughout Europe and the Americas (Hall & Thomsen, 2025; Zhu, Fan, Shi, & Shi, 2019).

2.3.2 *Principles of Circular Economy*

The circular economy operates on three fundamental principles that collectively create a regenerative system: design out waste and pollution, keep products and materials in use, and regenerate natural systems (Collado-Ruano & Segovia Sarmiento, 2022; Girard, 2021). These principles represent a systemic approach to economic development designed to benefit businesses, society, and the environment by gradually decoupling growth from finite resource consumption.

2.3.2.1 Design Out Waste and Pollution

The first principle requires products, materials, and infrastructure to be designed for circularity from the outset, preventing waste creation rather than managing it post-production (Moorhouse, 2023). This approach moves beyond end-of-life waste management toward preventive design strategies that eliminate negative environmental impacts during the design phase (Law & Narayan, 2021; Markevičiūtė & Varžinskas, 2022). Products are conceived as components in closed-loop systems, where materials flow continuously without becoming waste (Bharadwaj et al., 2023; Camilleri, 2019).

Practical applications include using renewable inputs, non-toxic materials, and modular design principles that facilitate disassembly and component recovery (Biswas & John, 2022). Examples range from buildings designed as material banks where components can be systematically disassembled and reused to product-as-service models that maintain manufacturer responsibility for product lifecycles (Bragança et al., 2025). The Copenhagen Nordhavn District exemplifies this principle through waste-to-energy facilities that convert municipal waste into electricity and heat, eliminating waste while generating urban energy (Lucciarini & Galdini, 2024).

2.3.2.2 Keep Products and Materials in Use

The second principle focuses on maximizing material utility through maintenance, reuse, refurbishment, remanufacturing, and recycling, ensuring products and materials circulate at their highest value (Jensen, 2025a; Kandpal, Jaswal, Santibanez Gonzalez, & Agarwal, 2024; Psarommatis & May, 2025). This creates cascading value chains where materials maintain utility across multiple lifecycles rather than following linear paths to disposal (Campbell-Johnston, Vermeulen, Reike, & Brullot, 2020). Urban contexts offer particular advantages for implementing sharing economy models, product-as-service arrangements, and collaborative consumption systems due to high population density and resource concentration (Deal, 2020).

2.3.2.3 Regenerate Natural Systems

The third principle positions circular systems as actively restorative rather than merely sustainable, focusing on improving natural environments and building biodiversity (Gibbons, 2020; Morseletto, 2020). This involves using farming practices that restore soils, implementing urban greening that enhances ecosystem services, and adopting renewable energy systems that reduce environmental stress (Ogwu & Kosoe, 2025; Rawat et al., 2022). Circular cities integrate green infrastructure as functional components of urban metabolism rather than decorative additions (Perrotti & Stremke, 2020).

2.3.2.4 Comparison with Linear and Circular Economies

The fundamental distinction between linear and circular economic models lies in their approach to resource flows, value creation, and environmental impact (Table 2.1). Linear systems operate as open loops where materials flow unidirectionally from extraction to disposal, while circular systems function as closed loops where materials continuously circulate.

2.3.3 Key Differences and Shortcomings of Traditional Model

The traditional linear economy is characterized by resource intensity, environmental externalities, and inefficient waste management, which undermine urban sustainability and future security (Ragazou, Zournatzidou, Sklavos, & Sariannidis, 2024). Recent syntheses on waste management and circular economy further underline that these shortcomings are structurally embedded in how contemporary waste systems are designed and governed. They show that without a deliberate shift toward circular value chains emphasizing prevention, recovery, and regenerative use of materials, waste infrastructures tend to lock cities into high-emission, resource-intensive trajectories. Framing waste management explicitly within circular economy principles therefore becomes a critical lever for reconfiguring urban systems away from linear dependence and toward long-term sustainability (Sreenath, Dhanaraj, Rath, Bhardwaj, & Choudhary, 2025). Circular economies address these shortcomings by promoting systemic change, closing resource loops, and enabling cities to become regenerative hubs for people and nature (Cervantes Puma, Salles, & Bragança, 2024). This framework redefines stewardship in urban settings, suggesting that circular economic principles are essential for advancing both sustainability and resilience moving beyond mere mitigation toward long-term ecological vitality (Bühler, Hollenbach, Köhler, & Armstrong, 2024).

Table 2.1 Comparative analysis: linear vs. circular economic models

Linear economy limitations	*Impact description*	*Circular economy advantages*	*Solution description*
Resource depletion	Continuous extraction exhausts finite resources, jeopardizing future availability (Sillanpää & Ncibi, 2019; Skawińska & Zalewski, 2021)	Resource efficiency	Closed-loop systems maximize utility and reduce virgin resource extraction (Hagelüken & Goldmann, 2022; Mishra, Dutta, Jayasankar, Jain, & Mathiyazhagan, 2023)
Pollution generation	Manufacturing creates significant air, water, and soil contamination (Mihai & Minea, 2021)	Waste elimination	Design strategies prevent waste creation rather than managing disposal (Salmenperä, Pitkänen, Kautto, & Saikku, 2021)
Waste accumulation	Landfills overflow creating environmental hazards while wasting resources (Yang et al., 2023)	Regenerative impact	Systems actively restore and enhance natural environments (Čega, Drezgić, & Čišić, 2024)
Climate impact	Energy-intensive processes contribute substantially to greenhouse emissions (Abbas, Imran, & Sattar, 2025; Feldmann, 2023)	Economic resilience	Local resource cycles reduce dependency on external supply chains (Morseletto, 2020)
Economic inefficiency	Planned obsolescence forces repeated purchases while underutilizing materials (Figge, Dimitrov, Schlosser, & Chenavaz, 2022)	Innovation catalyst	Circular models drive technological and social innovation (Jensen, 2025b; Suchek, Fernandes, Kraus, Filser, & Sjögrén, 2021)

2.3.4 *Framework of Circular Economy*

The CE is underpinned by interdisciplinary theories that challenge the linear "*take-make-waste*" paradigm. These frameworks emphasize regeneration, systemic efficiency, and harmony with ecological limits, offering blueprints for sustainable urban transformation. We examine key theoretical foundations below, emphasizing their relevance to cities. The cradle-to-cradle (C2C) framework (Fig. 2.4), developed by Michael Braungart

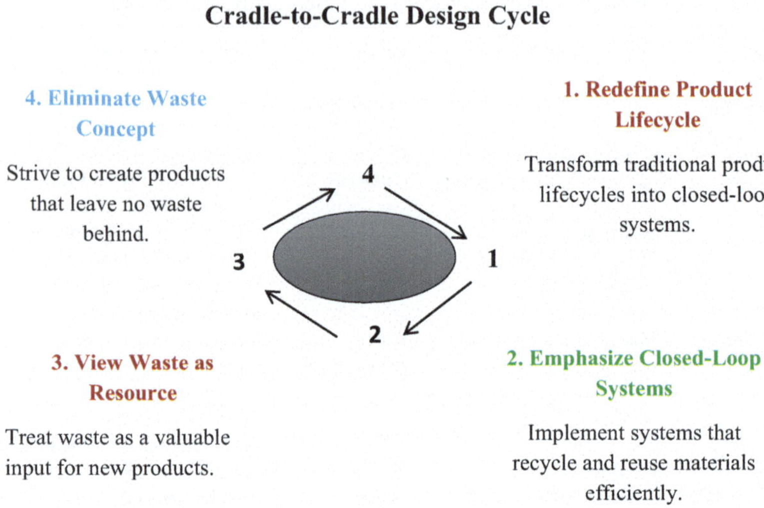

Fig. 2.4 Cradle-to-cradle framework

and William McDonough, redefines product and system design by emulating nature's regenerative cycles. Unlike the traditional linear "*cradle-to-grave*" model, C2C promotes a circular approach in which materials are continuously cycled as either biological or technical nutrients. Biological nutrients safely decompose and return to the environment, while technical nutrients are reused indefinitely without loss of quality. The framework emphasizes key design principles, including material health, renewable energy, product disassembly, and ecological intelligence (Sherratt, 2013). In urban contexts, applying C2C principles minimizes waste, optimizes resource efficiency, and increases regenerative cities. This approach directly supports SDG 12: Ensure sustainable consumption and production patterns, particularly Target 12.5, which aims to reduce waste generation substantially through prevention, reduction, recycling, and reuse. It also advances SDG 11: Make cities inclusive, safe, resilient, and sustainable, specifically Target 11.6, focused on reducing the adverse environmental impact of cities through improved waste management.

Industrial ecology (IE), pioneered by Robert U. Ayres, analyzes industrial systems with the goal of optimizing material and energy flows, much like natural ecosystems (Fig. 2.5). Rather than viewing industries in

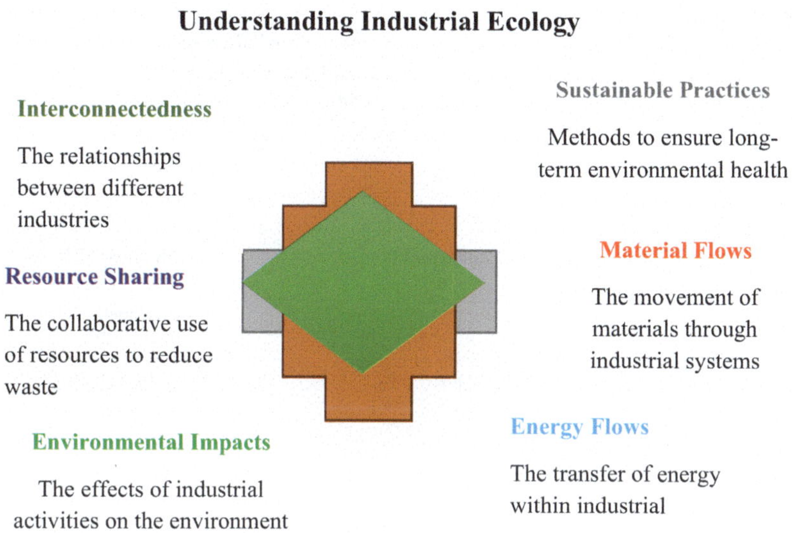

Fig. 2.5 Industrial ecology framework

isolation, IE encourages a systemic perspective where waste from one process serves as input for another, promoting resource efficiency and minimizing environmental impact. Central to IE is the idea that industrial activity should be embedded within the carrying capacity of ecosystems, mimicking nature's closed-loop cycles. Through methods such as material flow analysis (MFA) and life cycle assessment (LCA), IE provides powerful tools to redesign production processes, urban infrastructures, and supply chains toward sustainability (Klancko, 2003). In the context of urban sustainability and CEs, IE supports the development of eco-industrial parks, urban symbiosis, and integrated waste management systems. This framework directly advances SDG 8: Promote sustainable economic growth, productive employment, and decent work, Target 8.4, which calls for improving global resource efficiency. It also aligns with SDG 12: Ensure sustainable consumption and production patterns, specifically Target 12.2, emphasizing the sustainable management and efficient use of natural resources.

The Blue Economy (BE), conceptualized by Gunter Pauli, proposes an economic model that leverages locally available resources to drive

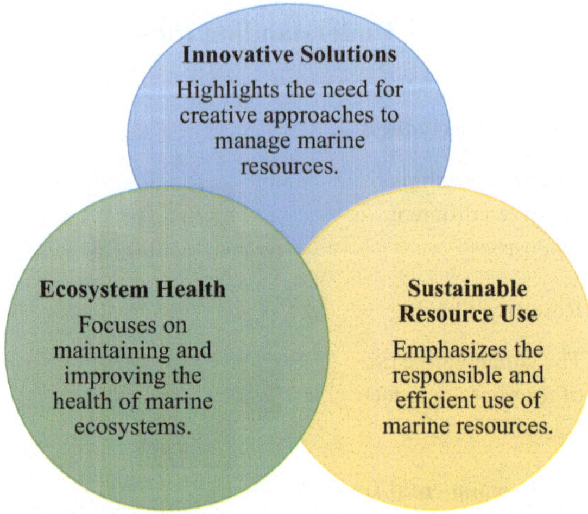

Fig. 2.6 Blue Economy framework

sustainable development, reduce environmental impact, and create employment opportunities (Fig. 2.6). Unlike traditional economic systems that often rely heavily on external inputs and generate significant waste, the BE emphasizes innovation inspired by nature's efficiency. It encourages the transformation of local, often underutilized, resources into value-added goods and services without causing environmental harm (Kumaraswamy, Choudhury, Ponappa, & Majumdar, 2025). By mimicking ecosystems where no waste exists, the BE seeks to establish self-sustaining industrial and urban systems (Pauli, 2010). In the context of urban sustainability, BE principles promote decentralized production, local job creation, waste-to-resource innovations, and closed-loop systems. This framework directly advances SDG 9: Build resilient infrastructure to promote sustainable industrialization, notably Target 9.4, which focuses on upgrading industries to enhance sustainability and resource-use efficiency. It also supports SDG 12: Responsible consumption and production, specifically Target 12.4, emphasizing the environmentally sound management of chemicals and waste throughout their life cycle. Emerging coastal case studies further illustrate how the Blue Economy can operationalize these principles through the sustainable use of living marine resources. Investigations of wild seaweeds along Nipah Beach in Indonesia, for example, have identified diverse macroalgal species with significant antioxidant potential, demonstrating that locally available marine biomass

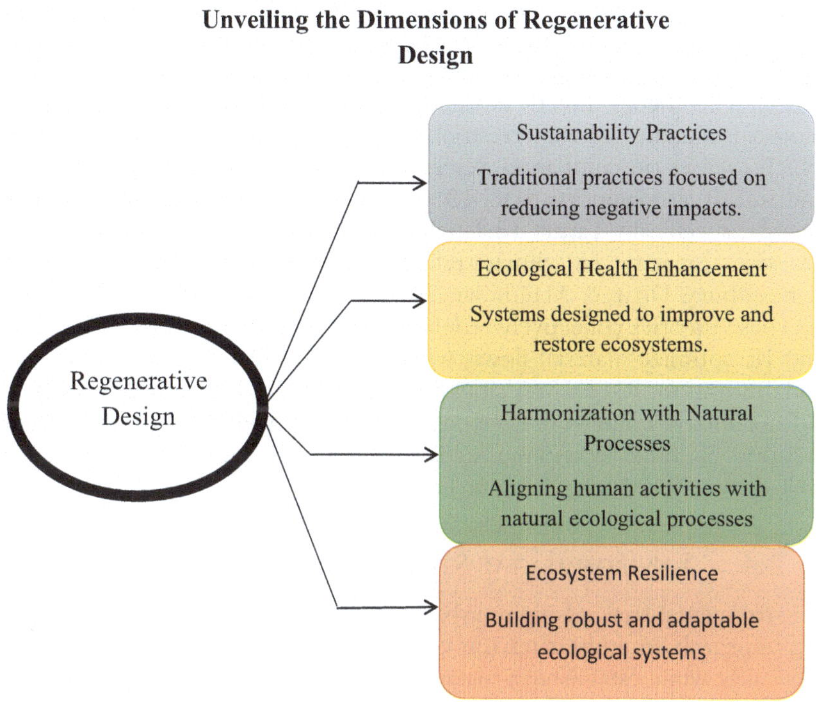

Fig. 2.7 Regenerative design framework

can be transformed into high-value bioactive products. Such nature-based value chains exemplify how coastal and peri-urban regions can couple ecosystem conservation with circular, place-based economic development under a Blue Economy paradigm (Cokrowati et al., 2025).

Regenerative design (RD), pioneered by John T. Lyle, envisions systems that not only sustain but actively restore, renew, and revitalize their sources of energy and materials (Fig. 2.7). Unlike conventional sustainable design, which often focuses on minimizing harm, RD seeks to create environments with a net-positive impact on ecological and human systems. It promotes closed-loop cycles of resource use, where waste becomes a resource, and landscapes are managed to enhance biodiversity, soil health, and water systems. In urban contexts, RD advocates for buildings, infrastructure, and public spaces that contribute positively to environmental health and community well-being (Lyle, 1994). By embedding RD principles into urban planning, cities can evolve into resilient ecosystems capable of adapting to climate change and resource scarcity (Bharathy et al., 2025). This theory directly advances SDG 15: Protect, restore, and promote sustainable use of terrestrial ecosystems, particularly Target 15.3, which focuses on combating desertification and restoring degraded land and soil. It also supports SDG 13: Take urgent action to combat climate change, specifically Target 13.1, which seeks to strengthen resilience and adaptive capacity to climate-related hazards and natural disasters (Choudhury, Dixit, & Majumdar, 2025).

These theories collectively redefine urban resource management. C2C and IE optimize material flows, while the BE prioritizes value retention and innovation. RD ensures cities emulate ecological resilience. They provide a cohesive framework for cities to transition from extraction to regeneration, aligning economic activity with environmental and social well-being (Korhonen, Nuur, & Feldmann, 2023).

2.4 Environmental Stewardship in Urban Settings

Environmental stewardship in urban contexts represents a vital and evolving approach to addressing the complex environmental challenges that cities face while establishing sustainable pathways for urban development (Mansourihanis, Maghsoodi Tilaki, Yousefian, & Zaroujtaghi, 2023; Pera, 2020). This concept embodies collaborative participation among diverse stakeholders, systematic responses to ecological pressures, and the creation of robust policy mechanisms that support effective environmental governance (Blühdorn & Deflorian, 2019). As cities keep getting bigger, stewardship gives them a way to balance their goals for growth with the need to protect natural systems that are important for long-term health and resilience (Conti, Battaglia, Calabrese, & Simone, 2021). In cities, environmental stewardship means that people, groups, and institutions are actively and responsibly involved in protecting, restoring, and improving urban ecosystems and the ecological and social functions that go along with them (Campbell, Svendsen, Johnson, & Landau, 2022; Murphy, Enqvist, & Tengö, 2019). This involvement is based on a common value of care and a shared goal of making cities more sustainable and fair for everyone. At its core, stewardship is proactive; it extends beyond simple compliance or conservation, emphasizing ongoing action, education, and innovation. Urban environmental stewardship acknowledges that even highly modified urban ecosystems continue to operate according to ecological principles and that their health and resilience depend on deliberate human intervention and care (Domínguez-Escrig, Mallén-Broch, Lapiedra-Alcamí, & Chiva-Gómez, 2019). Furthermore, stewardship in cities encompasses a broad range of activities such as conserving green spaces, managing community gardens, monitoring local ecosystems, and advocating for sustainable urban planning (Aiguobarueghian, Adanma, Ogunbiyi, & Solomon, 2024a; McCarthy & Russo, 2023). It also includes teaching people about the environment and creating a culture of shared responsibility for urban nature.

2.4.1 *Active Role of Citizens, Governments, and Businesses*

2.4.1.1 *Citizens as Environmental Stewards*

Citizens form the foundation of environmental stewardship in cities through direct participation in ecological management and civic engagement aimed at improving environmental and social well-being (Campbell et al., 2022; Kiss et al., 2022). Urban stewards may act as volunteers, community organizers, or professionals who dedicate time and expertise to conservation and sustainability projects (Lopez & Weaver, 2023). Activities such as neighborhood clean-ups, tree planting, and advocacy for green policies illustrate how citizens translate environmental concern into tangible action.

Studies indicate that civic stewardship networks frequently function within localized clusters while coordinating on a citywide level through partnerships with municipal agencies (Caggiano, Landau, Campbell, Johnson, & Svendsen, 2024; Jasny, Johnson, Campbell, Svendsen, & Redmond, 2019; Wolf, Brinkley, & Blahna, 2021). These networks help build multi-level capacity for managing ecosystem services in cities. Through such collaborations, communities and governments co-develop flexible institutional arrangements that respond to both social and ecological needs, fostering experimentation and shared management practices within urban landscapes.

2.4.1.2 *Government Environmental Leadership*

Governments at all levels play a central role in shaping and sustaining environmental stewardship through regulation, resource allocation, and policy innovation (Miao & Nduneseokwu, 2024; Shan & Ji, 2024). Government stewardship responsibilities include enforcing environmental regulations, developing comprehensive policies aligned with sustainability goals, and providing funding and technical assistance for community initiatives (Bennett et al., 2018). Equally important is the government's role in coordinating partnerships among civic groups, businesses, and other institutions to ensure that stewardship efforts are inclusive, effective, and scalable.

2.4.1.3 Business Sector Participation

Businesses also play a pivotal role in urban environmental stewardship by integrating sustainability into their operations and community relationships (Lăzăroiu, Ionescu, Andronie, & Dijmărescu, 2020). Corporate stewardship encompasses environmentally responsible management of assets and resources through pollution prevention, habitat restoration, and the pursuit of continuous improvement (Ordonez-Ponce, Clarke, & MacDonald, 2021; Wahyuni, Ulfa, Novianti, & Rosalinda, 2024). Common business practices that support stewardship include implementing ISO 14001-certified environmental management systems (Bravi, Santos, Pagano, & Murmura, 2020; Johnstone, 2022), adopting sustainable procurement standards, partnering with local conservation initiatives, and investing in clean technologies and circular business models (Waxin, Knuteson, & Bartholomew, 2019). These actions not only mitigate environmental impacts but also strengthen corporate reputation and community trust.

2.4.2 Challenges in Urban Environments

Urban environments face interconnected environmental challenges that demand systemic and coordinated responses. Among the most pressing are resource overconsumption, waste management, and ecological degradation.

2.4.2.1 Resource Overconsumption

Cities consume a disproportionately large share of global resources over 75% of all materials and energy while generating more than half of global waste and around 60% of greenhouse gas emissions, despite covering only about 2% of the Earth's surface (Ramírez-Márquez, Posadas-Paredes, & Ponce-Ortega, 2025). High-income nations, in particular, consume roughly six times more materials per person and produce ten times more climate impacts compared to low-income countries. This overconsumption manifests in various ways: the heavy material and energy demand of urban infrastructure, consumption-oriented lifestyles that drive overuse of resources, and inefficient utilization of assets such as vehicles, food, and office spaces (Guo & Klein, 2025; Kälviäinen, 2022). For example, in many European cities, cars remain parked about 92% of the time, approximately one-third of food is wasted, and office buildings often operate at less than half capacity (Ellen MacArthur Foundation, 2019).

2.4.2.2 Waste Management

Urban waste management remains one of the most critical challenges for city governments (Ebekozien, Aigbavboa, Nwaole, Aginah, & Aigbedion, 2022; Lissah, Ayanore, Krugu, Aberese-Ako, & Ruiter, 2021). Globally, cities produce around 2.01 billion tonnes of municipal solid waste each year, and at least one-third of it is not managed in environmentally safe ways (Kumar & Ram, 2021; Sharma & Jain, 2020). In many developing regions, waste management can consume up to half of municipal budgets while still leaving health and environmental risks unaddressed. Circular economy strategies are increasingly being adopted to counter this issue by focusing on waste prevention, material recovery, and resource recirculation (Purchase et al., 2021). Mexico City's Comprehensive Waste Management Program provides a strong example; it diverts over 2000 tonnes of waste from landfills daily while recycling more than 1700 tonnes, demonstrating the potential of circular models in urban waste systems (Villarroel & Yanez, 2024). Beyond municipal mixed-waste streams, food and agro-industrial residues also represent a critical frontier for waste minimization and circular valorization. Detailed case-study evidence shows that these organic waste fractions can be converted into biofuels, simultaneously diverting biodegradable material from landfills and contributing to low-carbon energy portfolios. Such biofuel pathways operationalize circular economy principles by transforming unavoidable organic wastes into value-added energy carriers that support urban and regional sustainability agendas (Choudhary, Harne, Dhar, Alqahtani, & Patel, 2025).

2.4.2.3 Pollution and Ecological Degradation

Urban degradation manifests through deteriorating air, water, and soil quality, as well as through loss of biodiversity (Mishra, 2025). Vehicle emissions, industrial output, and intensive energy use generate pollutants such as carbon monoxide and sulfur dioxide, which contribute to smog formation and respiratory illnesses (Zipperer, Northrop, & Andreu, 2020). Similarly, urban water bodies often suffer from contamination due to industrial discharge, sewage, and stormwater runoff, leading to disrupted hydrological cycles and heightened flood risks. Urban expansion also fragments habitats and reduces biodiversity, while soil sealing and contamination limit natural regeneration and reduce urban ecosystem functionality (Celletti et al., 2025).

2.4.3 Role of Policy and Governance

Effective environmental stewardship depends on strong governance frameworks that integrate regulatory, institutional, and participatory approaches. Environmental policies such as the National Environmental Policy Act, Clean Air Act, Clean Water Act, and Resource Conservation and Recovery Act establish foundational regulations governing pollution control, waste management, and ecosystem conservation. Agencies such as the Environmental Protection Agency (EPA) oversee implementation, monitoring, and enforcement. State and local governments then adapt these frameworks to regional contexts, often introducing innovative approaches to land use, green infrastructure, and environmental justice. For example, local zoning laws may incentivize green roofs, rain gardens, and urban forests, while conservation easements allow for the protection of ecologically valuable land through partnerships between public agencies, NGOs, and private landowners.

2.4.3.1 Community-Based Stewardship Initiatives

Grassroots environmental stewardship initiatives are equally crucial. These efforts empower local communities to take ownership of environmental issues through participatory decision-making, capacity-building, and knowledge sharing (Oyamo, Etan, Offiong, & Osang, 2025). Effective community initiatives are characterized by inclusivity, transparency, and the integration of traditional ecological knowledge (Osazuwa, Olowonihi, Oghogho, Mboto, & Gogogwute, 2024; Suriyankietkaew, Krittayaruangroj, Thinthan, & Lumlongrut, 2025). Evidence from rural communities in the Poltava Region of Ukraine illustrates how the design and placement of community bins, combined with locally tailored engagement strategies, can significantly influence citizen participation in waste segregation and the overall performance of municipal solid waste systems. The case study shows that regional socio-cultural conditions, institutional arrangements, and spatial organization around shared collection points strongly shape the success of community-based waste initiatives. These insights underscore the need for urban and peri-urban stewardship programs to adapt circular waste management instruments such as community bins to local social and ecological contexts rather than relying on uniform technical solutions (Bredun, Choudhary, & Kumar, 2024). The Malpai Borderlands Group offers a successful model: a coalition of private landowners and ranchers working collaboratively to restore and maintain natural ecological processes across shared landscapes (Martin, 2024). Their philosophy of undertaking projects *with* communities rather than *for* them ensures mutual respect and sustained engagement. Therefore, community-based stewardship thrives when supported by strong infrastructure, including training programs, financial resources, and collaborative networks that connect communities with government agencies and researchers. Platforms for sharing best practices and lessons learned further enhance collective learning and long-term impact.

2.4.3.2 Circular Economy Implementation in Global Cities

The following offers an in-depth examination of how five leading global cities are implementing circular economy strategies to transform their urban systems from linear "*take-make-dispose*" models to regenerative, sustainable frameworks. Amsterdam leads globally with its ambition to achieve 100% circularity by 2050, guided by Raworth's Doughnut Economics framework. Its *Circular Strategy 2020–2025* prioritizes food systems, consumer goods, and construction sectors, aiming to halve raw material use by 2030 and ensure all municipal procurement follows circular principles. The city employs a circular scan to track material flows, and over 200 companies and projects have received support to promote circular economy initiatives (Calisto Friant, Reid, Boesler, Vermeulen, & Salomone, 2023). Similarly, Barcelona integrates circular economy objectives into its Zero Waste Plan, targeting a 60% recycling rate by 2024 and encouraging waste prevention through tax incentives, community networks, and adaptive reuse of vacant spaces under the PLA BUITS program. The city also hosts annual circular city events and sectoral workshops that connect stakeholders to co-create sustainable solutions (Pablo-Romero, Sánchez-Braza, & Torreblanca, 2024). In Asia, Singapore exemplifies technology-driven circularity through its *Zero Waste Masterplan*, which integrates policy and innovation to address severe land constraints. The nation has achieved a 99% recycling rate in construction waste and established the Tuas Nexus facility that combines water and waste treatment, reducing emissions by 200,000 tonnes annually. Furthermore, the iconic Gardens by the Bay recycles 70% of green waste into energy (Carrière, Weigend Rodríguez, Pey, Pomponi, & Ramakrishna, 2020). Copenhagen also demonstrates strong alignment between circularity and climate neutrality, with its Circular Strategy supporting the city's carbon-neutral 2025 target. It aims for 70% waste recycling by 2024 through robotics-based sorting, electric collection vehicles, and the Amager Bakke waste-to-energy plant, which provides heat and power to more than 600,000 residents (Pirenne, 2023). In the United States, San Francisco stands as a zero-waste pioneer, diverting about 80% of its municipal waste through mandatory recycling and composting initiatives. Partnering with Recology, the city composts approximately 650 tons of food waste daily, complemented by public education and financial incentives that encourage community participation in sustainable waste practices (Abruzzo, 2019; Ashiagbor, 2025).

2.5 Mechanisms of Circular Economies in Urban Contexts

CEs reconfigure urban systems by prioritizing resource efficiency, waste valorization, and systemic innovation. These mechanisms enable cities to transition from linear consumption patterns to regenerative loops, reducing environmental strain while fostering economic and social resilience. Below, key CE mechanisms are explored through their application in urban settings.

2.5.1 Resource Efficiency: Urban Mining and Circular Supply Chains

Cities are reservoirs of underutilized materials, from construction debris to discarded electronics. Urban mining, extracting valuable resources from waste streams, has emerged to offset virgin material extraction (Bhardwaj, Rath, Jain, & Choudhury, 2025). For example, the Netherlands Circular Innovation Partnership takes metals, plastics, and textiles from municipal waste and puts them back into local manufacturing (Ashraf, Adeniyi, & van Seters, 2024). Building on such examples, recent analyses of urban resource recovery emphasize how important it is to view these initiatives as components of deliberately designed circular supply chains rather than isolated projects. By mapping material flows, identifying key nodal actors, and structuring logistics around recovery, refurbishment, and secondary markets, these studies show how urban systems can systematically retain value that would otherwise be lost. This supply-chain perspective is crucial for translating resource recovery pilots into integrated, citywide circular infrastructures that support long-term urban sustainability (Narayan et al., 2025). Similarly, circular supply chains prioritize modular design and standardized components, enabling easier disassembly and reuse. A 2023 study found that cities adopting such practices reduced raw material demand by 22% while cutting costs (Yang et al., 2023) (Fig. 2.8). In parallel to material-focused urban mining, biological transformation of food and agro-industrial wastes is emerging as a critical pillar of resource-efficient circular supply chains. Comprehensive analyses of these organic streams demonstrate how targeted bioconversion pathways can generate bioenergy, biochemicals, and bio-based materials, thereby retaining value that would otherwise be lost through disposal. By embedding such bio-transformation routes into urban metabolisms, cities can substantially reduce dependence on virgin resources while simultaneously advancing waste prevention and value recovery objectives (Kumar et al., 2025).

2.5.2 Waste Minimization: Recycling, Upcycling, and Zero-Waste Initiatives

CE principles reframe waste as a resource (Fig. 2.9). Advanced recycling systems, such as Singapore's Semakau Landfill, integrate waste-to-energy technologies to convert non-recyclables into electricity, construction debris, and ferrous metal, achieving high recycling rates of 99% (National Environment Agency, 2015). Upcycling, transforming waste into higher-value products, is exemplified by initiatives like San Francisco's Recycle Art, which repurposes plastic waste into public art installations (Choudhury & Roy, 2025; Hod, 2025).

Upcycling also extends to the development of advanced bio-based materials from organic waste streams. Recent case-study work demonstrates how food and agro-industrial residues can be transformed into biopolymers with functional properties suitable for packaging and other material applications, thus substituting fossil-based plastics. Such biopolymer pathways exemplify high-value circular valorization, simultaneously reducing landfill burdens and supporting sustainable material cycles within

Fig. 2.8 Members from Voice of Environment (VoE) are collecting empty packaged drinking water bottles at Kamakhya Temple premise in Guwahati city of Assam, India, during the mega event Ambubachi Mela as per its mission of the clean and eco-friendly Kamakhya Temple initiative (Photo by Moharana Choudhury)

Fig. 2.9 A photo showing sewage treatment plant (Photo by: Moharana Choudhury)

urban and regional economies (Choudhary et al., 2025). Meanwhile, zero-waste cities like Ljubljana leverage policy mandates and citizen engagement to achieve 90% landfill diversion rates, demonstrating scalability (Kyrkou, 2024).

2.5.3 Emission Reduction: Circular Energy and Construction

The built environment accounts for 40% of global CO_2 emissions, making CE practices critical. Circular construction employs recycled steel, low-carbon concrete, and modular designs to reduce embodied carbon. Rotterdam's Circularity Hub uses 90% recycled materials in infrastructure projects, reducing CO_2 emissions by 68% (Circle Economy Foundation, 2025). In energy systems, district heating networks powered by bioenergy from organic waste (e.g., Copenhagen's Amager

Bakke plant) exemplify closed-loop energy systems, cutting fossil fuel reliance (Edo, 2021).

2.5.4 *Socio-economic Inclusion: Equitable Access and Job Creation*

CEs democratize resource access while generating green jobs. Sharing economies, such as Berlin's Next Bike bicycle-sharing program, reduce private ownership and improve mobility for low-income residents (Heumann et al., 2025). Meanwhile, CE sectors create employment: the EU's circular transition is projected to generate 700,000 jobs by 2030, particularly in repair, recycling, and remanufacturing (European Commission, 2020). Community-led initiatives, like Nairobi's Mukuru Recycling Centre, empower marginalized groups by formalizing waste-picker roles (Sverdlik, Mitlin, & Dodman, 2019).

2.5.5 *Digital Enablers: IoT, AI, and Blockchain*

Digital tools optimize resource flows and transparency. In the urban water sector, spatial intelligence and AI-driven analytics are increasingly being deployed to optimize wastewater collection and treatment as part of circular resource management. Recent work on smart sewer networks demonstrates how integrated sensing, geospatial data, and predictive control can reduce overflow events, improve hydraulic performance, and enhance energy efficiency across wastewater infrastructure. By turning traditionally passive sewer systems into actively managed, data-rich networks, such approaches embed circular economy principles directly into urban water cycles and strengthen the resilience of sanitation services (Choudhary, Tigga, et al., 2025). Smart bins with IoT sensors in Seoul track waste generation patterns, slashing collection costs by 83% (Tomás, 2017). AI-driven platforms, like Amsterdam's Circular IQ, match industrial by-products with potential users, boosting industrial symbiosis (Circular IQ, n.d.). Blockchain further ensures traceability, as seen in Hong Kong's Green Blockchain Initiative, which certifies recycled material origins to prevent greenwashing (Kitano, 2024). Collectively, these mechanisms demonstrate how CEs reconfigure urban systems holistically. By aligning technological innovation with equitable policies, cities can decouple growth from resource depletion, advancing sustainability while addressing social inequities.

2.6 Intersections of Circular Economy and Urban Sustainability

Circular economy practices directly support urban sustainability by reducing waste, improving efficiency, and fostering innovation (Aiguobarueghian, Adanma, Ogunbiyi, & Solomon, 2024b; Klein, Ramos, & Deutz, 2020).

Waste Reduction: CE initiatives prioritize prevention and resource recovery through material recovery facilities, industrial symbiosis, and community-led recycling (Ngwemba, 2024; Nirmani, 2024). A critical yet often underutilized component of this waste reduction agenda is the valorization of food and agro-industrial residues into biofuels. Synthesizing multiple technological pathways, recent work has shown how diverse organic waste streams can be converted into liquid and gaseous biofuels, thereby simultaneously reducing biodegradable waste loads and contributing to low-carbon energy systems. These biofuel routes exemplify how circular waste management can directly reinforce urban sustainability by coupling resource recovery with climate mitigation objectives (Choudhary et al., 2025).

Resource Efficiency: Smart technologies like IoT and AI enhance recycling and optimize resource flows (Jogarao, Lakshmanna, & Naidu, 2024; Lakhouit, 2025). Denmark's reuse of bricks saves production energy (Christensen, Johansen, Buchard, & Glarborg, 2022), while projects such as Circle House demonstrate that 90% of building materials can be reused (Christensen, Hjul-Nielsen, Moalem, & Johansen, 2021).

Circular Urban Design: Cities integrate green infrastructure, modular construction, and regenerative landscapes that provide ecosystem services such as cooling, stormwater management, and carbon storage (Calheiros & Stefanakis, 2021; Deksissa, Trobman, Zendehdel, & Azam, 2021). Espoo's Kera District in Finland showcases a circular neighborhood using recyclable materials and sustainable mobility (Paavola, 2020; Reyes, 2023).

Business and Innovation: New circular business models, resource recovery, sharing, and product-as-a-service drive economic transformation (Han, Heshmati, & Rashidghalam, 2020).

2.7 Case Studies: Circular Economy Best Practices in Cities

2.7.1 Netherlands Circular 2050 Plan

The Dutch government's "A Circular Economy in the Netherlands by 2050" program outlines a national strategy to transition to a sustainable CE. It sets an interim goal of reducing the use of primary raw materials (minerals, fossil fuels, and metals) by 50% by 2030, aiming for full circularity by 2050. The strategy focuses on five priority sectors: biomass and food, plastics, manufacturing, construction, and consumer goods. Key interventions include fostering legislation, market incentives, financing, innovation, and international cooperation. The program emphasizes stakeholder collaboration and aims to decouple economic growth from resource consumption, ensuring environmental sustainability and economic resilience (Circolife Team, 2025; Government of the Netherlands, 2016).

2.7.2 Tokyo's Urban Mining Initiative

In preparation for the Tokyo 2020 Olympics, Japan launched the "*Tokyo 2020 Medal Project*," a groundbreaking urban mining initiative that transformed electronic waste into Olympic medals. Between April 2017 and March 2019, Japanese citizens donated approximately 78,985 tonnes of discarded electronics, including over 6 million mobile phones. From this collection, organizers extracted 30.3 kg of gold, 4100 kg of silver, and 2700 kg of bronze, sufficient to produce around 5000 medals for the Olympic and Paralympic Games. This endeavor conserved an estimated 4.5–5.1 terajoules of energy and reduced CO_2 emissions by approximately 420 metric tons. The project exemplified a successful CE model, engaging over 90% of Japanese municipalities and raising public awareness about sustainable practices. By repurposing e-waste into symbols of athletic excellence, Tokyo set a precedent for future host cities to integrate environmental responsibility into global events (DW News, 2021).

2.7.3 Medellín's Green Corridors

Medellín, Colombia, has transformed its urban landscape by implementing over 30 interconnected green corridors, a key component of its

2024–2027 urban development plan. These corridors extend from the city's peripheries to its center, creating a green belt that reconnects fragmented neighborhoods and revitalizes lost public spaces (Restrepo Montoya, 2024). The initiative has led to the planting of approximately 893,000 trees and 2.6 million plants, reducing average city temperatures by 2 °C and improving air quality. Beyond environmental benefits, the project emphasizes social inclusion by involving local communities in decision-making through participatory budgeting, ensuring that the development aligns with residents' needs (Ritter, 2023). By integrating ecological restoration with community engagement, Medellín's green corridors exemplify how urban planning can promote environmental sustainability and social equity, serving as a sustainable model for cities worldwide.

2.7.4 Chennai's Circular Economy Approach to Construction Waste

Chennai processes approximately 1000 tonnes of construction and demolition (C&D) waste daily. The Greater Chennai Corporation (GCC) has established recycling facilities at Perungudi and Kodungaiyur to address this. These plants convert debris into materials like M-sand, P-sand, and aggregates using vertical shaft impactor technology, producing sand that is 30% cheaper than conventional options. Since January 2025, over 1 lakh tonnes of debris has been processed, supplying recycled materials to major MNCs like L&T. The initiative includes strict regulations, with penalties of up to ₹25,000 for illegal dumping. It promotes using recycled products in construction, exemplifying a sustainable CE model (Express News Service, 2025).

2.8 Lessons Learned

The transition to circular urban systems demands more than technological innovation; it necessitates scalable models, inclusive collaboration, and flexible governance. Drawing from case studies and applied research, three critical dimensions scalability, stakeholder collaboration, and adaptive governance emerge as pivotal to success (Table 2.2). The following table synthesizes these lessons, offering a framework for policymakers, practitioners, and researchers to institutionalize CE principles within complex urban ecosystems.

Table 2.2 Enablers of circular city development

Dimension	*Key insights*
Scalability	Localized circular initiatives, when structured around modular designs and resource-efficient technologies, exhibit strong scalability potential. Successful scaling requires not only technological and financial readiness but also robust policy frameworks and standardized metrics for replication. Pilot programs serve as critical proofs of concept, necessitating agility in adaptation across different urban contexts. Recent methodological work on circular economy performance measurement has shown that metric design is not a purely technical exercise but a strategic governance tool. By comparing alternative indicator sets and evaluation techniques, these studies demonstrate how carefully constructed metrics can capture environmental, economic, and social dimensions of circularity, enabling cities to monitor progress, benchmark initiatives, and recalibrate interventions over time. Embedding such multidimensional performance frameworks into urban circular strategies is therefore essential for turning pilot projects into scalable, evidence-based policy pathways (Narayan, Pathak, et al., 2025)
Stakeholder collaboration	Systemic circularity is unattainable without deep collaboration among governments, private enterprises, academia, and civil society. Multi-stakeholder platforms foster trust, shared ownership, and the pooling of expertise, while participatory governance structures enhance legitimacy and public acceptance. Strategic alliances catalyze innovation and resilience within evolving urban ecosystems
Adaptive governance	Fixed, traditional governance approaches are ill-suited to the dynamic requirements of circular cities. Adaptive governance, characterized by experimental policymaking, continuous feedback loops, and institutional flexibility, emerged as a critical enabler. Cities that embedded iterative learning processes and allowed regulatory recalibration proved more capable of sustaining long-term circular transformations

2.9 Social and Economic Impacts

The shift toward a CE in urban settings delivers broad social and economic benefits by creating new jobs, strengthening community participation, and encouraging behavioral change through education and awareness (Vanhuyse, Fejzić, Ddiba, & Henrysson, 2021).

2.9.1 Job Creation and Economic Resilience

Circular strategies drive green employment and build economic resilience by localizing production and reducing reliance on global supply chains

(Mahmood, Misra, Sun, Luqman, & Papa, 2024). According to the Ellen MacArthur Foundation, a global circular transition could generate USD 4.5 trillion in benefits and create over six million new jobs by 2030 (Cherrafi, Chiarini, Belhadi, El Baz, & Chaouni Benabdellah, 2022; Eduljee & Ghosh, 2023). Key opportunities emerge in circular construction, reuse and repair industries, recycling and resource recovery, and digital sharing platforms. These sectors employ engineers, designers, technicians, and service operators, while localized material cycles reduce exposure to raw material price volatility and supply-chain disruptions, fostering stable urban economies.

2.9.2 Community Engagement and Equity

Inclusive circular initiatives ensure equitable access to opportunities, particularly for marginalized communities (Ashton, Fratini, Isenhour, & Krueger, 2022). Programs such as Earth5R's Circular Waste Economy in Mumbai empower women and low-income residents to run upcycling and composting cooperatives that generate income and improve local conditions. In cities like New York and Barcelona, civic groups collaborate with municipal governments to integrate community-led greening and waste reduction projects, building social cohesion and shared responsibility (Oikonomaki, Papadaki, & Kakderi, 2024; Saez Ujaque, Roca, De Balanzó Joue, Fuertes, & Garcia-Almirall, 2021). Policy tools such as participatory budgeting and community benefit agreements further ensure that circular infrastructure investments address environmental justice and local empowerment.

2.9.3 Behavioral Change and Awareness

Long-term circular transformation depends on cultural and behavioral adaptation supported by education and communication (Gonella, Godinho Filho, Campos, & Ganga, 2024; Romero-Luis et al., 2022). Cities including San Francisco and Copenhagen introduced extensive public campaigns on zero waste and composting well before policy rollout, achieving participation and minimizing recycling contamination. Schools and universities increasingly integrate circular economy principles into curricula, while community maker spaces and repair cafés engage citizens in practical resource stewardship (Table 2.3). These interventions

Table 2.3 Challenges to implementing circular cities

Barrier	*Description*	*Impact on urban sustainability*
Infrastructure limitations	Outdated waste systems, limited space, insufficient technology and logistics for reuse and recycling (Mahpour, 2018)	Slows circular economy adoption and prevents efficient resource recovery (Dong, Wang, Scipioni, Park, & Ren, 2017)
Lack of regulatory support	Weak or inconsistent environmental policies, subsidies favoring linear models, unclear governance (Bourdin & Jacquet, 2025; Williams, 2019)	Reduces incentives for businesses and cities to invest in circular practices (Williams, 2021)
Market and consumer resistance	Perceived costs, habit-driven behavior, doubts about quality and convenience of circular products/ services	Limits participation in circular initiatives and slows cultural transition (Vanhuyse et al., 2021)
Data and measurement challenges	Lack of standardized metrics, poor supply-chain transparency, limited digital tracking tools (Fatorachian, Kazemi, & Pawar, 2025)	Makes evaluating progress difficult and leads to ineffective decision-making (Dizdaroglu, 2017)

normalize sustainable behavior and embed circular values into urban culture.

2.10 Challenges to Implementing Circular Cities

2.11 The Path Forward

As we envision the future of sustainable urbanism, an integrated, multi-pronged strategy is essential. Governance must prioritize multi-stakeholder platforms and circular urban planning, ensuring that public, private, and civil society actors collaborate dynamically. Economically, cities must deploy tax incentives, enforce extended producer responsibility (EPR), and leverage green bonds to finance the circular transition. Simultaneously, fostering innovation ecosystems through public-private R&D partnerships and nurturing circular startups will catalyze scalable solutions. Critically, genuine community engagement is indispensable: cities must invest in education, embrace participatory design, and employ behavioral nudges to normalize circular practices among citizens. In the legal and policy arenas, adaptive regulatory frameworks must dismantle barriers favoring linear models and institutionalize circularity as a baseline expectation. Ultimately, the path forward demands that we reimagine urban economies not as extractive but as regenerative systems deeply interconnected with ecological integrity and social equity. In this pivotal moment, this chapter not only calls for innovation but also to govern with vision, incentivize with justice, and engage with humility. The circular city is not merely a possibility; it is an imperative.

2.12 Strategies for Circular Economy and Environmental Stewardship

The effective deployment of CE and environmental stewardship in cities depends on coordinated policy, technological, educational, and collaborative measures that work together to drive systemic urban transformation.

2.12.1 Policy Recommendations

Governments should support circular innovation through incentives and regulation. Tax credits, low-interest loans, and grants can encourage businesses to adopt practices such as remanufacturing, modular construction, and material recovery. Reduced fees or fast-tracked permits should be offered for projects demonstrating closed-loop resource systems.

Extended producer responsibility (EPR) laws can ensure manufacturers reclaim end-of-life products, keeping materials within local cycles (Brown, Laubinger, & Börkey, 2023). Municipalities should also set mandatory waste reduction targets for recycling by 2030 linked to performance-based funding. Harmonizing zoning, building codes, and procurement rules can further enable adaptive reuse, waste-to-energy systems, and transparent data reporting on material flows.

2.12.2 Technological Innovations

Smart technologies can make circular systems more efficient. IoT sensors and AI-driven sorting robots improve waste collection and recycling accuracy, reducing contamination and operational costs. Similar advances are reshaping urban air quality management in heavily industrialized cities. Recent work by Kryvyi Rih in Ukraine demonstrates how AI-based modeling, coupled with dense monitoring networks, can generate high-resolution spatiotemporal forecasts of pollutant concentrations and support more proactive regulatory and operational decisions. By integrating predictive analytics into routine environmental governance, such systems strengthen urban environmental stewardship and help align industrial development with air quality and public health objectives (Halaktionov et al., 2025). Blockchain platforms enhance transparency by tracking material origins and facilitating automated compliance through smart contracts. Digital sharing platforms and open data ecosystems can optimize asset use, extend product life, and inform policy through real-time data.

2.12.3 *Education and Awareness*

Embedding circular principles in education fosters long-term behavioral change (Altassan, 2023). Integrating CE concepts into K–12 and university curricula cultivates systems thinking and sustainable design skills. Public campaigns across digital and community platforms can promote composting, repair, and reuse habits. Maker spaces, repair cafés, and upcycling workshops engage citizens directly, while cultural programs and art initiatives can reinforce circular values through creative participation (Christou et al., 2024).

2.12.4 *Collaboration Models*

Collaboration between public institutions, private sectors, and communities is essential. Public-private partnerships (PPPs) can co-invest in material recovery, waste-to-energy, and industrial symbiosis projects (Nwokolo, 2025). Multi-stakeholder networks and city alliances (such as C40 and OECD) facilitate knowledge exchange and shared governance (Piazza, 2024; Rydin, 2021). Cross-sector coalitions connecting construction, manufacturing, and waste management can pilot integrated circular systems, while innovation challenges and hackathons encourage startups and researchers to co-create localized solutions.

2.13 The Future of Urban Sustainability Through Circular Economies

The future of urban sustainability through circular economies envisions transformative cities by 2050, where waste is minimized through regenerative design, products are continuously reused, and urban systems operate with closed resource loops. In these visionary scenarios, digitalization and artificial intelligence play a crucial role by optimizing material flows, improving accountability, and enabling real-time monitoring of environmental performance. As climate challenges intensify, circular strategies also strengthen climate adaptation and resilience, helping cities reduce emissions, enhance resource security, and better withstand environmental stresses such as heat waves, flooding, and resource scarcity. Ultimately, circular economies offer a blueprint for smarter, greener, and more resilient urban living.

2.14 Conclusion

The transition to a CE offers cities a powerful pathway to achieve environmental stewardship, economic resilience, and social equity. Supported by bibliometric analysis, which highlights the accelerating global research focus on CEs and urban sustainability, this chapter emphasizes the transformative potential of moving beyond the traditional linear model. Urban systems can regenerate resources, reduce emissions, and foster inclusive growth by incorporating frameworks such as C2C, IE, BE, and RD. Real-world examples from the Netherlands' urban mining to Medellín's green corridors demonstrate that circular principles are practical, scalable solutions. Nonetheless, the journey requires overcoming institutional inertia, technological barriers, and social resistance. Through strategic governance, innovative financing, public-private collaboration, and profound community engagement, cities can evolve into regenerative ecosystems. Embracing CEs is not simply an option; it is essential for building future-ready, resilient urban landscapes capable of thriving within the boundaries of a finite planet.

References

Abbas, Q., Imran, M., & Sattar, A. (2025). From linear to circular: The impact of economic policies and technological innovations on greenhouse gas emissions in The Netherlands. *Carbon Balance and Management, 20*(1), 11. https://doi.org/10.1186/s13021-025-00297-1

Abruzzo, Sondra. (2019). "*Best Practices to Encourage Landfill Diversion in Waste Management Programs*" . Scripps Senior Theses. 1329. https://scholarship.claremont.edu/scripps_theses/1329

Ahmed, N., Xinagyu, G., Alnafissa, M., et al. (2025). Linear and non-linear impact of key agricultural components on greenhouse gas emissions. *Scientific Reports, 15*, 5314. https://doi.org/10.1038/s41598-025-88159-1

Aiguobarueghian, I., Adanma, U. M., Ogunbiyi, E. O., & Solomon, N. O. (2024a). An overview of initiatives and best practices in resource management and sustainability. *World Journal of Advanced Research and Reviews, 22*(2), 1734–1745. https://doi.org/10.30574/wjarr.2024.22.2.1519

Aiguobarueghian, I., Adanma, U. M., Ogunbiyi, E. O., & Solomon, N. O. (2024b). Waste management and circular economy: A review of sustainable practices and economic benefits. *World Journal of Advanced Research and Reviews, 22*(2), 1708–1719. https://doi.org/10.30574/wjarr.2024.22.2.1517

Altassan, A. (2023). Sustainable integration of solar energy, behavior change, and recycling practices in educational institutions: A holistic framework for environmental conservation and quality education. *Sustainability, 15*(20), 15157. https://doi.org/10.3390/su152015157

Ashiagbor, E. (2025). *Effectiveness of Urban composting and waste management in the San Francisco Bay Area*. https://repository.usfca.edu/capstone/1896

Ashraf, N., Adeniyi, D., & van Seters, J. (2024). *External implications of the circular economy transition of The Netherlands and the EU. ECDPM discussion paper 362. Maastricht: ECDPM.* https://www.eumonitor.eu/9353000/1/j4nvgs5kjg27kof_j9vvik7m1c3gyxp/vmarpv1caozm/f=/blg1127540.pdf

Ashton, W. S., Fratini, C. F., Isenhour, C., & Krueger, R. (2022). Justice, equity, and the circular economy: Introduction to the special double issue. *Local Environment, 27*(10–11), 1173–1181. https://doi.org/10.1080/13549839.2022.2118247

Bennett, N. J., Whitty, T. S., Finkbeiner, E., Pittman, J., Bassett, H., Gelcich, S., & Allison, E. H. (2018). Environmental stewardship: A conceptual review and analytical framework. *Environmental Management, 61*(4), 597–614. https://doi.org/10.1007/s00267-017-0993-2

Bharadwaj, G., Chopde, S., Taluja, R., Lalitha, G., Chandrashekar, R., & Dhahi, H. A. (2023). Closing the loop: Advances in materials, energy, and waste management. *E3S Web of Conferences, 453*, 01024. https://doi.org/10.1051/e3sconf/202345301024

Bharathy, N., Kowsalya, P., David, Hullebusch, E. D. M., Choudhury, M., Khan, P. A., Aidoo, M. K., & Parthasarathi, T. (2025). Breaking barriers: Innovative approaches and Technologies for Climate Resilience: A case study from European countries. In *Sustainable synergy: Harnessing ecosystems for climate resilience* (pp. 177–190). Springer Nature Switzerland. https://doi.org/10.1007/978-3-031-77957-2_12

Bhardwaj, L. K., Rath, P., Jain, H., & Choudhury, M. (2025). Exploring the effects of e-waste on soil, water quality and human health. *Discover Civil Engineering, 2*(1), 1–16. https://doi.org/10.1007/s44290-025-00167-2

Biswas, W. K., & John, M. (2022). *Engineering for sustainable development: Theory and practice*. Wiley. https://doi.org/10.1002/9781119721000

Blühdorn, I., & Deflorian, M. (2019). The collaborative management of sustained unsustainability: On the performance of participatory forms of environmental governance. *Sustainability, 11*(4), 1189. https://doi.org/10.3390/su11041189

Bourdin, S., & Jacquet, N. (2025). Closing the loop at the local scale: Investigating the drivers of and barriers to the implementation of the circular economy in cities and regions. *Ecological Economics, 231*, 108542. https://doi.org/10.1016/j.ecolecon.2025.108542

Bragança, L., Griffiths, P., Askar, R., Salles, A., Ungureanu, V., Tsikaloudaki, K., Bajare, D., Zsembinszki, G., & Cvetkovska, M. (2025). *Circular economy design and management in the built environment: A critical review of the state of the art*. Springer Nature. https://doi.org/10.1007/978-3-031-73490-8

Bravi, L., Santos, G., Pagano, A., & Murmura, F. (2020). Environmental management system according to ISO 14001:2015 as a driver to sustainable development. *Corporate Social Responsibility and Environmental Management, 27*(6), 2599–2614. https://doi.org/10.1002/csr.1985

Bredun, V., Choudhary, R., & Kumar, A. (2024). Regional specifics of using community bins in waste management: A case study of rural communities in Poltava region (Ukraine). *Trends in Ecological and Indoor Environmental Engineering, 2*(4), 10–17. https://doi.org/10.62622/TEIEE.024.2.4.10-17

Brown, A., Laubinger, F., & Börkey, P. (2023). *New aspects of EPR: Extending producer responsibility to additional product groups and challenges throughout the product lifecycle. OECD environment working papers*. Organisation for Economic Co-operation and Development (OECD).

Bühler, M., Hollenbach, P., Köhler, L., & Armstrong, R. (2024). Unlocking resilience and sustainability with earth-based materials: A principled framework for urban transformation. *Frontiers in Built Environment, 10*, 1385116. https://doi.org/10.3389/fbuil.2024.1385116

Caggiano, H., Landau, L. F., Campbell, L. K., Johnson, M. L., & Svendsen, E. S. (2024). Civic stewardship and urban climate governance: Opportunities

for transboundary planning. *Journal of Planning Education and Research, 44*(3), 1620–1633. https://doi.org/10.1177/0739456X221104010
Calheiros, C. S. C., & Stefanakis, A. I. (2021). Green roofs towards circular and resilient cities. *Circular Economy and Sustainability, 1*(1), 395–411. https://doi.org/10.1007/s43615-021-00033-0
Calisto Friant, M., Reid, K., Boesler, P., Vermeulen, W. J. V., & Salomone, R. (2023). Sustainable circular cities? Analysing urban circular economy policies in Amsterdam, Glasgow, and Copenhagen. *Local Environment, 28*(10), 1331–1369. https://doi.org/10.1080/13549839.2023.2206643
Camilleri, M. A. (2019). The circular economy's closed loop and product service systems for sustainable development: A review and appraisal. *Sustainable Development, 27*(3), 530–536. https://doi.org/10.1002/sd.1909
Campbell, L. K., Svendsen, E., Johnson, M., & Landau, L. (2022). Activating urban environments as social infrastructure through civic stewardship. *Urban Geography, 43*(5), 713–734. https://doi.org/10.1080/02723638.2021.1920129
Campbell-Johnston, K., Vermeulen, W. J. V., Reike, D., & Brullot, S. (2020). The circular economy and cascading: Towards a framework. *Resources, Conservation and Recycling: X, 7*, 100038. https://doi.org/10.1016/j.rcrx.2020.100038
Carrière, S., Weigend Rodríguez, R., Pey, P., Pomponi, F., & Ramakrishna, S. (2020). Circular cities: The case of Singapore. *Built Environment Project and Asset Management, 10*(4), 491–507. https://doi.org/10.1108/BEPAM-12-2019-0137
Čega, R. S., Drezgić, S., & Čišić, D. (2024). Exploring the regenerative economy: A comprehensive review of literature. *Journal of Regenerative Economics, 1*(1), 1–40.
Celletti, S., Poreba, L., Wawer, R., Padoan, E., Comis, S., Bartosiewicz, B., & Schiavon, M. (2025). The potential of nature-based solutions for urban soils: Focus on green infrastructure and bioremediation. *Frontiers in Environmental Science, 13*, 1634662. https://doi.org/10.3389/fenvs.2025.1634662
Cervantes Puma, G. C., Salles, A., & Bragança, L. (2024). Nexus between urban circular economies and sustainable development goals: A systematic literature review. *Sustainability, 16*(6), 2500. https://doi.org/10.3390/su16062500
Cherrafi, A., Chiarini, A., Belhadi, A., El Baz, J., & Chaouni Benabdellah, A. (2022). Digital technologies and circular economy practices: Vital enablers to support sustainable and resilient supply chain management in the post-COVID-19 era. *The TQM Journal, 34*(7), 179–202. https://doi.org/10.1108/TQM-12-2021-0374
Choudhary, R., Bhardwaj, L. K., Nguyen Tan, T., Amesho, K. T. T., Chegini, S., Boiko, O., & Sillanpää, M. (2025). Spatial intelligence integration in smart wastewater systems: Advancing efficiency and sustainability in urban sewer networks. In M. Choudhury, S. Majumdar, S. Goswami, & M. Sillanpää (Eds.),

Smart wastewater systems and climate change: Innovations through spatial intelligence. Royal Society of Chemistry. https://doi.org/10.1039/9781837676880-00155

Choudhary, R., Harne, K., Dhar, S., Alqahtani, F. M., & Patel, K. (2025). Case study on biofuels derived from food and agro-industrial wastes. In M. Sillanpää, M. Choudhury, & S. Goswami (Eds.), *Food and agro-industrial wastes: Sustainable impacts, transformation and added value of by-products*. Academic Press. https://doi.org/10.1016/B978-0-443-31654-8.00030-8

Choudhary, R., Kumar, A., Kumar, V., Elmnifi, M. H., Dhar, S., Alqahtani, F. M., & Patel, K. (2025). Case study on biopolymers derived from food and agro-industrial wastes. In M. Sillanpää, M. Choudhury, & S. Goswami (Eds.), *Food and agro-industrial wastes: Sustainable impacts, transformation and added value of by-products*. Academic Press. https://doi.org/10.1016/B978-0-443-31654-8.00026-6

Choudhary, R., Tigga, S., Bhardwaj, L. K., Goswami, S., Nguyen Tan, T., & Chaudhari, S. K. (2025). Biofuels from food and agro-industrial wastes. In M. Sillanpää, M. Choudhury, & S. Goswami (Eds.), *Food and agro-industrial wastes: Sustainable impacts, transformation and added value of by-products*. Academic Press. https://doi.org/10.1016/B978-0-443-31654-8.00004-7

Choudhury, M., Dixit, G., & Majumdar, S. (2025). *Sustainable synergy: Harnessing ecosystems for climate resilience*. Springer Nature. https://doi.org/10.1007/978-3-031-77957-2

Choudhury, M., & Roy, P. (2025). Challenges with microplastic pollution in the regime of UN sustainable development goals. *World Development Sustainability, 6*, 100216. https://doi.org/10.1016/j.wds.2025.100216

Christensen, D., Hjul-Nielsen, J., Moalem, R. M., & Johansen, B. (2021). Circular economy in Denmark: Bornholm's vision to achieve 100 percent reuse and recycling. In S. K. Ghosh & S. K. Ghosh (Eds.), *Circular economy: Recent trends in global perspective* (pp. 385–424). Springer Nature. https://doi.org/10.1007/978-981-16-0913-8_13

Christensen, T. B., Johansen, M. R., Buchard, M. V., & Glarborg, C. N. (2022). Closing the material loops for construction and demolition waste: The circular economy on the Island Bornholm, Denmark. *Resources, Conservation Recycling Advances, 15*, 200104. https://doi.org/10.1016/j.rcradv.2022.200104

Christou, O., Manou, D. B., Armenia, S., Franco, E., Blouchoutzi, A., & Papathanasiou, J. (2024). Fostering a whole-institution approach to sustainability through systems thinking: An analysis of the state-of-the-art in sustainability integration in higher education institutions. *Sustainability, 16*(6), 2508. https://doi.org/10.3390/su16062508

Circle Economy Foundation. (2025). *BlueCity: Repurposing a water park into a circular model city*. Knowledge Hub. https://knowledge-hub.circle-economy.com/article/30011

Circolife Team. (2025). *Amsterdam: A circular city leading the way*. Circolife. https://circolife.com/blog/amsterdam-a-circular-city-leading-the-way/#:~:text=In%202020%2C%20the%20city%20introduced,for%20processing%20and%20reducing%20waste

Circular economy: Definition, Importance and Benefits. (2023). Topics | European Parliament. https://www.europarl.europa.eu/topics/en/article/20151201STO05603/circular-economy-definition-importance-and-benefits

Circular IQ. (n.d.). *Circular IQ home*. Retrieved April 28, 2025, from https://circular-iq.com/circular-iq-home/

Cokrowati, N., Choudhary, R., Yatin, N., Apriliyanti, F. J., Amanda, R. A., Ramadhan, A. F., Kumar, A., & Hrdzelidze, S. (2025). Identification of wild seaweeds at Nipah Beach, Indonesia, and study of their antioxidant potential. *Trends in Ecological and Indoor Environmental Engineering, 3*(3), 1–7. https://doi.org/10.62622/TEIEE.025.3.3.01-07

Collado-Ruano, J., & Segovia Sarmiento, J. (2022). Ecological economics foundations to improve environmental education practices: Designing regenerative cultures*. *World Futures, 78*(7), 456–483. https://doi.org/10.1080/02604027.2022.2072158

Conti, M. E., Battaglia, M., Calabrese, M., & Simone, C. (2021). Fostering sustainable cities through resilience thinking: The role of nature-based solutions (Nbss): Lessons learned from two Italian case studies. *Sustainability, 13*(22), 12875. https://doi.org/10.3390/su132212875

Das, S., Choudhury, M. R., Chatterjee, B., Das, P., Bagri, S., Paul, D., Bera, M., & Dutta, S. (2024). Unraveling the urban climate crisis: Exploring the nexus of urbanization, climate change, and their impacts on the environment and human well-being—A global perspective. *AIMS Public Health, 11*(3), 963–1001. https://doi.org/10.3934/publichealth.2024050

Deal, G. (2020). *Circular economy action plan*. https://www.pac.gr/bcm/uploads/new_circular_economy_action_plan.pdf

Deksissa, T., Trobman, H., Zendehdel, K., & Azam, H. (2021). Integrating urban agriculture and stormwater management in a circular economy to enhance ecosystem services: Connecting the dots. *Sustainability, 13*(15), 8293. https://doi.org/10.3390/su13158293

Dizdaroglu, D. (2017). The role of indicator-based sustainability assessment in policy and the decision-making process: A review and outlook. *Sustainability, 9*(6), 1018. https://doi.org/10.3390/su9061018

Domínguez-Escrig, E., Mallén-Broch, F. F., Lapiedra-Alcamí, R., & Chiva-Gómez, R. (2019). The influence of leaders' stewardship behavior on innovation success: The mediating effect of radical innovation. *Journal of Business Ethics, 159*(3), 849–862. https://doi.org/10.1007/s10551-018-3833-2

Dong, L., Wang, Y., Scipioni, A., Park, H.-S., & Ren, J. (2017). Recent progress on innovative urban infrastructures system towards sustainable resource management. *Resources, Conservation and Recycling, 122*, 43–54. https://doi.org/10.1016/j.resconrec.2017.02.020

DW News. (2021). *Tokyo 2020 Olympic medals made from old smartphones, laptops.* DW. https://www.dw.com/en/tokyo-2020-olympic-medals-made-from-old-smartphones-laptops/a-58325432

Ebekozien, A., Aigbavboa, C., Nwaole, A. N. C., Aginah, I. L., & Aigbedion, M. (2022). Sustainable cities through household waste management: An unexplored approach to challenges confronting private solid waste management. *Facilities, 40*(5/6), 365–379. https://doi.org/10.1108/F-09-2021-0078

Edo, M. (2021). *Waste-to-energy and social acceptance: Copenhill waste-to-energy Plant in Copenhagen (IEA bioenergy: Task 36)*. IEA Bioenergy. https://www.ieabioenergy.com/wp-content/uploads/2021/03/T36_WtE-and-Social-Acceptance_Copnehhill-WtE-plant-in-Copenhagen.pdf

Eduljee, G., & Ghosh, S. K. (2023). *The circular economy: Meeting sustainable development goals.* Royal Society of Chemistry. https://doi.org/10.1039/9781837670697

Ellen MacArthur Foundation. (2019). How the circular economy tackles climate change. Ellen MacArthur Foundation. https://www.hoop-hub.eu/virtual_images/134-6254016ea43c113bc152bb9f06f1ec02.pdf

European Commission. (2020). Q&A: A new circular economy action plan for a cleaner and more competitive Europe. https://ec.europa.eu/commission/presscorner/detail/en/qanda_20_419

Express News Service. (2025). *Corporation's plants turn construction waste into usable sand, gravel, paver blocks at Perungudi and Kodungaiyur*. The New Indian Express. https://www.newindianexpress.com/cities/chennai/2025/Apr/17/corporations-plants-turn-construction-waste-into-usable-sand-gravel-paver-blocks-at-perungudi-and-kodungaiyur

Fatorachian, H., Kazemi, H., & Pawar, K. (2025). Digital technologies in food supply chain waste management: A case study on sustainable practices in smart cities. *Sustainability, 17*(5), 1996. https://doi.org/10.3390/su17051996

Feldmann, A. (2023). Circular economy for energy-intensive industries. In F. Urban & J. Nordensvärd (Eds.), *Handbook on climate change and technology* (pp. 228–236). Edward Elgar Publishing. https://doi.org/10.4337/9781800882119.00024

Figge, F., Dimitrov, S., Schlosser, R., & Chenavaz, R. (2022). Does the circular economy fuel the throwaway society? The role of opportunity costs for products that lose value over time. *Journal of Cleaner Production, 368*, 133207. https://doi.org/10.1016/j.jclepro.2022.133207

Fischer-Kowalski, M., Rovenskaya, E., Krausmann, F., Pallua, I., & Mc Neill, J. R. (2019). Energy transitions and social revolutions. *Technological Forecasting and Social Change, 138*, 69–77. https://doi.org/10.1016/j.techfore.2018.08.010

Fratini, C. F., Georg, S., & Jørgensen, M. S. (2019). Exploring circular economy imaginaries in European cities: A research agenda for the governance of urban sustainability transitions. *Journal of Cleaner Production, 228*, 974–989. https://doi.org/10.1016/j.jclepro.2019.04.193

Fusco Girard, L., & Nocca, F. (2019). Moving towards the circular economy/city model: Which tools for operationalizing this model? *Sustainability, 11*(22), 6253. https://doi.org/10.3390/su11226253

Ghisellini, P., Cialani, C., & Ulgiati, S. (2016). A review on circular economy: The expected transition to a balanced interplay of environmental and economic systems. *Journal of Cleaner Production, 114*, 11–32. https://doi.org/10.1016/j.jclepro.2015.09.007

Gibbons, L. V. (2020). Regenerative—The new sustainable? *Sustainability, 12*(13), 5483. https://doi.org/10.3390/su12135483

Girard, L. F. (2021). The evolutionary circular and human centered city: Towards an ecological and humanistic "re-generation" of the current city governance. *Human Systems Management, 40*(6), 753–775. https://doi.org/10.3233/HSM-211218

Gonella, J. D. S. L., Godinho Filho, M., Campos, L. M. D. S., & Ganga, G. M. D. (2024). People's awareness and behaviours of circular economy around the world: Literature review and research agenda. *Sustainability Accounting, Management and Policy Journal, 15*(5), 1118–1154. https://doi.org/10.1108/SAMPJ-08-2022-0413

Government of the Netherlands. (2016). A circular economy in The Netherlands by 2050: Government-wide programme for a circular economy. Ministry of Infrastructure and the Environment Ministry of Economic Affairs. https://circulareconomy.europa.eu/platform/sites/default/files/17037circulaireeconomie_en.pdf

Guo, J., & Klein, K. (2025). *Living well in a finite world: A systems approach to degrowth and public health in cities.* Schumacher Institute. https://schumacherinstitute.org.uk/wp-content/uploads/2025/07/202505-Living-Well-in-a-Finite-World-Jiahui-Guo-and-Kristine-Klein.pdf

Hagelüken, C., & Goldmann, D. (2022). Recycling and circular economy—Towards a closed loop for metals in emerging clean technologies. *Mineral Economics, 35*(3–4), 539–562. https://doi.org/10.1007/s13563-022-00319-1

Halaktionov, M., Bredun, V., Choudhary, R., Goroneskul, M., Kumar, A., Ouiya, F., Sydorenko, V., & Markina, L. (2025). AI-enhanced air quality assessment and prediction in industrial cities: A case study of Kryvyi Rih, Ukraine. *Ecological Engineering Environmental Technology, 26*(6), 45–56. https://doi.org/10.12912/27197050/203725

Hall, M. J., & Thomsen, M. R. (2025). *The circular economy in the EU and China: Comparing socio-technical systems and examining how the EU and China utilise policy to facilitate circular transitions* (master's thesis). Roskilde University. https://rucforsk.ruc.dk/ws/files/111143164/MA_Thesis_Michael_and_Marius_The_Circular_Economy_in_the_EU_and_China___Comparing_Socio_Technical_Systems_and_Examining_How_the_EU_and_China_Utilise_Policy_to_Facilitate_Circular_Transitions.pdf.

Han, J., Heshmati, A., & Rashidghalam, M. (2020). Circular economy business models with a focus on servitization. *Sustainability, 12*(21), 8799. https://doi.org/10.3390/su12218799

Henderson, P. (2023). *The circular economy: A missing piece in city climate action plans?* Ellen MacArthur Foundation. https://www.ellenmacarthurfoundation.org/articles/the-circular-economy-a-missing-piece-in-city-climate-action-plans

Heumann, M., Kraschewski, T., Otto, P., Tilch, L., Brauner, T., & Breitner, M. H. (2025). Factors influencing the usage of shared micromobility: Implications from Berlin. *Journal of Cycling and Micromobility Research, 4*, 100063. https://doi.org/10.1016/j.jcmr.2025.100063

Hod, I. (2025). *From trash to treasure: San Francisco artist turns garbage into masterpieces.* CBS News. https://www.cbsnews.com/sanfrancisco/news/from-trash-to-treasure-san-francisco-artist-turns-garbage-into-masterpieces/

Jasny, L., Johnson, M., Campbell, L. K., Svendsen, E., & Redmond, J. (2019). Working together: The roles of geographic proximity, homophilic organizational characteristics, and neighborhood context in civic stewardship collaboration networks in Philadelphia and New York City. *Ecology and Society, 24*(4), art8. https://doi.org/10.5751/ES-11140-240408

Jensen, H. H. (2025a). 4th and 5th r—Maximizing product lifecycles: Integrating refurbishment and remanufacturing strategies. In H. H. Jensen (Ed.), *Circular economy opportunities and pathways for manufacturers* (pp. 161–199). Springer Nature. https://doi.org/10.1007/978-3-031-75279-7_7

Jensen, H. H. (2025b). Circular economy as a catalyst for innovation and growth. In H. H. Jensen (Ed.), *Circular economy opportunities and pathways for manufacturers* (pp. 429–442). Springer Nature. https://doi.org/10.1007/978-3-031-75279-7_18

Jogarao, M., Lakshmanna, B. C., & Naidu, S. T. (2024). Ai-enabled circular economy management for sustainable smart cities: Integrating artificial intelligence in resource optimization and waste reduction. In V. Kandpal, E. D. Santibanez-Gonzalez, P. Chatterjee, & M. K. Nallapaneni (Eds.), *Smart cities and Circular economy* (pp. 83–96). Emerald Publishing Limited. https://doi.org/10.1108/978-1-83797-957-820241008

Johnstone, L. (2022). The means to substantive performance improvements—Environmental management control systems in ISO 14001– Certified SMEs.

Sustainability Accounting, Management and Policy Journal, 13(5), 1082–1108. https://doi.org/10.1108/SAMPJ-11-2021-0456

Kälviäinen, M. (2022). User-driven service design for environmentally responsible consumption [fi=Publisher's version|sv=Publisher's version|en=Publisher's version|]. http://www.theseus.fi/handle/10024/755960

Kandpal, V., Jaswal, A., Santibanez Gonzalez, E. D. R., & Agarwal, N. (2024). Circular economy principles: Shifting towards sustainable prosperity. In I. V. Kandpal, A. Jaswal, E. D. R. Santibanez Gonzalez, & N. Agarwal (Eds.), *Sustainable energy transition* (pp. 125–165). Springer Nature. https://doi.org/10.1007/978-3-031-52943-6_4

Kara, S., Hauschild, M., Sutherland, J., & McAloone, T. (2022). Closed-loop systems to circular economy: A pathway to environmental sustainability? *CIRP Annals, 71*(2), 505–528. https://doi.org/10.1016/j.cirp.2022.05.008

Kiss, B., Sekulova, F., Hörschelmann, K., Salk, C. F., Takahashi, W., & Wamsler, C. (2022). Citizen participation in the governance of nature-based solutions. *Environmental Policy and Governance, 32*(3), 247–272. https://doi.org/10.1002/eet.1987

Kitano, Y. (2024). *Hong Kong harnesses blockchain in green finance push.* Nomura Connects. https://www.nomuraconnects.com/focused-thinking-posts/hong-kong-harnesses-blockchain-in-green-finance-push/

Klancko, R. J. (2003). A handbook of industrial ecology. Robert U. Ayres and Leslie W. Ayres, eds. 2002. Edward Elgar publishing, Northampton, MA. 680 pp. $285 hardcover. *Environmental Practice, 5*(2), 183–184. https://doi.org/10.1017/S1466046603261123

Klein, N., Ramos, T., & Deutz, P. (2020). Circular economy practices and strategies in public sector organizations: An integrative review. *Sustainability, 12*(10), 4181. https://doi.org/10.3390/su12104181

Korhonen, J., Nuur, C., & Feldmann, A. (2023). Circular economy as an integrative framework: A synthesis of theoretical convergences. *Ecological Economics, 204*, 107667. https://doi.org/10.1016/j.ecolecon.2022.107667

Kovacic, Z., Strand, R., & Völker, T. (2019). *The circular economy in Europe: Critical perspectives on policies and imaginaries* (1st ed.). https://doi.org/10.4324/9780429061028

Kumar, A., & Ram, C. (Eds.). (2021). *Nanobiotechnology for green environment.* CRC Press. https://doi.org/10.1201/9781000342505

Kumar, P., Chauhan, S., Addai, G., Choudhary, R., Sudan, P., & Kumar, P. (2025). Valorization of food and agroindustrial wastes—Biological transformation. In M. Sillanpää, M. Choudhury, & S. Goswami (Eds.), *Food and agro-industrial wastes: Sustainable impacts, transformation and added value of by-products.* Academic Press. https://doi.org/10.1016/B978-0-443-31654-8.00008-4

Kumaraswamy, T. R., Choudhury, M., Ponappa, S., & Majumdar, S. (2025). A significance of climate resilience: Strategies for local, regional, and national levels. In *Sustainable synergy: Harnessing ecosystems for climate resilience* (pp. 193–205). Springer Nature. https://doi.org/10.1007/978-3-031-77957-2_13

Kyrkou, M. (2024). *Zero-waste cities: Case studies and future prospects.* WattCrop. https://wattcrop.com/zero-waste-cities-case-studies-and-future-prospects/

Lakhouit, A. (2025). Revolutionizing urban solid waste management with AI and IoT: A review of smart solutions for waste collection, sorting, and recycling. *Results in Engineering, 25*, 104018. https://doi.org/10.1016/j.rineng.2025.104018

Law, K. L., & Narayan, R. (2021). Reducing environmental plastic pollution by designing polymer materials for managed end-of-life. *Nature Reviews Materials, 7*(2), 104–116. https://doi.org/10.1038/s41578-021-00382-0

Lăzăroiu, G., Ionescu, L., Andronie, M., & Dijmărescu, I. (2020). Sustainability management and performance in the urban corporate economy: A systematic literature review. *Sustainability, 12*(18), 7705. https://doi.org/10.3390/su12187705

Lissah, S. Y., Ayanore, M. A., Krugu, J. K., Aberese-Ako, M., & Ruiter, R. A. C. (2021). Managing urban solid waste in Ghana: Perspectives and experiences of municipal waste company managers and supervisors in an urban municipality. *PLoS One, 16*(3), e0248392. https://doi.org/10.1371/journal.pone.0248392

Lopez, C. W., & Weaver, R. (2023). What influences where volunteers practice environmental stewardship? The role of scale(S) in sorting stewards. *Journal of Environmental Planning and Management, 66*(9), 1983–2008. https://doi.org/10.1080/09640568.2022.2049596

Lucciarini, S., & Galdini, R. (2024). Is an urban waste-to-energy plant a "green" megaproject? The power of narratives in shaping the city: A Danish case study. *Environmental Sociology, 10*(3), 359–370. https://doi.org/10.1080/23251042.2024.2357691

Lyle, J. T. (1994). *Regenerative design for sustainable development.* Wiley.

Mahmood, S., Misra, P., Sun, H., Luqman, A., & Papa, A. (2024). Sustainable infrastructure, energy projects, and economic growth: Mediating role of sustainable supply chain management. *Annals of Operations Research, 355*, 1099–1130. https://doi.org/10.1007/s10479-023-05777-6

Mahpour, A. (2018). Prioritizing barriers to adopt circular economy in construction and demolition waste management. *Resources, Conservation and Recycling, 134*, 216–227. https://doi.org/10.1016/j.resconrec.2018.01.026

Mansourihanis, O., Maghsoodi Tilaki, M. J., Yousefian, S., & Zaroujtaghi, A. (2023). A computational geospatial approach to assessing land-use compatibility in urban planning. *Land, 12*(11), 2083. https://doi.org/10.3390/land12112083

Markevičiūtė, Z., & Varžinskas, V. (2022). Smart material choice: The importance of circular design strategy applications for bio-based food packaging preproduction and end-of-life life cycle stages. *Sustainability, 14*(10), 6366. https://doi.org/10.3390/su14106366

Martin, J. V. (2024). Conservation and conviviality in the American west. *Elementa: Science of the Anthropocene, 12*(1), 00073. https://doi.org/10.1525/elementa.2023.00073

McCarthy, L. J., & Russo, A. (2023). Exploring the role of nature-based typologies and stewardship schemes in enhancing urban green spaces: Citizen perceptions of landscape design scenarios and ecosystem services. *Journal of Environmental Management, 346*, 118944. https://doi.org/10.1016/j.jenvman.2023.118944

Miao, Q., & Nduneseokwu, C. (2024). Environmental leadership and governance: Strategies, mechanisms and multi-levels. In I. Q. Miao & C. Nduneseokwu (Eds.), *Environmental leadership in a VUCA era* (pp. 159–206). Springer Nature. https://doi.org/10.1007/978-981-96-0324-4_4

Mihai, F.-C., & Minea, I. (2021). Sustainable alternative routes versus linear economy and resources degradation in eastern Romania. *Sustainability, 13*(19), 10574. https://doi.org/10.3390/su131910574

Mishra, A., Dutta, P., Jayasankar, S., Jain, P., & Mathiyazhagan, K. (2023). A review of reverse logistics and closed-loop supply chains in the perspective of circular economy. *Benchmarking: An International Journal, 30*(3), 975–1020. https://doi.org/10.1108/BIJ-11-2021-0669

Mishra, H. (2025). Environmental degradation and impacts on agricultural production: A challenge to urban sustainability. In J. A. Parray, N. Shameem, & A. K. Haghi (Eds.), *Sustainable Urban environment and waste management: Theory and practice* (pp. 53–92). Springer Nature. https://doi.org/10.1007/978-981-96-1140-9_3

Moorhouse, D. (2023). *An investigation of the theory and practice of zero-waste fashion design for the luxury sector incorporating a circular design model* [doctoral dissertation, University of Huddersfield]. University of Huddersfield Repository. https://pure.hud.ac.uk/ws/portalfiles/portal/70819650/19_final_thesis.pdf

Morseletto, P. (2020). Restorative and regenerative: Exploring the concepts in the circular economy. *Journal of Industrial Ecology, 24*(4), 763–773. https://doi.org/10.1111/jiec.12987

Möslinger, M., Ulpiani, G., & Vetters, N. (2023). Circular economy and waste management to empower a climate-neutral urban future. *Journal of Cleaner Production, 421*, 138454. https://doi.org/10.1016/j.jclepro.2023.138454

Murphy, A., Enqvist, J. P., & Tengö, M. (2019). Place-making to transform urban social–ecological systems: Insights from the stewardship of urban lakes in

Bangalore, India. *Sustainability Science, 14*(3), 607–623. https://doi.org/10.1007/s11625-019-00664-1

Narayan, S., Khatun, B., Munasinghe, S. H., Pandey, S. R., Choudhary, R., Troilo, M., & Birje, T. S. (2025). Performance metrics for circular economy: Techniques for evaluating effectiveness and measuring progress. In S. Majumdar, M. Choudhury, & A. Cheshmehzangi (Eds.), *Sustainable urban future: Embracing circular economy in waste management* (pp. 249–270). Springer. https://doi.org/10.1007/978-981-95-2976-6_11

Narayan, S., Pathak, P., Khatun, B., Choudhary, R., Troilo, M., Raj, C., & Vambol, V. (2025). Resource recovery in urban settings: Developing circular supply chains for sustainable cities. In S. Majumdar, M. Choudhury, & A. Cheshmehzangi (Eds.), *Sustainable urban future: Embracing circular economy in waste management* (pp. 225–247). Springer. https://doi.org/10.1007/978-981-95-2976-6_10

Narayan, S., Yadav, D., Jha, S., Khatun, B., Choudhary, R., Verma, U. K., & Kumar, D. (2025). From past to present: The evolution of waste management and the rise of circular economy principles. In S. Majumdar, M. Choudhury, & A. Cheshmehzangi (Eds.), *Sustainable urban future: Embracing circular economy in waste management* (pp. 23–53). Springer. https://doi.org/10.1007/978-981-95-2976-6_2

National Environment Agency. (2015). Solid waste management technology roadmap. https://www.nccs.gov.sg/files/docs/default-source/default-document-library/solid-waste-management-technology-roadmap.pdf

Ngwemba, E. B. (2024). *Contributions of circular economy practices to waste management in Mzuzu City, Malawi* [master's thesis, Mzuzu University]. Mzuzu University Digital Repository. https://repository.mzuni.ac.mw/handle/123456789/614

Nirmani, I. A. P. (2024). Understanding the role of community participation in enhancing circular economy outcomes in developing regions. *World Journal of Advanced Research and Reviews, 23*(3), 3034–3043. https://doi.org/10.30574/wjarr.2024.23.3.2935

Nwokolo, S. C. (2025). From policy to practice: Evaluating the role of private sector champions like Elon musk in shaping trump's 2.0 climate agenda. *Trends in Renewable Energy, 11*(1), 122–154. https://doi.org/10.17737/tre.2025.11.1.00190

Ogwu, M. C., & Kosoe, E. A. (2025). Integrating green infrastructure into sustainable agriculture to enhance soil health, biodiversity, and microclimate resilience. *Sustainability, 17*(9), 3838. https://doi.org/10.3390/su17093838

Oikonomaki, E., Papadaki, I., & Kakderi, C. (2024). Promoting green transformations through smart engagement: An assessment of 100 citizen-led urban greening projects. *Land, 13*(4), 556. https://doi.org/10.3390/land13040556

Ordonez-Ponce, E., Clarke, A., & MacDonald, A. (2021). Business contributions to the sustainable development goals through community sustainability partnerships. *Sustainability Accounting, Management and Policy Journal, 12*(6), 1239–1267. https://doi.org/10.1108/SAMPJ-03-2020-0068

Osazuwa, C. M., Olowonihi, A. P., Oghogho, M. Y., Mboto, A. A., & Gogogwute, O. (2024). Role of community-based conservation in Ekiti state. *The American Journal of Interdisciplinary Innovations and Research, 6*(11), 97–112. https://doi.org/10.37547/tajiir/Volume06Issue11-07

Oyamo, V. I., Etan, M. O., Offiong, A. E. A., & Osang, G. O. (2025). Examining the effectiveness of community-based projects in fostering environmental stewardship and promoting local sustainability initiatives in cross river state, Nigeria. *Transdisciplinary Journal of Education Sustainable Development Studies, 1*(2), 258–279. https://doi.org/10.60787/TJESDS.VOL1NO2.19

Paavola, M. (2020). Transport and mobility in an urban circular economy. https://lutpub.lut.fi/handle/10024/160766

Pablo-Romero, M. D. P., Sánchez-Braza, A., & Torreblanca, C. (2024). Implementing the circular economy in the European Union and Spain: Links to the low-carbon transition. *Energies, 17*(21), 5255. https://doi.org/10.3390/en17215255

Pauli, G. (2010). *The blue economy: 10 years, 100 innovations, 100 million jobs.* Paradigm Publications.

Pera, A. (2020). Assessing sustainability behavior and environmental performance of urban systems: A systematic review. *Sustainability, 12*(17), 7164. https://doi.org/10.3390/su12177164

Perrotti, D., & Stremke, S. (2020). Can urban metabolism models advance green infrastructure planning? Insights from ecosystem services research. *Environment and Planning B: Urban Analytics and City Science, 47*(4), 678–694. https://doi.org/10.1177/2399808318797131

Piazza, R. (2024). Learning cities: Catalysts for sustainable urban development through collaborative partnerships. *Form@re - Open Journal per La Formazione in Rete, 24*(2), 228–242. https://doi.org/10.36253/form-16069

Pirenne, F. (2023). *A comparative study on the adoption of the circular economy in European port cities* [master's thesis, HEC School of management, University of Liège]. *MatheO repository.* http://hdl.handle.net/2268.2/17531

Psarommatis, F., & May, G. (2025). A cost–benefit model for sustainable product reuse and repurposing in circular remanufacturing. *Sustainability, 17*(1), 245. https://doi.org/10.3390/su17010245

Purchase, C. K., Al Zulayq, D. M., O'Brien, B. T., Kowalewski, M. J., Berenjian, A., Tarighaleslami, A. H., & Seifan, M. (2021). Circular economy of construction and demolition waste: A literature review on lessons, challenges, and benefits. *Materials, 15*(1), 76. https://doi.org/10.3390/ma15010076

Ragazou, K., Zournatzidou, G., Sklavos, G., & Sariannidis, N. (2024). Integration of circular economy and urban metabolism for a resilient waste-based sustainable urban environment. *Urban Science, 8*(4), 175. https://doi.org/10.3390/urbansci8040175

Ramírez-Márquez, C., Posadas-Paredes, T., & Ponce-Ortega, J. M. (2025). From resource abundance to responsible scarcity: Rethinking natural resource utilization in the age of hyper-consumption. *Resources, 14*(8), 118. https://doi.org/10.3390/resources14080118

Rashid, S., & Malik, S. H. (2023). Transition from a linear to a circular economy. In S. A. Bandh, F. A. Malla, & A. T. Hoang (Eds.), *Renewable energy in Circular economy* (pp. 1–20). Springer International Publishing. https://doi.org/10.1007/978-3-031-42220-1_1

Rawat, L. S., Maikhuri, R. K., Bahuguna, Y. M., Jugran, A. K., Maletha, A., Jha, N. K., Phondani, P. C., Dhyani, D., Pharswan, D. S., & Chamoli, S. (2022). Rejuvenating ecosystem services through reclaiming degraded land for sustainable societal development: Implications for conservation and human wellbeing. *Land Use Policy, 112*, 105804. https://doi.org/10.1016/j.landusepol.2021.105804

Restrepo Montoya, A. (2024). *Environmental urbanism and urban geographies: Medellín 2024–2027 urban plan.* ArchDaily. https://www.archdaily.com/1019444/environmental-urbanism-and-urban-geographies-medellin-2024-2027-urban-plan

Reyes, D. J. (2023). Sustainable development planning in the built environment: Towards a circular economy : A case study in Lahti, Finland. [fi=Ylempi AMK-opinnäytetyö|sv=Högre YH-examensarbete|en=master's thesis|]. http://www.theseus.fi/handle/10024/808418

Ritter, M. (2023). *The green corridors of Medellin: A case of successful adaptation?* IMAGINE Adaptation. https://imagineadapt.bc3research.org/news/the-green-corridors-of-medellin-a-case-of-successful-adaptation/

Romero-Luis, J., Carbonell-Alcocer, A., Gertrudix, M., Gertrudis Casado, M. D. C., Giardullo, P., & Wuebben, D. (2022). Recommendations to improve communication effectiveness in social marketing campaigns: Boosting behavior change to foster a circular economy. *Cogent Social Science, 8*(1), 2147265. https://doi.org/10.1080/23311886.2022.2147265

Rydin, Y. (2021). Governance theories: Stakeholders, networks and collaboration. In I. Y. Rydin (Ed.), *Theory in planning research* (pp. 83–104). Springer. https://doi.org/10.1007/978-981-33-6568-1_5

Saez Ujaque, D., Roca, E., De Balanzó Joue, R., Fuertes, P., & Garcia-Almirall, P. (2021). Resilience and urban regeneration policies. Lessons from community-led initiatives. The case study of canfugarolas in mataro(Barcelona). *Sustainability, 13*(22), 12855. https://doi.org/10.3390/su132212855

Salmenperä, H., Pitkänen, K., Kautto, P., & Saikku, L. (2021). Critical factors for enhancing the circular economy in waste management. *Journal of Cleaner Production, 280*, 124339. https://doi.org/10.1016/j.jclepro.2020.124339

Sarna, B., Kortelainen, S., Plotnikova, E., & Saqib, S. (2025). Foundations and catalysts for circular economy: A Coursebook on Circular Economy—Course 1.

Saxena, V. (2025). Water quality, air pollution, and climate change: Investigating the environmental impacts of industrialization and urbanization. *Water, Air Soil Pollution, 236*(2), 73. https://doi.org/10.1007/s11270-024-07702-4

Scheel, C., Aguiñaga, E., & Bello, B. (2020). Decoupling economic development from the consumption of finite resources using circular economy: A model for developing countries. *Sustainability, 12*(4), 1291. https://doi.org/10.3390/su12041291

Shan, C., & Ji, X. (2024). Environmental regulation and green technology innovation: An analysis of the government subsidy policy's role in driving corporate green transformation. *Industrial Engineering and Innovation Management, 7*(1). https://doi.org/10.23977/ieim.2024.070106

Sharma, K. D., & Jain, S. (2020). Municipal solid waste generation, composition, and management: The global scenario. *Social Responsibility Journal, 16*(6), 917–948. https://doi.org/10.1108/SRJ-06-2019-0210

Sherratt, A. (2013). Cradle to cradle. In S. O. Idowu, N. Capaldi, L. Zu, & A. D. Gupta (Eds.), *Encyclopedia of corporate social responsibility* (pp. 630–638). Springer. https://doi.org/10.1007/978-3-642-28036-8_165

Sillanpää, M., & Ncibi, C. (2019). *The circular economy: Case studies about the transition from the linear economy*. Academic Press. https://doi.org/10.1016/C2018-0-01363-3

Skawińska, E., & Zalewski, R. (2021). *Contemporary challenges for sustainable development: Linear versus circular economy*. Polskie Towarzystwo Ekonomiczne w Zielonej Górze.

Sreenath, A. B., Dhanaraj, K., Rath, P., Bhardwaj, L. K., & Choudhary, R. (2025). Waste management and circular economy. In S. Kulkarni & H. Akolkar (Eds.), *Sustainable chemistry and pioneering green engineering solutions (chap. 11)*. IGI Global. https://doi.org/10.4018/979-8-3373-1409-9.ch011

Stahel, W. R. (2020). History of the circular economy. The historic development of circularity and the circular economy. In S. Eisenriegler (Ed.), *The Circular economy in the European Union* (pp. 7–19). Springer International Publishing. https://doi.org/10.1007/978-3-030-50239-3_2

Stahel, W. R. (2021). The circular industrial economy of the anthropocene and its benefits to society. In K. B. Misra (Ed.), *Handbook of advanced Performability engineering* (pp. 249–270). Springer International Publishing. https://doi.org/10.1007/978-3-030-55732-4_11

Stahel, W. R., & Clift, R. (2016). Stocks and flows in the performance economy. In R. Clift & A. Druckman (Eds.), *Taking stock of industrial ecology* (pp. 137–158). Springer. https://doi.org/10.1007/978-3-319-20571-7_7

Suchek, N., Fernandes, C. I., Kraus, S., Filser, M., & Sjögrén, H. (2021). Innovation and the circular economy: A systematic literature review. *Business Strategy and the Environment, 30*(8), 3686–3702. https://doi.org/10.1002/bse.2834

Suriyankietkaew, S., Krittayaruangroj, K., Thinthan, S., & Lumlongrut, S. (2025). Creative tourism as a driver for sustainable development: A model for advancing SDGs through community-based tourism and environmental stewardship. *Environmental and Sustainability Indicators, 27*, 100828. https://doi.org/10.1016/j.indic.2025.100828

Sverdlik, A., Mitlin, D., & Dodman, D. (2019). *Realising the multiple benefits of climate resilience and inclusive development in informal settlements.* Cities Alliance. https://www.citiesalliance.org/sites/default/files/2019-12/JWP%20%282019%29%20Realising%20the%20Multiple%20Benefits.pdf

Tomás, J. P. (2017). *How Seoul used the IoT to improve waste management and collection.* RCR Wireless News. https://www.rcrwireless.com/20171009/fundamentals/how-seoul-improved-waste-collection-via-smart-waste-management-tool-tag23-tag99

United Nations. (2023). Goal 11: Make cities and human settlements inclusive, safe, resilient and sustainable. In The Sustainable Development Goals Report 2023. United Nations. https://unstats.un.org/sdgs/report/2023/goal-11/

United Nations Environment Programme. (2024). Global waste management outlook 2024. United Nations Environment Programme. https://www.unep.org/resources/global-waste-management-outlook-2024

Vanhuyse, F., Fejzić, E., Ddiba, D., & Henrysson, M. (2021). The lack of social impact considerations in transitioning towards urban circular economies: A scoping review. *Sustainable Cities and Society, 75*, 103394. https://doi.org/10.1016/j.scs.2021.103394

Villarroel, K., & Yanez, Á. (2024). *Urban waste management for cleaner cities: Towards a circular economy in LAC* [blog post]. Inter-American Development Bank Blogs. https://blogs.iadb.org/ciudades-sostenibles/en/urban-waste-management-in-lac-cities/

Wahyuni, W., Ulfa, F., Novianti, A., & Rosalinda, R. (2024). Sustainable business ethics: Fostering corporate responsibility and environmental stewardship. *Advances: Jurnal Ekonomi Bisnis, 2*(4). https://doi.org/10.60079/ajeb.v2i4.351

Waxin, M.-F., Knuteson, S. L., & Bartholomew, A. (2019). Outcomes and key factors of success for iso 14001 certification: Evidence from an emerging Arab gulf country. *Sustainability, 12*(1), 258. https://doi.org/10.3390/su12010258

Williams, J. (2019). Circular cities: Challenges to implementing looping actions. *Sustainability, 11*(2), 423. https://doi.org/10.3390/su11020423

Williams, J. (2021). Circular cities: What are the benefits of circular development? *Sustainability, 13*(10), 5725. https://doi.org/10.3390/su13105725

Wolf, K. L., Brinkley, W., & Blahna, D. (2021). Civic environmental stewardship: Aligning organizational and participant motivations. *Cities and the Environment, 14*(2). https://doi.org/10.15365/cate.2021.140203

Yang, M., Chen, L., Wang, J., Msigwa, G., Osman, A. I., Fawzy, S., Rooney, D. W., & Yap, P.-S. (2023). Circular economy strategies for combating climate change and other environmental issues. *Environmental Chemistry Letters, 21*(1), 55–80. https://doi.org/10.1007/s10311-022-01499-6

Zhu, J., Fan, C., Shi, H., & Shi, L. (2019). Efforts for a circular economy in China: A comprehensive review of policies. *Journal of Industrial Ecology, 23*(1), 110–118. https://doi.org/10.1111/jiec.12754

Zipperer, W. C., Northrop, R., & Andreu, M. (2020). Urban development and environmental degradation. In W. C. Zipperer, R. Northrop, & M. Andreu (Eds.), *Oxford research Encyclopedia of environmental science*. Oxford University Press. https://doi.org/10.1093/acrefore/9780199389414.013.97

CHAPTER 3

Integrating Innovations: Smart Technologies and Circular Economy Synergies

Prachi Wasnik, Sushobhan Majumdar, and Ravi Rajak

3.1 Introduction

3.1.1 The Digital Revolution Meets Sustainability

The convergence of smart technologies with circular economy principles has rapidly shifted from conceptual debate to real-world implementation. Digital solutions are now driving measurable economic value while helping industries respond to climate pressures, resource scarcity, and waste challenges (Choudhury, Cheshmehzangi, & Majumdar, 2025; Majumdar, Choudhury, & Cheshmehzangi, 2025). The global digital circular economy market was valued at USD 2.34 billion in 2024 and is projected to

P. Wasnik
Dyson School of Design Engineering, Imperial College London, London, UK

S. Majumdar
Department of Geography, Pandit Raghunath Murmu Smriti Mahavidyalaya, Bankura, West Bengal, India

R. Rajak (✉)
Department of Economics, University of Bristol, Bristol, UK
e-mail: ravi.india.2019@bristol.ac.uk

S. Majumdar, M. Choudhury (eds.), *City of the Future*,
https://doi.org/10.1007/978-3-032-18026-1_3

reach USD 13.84 billion by 2032, growing at a compound annual growth rate of 24.88% (DataM Intelligence, 2025). This strong trajectory reflects a fundamental change in how organizations view digital innovation, not simply as a support tool, but as a core engine of circular transformation (González, Kandpal, Machado, Martens, & Majumdar, 2023).

Investment momentum across sectors reinforces this shift. The global electronics recycling industry, for instance, is expected to grow from USD 37 billion in 2022 to more than USD 108 billion by 2030 (UNCTAD, 2024). Artificial intelligence (AI)-enabled waste sorting, advanced material recovery systems, and smart logistics technologies are forecast to expand at over 20% annually, driven by automation in manufacturing and municipal waste infrastructure. Leading firms such as AMP Robotics, Circularise, and Philips Lighting are already using AI, blockchain, and IoT platforms to track materials, extend product life, and adopt product-as-a-service models. These advancements signal a decisive shift toward technology-enabled circularity, in which digital systems underpin regenerative, resource-efficient economic models (Kowsalya et al., 2025).

The global electronics recycling industry alone is predicted to increase from $37 billion in 2022 to an estimated $108 billion by 2030, illustrating the potential for economic growth that arises when smart technologies align with circular principles (UNCTAD, 2024). This convergence is transforming how businesses approach sustainability, generating new business models, and optimizing resource flows.

3.1.2 Evolution of Circular Economy Concepts

Academic discourse around circularity has undergone substantial transformation in recent years, evolving from traditional waste management approaches to comprehensive sustainability paradigms. The circular economy now represents a new sustainability paradigm and a major driver of innovation in industrial firms, with digital sterilization enabled through smart products and their embeddedness in Internet systems (Alcayaga & Hansen, 2025; Majumdar et al., 2025). Contemporary scholarship reveals increasingly sophisticated frameworks that emphasize systemic regeneration, lifecycle optimization, and value-retention strategies. Modern approaches integrate digital twin technology into circular economy frameworks, enabling continuous monitoring, predictive analytics, and AI-driven decision-making to optimize operations, extend product lifespans, and achieve sustainability goals (Wang, Liu, & Zhang, 2025).

The smart circular economy, which represents the intersection of circular economy principles and digital technologies such as AI, blockchain, cyber-physical systems, big data, and IoT, has proven helpful in addressing global sustainability challenges and improving organizational performance (Majumdar et al., 2025; Truant, Borlatto, & Corazza, 2024). Modern circular models integrate advanced design philosophies that prioritize durability, modularity, and material recovery while simultaneously creating new revenue streams through product-as-a-service and sharing-economy business models. Research demonstrates that achieving meaningful scale requires supply chain integration to mediate the relationship between smart manufacturing technologies and circular economy implementation through enhanced collaborations within and between companies (Di Maria, De Marchi, & Galeazzo, 2022; Neri et al., 2024). The integration of IoT, blockchain, and digital identification systems provides closed-loop information systems that are crucial for sustainable material circulation, enabling real-time monitoring of product lifecycles and improved data dissemination among stakeholders (DataM Intelligence, 2025).

3.1.3 Addressing Sustainability in Rapidly Growing Cities

Rapid urbanization is intensifying pressure on cities, making sustainability an urgent priority for governments, communities, and planners. As populations grow, demand for housing, energy, transportation, and resources accelerates, often outpacing the capacity of existing infrastructure (Chatterjee & Majumdar, 2022; Gough, 2015). This expansion can lead to increased waste, pollution, and social inequality, if not managed carefully. With climate change amplifying risks such as heatwaves, flooding, and resource scarcity, cities must adopt sustainable strategies that reduce environmental impacts while enhancing resilience (Choudhury, Dixit, & Majumdar, 2025). Addressing sustainability is no longer optional; it is essential for ensuring livable, equitable, and economically stable urban futures (Sanchez & Ye, 2023).

3.1.4 Smart Technologies and Circular Economy Principles

Smart technologies and circular economy principles are increasingly intertwined as cities seek solutions that optimize resource use while reducing environmental footprints (Majumdar & Chatterjee, 2020;

Table 3.1 Smart technologies and circular economy principles

Aspect	*Smart technologies*	*Circular economy principles*	*Integration potential*
Primary focus	Efficiency, data-driven decision-making	Resource regeneration, waste reduction	Optimizing resource flows using data analytics
Key tools	IoT, AI, big data, digital platforms	Reduce, reuse, repair, remanufacture, recycle	IoT-enabled waste sorting, predictive maintenance
Outcome	Real-time monitoring and automation	Closed-loop systems and extended product life	Enhanced sustainability and resilience

Source: Compiled by the Authors

Mohamed, 2023). Digital tools such as IoT sensors, AI, and data analytics enable cities to monitor consumption patterns, identify inefficiencies, and manage resources in real time, critical capabilities for achieving circularity (Majumdar et al., 2025). At the same time, circular economy frameworks provide the sustainable foundation that ensures technological advancements do more than improve efficiency; they support long-term regeneration and reduced waste (Choudhury et al., 2025; Wu et al., 2025; Zulkifli, Napu, Hakim, & Murini, 2024). Together, these approaches create a synergy in which data-driven intelligence enhances circular resource loops, and circular principles guide the responsible deployment of technology. Table 3.1 shows smart technologies and circular economy principles.

This chapter is guided by key questions that explore how cities can merge smart technologies with circular economy strategies to build resilient, regenerative urban systems. These questions include the following: How can digital tools enhance circular material and energy flows? What governance models and policies are needed to support integrated innovation? How can cities overcome social, economic, and technological barriers to adoption! Moreover, what risks must be addressed to ensure that smart solutions remain ethical, equitable, and sustainable! These guiding inquiries shape the chapter's exploration of practical pathways for creating future-ready cities (Dixit, Majumdar, & Choudhury, 2025).

3.2 Role of AI, IoT, and Blockchain in Circular Economy

3.2.1 Artificial Intelligence Applications

Artificial intelligence serves as the driving force to achieve circular economy goals, transforming every aspect of business from product design and operational management to service delivery (Jensen, 2025). AI is an interdisciplinary domain that establishes relationships between economic growth and its surroundings to enable the sustainable use of natural resources, with the capability to perform functions associated with human intelligence, such as reasoning and learning (Liao et al., 2024; Majumdar, 2023; Raut, Hossain, Kouhizadeh, & Fazio, 2025). Contemporary implementations demonstrate AI's effectiveness in enabling circular economy adoption through socio-technical mechanisms, particularly in predictive maintenance systems that leverage machine learning algorithms to analyze IoT data in real time (Bag & Mangla, 2025; Hwang, Kim, & Park, 2025).

Machine learning-assisted iterative design processes enable faster prototyping and testing of recyclable, biodegradable, or renewable products suitable for circular economy applications (Intesa Sanpaolo Innovation Center, 2024). Advanced data-driven techniques in circular economy applications demonstrate AI's capacity to promote waste-to-energy conversion and support cascade reuse paradigms, reverse supply chains by recovering end-of-life materials, and extend lifecycles through repair and reuse (Chen, Zhang, & Wang, 2025). Integration of machine learning technology with green supply chain management in smart cities optimizes resource use through gravitational algorithm-enhanced support vector machine models (Liu & Wang, 2024).

3.2.2 Internet of Things Infrastructure

IoT applications enhance efficiency, reduce waste, and prolong product lifecycles through data analytics, real-time tracking, and automation, while fostering inventive circular business models such as product-as-a-service and sharing economies (Hassan & Rahman, 2024). The global IoT platforms market is experiencing robust growth, driven by sustainability trends such as circular economy initiatives and real-time environmental monitoring enabled by organizations that track environmental changes through IoT devices and sensors (Precedence Research, 2025). IoT-driven waste management solutions optimize collection routes, improve recycling rates, and monitor landfill conditions, reducing landfill burden and promoting circular economy principles by enabling better tracking and management of materials (Stats and Insights, 2025).

Smart sensors embedded in systems monitor fill levels in real time, automatically alerting collection services and eliminating unnecessary trips, thereby reducing fuel consumption and operational costs by up to 30% in municipalities (Okon Recycling, 2025). The implementation of Time-of-Flight sensing technology with ESP32-S3 microcontrollers enables IoT connectivity for real-time monitoring systems, providing data that facilitates efficient resource management and supports informed decision-making in circular economy applications (Sensors Journal, 2025). IoT systems incorporating sensors for automatic waste detection and classification ensure recyclable materials are correctly identified, minimizing contamination and improving resource recovery efficiency (Ritzkal et al., 2025).

3.2.3 Blockchain Trust Mechanisms

Blockchain technology enables monitoring throughout the entire product lifecycle, ensuring compliance and sustainability through enhanced transparency and automation via smart contracts. At the same time, tokenization facilitates traceability and value distribution in circular supply chains (Carlos, Silva, & Gonçalves, 2025). Blockchain addresses critical challenges in supply chain management by bringing transparency to the entire product lifecycle, enabling monitoring of adherence to standards in fair trade and circular economy models through verifiable records (Hübschke, Buss, Holschbach, & Lier, 2025). Blockchain provides an immutable, transparent, and traceable system for gathering, storing, and retrieving data across networks, with timestamps and controls that confirm the authenticity and provenance of materials (Rajput, More, Adhikari, & Arya, 2025).

The blockchain for sustainable supply chains market reached USD 827.6 million in 2024, with applications that ensure authentic sourcing, track product journeys from raw materials to end users, and enable proper recycling, waste tracking, and circular economy practices (GM Insights, 2025). Research demonstrates a 400% increase in publications after 2020, with dominant topics including smart contracts, supply chain traceability, and sustainability, alongside emerging themes such as decentralization, privacy, and circular economy integration (MDPI Supply Chain, 2025).

Blockchain enhances trust and collaboration among partners in complex supply networks, with implementations including blockchain-based recycling systems that trace material use and improve compliance in closed-loop resource flows (Cai, Choi, & Zhang, 2025).

3.3 Foundations of Integration: Understanding the Two Domains

3.3.1 Smart Technologies in Urban Systems

3.3.1.1 Role of Smart Technologies

Smart technologies refer to digitally connected systems that collect, process, and use real-time data to enhance the efficiency, responsiveness, and sustainability of urban environments. Their defining characteristics include interoperability between devices and platforms, automation enabled by intelligent algorithms, seamless communication across networks, and the ability to learn and adapt over time (Apanavičienė & Shahrabani, 2023; Majumdar, Kandpal, & Anthopoulos, 2024). These technologies integrate hardware, software, and data infrastructure to improve decision-making, optimize resource use, and deliver more personalized and effective public services. In cities, smart technologies are designed to create environments that are not only technologically advanced but also more livable, resilient, and climate-responsible (Bauermann, Bussador, Bauermann, & Matrakas, 2024).

3.3.1.2 Role of IoT, AI, Big Data Analytics, and Digital Platforms

IoT, AI, big data analytics, and digital platforms form the backbone of smart urban systems, enabling cities to gather insights, predict trends, and coordinate complex operations. IoT sensors embedded in transportation networks, buildings, and utilities continuously monitor conditions such as traffic flow, energy consumption, and air quality (Mammadov, Asgarov, & Mammadova, 2024). AI processes this data to automate functions, optimize systems, and support predictive maintenance, while big data analytics uncovers patterns that inform long-term planning and policy decisions. Digital platforms integrate these capabilities, enabling real-time communication between city agencies, businesses, and residents. Together, these technologies create intelligent ecosystems that enhance efficiency, support sustainability goals, and improve overall quality of life in urban settings (Korczak, Dudycz, Nita, & Oleksyk, 2017).

3.3.2 *Core Principles of the Circular Economy*

3.3.2.1 Circular Value Loops: Reduce, Reuse, Repair, Remanufacture, Recycle

Circular value loops emphasize keeping materials, products, and resources in use for as long as possible through strategies such as reducing consumption, reusing existing goods, repairing damaged items, remanufacturing components, and recycling materials at the end of their lifecycle (Khan, Ali, & Singh, 2022). These loops aim to slow, narrow, and close resource flows, reducing the need for virgin materials while minimizing waste generation. By extending product lifespans and enabling continuous value recovery, circular value loops support more sustainable production and consumption patterns and form the foundation of circular economy thinking in urban systems (Smol, Marcinek, & Duda, 2024).

3.3.2.2 Design Thinking for Closed-loop Systems

Design thinking for closed-loop systems involves creating products, buildings, and services that anticipate and minimize waste from the outset. This approach prioritizes modularity, durability, ease of repair, and the use of materials that can be safely cycled back into the economy (Rosário, Lopes, & Rosário, 2024). It requires designers, engineers, and urban planners to consider the entire lifecycle of a product or system from material sourcing and manufacturing to end-of-life recovery ensuring that every stage contributes to regeneration rather than disposal. In cities, applying closed-loop design principles supports the development of resilient infrastructure and sustainable resource-use patterns (Forczek-Brataniec & Jamioł, 2024).

3.3.2.3 Economic and Environmental Benefits of Transitioning from Linear to Circular Models

Transitioning from linear "take–make–dispose" models to circular economies offers substantial economic and environmental benefits for cities. Economically, circular strategies can reduce costs by lowering material demand, extending product lifespans, and generating new business opportunities in repair, remanufacturing, and resource recovery sectors (Gupta, Dayal, Horiguchi, & Ghosh, 2025). Environmentally, circular models significantly reduce waste, pollution, and carbon emissions, while decreasing the extraction of finite natural resources. By promoting resource efficiency and fostering innovation, circular economies enhance urban resilience and support long-term sustainability goals, making cities more adaptable and competitive in the face of global challenges (Sousa, Bragança, Silva, & Oliveira, 2024).

3.4 Circular Principles Shaping Smart Technology Deployment

3.4.1 *Designing Smart Solutions with Lifecycle Thinking*

3.4.1.1 Minimizing Energy Consumption and E-Waste

Minimizing energy consumption and e-waste is central to lifecycle thinking in smart solution design. This approach requires developers to evaluate the environmental impacts of technologies from manufacturing through disposal, ensuring that systems use energy-efficient components, low-power protocols, and renewable energy sources wherever possible (Müller, Pradhananga, & ElZomor, 2024). It also emphasizes reducing electronic waste by designing devices with longer lifespans, optimizing software to extend hardware usability, and enabling responsible end-of-life recovery (Agarwal, Darbar, Saha, Choudhury, & Singh, 2023). By aligning smart technologies with lifecycle considerations, cities can curb resource depletion, lower greenhouse gas emissions, and prevent the accumulation of hazardous e-waste (Krumay & Brandtweiner, 2016).

3.4.1.2 Durable, Repairable, and Upgradable Tech Components

Creating durable, repairable, and upgradable tech components allows urban smart systems to remain functional and relevant over extended periods, reducing the need for frequent replacements. Modular hardware designs enable easy repairs and component swaps, while open standards support interoperability and long-term maintenance (Zimmerman, Ofori, Barrett, Soler, & Harriman, 2018). Upgradable software and firmware ensure that devices can evolve with emerging technologies without requiring full system replacement. This design philosophy not only reduces material waste and lifecycle costs but also enhances the resilience and adaptability of digital urban infrastructure (Badenko, Bolshakov, Celani, & Puglisi, 2024).

3.4.2 *Social and Ethical Considerations*

3.4.2.1 Ensuring Technology Supports Equity and Community Resilience

Ensuring technology supports equity and community resilience requires cities to prioritize inclusive design and deployment strategies that benefit all residents. Smart systems must be accessible to diverse populations, accounting for differences in digital literacy, socioeconomic status, and cultural contexts (Nair & Schlumbrecht, 2024). Equitable technology use means distributing benefits such as improved mobility, cleaner environments, and better public services fairly across communities while avoiding unintended biases in algorithms or data use. Strengthening community resilience also involves engaging residents in decision-making, ensuring that technological solutions are transparent, trustworthy, and responsive to local needs (Garzón-Galvis, Richardson, & Solomon, 2022).

3.4.2.2 Avoiding Digital Waste and Ensuring Inclusive Access

Avoiding digital waste and ensuring inclusive access are essential for aligning smart city innovations with sustainable and ethical values. Digital waste occurs not only through discarded hardware but also through abandoned applications, redundant data systems, and poorly integrated platforms that burden cities with inefficiencies (Bregoli, Liscio, Nabiti, & Sospiro, 2024). To prevent this, planners should prioritize streamlined digital ecosystems, long-term interoperability, and responsible procurement practices. At the same time, inclusive access demands that all residents have affordable connectivity, accessible digital services, and the tools needed to participate fully in smart urban life. By addressing these concerns, cities can promote both environmental responsibility and social inclusion. Table 3.2 shows emerging technologies enabling circular cities (Kameni & Tekouabou, 2023).

Table 3.2 Emerging technologies enabling circular cities

Technology	*Description*	*Circular application*
AI-driven circular design	Algorithms optimizing materials, design, and lifecycle	Predicting material reuse, extending product life
Digital twins	Virtual city models for simulation and planning	Optimizing energy, waste, and mobility systems
Autonomous circular logistics	Robotics and autonomous vehicles for material movement	Efficient collection, sorting, and redistribution

Source: Compiled by the Authors

3.5 Integrative Frameworks for City-Level Implementation

3.5.1 Systems Thinking for Urban Innovation

3.5.1.1 Cities as Interconnected Systems

Cities function as complex, interconnected systems where ecological, social, and technological elements continuously interact. Urban planning that recognizes these interdependencies can better anticipate cascading effects, such as how transportation systems influence energy use, emissions, and public health, or how social behaviors shape resource consumption (Kearns, Beaty, & Barnett, 2007). By adopting a systems perspective, planners and policymakers can design interventions that optimize multiple outcomes simultaneously improving environmental sustainability, social equity, and technological efficiency rather than focusing narrowly on a single sector (Omole, Olajiga, & Olatunde, 2024).

3.5.1.2 Identifying Leverage Points for Integration

Identifying leverage points for integration involves pinpointing areas where small, targeted interventions can create outsized impacts across urban systems. Examples include optimizing energy grids through smart sensors to reduce emissions citywide, introducing circular material flows in construction to minimize waste while supporting economic growth, or deploying data-driven mobility solutions that improve accessibility and reduce congestion (Folorunsho, Adenekan, Ezeigweneme,

Somadina, & Okeleke, 2024). By focusing on these high-impact nodes, cities can implement integrated strategies that align smart technologies with circular economy principles, accelerating transformation while conserving resources and reducing systemic inefficiencies (Horn & Proksch, 2022).

3.5.2 Governance Models and Policy Enablers

3.5.2.1 Regulatory Frameworks Encouraging Circular Innovation

Regulatory frameworks play a critical role in promoting circular innovation by setting the rules, incentives, and standards that guide sustainable urban development (Choudhury, Cheshmehzangi, & Majumdar, 2025). Policies can encourage the adoption of resource-efficient technologies, mandate recycling and reuse, establish green procurement requirements, and provide tax or funding incentives for circular initiatives (Lopes, Gomes, & Nogueira, 2025). Effective frameworks also ensure accountability, enforce environmental standards, and create a predictable environment for businesses and innovators. By aligning regulations with circular economy goals, cities can accelerate the transition from linear systems to regenerative, resource-conscious models (Rončević & Cepoi, 2021).

3.5.2.2 Public–Private Partnerships and Cross-sector Collaboration

Public–private partnerships (PPPs) and cross-sector collaboration are essential for implementing large-scale circular and smart city initiatives. Complex urban challenges often require pooled expertise, shared resources, and coordinated action across government agencies, private companies, research institutions, and civil society (Katah & Ozturk, 2024). PPPs can finance infrastructure projects, pilot innovative technologies, and scale successful solutions efficiently. Cross-sector collaboration also facilitates knowledge exchange, ensures solutions address diverse stakeholder needs, and strengthens community engagement. Together, these approaches create a cooperative ecosystem that enables cities to integrate smart technologies and circular strategies more effectively (Lende & Ambadkar, 2024).

3.5.3 *Financing Synergistic Projects*

Financing synergistic projects that combine smart technologies with circular economy principles is crucial for cities seeking long-term sustainability and resilience. Effective financing ensures that innovative solutions are not only piloted but scaled, maintained, and adapted over time (Sharma, Choudhury, Agarwal, Sharma, & Sharma, 2023). Strategic investment in both green and digital infrastructure, alongside dedicated funding mechanisms for circular technology ecosystems, enables cities to implement integrated approaches that maximize environmental, social, and economic benefits (Putra & Knaap, 2018).

3.5.3.1 *Green and Digital Infrastructure Investment Strategies*

Investing in green and digital infrastructure requires a long-term, holistic approach that considers both environmental impact and technological advancement. Green infrastructure investments include renewable energy systems, sustainable water and waste management, energy-efficient buildings, and urban greening initiatives (Sahín, Kaplan, & Yılmaz, 2025). Digital infrastructure investments such as IoT networks, data platforms, and smart grids enhance the efficiency and adaptability of urban services. Prioritizing projects that combine both green and digital elements, like sensor-enabled energy-efficient buildings or smart waste management systems, can maximize synergies, reduce operational costs, and support circular resource flows (Apanavičienė & Shahrabani, 2023).

3.5.3.2 *Funding Mechanisms Supporting Circular Technology Ecosystems*

Funding mechanisms to support circular technology ecosystems include grants, public–private partnerships, green bonds, and impact investment programs specifically targeting sustainable innovation. These mechanisms help finance research, development, and deployment of technologies that enable recycling, remanufacturing, and resource optimization (Kuzior, Ляшенко, Petrova, & Serdiuk, 2023). Incentives such as tax credits, subsidies, and low-interest loans encourage businesses and startups to develop circular solutions, while municipal funds or blended financing models ensure alignment with citywide sustainability and resilience goals. By strategically directing financial resources, cities can create a supportive environment that nurtures scalable, regenerative, and technology-driven circular initiatives (Charoontham & Amornpetchkul, 2022).

3.6 Real-World Case Studies of Integrated Innovations

3.6.1 Smart Circular Neighborhoods: Urban Districts that Blend Digital Systems with Circular Resource Cycles

Smart circular neighborhoods are urban districts designed to integrate digital technologies with circular resource cycles, creating self-sustaining, resilient communities. These neighborhoods leverage IoT sensors, data platforms, and AI-driven systems to monitor and optimize energy use, water management, mobility, and waste flows in real time (Rane, Rane, Paramesha, & Kaya, 2024). At the same time, circular principles guide the design and operation of buildings, public spaces, and local industries to maximize material reuse, extend product lifespans, and minimize waste. Examples include districts with smart energy grids paired with renewable sources, shared mobility networks, modular and adaptable buildings, and digital platforms that facilitate material exchanges and product-as-a-service models (Keena, Rondinel-Oviedo, Pomasonco-Alvis, & Bouffard, 2024). By combining technological intelligence with circular strategies, smart circular neighborhoods not only improve efficiency and sustainability but also enhance livability, social equity, and community resilience.

3.6.2 Smart Manufacturing for Circular Material Use

Smart manufacturing for circular material use combines advanced digital technologies with circular economy principles to optimize production, minimize waste, and extend the lifecycle of materials (Table 3.3). By integrating IoT sensors, AI, robotics, and real-time data analytics, manufacturers can monitor material flows, predict maintenance needs, and optimize energy and resource efficiency throughout the production process (Blinova, Pant, Nijhawan, Prakash, & Sharma, 2024). Circular practices such as remanufacturing, component reuse, and closed-loop recycling are embedded into operations, ensuring that waste is minimized and materials

Table 3.3 Governance and financing mechanisms for smart circular projects

Mechanism	*Role*	*Benefits*
Regulatory frameworks	Encourage innovation and enforce standards	Predictable environment for innovation, reduced environmental impact
Public–private partnerships	Pool resources and expertise	Accelerates implementation, shares risk
Green and digital infrastructure investment	Fund integrated projects	Synergistic environmental and technological benefits
Circular technology funding	Support startups and ecosystem growth	Stimulates innovation, promotes scalable solutions

Source: Compiled by the Authors

are continually reintegrated into the value chain. For example, factories can track and recover scrap materials automatically, redesign products for easier disassembly, or implement digital platforms that connect suppliers and recyclers (Hashemi, Esrafilian, Hadavand, & Zeraatkar, 2019). This approach not only reduces environmental impact but also enhances economic efficiency, lowers costs, and supports scalable, sustainable industrial ecosystems.

3.7 Challenges, Risks, and Limitations

3.7.1 *Technological Lock-in and Digital Obsolescence*

Technological lock-in and digital obsolescence pose significant challenges to cities seeking long-term sustainability. Rapid innovation can lead to infrastructure and systems becoming outdated before their full values are realized, creating financial strain and unnecessary electronic waste (Goswami, Choudhury, & Agarwal, 2025). Lock-in occurs when cities become dependent on proprietary platforms or vendors, limiting flexibility, interoperability, and the ability to adopt better solutions in the future (Pereira & Azambuja, 2021). These risks make it crucial for cities to prioritize open standards, modular systems, and scalable technologies that can evolve with changing needs and avoid long-term inefficiencies.

3.7.2 *Privacy, Cybersecurity, and Ethical Dilemmas*

The increasing use of sensors, AI, and data platforms raises major privacy, cybersecurity, and ethical concerns. Vast amounts of personal and environmental data are collected to optimize urban services, making cities vulnerable to breaches, surveillance misuse, and algorithmic bias (Refaei, 2024). Ensuring data protection, transparency in data governance, and accountability in automated decision-making is essential for maintaining public trust. Ethical dilemmas also emerge around who controls data, how it is used, and whether it reinforces inequalities. Cities must therefore adopt strong cybersecurity standards, clear consent mechanisms, and ethical frameworks for responsible digital governance (Fayayola & Olorunfemi, 2023).

3.7.3 *Economic and Social Barriers*

Economic and social barriers can limit the adoption of smart and circular innovations, particularly in cities with financial constraints or uneven access to technology. High upfront costs for digital infrastructure, workforce capacity gaps, and limited funding can slow progress. Socially, communities may resist changes if they feel excluded from decision-making or fear job displacement due to automation (Liu, Wang, & Hoo, 2022). Addressing these barriers requires equitable investment strategies, inclusive stakeholder engagement, and workforce development programs that ensure all residents benefit from the transition to smarter and more sustainable systems (Hassebo, Tealab, & Hamouda, 2024).

3.7.4 *Policy and Regulatory Constraints*

Policy and regulatory constraints can hinder the integration of smart technologies and circular economy principles in urban environments. Outdated regulations, fragmented governance structures, and slow policy adaptation can prevent cities from implementing innovative solutions at scale (Batista, Carrara, & Olivi, 2025). Furthermore, conflicting standards across sectors such as waste management, energy, transportation, and data governance create bottlenecks that impede cross-sector collaboration. To overcome these limitations, governments need adaptive regulatory frameworks, clear guidelines for data sharing, and policies that encourage innovation while safeguarding environmental and social outcomes (Jiang, Zhu, & Pan, 2024).

3.8 Roadmap for Future Integration

3.8.1 Strategic Priorities for Cities

3.8.1.1 Steps for City Planners and Policymakers

City planners and policymakers should begin by setting clear sustainability and digital transformation goals that align with broader climate and development commitments. They need to establish cross-departmental coordination structures, invest in data infrastructure, and adopt evidence-based planning tools that track resource flows, environmental impacts, and system efficiencies (Judijanto, Erkamim, Dolphina, & Utama, 2023). Updating regulations such as procurement policies, building codes, zoning rules, and waste management standards to incentivize circular practices is crucial, as is creating funding mechanisms that support innovative pilot projects in mobility, energy, housing, and waste systems. Meaningful public engagement and collaboration with private-sector, academic, and civil society partners ensure that solutions are equitable, transparent, and responsive to local needs (Pasupuleti, 2024).

3.8.1.2 Integrating Circular and Digital Strategies into Long-Term Urban Plans

Embedding circular and digital strategies into long-term urban plans requires cities to treat technology and sustainability as interconnected pillars rather than separate initiatives. This involves incorporating lifecycle thinking into urban design, promoting resource-efficient land use, and using digital tools such as GIS mapping, digital twins, and predictive analytics to understand future demands and optimize resource loops (Rajpal, Choudhury, Goswami, Chakravorty, & Raghavan, 2024). Smart infrastructure should be planned to support circular systems, including shared mobility networks, intelligent waste recovery facilities, and buildings designed for adaptability and material reuse (Ghobakhloo et al., 2024). Long-term plans must also prioritize resilience by ensuring that digital systems reduce vulnerability to climate impacts and social inequalities. Ultimately, integrating these strategies enables cities to evolve into regenerative, efficient, and future-ready environments (Bouramdane, 2023).

3.8.2 Emerging Technologies and Opportunities

3.8.2.1 AI-Driven Circular Design

AI-driven circular design leverages advanced algorithms to help cities create products, buildings, and systems that minimize waste and maximize resource recovery. By analyzing large datasets on material flows, product lifecycles, and environmental impacts, AI can identify opportunities to extend asset longevity, optimize material choices, and improve repairability or recyclability (Cognoli, Cocco, & Ruggiero, 2024). These tools support designers, engineers, and policymakers in developing closed-loop solutions that reduce emissions and resource consumption while enhancing economic efficiency (Asres, Ghebrab, & Ekwaro-Osire, 2021). In an urban context, AI can also enable adaptive infrastructure planning, predicting future demand and guiding decisions that align with circular principles from the outset.

3.8.2.2 Digital Twins for Resource-Efficient City Planning

Digital twins, virtual replicas of physical urban systems, offer powerful opportunities for resource-efficient city planning. By simulating energy use, mobility patterns, waste generation, and environmental impacts, digital twins allow planners to test scenarios before implementation, reducing costly mistakes and accelerating sustainable innovation (Chande, 2024). They enable real-time monitoring and predictive management of infrastructure, helping cities optimize resource flows, improve resilience, and identify efficiencies across building performance, transportation networks, and utilities (Kushawaha, Gupta, & Singh, 2024). When integrated with circular economy strategies, digital twins support long-term planning that prioritizes reuse, regeneration, and intelligent resource allocation.

3.8.2.3 Autonomous Circular Logistics

Autonomous circular logistics combines autonomous vehicles, robotics, and intelligent routing systems to create efficient, low-emission circular material flows within cities. These technologies can streamline the collection, sorting, and redistribution of materials such as reusable packaging, repairable electronics, or shared goods thereby minimizing waste and reducing transport-related emissions (Nagy & Szentesi, 2024). Autonomous delivery systems can also enhance urban sharing models and product-as-a-service frameworks by ensuring timely movement of goods with minimal human intervention. By improving precision, efficiency, and scalability, autonomous circular logistics helps cities build more flexible and sustainable supply chains that support circular economies (Betts, Gutiérrez-Franco, & Ponce-Cueto, 2022).

3.9 Conclusion

The integration of smart technologies with circular economy principles represents a transformative pathway for the cities of the future. By leveraging data-driven systems, AI, IoT, and digital platforms, urban areas can optimize resource flows, reduce waste, and enhance operational efficiency. At the same time, circular strategies focused on reducing, reusing, repairing, remanufacturing, and recycling ensure that technological innovations contribute to long-term sustainability rather than short-term gains. This chapter has highlighted the critical interdependencies between technology and circularity, the governance models and financing strategies needed to support their convergence, and the emerging opportunities offered by AI-driven design, digital twins, and autonomous logistics. It has also addressed the challenges, including technological lock-in, ethical concerns, economic and social barriers, and regulatory constraints, emphasizing the need for careful planning and inclusive implementation. Ultimately, achieving smart, circular cities requires a holistic approach that views urban systems as interconnected ecological, social, and technological networks. By identifying leverage points, fostering cross-sector collaboration, and aligning investment with sustainability goals, cities can build resilient, equitable, and regenerative urban environments. The journey toward such integrated urban futures is complex, but the potential rewards in environmental, social, and economic terms make it both urgent and essential.

References

Agarwal, S., Darbar, S., Saha, S., Choudhury, M., & Singh, R. P. (2023). E-waste management using different cost-effective, eco-friendly biological techniques: An overview. In *Waste management and resource recycling in the developing world* (pp. 205–235). Elsevier. https://doi.org/10.1016/B978-0-323-90463-6.00026-9

Alcayaga, A., & Hansen, E. G. (2025). Smart circular economy as a service business model: An activity system framework and research agenda. *R&D Management, 55*(1), 49–68.

Apanavičienė, R., & Shahrabani, M. (2023). Key factors affecting smart building integration into smart city: Technological aspects. *Smart Cities, 6*(4), 1832–1857. https://doi.org/10.3390/smartcities6040085

Asres, E., Ghebrab, T., & Ekwaro-Osire, S. (2021). Framework for design of sustainable flexible pavement. *Infrastructures, 7*(1), 6. https://doi.org/10.3390/infrastructures7010006

Badenko, V., Bolshakov, N., Celani, A., & Puglisi, V. (2024). Principles for sustainable integration of bim and digital twin technologies in industrial infrastructure. *Sustainability, 16*(22), 9885. https://doi.org/10.3390/su16229885

Bag, S., & Mangla, S. K. (2025). Investigating the role of smart and resilient supplier management practices in circular economy: A supply chain practice view perspective. *Business Strategy and the Environment, 34*, 3919–3939.

Batista, L., Carrara, J., & Olivi, L. (2025). Are smart cities truly sustainable? Analysis of circular economy and waste management practices with proposals for the Brazilian context. *IOP Conference Series: Earth and Environmental Science, 1536*(1), 012021. https://doi.org/10.1088/1755-1315/1536/1/012021

Bauermann, B., Bussador, A., Bauermann, H., & Matrakas, M. (2024). Connecting the green to the digital: Integrating eco cities and smart regions. *Eco Cities, 5*(1), 2755. https://doi.org/10.54517/ec.v5i1.2755

Betts, K., Gutiérrez-Franco, E., & Ponce-Cueto, E. (2022). Key metrics to measure the performance and impact of reusable packaging in circular supply chains. *Frontiers in Sustainability, 3*. https://doi.org/10.3389/frsus.2022.910215

Blinova, T., Pant, R., Nijhawan, G., Prakash, A., & Sharma, A. (2024). Innovations in smart manufacturing: An experimental assessment of emerging technologies. *Bio Web of Conferences, 86*, 01064. https://doi.org/10.1051/bioconf/20248601064

Bouramdane, A. (2023). Shaping resilient buildings and cities: Climate change impacts, metrics, and strategies for mitigation and adaptation. *Information System and Smart City, 3*(1), 190. https://doi.org/10.59400/issc.v3i1.190

Bregoli, D., Liscio, M., Nabiti, N., & Sospiro, P. (2024). *Exploring the evolution of smart cities: A review on energy, mobility, and waste management aspects*. IntechOpen. https://doi.org/10.5772/intechopen.114330

Cai, Y., Choi, T. M., & Zhang, J. (2025). Building resilient circular supply chains with blockchain: Exploring the impact of relationships and capabilities. *Technovation, 137*, 105920.

Carlos, R. M., Silva, M. E., & Gonçalves, J. (2025). Blockchain-enabled integration of circular economy practices in supply chains: A conceptual framework. *Business Strategy & Development, 8*(1), 70222. https://doi.org/10.1002/bsd2.70222

Chande, S. (2024). An exploratory view of application of digital twin technology in education. *International Journal of Advanced Trends in Computer Science and Engineering, 13*(2), 53–58. https://doi.org/10.30534/ijatcse/2024/031322024

Charoontham, K., & Amornpetchkul, T. (2022). Startup accelerator analysis: Strategic decision on effort exertion and information disclosure regime. *Journal of Entrepreneurship in Emerging Economies, 16*(2), 418–445. https://doi.org/10.1108/jeee-06-2020-0188

Chatterjee, U., & Majumdar, S. (2022). Impact of land use change and rapid urbanization on urban heat Island in Kolkata city: A remote sensing based perspective. *Journal of Urban Management, 11*(1), 59–71. https://doi.org/10.1016/j.jum.2021.09.002

Chen, L., Zhang, Y., & Wang, H. (2025). AI-driven digital circular economy with material and energy sustainability for industry 4.0. *Energy Reports, 11*, 2468–2485.

Choudhury, M., Cheshmehzangi, A., & Majumdar, S. (2025). The circular economy paradigm: Core concepts and value. In S. Majumdar, M. Choudhury, & A. Cheshmehzangi (Eds.), *Sustainable urban future. Urban sustainability.* Springer. https://doi.org/10.1007/978-981-95-2976-6

Choudhury, M., Dixit, G., & Majumdar, S. (Eds.). (2025). *Sustainable synergy: Harnessing ecosystems for climate resilience.* Springer Nature. https://doi.org/10.1007/978-3-031-77957-2

Choudhury, M., Suhag, M., Jha, P., Harne, K. R., Samanta, P., Choudhury, M., Borgohain, R. B., Majumdar, S., Kravchenko, Y., & Choudhary, R. (2025). Urban waste management and challenges in emerging smart city systems governance and circular economy perspectives. *Ecological Engineering & Environmental Technology, 26*(10), 243–266. https://doi.org/10.12912/27197050/211014

Cognoli, R., Cocco, P., & Ruggiero, R. (2024). Innovative timber upcycling: Digital strategies for prolonging timber lifespan and promoting reuse. *IOP Conference Series: Earth and Environmental Science, 1402*(1), 012036. https://doi.org/10.1088/1755-1315/1402/1/012036

DataM Intelligence. (2025). *Digital circular economy market size, growth report 2025–2032.* https://www.datamintelligence.com/research-report/digital-circular-economy-market

Di Maria, E., De Marchi, V., & Galeazzo, A. (2022). Industry 4.0 technologies and circular economy: The mediating role of supply chain integration. *Journal of Cleaner Production, 348*, 131384.

Dixit, G., Majumdar, S., & Choudhury, M. (2025). Introduction to ecosystem Services for Climate Resilience: A theoretical framework. In *Sustainable synergy: Harnessing ecosystems for climate resilience* (pp. 3–16). Springer Nature. https://doi.org/10.1007/978-3-031-77957-2_1

Fayayola, O., & Olorunfemi, O. (2023). Ethical decision-making in it governance: A review of models and frameworks. *International Journal of Science and Research Archive, 11*(2), 130–138. https://doi.org/10.30574/ijsra.2024.11.2.0373

Folorunsho, S., Adenekan, O., Ezeigweneme, C., Somadina, I., & Okeleke, P. (2024). Developing smart cities with telecommunications: Building connected and sustainable urban environments. *Engineering Science & Technology Journal, 5*(8), 2492–2519. https://doi.org/10.51594/estj.v5i8.1441

Forczek-Brataniec, U., & Jamioł, K. (2024). The role of public space in building the resilience of cities: Analysis of representative projects from IFLA Europe exhibitions. *Sustainability, 16*(12), 5105. https://doi.org/10.3390/su16125105

Garzón-Galvis, C., Richardson, M., & Solomon, G. (2022). Tracking environmental and health disparities to strengthen resilience before the next crisis. *Environmental Justice, 15*(5), 319–329. https://doi.org/10.1089/env.2020.0025

Ghobakhloo, M., Iranmanesh, M., Fathi, M., Rejeb, A., Foroughi, B., & Nikbin, D. (2024). Beyond industry 4.0: A systematic review of industry 5.0 technologies and implications for social, environmental and economic sustainability. *Asia-Pacific Journal of Business Administration, 17*, 889–914. https://doi.org/10.1108/apjba-08-2023-0384

GM Insights. (2025). *Blockchain for sustainable supply chains market size to hit USD 9.2 billion by 2034.* Retrieved from https://www.gminsights.com/industry-analysis/blockchain-for-sustainable-supply-chains-market

González, E., Kandpal, V., Machado, M., Martens, M., & Majumdar, S. (2023). A bibliometric analysis of circular economies through sustainable smart cities. *sustainability, 15*(22), 15892. https://doi.org/10.3390/su152215892

Goswami, S., Choudhury, M., & Agarwal, S. (Eds.). (2025). *Electronic waste: Impact on health, animals, and the environment* (1st ed.). CRC Press. https://doi.org/10.1201/9781003582311

Gough, M. (2015). Reconciling livability and sustainability. *Journal of Planning Education and Research, 35*(2), 145–160. https://doi.org/10.1177/0739456x15570320

Gupta, M., Dayal, U., Horiguchi, S., & Ghosh, D. (2025). Demand and replacement forecasting for remanufactured parts of industrial products. *Journal of Advanced Manufacturing and Processing, 7*(3). https://doi.org/10.1002/amp2.70028

Hashemi, A., Esrafilian, R., Hadavand, S., & Zeraatkar, M. (2019). Application of fuzzy topsis for evaluation of green supply chain management practices (case study: Zanjan sepehr khodro; Iran). *International Journal of Engineering and Technology, 11*(6S), 13–24. https://doi.org/10.21817/ijet/2019/v11i6/191106084

Hassan, M., & Rahman, A. (2024). IoT-driven transformation of circular economy efficiency: An overview. *Mathematical and Computational Applications, 29*(4), 49. https://doi.org/10.3390/mca29040049

Hassebo, A., Tealab, M., & Hamouda, M. (2024). From a traditional city to a smart city: The measurement of cities' readiness for transition, Egypt as a case study. *Urban Science, 8*(4), 212. https://doi.org/10.3390/urbansci8040212

Horn, E., & Proksch, G. (2022). Symbiotic and regenerative sustainability frameworks: Moving towards circular city implementation. *Frontiers in Built Environment, 7*. https://doi.org/10.3389/fbuil.2021.780478

Hübschke, M., Buss, E., Holschbach, E., & Lier, S. (2025). Blockchain in supply chain management: A comprehensive review of success measurement methods. *Management Review Quarterly, 75*(3), 1245–1280. https://doi.org/10.1007/s11301-025-00546-0

Hwang, J., Kim, H., & Park, S. (2025). AI-driven circular transformation: Unlocking sustainable startup success through co-creation dynamics in circular economy ecosystems. *Sustainable Development, 33*(2), 70001. https://doi.org/10.1002/sd.70001

Intesa Sanpaolo Innovation Center. (2024). Why artificial intelligence enables the circular economy. https://www.intesasanpaoloinnovationcenter.com/en/news-and-events/news/2024/04/

Jensen, H. H. (2025). Mastering the circular economy and AI to stay competitive by 2030. World Economic Forum. https://www.weforum.org/stories/2025/08/circular-economy-ai-competitive/

Jiang, Z., Zhu, Z., & Pan, S. (2024). Data governance in multimodal behavioral research. *Multimodal Technologies and Interaction, 8*(7), 55. https://doi.org/10.3390/mti8070055

Judijanto, L., Erkamim, M., Dolphina, E., & Utama, I. (2023). Implementation of digitalization of city infrastructure for improved sustainability: Case study on smart city project in Surabaya, Indonesia. *West Science Nature and Technology, 1*(02), 64–72. https://doi.org/10.58812/wsnt.v1i02.488

Kameni, E., & Tekouabou, S. (2023). The role of inclusive educational technologies in transforming African cities into inclusive smart cities. *E3s Web of Conferences, 418*, 03003. https://doi.org/10.1051/e3sconf/202341803003

Katah, I., & Ozturk, O. (2024). Smart cities and sustainability: Comparative analysis and strategic insights for Qatar. https://doi.org/10.20944/preprints202410.1106.v1.

Kearns, A., Beaty, M., & Barnett, G. (2007). A social-ecological perspective on health in urban environments. *New South Wales Public Health Bulletin, 18*(4), 48. https://doi.org/10.1071/nb07031

Keena, N., Rondinel-Oviedo, D., Pomasonco-Alvis, M., & Bouffard, A. (2024). Beyond green: Integrating economic and social aspects to environmental life cycle assessments in Canadian housing. *IOP Conference Series: Earth and*

Environmental Science, 1363(1), 012017. https://doi.org/10.1088/1755-1315/1363/1/012017

Khan, S., Ali, S., & Singh, R. (2022). Determinants of remanufacturing adoption for circular economy: A causal relationship evaluation framework. *Applied System Innovation, 5*(4), 62. https://doi.org/10.3390/asi5040062

Korczak, J., Dudycz, H., Nita, B., & Oleksyk, P. (2017). *Towards process-oriented ontology for financial analysis., 11*, 981–987. https://doi.org/10.15439/2017f181

Kowsalya, P., Jennifer Flora, G., Kousalya, A., Toha, M. A., Khan, P. A., Choudhury, M., & Parthasarathi, T. (2025). Circular systems unveiled: Essential components and their impact on Recycling, reuse, and resource recovery. In S. Majumdar, M. Choudhury, & A. Cheshmehzangi (Eds.), *Sustainable Urban Future. Urban sustainability* (pp. 55–71). Springer. https://doi.org/10.1007/978-981-95-2976-6_3

Krumay, B., & Brandtweiner, R. (2016). *Companies' efforts towards reduction, reuse, recycling and recovery (4rs) of e-waste.* WIT Press. https://doi.org/10.2495/wm160011

Kushawaha, V., Gupta, G., & Singh, L. (2024). Enhancing energy efficiency: Advances in smart grid optimization. *International Journal of Innovative Research in Engineering & management, 11*(2), 100–105. https://doi.org/10.55524/ijirem.2024.11.2.20

Kuzior, A., Ляшенко, B., Petrova, I., & Serdiuk, O. (2023). Integrated models of the combination of eu grant funding and private funding in the energy sector of Ukraine based on public-private partnership. *Polityka Energetyczna—Energy Policy Journal, 26*(4), 165–194. https://doi.org/10.33223/epj/171325

Lende, P., & Ambadkar, S. (2024). Factors affecting the consideration of public-private partnerships (ppps) for smart city projects in Indore city. https://doi.org/10.21203/rs.3.Rs-3991568/v1.

Liao, Y., Liao, W., Wu, D., Zhang, C., Xu, J., & Majumdar, S. (2024). Inspired applications of artificial Intelligence in disaster-resilient regional construction for climate risk. In S. Ghosh, S. Majumdar, & A. Cheshmehzangi (Eds.), *Cities of tomorrow: Urban resilience and climate change preparedness. Urban sustainability.* Springer. https://doi.org/10.1007/978-981-97-9658-8_16

Liu, T., & Wang, Z. (2024). Optimizing green supply chain circular economy in smart cities with integrated machine learning technology. *Heliyon, 10*(9), e29825. https://doi.org/10.1016/j.heliyon.2024.e29825

Liu, X., Wang, F., & Hoo, W. (2022). Impact of traditional behavior of customers, employees, and social enterprises on the fear of change and resistance to innovation. *Frontiers in Psychology, 13.* https://doi.org/10.3389/fpsyg.2022.923094

Lopes, J., Gomes, S., & Nogueira, E. (2025). Pathways to circularity: Engagement patterns of European SMEs in the circular economy. *Business Strategy and the Environment, 34*(3), 3848–3864. https://doi.org/10.1002/bse.4192

Majumdar, S. (2023). Uses of artificial intelligence and GIS technologies in healthcare services. In R. S. Tomar, R. K. Chauhan, & S. K. Nayak (Eds.), *Artificial intelligence and data science in contemporary healthcare (chapter 18)*. CRC Press. https://doi.org/10.1201/9781003363606-18

Majumdar, S., & Chatterjee, U. (2020). Modelling urban growth using urban growth deterministic model in Kolkata metropolitan area: A geo-statistical approach. *Modeling Earth Systems and Environment, 7*(4), 2241–2249. https://doi.org/10.1007/s40808-020-00985-6

Majumdar, S., Choudhury, M., & Cheshmehzangi, A. (Eds.). (2025). *Sustainable urban future: Embracing circular economy in waste management*. Springer. https://doi.org/10.1007/978-981-95-2976-6

Majumdar, S., Kandpal, V., & Anthopoulos, L. G. (2024). *Smart cities: Innovations, challenges and future perspectives*. Springer Nature. https://doi.org/10.1007/978-3-031-59846-3

Mammadov, E., Asgarov, A., & Mammadova, A. (2024). Applications of iot in civil engineering: From smart cities to smart infrastructure. *Luminis Applied Science and Engineering, 1*(1), 13–28. https://doi.org/10.69760/lumin.202400003

MDPI Supply Chain. (2025). Mapping the landscape of blockchain for transparent and sustainable supply chains: A bibliometric and thematic analysis. *Logistics, 9*(3), 86.

Mohamed, A. (2023). Nurturing sustainable urban space: Integrating smart city innovations and earthship design principles for eco-friendly futures. *International Journal of Architectural Engineering and Urban Research, 6*(2), 376–400. https://doi.org/10.21608/ijaeur.2024.265873.1066

Müller, C., Pradhananga, P., & ElZomor, M. (2024). Pathways to decarbonization, circular construction, and sustainability in the built environment. *International Journal of Sustainability in Higher Education, 25*(6), 1315–1332. https://doi.org/10.1108/ijshe-09-2023-0400

Nagy, G., & Szentesi, S. (2024). Green logistics: Transforming supply chains for a sustainable future. *Advanced Logistic systems - Theory and Practice, 18*(3), 29–42. https://doi.org/10.32971/als.2024.026

Nair, N., & Schlumbrecht, M. (2024). Existing health inequities in the treatment of advanced and metastatic cancers. *Current Oncology Reports, 26*(12), 1553–1562. https://doi.org/10.1007/s11912-024-01617-3

Neri, A., Negri, M., Cagno, E., Franzò, S., Kumar, V., Lampertico, T., & Bassani, C. A. (2024). The relationship between digital technologies and the circular economy: A systematic literature review and a research agenda. *R&D Management, 54*(5), 1557–1588. https://doi.org/10.1111/radm.12715

Okon Recycling. (2025). The future of waste management: AI, IoT, and circular economy solutions explained. https://www.okonrecycling.com/sustainable-waste-management-solutions/

Omole, F., Olajiga, O., & Olatunde, T. (2024). Sustainable urban design: A review of eco-friendly building practices and community impact. *Engineering Science & Technology Journal, 5*(3), 1020–1030. https://doi.org/10.51594/estj.v5i3.955

Pasupuleti, M. (2024). Evaluating housing development through the lens of climate vulnerability and energy sustainability. *International Journal of Academic and Industrial Research Innovations, 04*(01), 135–146. https://doi.org/10.62311/nesx/rp-8-jan-2024

Pereira, G., & Azambuja, L. (2021). Smart sustainable city roadmap as a tool for addressing sustainability challenges and building governance capacity. *Sustainability, 14*(1), 239. https://doi.org/10.3390/su14010239

Precedence Research. (2025). Digital circular economy market size to hit USD 25.40 billion by 2034. https://www.precedenceresearch.com/digital-circular-economy-market

Putra, Z., & Knaap, W. (2018). Urban innovation system and the role of an open web-based platform: The case of Amsterdam smart city. *Journal of Regional and City Planning, 29*(3), 234. https://doi.org/10.5614/jrcp.2018.29.3.4

Rajpal, A., Choudhury, M., Goswami, S., Chakravorty, A., & Raghavan, V. (2024). *Waste management and treatment: Advances and innovations* (1st ed.). CRC Press. https://doi.org/10.1201/9781003258377

Rajput, D. V., More, P. R., Adhikari, P. A., & Arya, S. S. (2025). Blockchain technology in the food supply chain: A way towards circular economy and sustainability. *Sustainable Food Technology, 3*(2), 456–478. https://doi.org/10.1039/D5FB00065C

Rane, N., Rane, J., Paramesha, M., & Kaya, Ö. (2024). Artificial intelligence, machine learning, and deep learning for enabling smart and sustainable cities and infrastructure. In *Artificial Intelligence and Industry in Society 5.0*. Deep Science Publishing. https://doi.org/10.70593/978-81-981271-1-2_2

Raut, S., Hossain, N. U. I., Kouhizadeh, M., & Fazio, S. A. (2025). Application of artificial intelligence in circular economy: A critical analysis of the current research. *Expert Systems with Applications, 251*, 123812.

Refaei, M. (2024). AI governance in neom city: Exploring legal personality for smart robots and a framework for ethical innovation. Kuey. https://doi.org/10.53555/kuey.v30i5.2443.

Ritzkal, Prakosa, B. A., Rifki, M., Setiawan, E., Pramuko, A. E. K., Suhadi, S., & Syafrial. (2025). Using circular economy to manage organic and inorganic waste with internet of things-based monitoring system. *Instrumentation Mesure Métrologie, 24*(1), 45–52. https://doi.org/10.18280/i2m.240106

Rončević, B., & Cepoi, V. (2021). *Technologies and innovations in regional development: European Union and its strategies.* OAPEN. https://doi.org/10.3726/b17733

Rosário, A., Lopes, P., & Rosário, F. (2024). Sustainability and circular economy business development. https://doi.org/10.20944/preprints202406.1697.v1

Sahín, S., Kaplan, E., & Yılmaz, Ö. (2025). Green bonds and sustainable finance: a multidimensional analysis of macroeconomic and environmental determinants across nine countries. DOI: 10.21203/rs.3.rs-6065016/v1.

Sanchez, T., & Ye, X. (2023). The implications of human mobility and accessibility for transportation and livable cities. *Urban Science, 7*(4), 107. https://doi.org/10.3390/urbansci7040107

Sensors Journal. (2025). IoT-enabled real-time monitoring of urban garbage levels using time-of-flight sensing technology. *Sensors, 25*(7), 2152. https://doi.org/10.3390/s25072152

Sharma, A., Choudhury, M., Agarwal, S., Sharma, R., & Sharma, R. (2023). Corporate social responsibility and roles of developers for sustainability in companies. In *The route towards global sustainability: Challenges and management practices* (pp. 313–332). Springer International Publishing. https://doi.org/10.1007/978-3-031-10437-4_16

Smol, M., Marcinek, P., & Duda, J. (2024). Circular business models (cbms) in environmental management—Analysis of definitions, typologies and methods of creation in organizations. *Sustainability, 16*(3), 1209. https://doi.org/10.3390/su16031209

Sousa, R., Bragança, L., Silva, D., & Oliveira, R. (2024). Challenges and solutions for sustainable food systems: The potential of home hydroponics. *Sustainability, 16*(2), 817. https://doi.org/10.3390/su16020817

Stats and Insights. (2025). *IoT for sustainability: Powering a greener future—Stats and insights.* https://statsandinsights.com/2025/01/13/iot-for-sustainability-greener-future/

Truant, E., Borlatto, C., & Corazza, L. (2024). Drivers and barriers of smart technologies for circular economy: Leveraging smart circular economy implementation to nurture companies' performance. *Technological Forecasting and Social Change, 201*, 123223.

UNCTAD. (2024). *Digital economy report 2024: Shaping an environmentally sustainable and inclusive digital future.* United Nations Conference on Trade and Development.

Wang, Y., Liu, H., & Zhang, J. (2025). A review of digital twin integration in circular manufacturing for sustainable industry transition. *Sustainability, 17*(16), 7316. https://doi.org/10.3390/su17167316

Wu, D., Liao, W., Yu, H., Majumdar, S., Xu, J., & Choudhury, M. (2025). Assessing traditional waste management practices in urban environments and recent trends. In S. Majumdar, M. Choudhury, & A. Cheshmehzangi (Eds.),

Sustainable Urban Future. Urban sustainability. Springer. https://doi.org/10.1007/978-981-95-2976-6_5

Zimmerman, P., Ofori, M., Barrett, D., Soler, J., & Harriman, A. (2018). Considerations and examples of a modular open systems approach in defense systems. *The Journal of Defense Modeling and Simulation Applications Methodology Technology, 16*(4), 373–388. https://doi.org/10.1177/1548512917751281

Zulkifli, Z., Napu, F., Hakim, A., & Murini, M. (2024). Bibliometric analysis on the role of sustainable entrepreneurship in addressing circular economy challenges. *West Science Interdisciplinary Studies, 2*(02), 458–466. https://doi.org/10.58812/wsis.v2i02.679

CHAPTER 4

From Barriers to Breakthroughs: Innovating for a Sustainable Urban Future

Suiven John Paul Tume, Clarius Kuyar, Yongka Bernard Nyuyse, Foka Sherifatu Kinyuy, Ngoran Larissa Tomla, Leinyuy Serge Nangsinnyuy, and Sushobhan Majumdar

4.1 Introduction

Innovating for a sustainable urban future requires addressing key challenges, including climate change, social inequality, and inadequate infrastructure. Investing in green buildings, renewable energy, and smart transportation systems can reduce carbon footprints and enhance the quality of life (Adigun, Ibuotenang, & Shaibu, 2025; Kumar et al., 2025; Majumdar, 2018). Leveraging technologies like IoT, big data analytics, and AI can improve urban planning, management, and services, making cities more efficient and responsive (Chatterjee, Antipova, Ghosh,

S. J. P. Tume (✉) • C. Kuyar • Y. B. Nyuyse • F. S. Kinyuy • N. L. Tomla • L. S. Nangsinnyuy
Department of Geography and Planning, The University of Bamenda, Bamenda, Cameroon

S. Majumdar
Department of Geography, Pandit Raghunath Murmu Smriti Mahavidyalaya, Bankura, West Bengal, India

S. Majumdar, M. Choudhury (eds.), *City of the Future*,
https://doi.org/10.1007/978-3-032-18026-1_4

Majumdar, & Setiawati, 2023; Chatterjee, Biswas, Mukherjee, & Majumdar, 2022; Majumdar, Kandpal, & Anthopoulos, 2024). This takes into account inclusive governance by implementing participatory governance models, engaging citizens in decision-making, and prioritising social equity, thereby fostering more inclusive and resilient cities (Cheshmehzangi, 2025). Resilient cities are built through climate-resilient infrastructure, the promotion of green spaces, and the implementation of climate-adaptive policies, which can help cities mitigate the impacts of climate change (Chowdhury et al., 2025; Ghosh, Majumdar, & Cheshmehzangi, 2024; Sabyrbekov & Overland, 2025). This is possible through fostering partnerships between governments, private sector entities, and civil society, which can mobilise resources, expertise, and funding to support sustainable urban development (Alizadeh & Meshkini, 2025).

4.1.1 Context and Relevance

The global urban population has grown, with more than half of the world's population now living in urban areas. This trend is expected to continue, with projections suggesting that by 2050, around 68% of the world's population will reside in cities (UN-Habitat, 2024). The emergence and growth of megacities (cities with populations exceeding 10 million) are notable (Majumdar & Sivaramakrishnan, 2015; Majumdar & Sivaramakrishnan, 2016; Majumdar & Sivaramakrishnan, 2017). Cities like Tokyo, Delhi, and Shanghai are expanding rapidly, posing challenges in infrastructure, housing, and services (Chin, Alias, & Alkharabsheh, 2025). Much of the world's urban growth is occurring in developing countries, particularly in Africa and Asia. These regions are experiencing rapid urbanisation driven by economic opportunities, rural-to-urban migration, and population growth. There is a growing emphasis on developing smart cities that leverage technology to improve urban living. This includes the use of data analytics, IoT (Internet of Things), and sustainable practices to enhance transportation, energy efficiency, and public services. With increasing awareness of climate change and environmental issues, there is a stronger focus on sustainable urban development (Kumar et al., 2025; Ripp, Gustafsson, Jones, Majumdar, & Ginzarly, 2025). Cities are adopting green building practices, expanding public transportation, and creating more green spaces to improve the quality of life. As cities face challenges such as climate change, natural disasters, and pandemics, there is a growing emphasis on building urban resilience (Liao et al., 2024). This includes improving infrastructure, disaster preparedness, and community engagement (Chatterjee & Majumdar, 2022; Ghosh et al., 2024).

In North America and Europe, there is a trend towards suburbanisation and the growth of exurban areas as people seek larger living spaces outside traditional urban centres. Urban areas are also experiencing gentrification, in which rising property values and investment displace lower-income residents. This trend raises concerns about social equity and access to affordable housing. As urban areas become increasingly reliant on technology, disparities in access to digital resources can exacerbate inequalities within cities. Addressing the digital divide is becoming a critical issue for urban planners (Majumdar & Chatterjee, 2020).

Insufficient infrastructure can pose significant challenges. Many cities need upgrades to their transportation systems, waste management facilities, and energy grids to support sustainable practices. This ageing infrastructure cannot contain urban sprawl because the expansion of urban areas into surrounding rural land can lead to increased resource consumption (Majumdar, 2018), habitat destruction, and reliance on automobiles, making sustainable development more challenging (Majumdar & Chatterjee, 2020). Furthermore, economic disparities can create barriers to sustainability. Vulnerable populations may lack access to resources, green spaces, and essential services, making it challenging to implement equitable sustainability practices (Majumdar, 2019; Majumdar, 2020). Established industries and stakeholders may resist changes that threaten their interests. This resistance can slow down the adoption of sustainable technologies and practices. Fragmented planning processes that do not consider environmental, social, and economic factors can lead to unsustainable outcomes. Integrated approaches are necessary for practical urban sustainability (Majumdar & Sivaramakrishnan, 2018a; Majumdar & Sivaramakrishnan, 2018b).

A citation analysis (Fig. 4.1) of research on the sustainable urban future from 2008 to 2025 reveals a significant increase in scholarly attention to interdisciplinary approaches linking urban planning, sustainability, and technological innovation. Early studies (2008–2013) primarily focused on conceptual frameworks of sustainable cities, emphasising environmental management, compact city models, and policy integration. Between 2014 and 2019, citation trends indicate a surge in research related to smart cities, resilient infrastructure, and climate adaptation, reflecting global policy

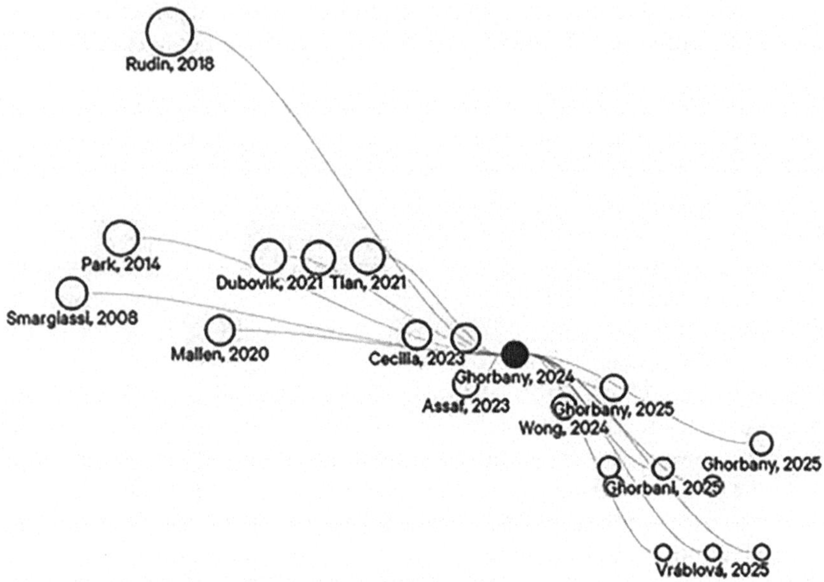

Fig. 4.1 Citation analysis for urban sustainability research

shifts following the adoption of the UN Sustainable Development Goals (SDGs) and the Paris Agreement.

Highly cited works during this period often centred on the nexus between digital technologies, urban governance, and ecological sustainability. From 2020 onwards, the literature shows increasing emphasis on nature-based solutions, data-driven urban management, and social inclusion, driven by the dual challenges of climate change and post-pandemic urban recovery. Influential authors and institutions from Europe and Asia dominate the citation landscape, with notable collaboration between academia, government, and industry. The analysis highlights a transition from theory-building to implementation-oriented research, demonstrating how cities are experimenting with integrated planning, renewable energy systems, and participatory innovation. Overall, the citation patterns underscore a maturing field characterised by cross-sectoral knowledge production and a growing recognition that sustainable urban futures depend on both technological advancement and inclusive governance frameworks.

Cities are often ill-prepared for the impacts of climate change, such as extreme weather events and rising sea levels. This vulnerability can undermine sustainability efforts and divert resources to immediate disaster response. While technology can facilitate sustainability, access to advanced technologies may be limited in certain regions. Additionally, the high costs of implementing new technologies can be a barrier for many cities. Cultural attitudes towards consumption, waste, and environmental stewardship can influence the success of sustainability initiatives. Changing long-standing behaviours and practices can be challenging. A lack of reliable data on urban systems, resource use, and environmental impacts can hinder effective decision-making and policy formulation for sustainable development (Majumdar & Sivaramakrishnan, 2018c, 2018d).

4.1.2 *Purpose of the Chapter*

This chapter aims to identify key barriers to urban sustainability and showcase innovations and transformative strategies that overcome them. Urban sustainability faces several barriers that can hinder progress towards creating environmentally friendly, socially inclusive, and economically viable cities. Many cities, especially in developing countries, struggle to secure adequate funding for sustainable infrastructure projects. Budget constraints can limit investments in public transportation, green spaces, and renewable energy. A lack of political commitment to sustainability initiatives can impede progress. Corruption, bureaucratic inefficiencies, and fragmented governance structures can also hinder the implementation of sustainable policies. Insufficient public awareness about sustainability issues can lead to resistance to change. Engaging communities in decision-making processes is crucial for fostering support for sustainability initiatives.

4.1.3 *Framing of Challenges*

Urbanisation challenges can be framed in various ways, depending on the context and focus. In the context of this chapter, the challenges include economic (infrastructure overload, housing affordability); social (social deprivation and emergence of slums in developing countries and homelessness in developed countries); environment (climate change and vulnerability of urban areas); and governance issues (institutional capacity, participatory governance, corruption, and transparency) (Fig. 4.2).

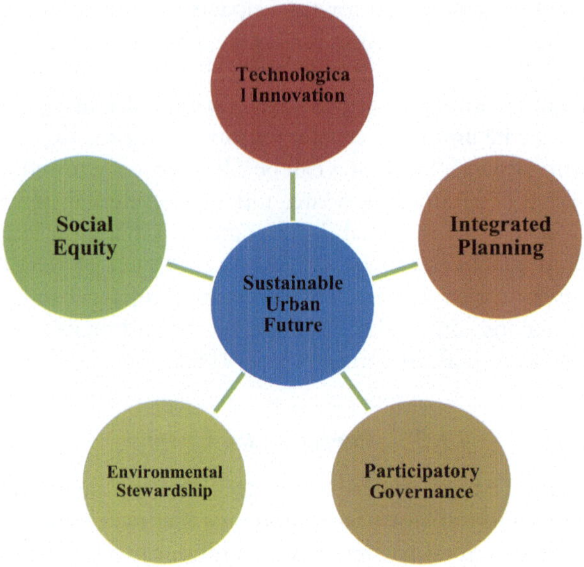

Fig. 4.2 Framework for sustainable urban innovation

4.2 Understanding the Barriers

4.2.1 Structural and Institutional Challenges

Urbanisation presents numerous structural and institutional challenges that can complicate the development of sustainable and resilient cities. These challenges often stem from rapid population growth, inadequate infrastructure, and the complexities of governance.

4.2.1.1 Structural Challenges

Rapid urban growth often outpaces the development of essential infrastructure, including transportation, water supply, sanitation, and waste management systems. This can lead to congestion, pollution, and a decline in quality of life. In addition, urbanisation frequently creates a significant housing demand that outstrips supply, leading to overcrowding and the proliferation of informal settlements, or slums. These areas often lack basic services and infrastructure (Chin et al., 2025). Urban housing scarcity is not only a problem in developing countries, as it is also common in developed economies. While urban housing problems in developed countries result in homelessness, the situation in less developed countries is characterised by the development of slums and socio-economic and environmental deprivation (Tume et al., 2025; Tume et al., 2025). Urbanisation can exacerbate social inequalities, leading to spatial segregation in which affluent populations reside in well-serviced areas. At the same time, marginalised communities are pushed to the peripheries with limited access to services. Cities around the world also struggle with traffic congestion, limited public transport options, and inadequate road networks. This can hinder mobility, increase travel times, and contribute to air pollution. Again, urbanisation can lead to the overexploitation of natural resources, the loss of green spaces, and increased pollution. Urban areas may also be more vulnerable to climate change impacts due to their high density and infrastructure vulnerabilities.

4.2.1.2 Institutional Challenges

Urban areas often encompass multiple jurisdictions, leading to fragmented governance that complicates coordinated planning and resource allocation. This can result in inconsistent policies and services, most often due to a lack of political will. Effective urban planning and sustainable development require strong political commitment (Kumar et al., 2025). However, political instability or lack of interest can hinder the implementation of necessary reforms and initiatives, leading to outdated or poorly enforced regulations that can stifle innovation and impede the development of sustainable urban practices. Bureaucratic hurdles may slow down project approvals and implementation.

Some local governments lack the technical expertise and resources needed for effective urban planning. This can result in poorly designed policies that fail to address the needs of growing populations. In some cases, corruption and mismanagement take centre stage. Corruption within urban governance can divert resources away from critical infrastructure projects and services. Mismanagement can further exacerbate existing challenges and undermine public trust.

Urban areas often face inadequate stakeholder engagement due to their cosmopolitan nature. Effective urban governance requires the involvement of various stakeholders, including community members, businesses, and civil society organisations. A lack of engagement can lead to policies that fail to reflect residents' needs and priorities. Institutional inertia may result from entrenched cultural attitudes towards urbanisation and development practices. Overcoming resistance to change is essential for implementing innovative solutions. There are also data deficiencies, as urban planners often rely on data to inform decision-making. Many cities face challenges in collecting and analysing data related to demographics, resource use, and environmental impacts, making it difficult to develop evidence-based policies.

4.3 Breakthroughs in Urban Sustainability

4.3.1 Governance and Policy Innovation

Governance and policy innovation encompass participatory urban planning and co-design, as well as decentralised governance models and community-led development. Participatory urban planning and co-design involve community members, stakeholders, and various interest groups in the planning and design of urban spaces. These methodologies aim to ensure that the voices of residents and users are heard and integrated into decision-making, leading to more inclusive, equitable, and responsive urban environments. Participatory urban planning is a collaborative approach that engages citizens in the planning process of urban development. It seeks to empower communities by involving them in discussions about their needs, priorities, and visions for their neighbourhoods. The principles of participatory urban planning include:

- Engaging a diverse range of stakeholders, including marginalised groups, to ensure all voices are represented.
- Providing clear information about planning processes, goals, and decisions to build trust among participants.
- Enabling community members to influence decisions that affect their lives, fostering a sense of ownership and responsibility.
- Encouraging partnerships between government agencies, community organisations, and residents to leverage resources and expertise.
- Establishing channels for continuous dialogue and feedback throughout the planning process.

Co-design extends the principles of participatory planning by actively involving community members in the design process itself. It emphasises collaboration between designers (planners, architects) and users to create spaces that meet the community's needs. Co-design includes collaboration (designers work alongside community members to co-create solutions rather than imposing top-down designs); creativity (encouraging innovative ideas from all participants, recognising that local knowledge can lead to unique solutions); iterative processes (engaging in cycles of design, feedback, and refinement to improve outcomes based on community input); empathy (understanding the lived experiences of community members to inform design decisions that resonate with their needs); ownership (fostering a sense of pride and belonging among community members by involving them in the creation of their environments) (Table 4.1).

Decentralised governance models and community-led development empower local communities to take ownership of their development processes. This approach has several benefits, including increased participation as community members are more likely to participate in decision-making processes that affect their lives. Again, decentralised governance models promote transparency and accountability, as decision-makers are more directly accountable to the community, and community-led development initiatives are more likely to be culturally relevant and responsive to local needs. With sustainability in mind, decentralised governance models can lead to more sustainable development outcomes, as community members are invested in the long-term success of their initiatives (Kumar et al., 2025).

This is possible when subsidiarity, participation, transparency, and capacity-building initiatives prevail. Decentralised governance models and community-led initiatives are champions of climate change adaptation,

Table 4.1 Framework for transforming barriers to breakthroughs

Barrier type	*Innovative intervention*	*Expected impact*
Policy fragmentation	Integrated urban governance models	Streamlined decision-making (e.g. Singapore)
Emission challenges	Renewable and efficient energy systems	Reduced carbon footprint (e.g. Copenhagen)
Financial constraints	Low-cost, high-impact design	Scalable sustainability (e.g. Curitiba)
Climate stress	Nature-based cooling strategies	Urban resilience and comfort (e.g. Melbourne)
Digital inequality	Inclusive digital literacy programmes	Equitable access and participation (e.g. Seoul)

Source: Compiled by the Authors

environmental crisis mitigation, and community resilience to climate-related shocks and stresses (Fig. 4.3).

4.3.2 *Financing Sustainable Cities*

Financing sustainable cities is built on green bonds, climate funds, and impact investing. It is also facilitated by public-private partnerships (PPPs) and circular economy financing (González, Kandpal, Machado, Martens, & Majumdar, 2023). Green bonds are fixed-income securities issued to finance projects with positive environmental impacts, such as renewable energy, energy efficiency, and sustainable water management (Kumar et al., 2025). They offer investors a stable return while supporting environmentally friendly projects. Like green bonds, climate funds are financial vehicles that invest in projects and initiatives to mitigate or adapt to climate change. Climate funds can support projects in emerging markets, promoting sustainable development and climate resilience (Beck et al., 2025; Chowdhury et al., 2025; Dixit, Majumdar, & Choudhury, 2025; Kumaraswamy, Choudhury, Ponappa, & Majumdar, 2025).

On the other hand, impact investing involves investing in companies, organisations, or projects that aim to generate both financial returns and

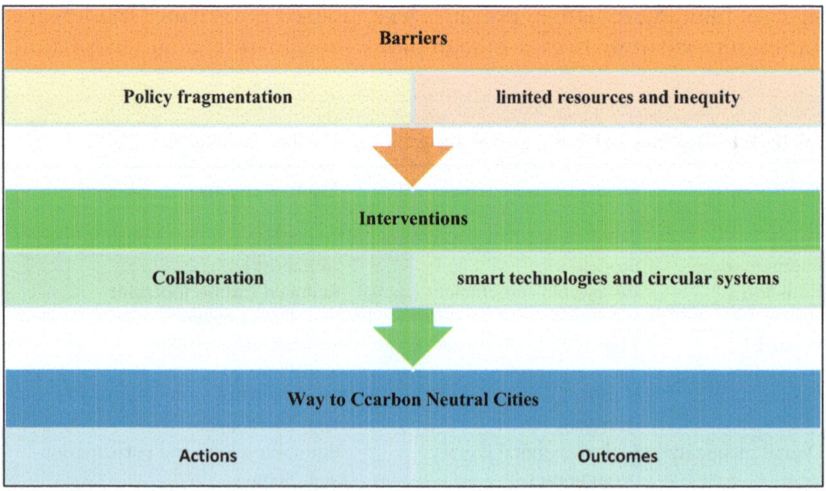

Fig. 4.3 The path from barriers to breakthroughs

positive social or environmental impacts. Impact investing can mobilise significant capital towards sustainable development and climate action. These financial tools are crucial for addressing climate change and promoting sustainable development. They offer investors opportunities to support environmentally friendly projects while generating returns.

Public-private partnerships (PPPs) and circular economy financing are innovative approaches to drive sustainable development and climate action. PPPs involve collaboration between governments and private sector entities to deliver public services or infrastructure projects (Majumdar, Choudhury, & Cheshmehzangi, 2025). PPPs can leverage private sector expertise, finance, and innovation to improve public services and infrastructure. PPPs in renewable energy, transportation, and waste management can support sustainable development and reduce greenhouse gas emissions (Chowdhury et al., 2025).

Circular economy financing involves investing in businesses, projects, or initiatives that promote the transition to a circular economy, where resources are used efficiently, and waste is minimised. Circular economy financing can support sustainable development, reduce waste, and promote resource efficiency. Investing in companies that develop sustainable products, services, or business models can generate returns while supporting the transition to a circular economy (Kumar et al., 2025). PPPs and circular economy financing can mobilise new sources of funding for sustainable development and climate action. They can facilitate collaboration and risk-sharing between governments and private sector entities, promoting more efficient and effective project delivery. PPPs and circular economy financing can support the achievement of the Sustainable Development Goals (SDGs) and the Paris Agreement.

4.3.3 Technological Transformation

The integration of 'advanced technologies', such as 'Internet of Things (IoT)' devices, 'smart sensors', and 'artificial intelligence (AI)', allows for 'predictive maintenance' that not only improves system uptime but also mitigates environmental risks (Kisić, Đurović, & Petrović, 2017). Continuous monitoring of system health helps detect anomalies early, preventing leaks, emissions, or hazardous failures that could result in environmental violations. Predictive analytics optimise maintenance schedules, thereby reducing unnecessary resource use and ensuring regulatory thresholds (emissions limits) are not exceeded (Suwignjo, Panjaitan, Baihaqy, & Rusdiansyah, 2023).

4.3.4 *Nature-Based and Regenerative Solutions*

Nature-based and regenerative solutions (NBRS) encompass urban greening, green corridors, blue infrastructure, climate-resilient urban design, and sponge cities. Urban greening, green corridors, and blue infrastructure are essential components of sustainable urban planning and climate resilience (Sabyrbekov & Overland, 2025). Urban greening involves incorporating green spaces, such as parks, gardens, and green roofs, into urban areas to improve environmental quality and human well-being. Urban greening can mitigate the urban heat island effect, improve air quality, and enhance biodiversity (Kumar et al., 2025). Green corridors, such as urban green spaces, are linear green spaces that connect urban areas, providing habitat for wildlife and promoting ecological connectivity. Green corridors can help maintain biodiversity, reduce urban fragmentation, and provide recreational spaces for residents. Both urban greening and green corridors are beneficial for carbon sequestration and temperature moderation in urban areas.

Blue infrastructure encompasses the use of natural or semi-natural systems, such as rivers, lakes, and wetlands, to manage urban water resources and promote ecosystem services (Chatterjee, Majumdar, & Bhunia, 2021). Blue infrastructure can help mitigate urban flooding, improve water quality, and enhance biodiversity. These strategies can help cities adapt to climate change by reducing the urban heat island effect, managing stormwater runoff, and improving air quality. They provide habitat for urban wildlife, improve ecological connectivity, enhance mental and physical health, and promote recreational opportunities and community engagement.

Sponge cities are a climate-resilient urban design approaches that incorporate green infrastructure to manage stormwater runoff, reduce flooding, and enhance urban sustainability. This concept uses nature-based solutions like parks, wetlands, and permeable pavements to absorb, filter, and store rainwater, mimicking natural hydrological processes (Majumdar, 2021a; Majumdar, Chatterjee, Shit, & Bhunia, 2021). Sponge cities are essentially permeable surfaces (allow rainwater to infiltrate the ground, reducing stormwater runoff); green roofs and bioswales (capture and filter rainwater, providing insulation and habitats for wildlife); natural waterways and wetlands (slow down and cleanse stormwater, supporting biodiversity and recreation); and multi-use landscapes (serve as parks, learning environments, and habitats for wildlife).

4.3.5 Inclusive and Equitable Innovation

Inclusive and equitable innovation can be tackled through affordable housing with sustainable materials and inclusive urban mobility. Affordable housing with sustainable materials is an innovative approach to building homes that are not only budget-friendly but also environmentally conscious (Kumar et al., 2025). Sustainable materials reduce environmental impact, promote energy efficiency, and create healthier living spaces. Sustainable materials in affordable housing minimise waste, conserve natural resources, and reduce greenhouse gas emissions (Chin et al., 2025). Sustainable materials, such as insulation, windows, and solar panels, reduce energy consumption and lower utility bills. They promote better indoor air quality, reducing health risks and improving occupant well-being, reducing maintenance and replacement costs over the building's lifespan, and increasing property value and appeal to environmentally conscious buyers. Examples of sustainable materials include recycled plastic, reclaimed wood, and repurposed materials, which reduce waste and conserve natural resources. Materials like bamboo, straw, and locally sourced wood promote sustainable forestry practices. Paints, adhesives, and finishes with low volatile organic compounds (VOCs) improve indoor air quality. Insulated concrete forms (ICFs) provide thermal mass, reducing energy consumption and improving building durability. Renewable energy systems, such as solar panels, reduce dependence on fossil fuels and lower energy costs.

Inclusive transport innovations encompass designing public transportation systems that cater to people with disabilities, the elderly, and those with limited mobility. It also includes car-sharing, bike-sharing, and ride-hailing services that promote shared transportation and reduce congestion; transitioning to eco-friendly vehicles to reduce emissions and improve air quality; intelligent transportation systems that optimise traffic flow, reduce congestion, and prioritise public transportation; and designing streets and sidewalks that prioritise pedestrian safety and accessibility.

4.4 Case Studies: Cities Leading the Way

4.4.1 Singapore: Smart City with Integrated Planning

Singapore stands as a global benchmark for smart and sustainable urban development, demonstrating how integrated planning and innovation can transform a city-state with limited natural resources into a thriving, resilient metropolis (Das, Lim, & Aravind, 2022). Guided by a long-term vision and strong governance, Singapore's urban strategy seamlessly integrates land use, transportation, housing, and environmental management through the coordinated efforts of agencies such as the Urban Redevelopment Authority (URA) and the Housing & Development Board (HDB) (Seidahmed, Lu, Chong, Ng, & Eltahir, 2018). The city's Smart Nation initiative exemplifies how data-driven technologies enhance urban efficiency, from intelligent traffic systems and real-time public transport monitoring to advanced waste and water management networks. Emphasis on sustainability is deeply embedded in policy frameworks, with green building standards, energy-efficient infrastructure, and extensive urban greenery such as the Park Connector Network and Gardens by the Bay contributing to climate resilience and liveability (Edwards et al., 2024). Singapore's use of digital twins and urban analytics allows planners to simulate and optimise city functions, while citizen engagement platforms ensure that innovation remains inclusive and responsive to community needs (Cui, Ferreira, Fung, & Matos, 2021). The integration of economic growth, technological advancement, and social equity reflects a holistic approach where urban planning is not a static blueprint but a dynamic, adaptive process. By aligning technological progress with environmental and social priorities, Singapore demonstrates how smart city innovation, underpinned by integrated planning, can foster sustainability, efficiency, and a high quality of life, serving as a model for cities worldwide aspiring to build a sustainable urban future (Górna & Górny, 2021).

4.4.2 Copenhagen: Carbon Neutral by 2025

Copenhagen has emerged as a global leader in sustainable urban transformation, pursuing the ambitious goal of becoming the world's first carbon-neutral capital by 2025 (Hansen & Agger, 2022). This vision is embedded in the 'Copenhagen Climate Plan', which integrates climate action across all sectors: energy, mobility, waste management, and urban design through a coherent, data-driven approach (Kohl & Andersen, 2022). The city's strategy emphasises a balanced combination of technological innovation, citizen engagement, and strong political commitment. More than 70% of Copenhagen's electricity is generated from renewable sources, particularly wind and biomass, while extensive district heating and cooling systems efficiently utilise waste heat and renewable energy (Damsø, Kjær, & Christensen, 2017). The city's investment in sustainable mobility, where over 40% of residents commute by bicycle, has significantly reduced transport-related emissions. Urban design initiatives such as green roofs, harbour cleaning projects, and climate-resilient infrastructure further enhance liveability and ecological balance. Importantly, Copenhagen's carbon neutrality plan aligns with social inclusion by promoting green jobs, community-based sustainability programmes, and public participation in climate governance (Lissandrello et al., 2025). The integration of smart technologies, including energy monitoring and climate data analytics, supports continuous performance evaluation and adaptive management. Although challenges remain especially in achieving complete decarbonisation of transportation and industry, the city's progress demonstrates the power of integrated planning, innovation, and collective responsibility. Copenhagen's approach provides a replicable model for cities worldwide, showing that economic prosperity, environmental stewardship, and high quality of life can coexist within a framework of carbon neutrality and sustainable urban growth (Jiang, Xu, Zhang, Zhang, & Wang, 2023).

4.4.3 Curitiba, Brazil: Low-Cost, High-Impact Transit and Urban Design

Curitiba, Brazil, is internationally recognised as a pioneer in sustainable urban planning and innovative public transportation. Facing rapid urban growth and limited financial resources in the 1970s, the city adopted a visionary, low-cost approach to urban design that prioritised accessibility, efficiency, and environmental responsibility (Lindau, Hidalgo, & Facchini, 2010). Under the leadership of then-mayor Jaime Lerner, Curitiba developed the Bus Rapid Transit (BRT) system, an affordable yet highly effective mass transit solution that has since been replicated worldwide (Miranda & Silva, 2012). The BRT system integrates dedicated bus lanes, prepaid boarding, and coordinated routes to deliver metro-level efficiency at a fraction of the cost, reducing traffic congestion, fuel consumption, and carbon emissions. Urban growth was strategically aligned along transit corridors, ensuring compact development and easy access to essential services. Complementing the transit system, Curitiba introduced extensive green spaces, recycling programmes, and innovative land-use policies that balance social equity with ecological sustainability. Community participation has been a cornerstone of Curitiba's success, with urban interventions designed to meet the needs of all socio-economic groups. The city's emphasis on multifunctional public spaces, pedestrian-friendly design, and environmental education fosters civic pride and a strong sense of belonging. Despite economic and political challenges, Curitiba's model remains a powerful example of how strategic planning, creativity, and governance can yield transformative urban outcomes (Krüger & Tamura, 2015). Its low-cost, high-impact solutions demonstrate that sustainable development is not dependent solely on wealth, but on visionary leadership, integrated planning, and a deep commitment to human-centred urban design (Bleviss, 2022).

4.4.4 Melbourne, Australia: Nature-Based Solutions and Urban Cooling

Melbourne, Australia, has emerged as a global leader in applying nature-based solutions to address climate change, urban heat, and environmental degradation (Chen et al., 2015). Confronted with rising temperatures, heatwaves, and water scarcity, the city adopted an integrated strategy that places green infrastructure at the heart of urban resilience and liveability (Jacobs, Gallant, Tapper, & Li, 2018). The Urban Forest Strategy and Climate Change Adaptation Plan aim to increase tree canopy cover to 40% by 2040, reduce the urban heat island effect, and enhance biodiversity within the metropolitan area (Furlong, Phelan, Dodson, & Considine, 2017). By investing in extensive street tree planting, green roofs, rain gardens, and revitalised waterways, Melbourne has redefined nature as an essential component of city infrastructure. These initiatives not only mitigate heat and improve air quality but also strengthen social cohesion by creating inviting public spaces and promoting community well-being (Radhakrishnan et al., 2019). The city's 'Cooling the City' programmes use advanced data analytics and spatial mapping to identify heat-vulnerable zones and guide targeted interventions, ensuring equitable distribution of cooling benefits. Collaboration among government agencies, researchers, and local communities has been key to Melbourne's success, demonstrating how participatory governance can support adaptive, evidence-based solutions. Moreover, nature-based approaches have delivered co-benefits, including enhanced stormwater management, carbon sequestration, and improved urban aesthetics (Kuller, Bach, Ramirez-Lovering, & Deletić, 2018). Melbourne's experience illustrates that integrating ecology with design and technology can transform cities into cooler, healthier, and more resilient environments, offering a blueprint for sustainable urban adaptation in a warming world (Sabri, Rajabifard, Ho, Amirebrahimi, & Bishop, 2016).

4.4.5 Seoul, South Korea: Digital Inclusion and Data-Driven Governance

Seoul, South Korea, stands at the forefront of smart urban transformation through its pioneering efforts in digital inclusion and data-driven governance (Cardoso & Ruiz, 2021). As one of the world's most connected cities, Seoul has harnessed technology not only to improve urban efficiency but also to enhance citizen engagement, transparency, and social equity (Lim, 2021). The city's Smart Seoul Strategy integrates digital platforms, open data systems, and artificial intelligence to optimise urban services, including public transportation, waste management, safety, and energy consumption. Through initiatives such as the Seoul Smart City Platform and the Big Data Campus, policymakers and researchers analyse vast datasets to inform evidence-based decision-making, improve public service delivery, and anticipate future urban challenges. Central to Seoul's approach is its commitment to bridging the digital divide. Programmes such as 'Smart Seoul Citizens' provide free digital literacy training and access to smart devices for elderly and low-income residents, ensuring that technological advancement benefits all segments of society (Jung, Jang, & Seo, 2016). The city also employs participatory tools, such as the mVoting and Oasis of Online Policy Suggestion platforms, enabling citizens to influence policy and urban planning decisions directly. These efforts reflect Seoul's philosophy of 'people-centred smart governance', in which technology serves to strengthen democracy, inclusivity, and sustainability. By combining advanced digital infrastructure with social innovation, Seoul demonstrates how data-driven governance can foster a transparent, responsive, and equitable urban ecosystem, setting a global benchmark for inclusive smart city development in the digital age (Križnik, 2013) (Table 4.2).

Table 4.2 Different case study models at a GLANCE

City	*Focus area*	*Key strategies*	*Main outcomes*	*Replicability potential*
Singapore	Integrated smart planning	Data-driven governance, circular resource management	High efficiency, resilient urban systems	Highly adaptable for dense urban centres
Copenhagen	Carbon neutrality	Renewable energy, sustainable mobility	Reduced emissions, enhanced liveability	High—with strong policy support
Curitiba	Low-cost transit	Bus Rapid Transit (BRT), compact city design	Affordable, accessible mobility	Very high—suitable for developing cities
Melbourne	Urban cooling	Urban forest, green roofs, water-sensitive design	Lower heat, improved well-being	Moderate to high—climate-dependent
Seoul	Digital inclusion	Open data, citizen participation	Transparent, equitable governance	Highly adaptable for smart city programmes

Source: Compiled by the Authors

4.5 Cross-sector Collaboration and Systems Thinking

Transforming cities towards sustainability cannot happen in silos. It demands partnerships that cut across institutions, academic disciplines, and economic sectors. When we view cities through a systems lens, we see them as complex, living organisms in which environmental conditions, social dynamics, economic forces, and technology interact (Cheshmehzangi, 2025). This perspective helps urban planners and policymakers identify pressure points (strategic interventions that can trigger broader change rather than just applying temporary fixes).

4.5.1 The Role of Academia and Research

Universities and research centres have become vital partners in creating knowledge alongside practitioners and developing policies grounded in evidence. Through urban living labs, climate modelling projects, and digital innovation centres, these institutions provide testing grounds for sustainability ideas before they scale up (Adigun et al., 2025; Lu et al., 2025; Majumdar, 2021b). Take, for example, how hydrological data and spatial analysis have shaped water reuse programmes and flood protection measures in multiple cities (Adigun et al., 2025). Researchers are also developing better ways to measure sustainability, which helps city governments turn broad goals like the SDGs into concrete, trackable actions (He, Guo, & Zhang, 2025).

4.5.2 *Private Sector as a Catalyst*

Businesses are not just bystanders in this transformation. They are driving much of the innovation. Companies are pouring investment into renewable energy projects, systems that convert waste into valuable resources, and technologies designed for climate adaptation (Kropelnytska, Yekimov, & Aleksandrovych, 2025). Through public-private partnerships, we are seeing scalable solutions emerge: green financing tools, sustainable supply chain practices, intelligent logistics systems, and electric vehicle networks (Alsaleh, 2025; Musida, Hanafi, & Sukardjo, 2025). What is more, corporations that have integrated climate considerations into their core governance find that sustainable business models can actually boost both environmental outcomes and market position (Lu et al., 2025).

4.5.3 *Community-Based Innovation*

Some of the most impactful work happens at the grassroots level, where local initiatives build social resilience and ensure no one gets left behind. Urban farming projects, for instance, do triple duty: they strengthen food security, create green spaces, and bring residents together around shared purpose (Nurulanis, Ahmad, Adnan, & Chuweni, 2025). Neighbourhood revitalisation efforts and socially minded housing developments give residents a voice in shaping spaces that actually work for their economic situations and daily lives (Chin et al., 2025; Marvi, Kalwar, Talpur, & Memon, 2025). When cities tap into local expertise and give communities digital tools to participate, people develop a lasting stake in sustainability (Zheng & Rahman, 2025).

4.5.4 *Global Networks and Knowledge Sharing*

Cities do not need to reinvent the wheel. International networks like C40 Cities, ICLEI, and UN-Habitat's Global Platform for Sustainable Cities create spaces where municipalities can learn from each other's successes and failures, share resources, and connect local work to global priorities. These platforms accelerate the spread of practical approaches to reducing carbon emissions, data-informed planning, and building resilience, with lessons flowing between cities in the Global South and North (Alizadeh & Meshkini, 2025). Weaving international frameworks into local governance structures builds capacity and ensures that innovations fit local contexts while staying globally relevant.

4.6 Future Trends and Opportunities

The cities of tomorrow will emerge from the intersection of technological advancement, social creativity, and environmental care. Several trends on the horizon could fundamentally reshape urban life.

4.6.1 *Decarbonisation of Cities*

Cutting urban carbon emissions means tackling energy, transport, and buildings simultaneously. Cities are electrifying bus and rail systems, integrating solar and wind power into their grids, and creating zones where only zero-emission vehicles can enter (Alsaleh, 2025; He et al., 2025; Majumdar, 2020a). Oslo and London show what is possible when cities combine smart regulation with infrastructure spending. Both have documented significant carbon reductions. Corporate climate frameworks are also helping by ensuring that private sector operations align with public sustainability goals (Lu et al., 2025).

4.6.2 *Circular Urbanism*

What if cities stopped treating resources as disposable? Circular urbanism reimagines the urban metabolism: materials are reused, waste is converted into energy, and old buildings provide raw materials for new construction (Guo, 2025). European and East Asian cities are already demonstrating how circular approaches to construction, combined with the right policy incentives, can close resource loops while creating employment in green industries (Balocco et al., 2025). Building waste management into city planning from the start gets us closer to the vision of zero-waste cities (Guo, 2025).

4.6.3 *Digital Twins and Virtual Planning Tools*

Digital twin technology has moved from experimental to essential. These virtual replicas of cities let planners simulate everything from energy consumption patterns to traffic flows to climate impacts, giving them predictive power they have never had before (Omar, 2025; Selvaraj et al., 2025). Modelling zero-energy buildings within these digital environments helps design smarter, climate-responsive cities. Singapore's Virtual Singapore platform demonstrates how digital governance tools can enhance resilience, improve transparency in decision-making, and facilitate collaboration among diverse stakeholders.

4.6.4 *Urban Resilience to Climate Extremes*

Climate hazards are intensifying. Heat waves last longer, floods come more frequently, and rainfall patterns become more extreme. Cities need proactive adaptation strategies. Green-blue infrastructure (parks, wetlands, and permeable surfaces), early warning systems, and climate-resilient building designs offer multiple benefits at once (Guo et al., 2025; Zahra, Tariq, & Jamil, 2025). Recent research on heat-related health risks and climate-adapted urban design underscores the need for health departments, environmental agencies, and planning offices to work together (Huang et al., 2025). Housing systems in climate-vulnerable regions present one of the most pressing challenges for urban resilience. In rapidly growing cities across the Global South, informal settlements and weak adaptation infrastructure leave residents particularly exposed to climate risks. Cameroon shows just how much unplanned housing and poor infrastructure amplify climate impacts, making resilient design and stronger policy frameworks urgent priorities (Tume, Nyuyse, et al., 2025; Tume, Zetem, et al., 2025). When cities build climate adaptation directly into their housing policies, they substantially reduce people's exposure to extreme weather while strengthening long-term sustainability.

4.6.5 *Youth and Grassroots Movements*

Young people and community organisers are changing the conversation around urban sustainability (Majumdar & Sivaramakrishnan, 2025). Using digital platforms and participatory design processes, youth activists are demanding accountability and pushing innovation forward (Mishra, 2025). Grassroots networks leverage digital tools to mobilise communities, elevate local knowledge, and build climate literacy. This transforms sustainability from something experts dictate into a collective civic endeavour.

4.7 Overcoming Resistance and Scaling Solutions

Innovation is happening fast, but cities still struggle with institutional resistance, cultural barriers, and disconnected policies. Breaking through requires aligning incentives, changing behaviours, and building governance systems people can trust (Table 4.3).

4.7.1 *Behavioural Change and Urban Culture Shifts*

Infrastructure alone won't deliver sustainability. People's daily choices matter just as much. Cities can encourage low-carbon living through behavioural nudges, environmental education, and designs that make sustainable choices the easy ones (Paul, 2025). Public art installations, storytelling campaigns, and education programmes can shift cultural attitudes

Table 4.3 Key barriers to sustainable urban innovation

Category	*Description*	*Possible solutions*
Institutional	Fragmented governance, lack of policy coordination	Establish integrated urban planning authorities
Financial	Limited funding for green infrastructure	Green financing, public-private partnerships
Technological	Unequal access to innovation and data	Open data policies, smart infrastructure investment
Social	Inequity and low public participation	Participatory governance, community-led innovation
Environmental	Resource depletion, climate vulnerability	Nature-based and circular economy solutions

Source: Compiled by the Authors

towards renewable energy, public transit, and car-sharing (Musida et al., 2025). When cities cultivate a culture of stewardship, sustainability shifts from something people feel forced to do to something they actually want.

4.7.2 Policy Levers and Incentives

Regulations can accelerate change when designed well. Tax breaks for clean energy, zoning reforms that promote mixed-use and transit-oriented development, and government purchasing policies that prioritise sustainability can drive widespread adoption of green practices (Al-Otaibi, Bowan, Alabdullatief, Albaiz, & Salah, 2025). Strong governance and institutional coordination, backed by data-driven decisions, create the conditions innovation needs to spread (Li, 2025; Lu et al., 2025).

4.7.3 Monitoring, Metrics, and Accountability

Measurement systems ensure commitments actually translate into action. Adapting the Sustainable Development Goals to local contexts, implementing ESG reporting, and deploying urban sustainability dashboards create feedback loops and accountability mechanisms (Huang et al., 2025). Cities using climate resilience indices and performance dashboards can transparently track their progress towards net-zero emissions and resilience goals (He et al., 2025; Omar, 2025).

4.8 Conclusion

The journey from barriers to breakthroughs in creating a sustainable urban future is both complex and transformative. As this chapter has shown, cities stand at the intersection of immense challenges and unprecedented opportunities. Overcoming structural, technological, and social barriers requires more than isolated interventions; it demands an integrated, systemic approach that aligns innovation with equity, environmental stewardship, and resilience. The transition to sustainable urbanism calls for reimagining how cities are planned, governed, and inhabited, placing people and the planet at the core of development strategies. Innovative solutions, whether driven by digital technologies, community engagement, or circular economy models, must be inclusive and adaptable, ensuring that progress benefits all urban residents rather than deepening existing inequalities. Policymakers, businesses, and citizens alike share responsibility for cultivating a culture of collaboration, transparency, and long-term vision. Ultimately, the sustainable city of the future is not defined solely by its infrastructure or technology, but by its capacity to nurture well-being, foster innovation, and sustain life within ecological limits. By transforming barriers into opportunities for creative and collective action, urban societies can chart a course towards more liveable, equitable, and regenerative cities. The challenge now is not to imagine whether such cities are possible but to act decisively to make them a reality.

References

Adigun, P. O., Ibuotenang, N. D., & Shaibu, S. E. (2025). Sustainable urban water management: Reuse, recycling and climate-resilient strategies. *International Journal of Scientific Research in Science, Engineering and Technology, 12*(1), 122–135. https://doi.org/10.32628/IJSRSET251217

Alizadeh, H., & Meshkini, A. (2025). On the road to urban sustainability: Identifying major barriers to urban sustainability in Iran. *Urban Ecology, 44*(4), 1–18. https://doi.org/10.1007/s10037-025-00238-y

Al-Otaibi, A., Bowan, P. A., Alabdullatief, A., Albaiz, M., & Salah, M. (2025). Barriers to sustainable building project performance in developing countries: A case of Ghana and the Kingdom of Saudi Arabia. *Sustainability, 17*(8), 3539. https://doi.org/10.3390/su17083539

Alsaleh, A. (2025). Electric and autonomous vehicles in Italian urban logistics: Sustainable solutions for last-mile delivery. *World Electric Vehicle Journal, 16*(7), 338. https://doi.org/10.3390/wevj16070338

Balocco, C., Pierucci, G., Baia, M., Borghi, C., Francini, S., Chirici, G., & Mancuso, S. (2025). Energy sustainability of urban areas by green systems: Applied thermodynamic entropy and strategic modeling means. *Atmosphere, 16*(8), 975. https://doi.org/10.3390/atmos16080975

Beck, D., Pal, A., Zaveri, P., Zheng, A., Tiwari, A. K., & Majumdar, S. (2025). Ecosystem resilience for climate resilience, strategies and innovations: A case study from developing countries. In M. Choudhury, G. Dixit, & S. Majumdar (Eds.), *Sustainable synergy: Harnessing ecosystems for climate resilience. Climate change management.* Springer. https://doi.org/10.1007/978-3-031-77957-2_4

Bleviss, D. (2022). The legacy of Jaime Lerner and Curitiba, Brazil. Wiley interdisciplinary reviews. *Energy and Environment, 11*(5). https://doi.org/10.1002/wene.436

Cardoso, I., & Ruiz, T. (2021). Smart tourism destinations - a case study of Seoul. *South Korea. Applied Tourism, 6*(1), 36–44. https://doi.org/10.14210/at.v6n1.p36-44

Chatterjee, U., Antipova, A., Ghosh, S., Majumdar, S., & Setiawati, M. (2023). *Urban environment and smart cities in Asian countries*. Springer. https://doi.org/10.1007/978-3-031-25914-2

Chatterjee, U., Biswas, A., Mukherjee, J., & Majumdar, S. (2022). *Advances in urbanism, smart cities, and sustainability*. Routledge. https://doi.org/10.1201/9781003126195

Chatterjee, U., & Majumdar, S. (2022). Impact of land use change and rapid urbanization on urban heat Island in Kolkata city: A remote sensing-based perspective. *Journal of Urban Management, 11*(1), 59–71. https://doi.org/10.1016/j.jum.2021.09.002

Chatterjee, U., Majumdar, S., & Bhunia, G. S. (2021). Land restoration strategies to mitigate the problems of the shrinkage of wetlands in West Bengal. In G. S. Bhunia, U. Chatterjee, A. Kashyap, & P. K. Shit (Eds.), *Modern cartography series* (Vol. 10, pp. 525–538). Academic Press. https://doi.org/10.1016/B978-0-12-823895-0.00019-1

Chen, D., Thatcher, M., Wang, X., Barnett, G., Kachenko, A., & Prince, R. (2015). Summer cooling potential of urban vegetation—A modeling study for Melbourne, Australia. *AIMS Environmental Science, 2*(3), 648–667. https://doi.org/10.3934/environsci.2015.3.648

Cheshmehzangi, A. (2025). Urban planning for the contemporary age: Navigating complexities and shaping urban futures. *Encyclopedia, 5*(1), 19. https://doi.org/10.3390/encyclopedia5010019

Chin, C. P., Alias, A., & Alkharabsheh, O. H. (2025). Urban housing regeneration strategies for sustainable city development: A systematic literature review. *Discover Sustainability, 6*, 1050. https://doi.org/10.21203/rs.3.Rs-6941972/v1

Chowdhury, B. R., et al. (2025). Building a resilient tomorrow: Ecosystem-based approaches to measurement, assessment, prediction, and future perspectives in climate resilience. In M. Choudhury, G. Dixit, & S. Majumdar (Eds.), *Sustainable synergy: Harnessing ecosystems for climate resilience. Climate change management*. Springer. https://doi.org/10.1007/978-3-031-77957-2_14

Cui, M., Ferreira, F., Fung, T., & Matos, J. (2021). Tale of two cities: How nature-based solutions help create adaptive and resilient urban water management practices in Singapore and Lisbon. *Sustainability, 13*(18), 10427. https://doi.org/10.3390/su131810427

Damsø, T., Kjær, T., & Christensen, T. (2017). Implementation of local climate action plans: Copenhagen—Towards a carbon-neutral capital. *Journal of Cleaner Production, 167*, 406–415. https://doi.org/10.1016/j.jclepro.2017.08.156

Das, D., Lim, N., & Aravind, P. (2022). Developing a smart and sustainable campus in Singapore. *Sustainability, 14*(21), 14472. https://doi.org/10.3390/su142114472

Dixit, G., Majumdar, S., & Choudhury, M. (2025). Introduction to ecosystem Services for Climate Resilience: A theoretical framework. In M. Choudhury, G. Dixit, & S. Majumdar (Eds.), *Sustainable synergy: Harnessing ecosystems for climate resilience. Climate change management.* Springer. https://doi.org/10.1007/978-3-031-77957-2_1

Edwards, P., Yean, S., Goh, Y., Simcock, R., Lee, B., Zhou, J., & Ping, H. (2024). Connecting human wellbeing, urban trees, and greenspace in Singapore: alternative novel approaches. https://doi.org/10.21203/rs.3.rs-4370989/v1.

Furlong, C., Phelan, K., Dodson, J., & Considine, R. (2017). Scoping the potential role of the water sector in urban greening and cooling: A case study of Melbourne. *WIT Transactions on the Built Environment, 1*, 85–95. https://doi.org/10.2495/cc170091

Ghosh, S., Majumdar, S., & Cheshmehzangi, A. (2024). *Cities of Tomorrow: Urban resilience and climate change preparedness.* Springer. https://doi.org/10.1007/978-981-97-9658-8

González, E., Kandpal, V., Machado, M., Martens, M., & Majumdar, S. (2023). A bibliometric analysis of circular economies through sustainable smart cities. *Sustainability, 15*(22), 15892. https://doi.org/10.3390/su152215892

Górna, A., & Górny, K. (2021). Singapore vs. the 'Singapore of Africa'—Different approaches to managing urban agriculture. *Land, 10*(9), 987. https://doi.org/10.3390/land10090987

Guo, J., Lu, L., Zha, F., Agathangelidis, I., Yao, Y., Li, Q., Shen, Q., & Li, Z. (2025). A grid-scale assessment framework for heat health risks: A case study of the Beijing–Tianjin–Hebei region, China. *Climate Services, 40*, 100603. https://doi.org/10.1016/j.cliser.2025.100603

Guo, Q. (2025). Waste management systems in urban planning for a sustainable future. *SDGs Review, 5*(5), e06595. https://doi.org/10.47172/2965-730X.SDGsReview.v5.n05.pe06595

Hansen, K., & Agger, A. (2022). Copenhagen CO2 neutrality in 2025? A polycentric analysis of urban climate governance in Copenhagen 2006–2020. *Environmental Policy and Governance, 33*(3), 288–300. https://doi.org/10.1002/eet.2030

He, W., Guo, X., & Zhang, C. (2025). Assessing the impact of climate-resilient city development on urban sustainability: Evidence from China. *Sustainability, 17*(10), 4381. https://doi.org/10.3390/su17104381

Huang, B., Qi, J., Pathak, M., Sharifi, A., Cheshmehzangi, A., Attia, S., Matzarakis, A., Ghaffarianhoseini, A., Yun, G. Y., Darko, A., Liu, X., & He, B.-J. (2025). Climate-driven transformations: A framework for the sustainable urban landscape system to enhance heat resilience. *Sustainable Cities and Society, 131*, 106684. https://doi.org/10.1016/j.scs.2025.106684

Jacobs, S., Gallant, A., Tapper, N., & Li, D. (2018). Use of cool roofs and vegetation to mitigate urban heat and improve human thermal stress in Melbourne,

Australia. *Journal of Applied Meteorology and Climatology, 57*(8), 1747–1764. https://doi.org/10.1175/jamc-d-17-0243.1

Jiang, W., Xu, B., Zhang, H., Zhang, H., & Wang, S. (2023). *Practice and application of BIM-based carbon-neutral temporary construction facility in UHV power grid project* (pp. 472–485). Atlantis Press. https://doi.org/10.2991/978-94-6463-262-0_50

Jung, K., Jang, H., & Seo, I. (2016). Government-driven social enterprises in South Korea: Lessons from the social enterprise promotion program in the Seoul metropolitan government. *International Review of Administrative Sciences, 82*(3), 598–616. https://doi.org/10.1177/0020852315586935

Kisić, E., Đurović, Ž., & Petrović, V. (2017). *Predictive maintenance based on control charts applied at a thermoelectric power plant* (Vol. 200). IntechOpen. https://doi.org/10.5772/intechopen.68685

Kohl, U., & Andersen, J. (2022). Copenhagen's struggle to become the world's first carbon-neutral capital: How corporatist power beats sustainability. *Urban Planning, 7*(3). https://doi.org/10.17645/up.v7i3.5327

Križnik, B. (2013). Changing approaches to urban development in South Korea. *International Development Planning Review, 35*(4), 395–418. https://doi.org/10.3828/idpr.2013.27

Kropelnytska, S., Yekimov, S., & Aleksandrovych, O. (2025). Barriers and opportunities for entrepreneurship development in Ukraine: Towards a sustainable and resilient economic future. *Journal of Vasyl Stefanyk Precarpathian National University, 12*(2), 116–130. https://doi.org/10.15330/jpnu.12.2.116-130

Krüger, E., & Tamura, C. (2015). Urban climate studies in a subtropical location: Literature review and current perspectives for Curitiba, Brazil. *Energy and Emission Control Technologies*, 55. https://doi.org/10.2147/eect.s60290

Kuller, M., Bach, P., Ramirez-Lovering, D., & Deletić, A. (2018). What drives the location choice for water sensitive infrastructure in Melbourne, Australia? *Landscape and Urban Planning, 175*, 92–101. https://doi.org/10.1016/j.landurbplan.2018.03.018

Kumar, P., Sahani, J., Corada, P. K., Ahlawat, A., Andrade, M. F., Athanassiadou, M., Cao, S.-J., Collins, L., Dey, S., Di Sabatino, S., Halios, C. H., Harris, F., Horton, C., Inostroza, L., Jones, L., Rodding, K. T., McCallan, B., McNabola, A., Kumar, M. R., Kumar, M. S., Sperandio, N. E. G., Owen, G., Ravindra, K., Reis, N. C., Soebarto, V., Wenk, J., Wood, H. S., & Yao, R. (2025). Urban greening for climate resilient and sustainable cities: Grand challenges and opportunities. *Frontiers in Sustainable Cities, 7*, 1595280. https://doi.org/10.3389/frsc.2025.1595280

Kumaraswamy, T. R., Choudhury, M., Ponappa, S., & Majumdar, S. (2025). A significance of climate resilience: Strategies for local, regional, and national levels. In M. Choudhury, G. Dixit, & S. Majumdar (Eds.), *Sustainable synergy:*

Harnessing ecosystems for climate resilience. Climate change management. Springer. https://doi.org/10.1007/978-3-031-77957-2_13

Li, Z. (2025). A literature review on sustainable urban development in China. *Advances in Economics Management and Political Sciences, 133*, 135–141. https://doi.org/10.54254/2754-1169/133/2025.19691

Liao, Y., Liao, W., Wu, D., Zhang, C., Xu, J., & Majumdar, S. (2024). Inspired applications of artificial intelligence in disaster-resilient regional construction for climate risk. In S. Ghosh, S. Majumdar, & A. Cheshmehzangi (Eds.), *Cities of tomorrow: Urban resilience and climate change preparedness. Urban sustainability*. Springer. https://doi.org/10.1007/978-981-97-9658-8_16

Lim, I. (2021). Exploring experience at the intersection of migration and digital democracy in South Korea. *Asiascape Digital Asia, 8*(3), 139–163. https://doi.org/10.1163/22142312-bja10019

Lindau, L., Hidalgo, D., & Facchini, D. (2010). Bus rapid transit in Curitiba, Brazil. *Transportation Research Record Journal of the Transportation Research Board, 2193*(1), 17–27. https://doi.org/10.3141/2193-03

Lissandrello, E., Papina, C., Elisei, P., Jensen, J., Dragomir, A., & Lazăr, A. (2025). Co-creating positive energy districts: experimental governance and transformative capacity in Bucharest and Copenhagen. https://doi.org/10.21203/rs.3.rs-6930084/v1.

Lu, Z., Bao, K., Hu, X., Gao, C., Wen, Y., & Zhang, T. (2025). Climate governance action and corporate environmental performance: Evidence from climate-resilient city construction. *Frontiers in Environmental Science, 13*, 1651610. https://doi.org/10.3389/fenvs.2025.1651610

Majumdar, S. (2018). A comparative study between the municipalities for delineating the sphere of urban influence: A case study of Kolkata's urban fringe. *SELP Journal of Social Science, 9*(36), 43–50.

Majumdar, S. (2019). Crisis for open space and loss of urban heritage around Kolkata city. *Research Journal of Humanities and Social Sciences, 10*(4), 969–972. https://doi.org/10.5958/2321-5828.2019.00158.X

Majumdar, S. (2020a). Assessment and detection of land cover changes in the southern fringe of Kolkata using remotely sensed data. *Geography, Environment, Sustainability, 13*(4), 121–132. https://doi.org/10.24057/2071-9388-2020-65

Majumdar, S. (2020b). Prospects of sustainable transport of automobile vehicles in Kolkata City. In G. R. Kanagachidambaresan, R. Maheswar, V. Manikandan, & K. Ramakrishnan (Eds.), *Internet of things in smart Technologies for Sustainable Urban Development. EAI/springer innovations in communication and computing*. Springer. https://doi.org/10.1007/978-3-030-34328-6_5

Majumdar, S. (2021a). Sustainability in urban expansion of a metropolitan city: Impacts of urban growth towards the outer fringe of Kolkata. In S. Mangla & M. Ram (Eds.), *Supply chain sustainability: Modeling and innovative research frameworks* (pp. 59–72). De Gruyter. https://doi.org/10.1515/9783110628593-003

Majumdar, S. (2021b). Analysis of groundwater level and groundwater modeling using GIS in Kolkata. In P. Kumar, M. Kumar, & V. Singh (Eds.), *Groundwater development and management: Issues and challenges in South Asia (chapter 13)*. CRC Press. https://doi.org/10.1201/9781003034506-13

Majumdar, S., & Chatterjee, U. (2020). Modelling urban growth using an urban growth deterministic model in Kolkata metropolitan area: A geo-statistical approach. *Modeling Earth Systems and Environment, 7*(4), 2241–2249. https://doi.org/10.1007/s40808-020-00985-6

Majumdar, S., Chatterjee, U., Shit, P. K., & Bhunia, G. S. (2021). Identification of groundwater potential areas using geospatial technologies: A case study of Kolkata, India. In P. P. Adhikary, P. K. Shit, P. Santra, G. S. Bhunia, A. K. Tiwari, & B. S. Chaudhary (Eds.), *Geostatistics and geospatial Technologies for Groundwater Resources in India*. Springer. https://doi.org/10.1007/978-3-030-62397-5_8

Majumdar, S., Choudhury, M., & Cheshmehzangi, A. (Eds.). (2025). *Sustainable urban future: Embracing circular economy in waste management*. Springer. https://doi.org/10.1007/978-981-95-2976-6

Majumdar, S., Kandpal, V., & Anthopoulos, L. G. (2024). *Smart cities: Innovations, challenges and future perspectives*. Springer. https://doi.org/10.1007/978-3-031-59846-3

Majumdar, S., & Sivaramakrishnan, L. (2015). Pattern of land use in and around Kolkata city: A spatio-temporal analysis. *Indian Cartographer, 35*, 218–223.

Majumdar, S., & Sivaramakrishnan, L. (2016). Rajpur Sonarpur municipality and its sphere of influence. *Geographical Review of India, 78*(4), 349–366.

Majumdar, S., & Sivaramakrishnan, L. (2017). Monitoring urban growth and land use change detection in the southern fringes of Kolkata metropolitan area. *The Indian Geographical Journal, 92*(2), 168–183.

Majumdar, S., & Sivaramakrishnan, L. (2018a). Demographic dynamics with changing land price in the southern part of Kolkata metropolitan area: A case study of Rajpur Sonarpur municipality. *Hill Geographer, 34*(1), 55–67.

Majumdar, S., & Sivaramakrishnan, L. (2018b). Spatial and temporal changes of land use and land cover in the southern fringes of Kolkata, West Bengal. *Indian Journal of Regional Science, 50*(2), 19–32.

Majumdar, S., & Sivaramakrishnan, L. (2018c). Land cover classification and change analysis in the southern part of Kolkata metropolitan area by spatio-temporal remote sensing. *Annals of the National Association of Geographers, India (NAGI), 38*(2), 301–315.

Majumdar, S., & Sivaramakrishnan, L. (2018d). Possibility of urbanization in the rural areas of Kolkata metropolitan area. *Transactions of the Institute of Indian Geographers, 40*(2), 225–239.

Majumdar, S., & Sivaramakrishnan, L. (2025). Impacts of vegetation and impervious surfaces on urban sustainability: A case study of Kolkata. In L. Sivaramakrishnan & S. Bandyopadhyay (Eds.), *Sustainable urban forms and communities: Urban geographies of eastern India (chapter 11)*. Routledge. https://doi.org/10.4324/9781003509400-11

Marvi, H., Kalwar, S., Talpur, A. M., & Memon, A. I. (2025). Barriers to socially sustainable residential neighbourhoods: Planning practice of a growing city—A case study of Qasimabad taluka, Hyderabad. *Journal of Research in Architecture & Planning, 35*(1), 35–49.

Miranda, H., & Silva, A. (2012). Benchmarking sustainable urban mobility: The case of Curitiba, Brazil. *Transport Policy, 21*, 141–151. https://doi.org/10.1016/j.tranpol.2012.03.009

Mishra, V. (2025). Rethinking urban future: Towards sustainable cities and communities. *Sustainable Cities and Communities, 4*, 91–98. https://doi.org/10.1007/s44177-025-00088-w

Musida, M., Hanafi, I., & Sukardjo, M. (2025). Electric vehicle shared services: A decade of innovation, challenges, and transformative impact on sustainable urban mobility—A systematic literature review. *Open Transportation Journal, 19*, e26671212380759. https://doi.org/10.2174/0126671212380759250429073257

Nurulanis, A., Ahmad, Z., Adnan, Y. M., & Chuweni, N. N. (2025). Urban agriculture: A pathway to sustainable urban development. *Planning Malaysia: Journal of the Malaysian Institute of Planners, 23*(2), 41–55.

Omar, O. (2025). Digital twins for climate-responsive urban development: Integrating zero-energy buildings into smart city strategies. *Sustainability, 17*(15), 6670. https://doi.org/10.3390/su17156670

Paul, J. (2025). Challenges and opportunities in implementing climate-resilient building codes in Nigeria (pp. 1–7). https://www.researchgate.net/profile/Joel-Paul-10/publication/390300584

Radhakrishnan, M., Löwe, R., Ashley, R., Gersonius, B., Arnbjerg-Nielsen, K., Pathirana, A., & Zevenbergen, C. (2019). Flexible adaptation planning process for urban adaptation in Melbourne, Australia. *Proceedings of the Institution of Civil Engineers: Engineering Sustainability, 172*(7), 393–403. https://doi.org/10.1680/jensu.17.00033

Ripp, M., Gustafsson, C., Jones, Z., Majumdar, S., & Ginzarly, M. (2025). Historic urban landscapes and heritage systems as the basis for sustainable urban development. *Land, 14*(9), 1783. https://doi.org/10.3390/land14091783

Sabri, S., Rajabifard, A., Ho, S., Amirebrahimi, S., & Bishop, I. (2016). Leveraging VGI integrated with 3d spatial technology to support urban intensification in Melbourne, Australia. *Urban Planning, 1*(2), 32–48. https://doi.org/10.17645/up.v1i2.623

Sabyrbekov, R., & Overland, I. (2025). Urban climate resilience in transition economies: The case of Central Asia. *Environmental Research Letters, 20*, 074057. https://doi.org/10.1088/1748-9326/addfee

Seidahmed, O., Lu, D., Chong, C., Ng, L., & Eltahir, E. (2018). Patterns of urban housing shape dengue distribution in Singapore at neighborhood and country scales. *Geohealth, 2*(1), 54–67. https://doi.org/10.1002/2017gh000080

Selvaraj, V., Nivethikha, R., Yi, L., Rao, M. S., Balaraman, R., & Jain, A. (2025). Digital twins and urban SDGs: Simulating smart cities for sustainable development. *International Journal of Environmental Sciences, 11*(19s), 849. https://theaspd.com/index.php

Suwignjo, P., Panjaitan, L., Baihaqy, A., & Rusdiansyah, A. (2023). Predictive analytics to improve inventory performance: A case study of an FMCG company. *Operations and Supply Chain Management, an International Journal, 16*, 293–310. https://doi.org/10.31387/oscm0530390

Tume, S. J. P., Nyuyse, Y. B., Kuyar, C., Ngoran, L. T., Ngoran, H. A., Foka, S. K., & Ngala, E. N. (2025). Vulnerability of urban housing to climate change in cameroon. In A. Cheshmehzangi, J. Zuo, A. Sharifi, R. Zhang, A. ZiafatiBafarasat, & J. Zhhao (Eds.), *Designing healthy buildings and communities: Urban sustainability* (pp. 49–89). Springer. https://doi.org/10.1007/978-981-96-3057-8_4

Tume, S. J. P., Zetem, C. C., Nyuyse, Y. B., Kuyar, C., Tomla, N. L., & Kinyuy, F. S. (2025). Vulnerability and resilience of urban housing to climate change in Cameroon. In M. Fouda & M. N. Essone (Eds.), *Collection Dynamiques Territoriales et Développement, Anthropiscence urbaine et durabilité: Regard pluridisciplinaire sur la coévolution à risque entre urbanisation et environnement au Cameroun.* (pp. 249–297). Éditions Cheick Anta Diop.

UN-Habitat. (2024). *Cities and climate action: World cities report 2024* (p. 342). United Nations Human Settlements Programme.

Zahra, F. T., Tariq, F., & Jamil, F. (2025). A study on climate-resilient housing in Pakistan: Identifying critical indicators for sustainable development. *Environmental Research Communications, 7*, 035022. https://doi.org/10.1088/2515-7620/adbe2d

Zheng, Y., & Rahman, M. M. (2025). Towards green and climate-resilient urbanization in Rajshahi City: Urban growth meets climate action in northern Bangladesh. *Chinese Journal of Urban and Environmental Studies, 13*(27). https://doi.org/10.1142/S2345748125500010

PART II

From Vision to Reality: Effective Implementation of Smart Solutions

CHAPTER 5

Smart Buildings: Integrating Energy Efficiency and Sustainable Design

Deepti Pargai, Basudeb Pyne, and Mamta Saharan

5.1 Introduction

In this fast-moving era where everything is becoming smarter, the need to remain sustainable is even greater. This is not the ancient time when people lived sustainably instead of just talking about it (Bibri, 2019). Today, not only is the talk to inspire, but also the incorporation of sustainable design principles, a need of the hour. The *World Cities Report 2022* emphasizes that the future of humanity is unequivocally urban, projecting that 68% of the global population will live in urban areas by 2050. The *World Cities Report 2022* envisions a future where cities are resilient, inclusive, green, and digitally empowered, driven by lessons from the

D. Pargai (✉)
Fashion and Textile Technology, National Institute of Fashion Technology, Srinagar, Jammu and Kashmir, India

B. Pyne
Department of Education, Ramakrishna Mission Sikshanamandira, Howrah, West Bengal, India

M. Saharan
SUNY Buffalo State Buffalo, Buffalo, NY, USA

S. Majumdar, M. Choudhury (eds.), *City of the Future*,
https://doi.org/10.1007/978-3-032-18026-1_5

COVID-19 pandemic and ongoing global challenges. Achieving this optimistic scenario requires transformative action through the implementation of the New Urban Agenda, emphasizing economic, social, environmental, and institutional resilience. Key priorities include reducing inequality and poverty, embracing climate action and smart technologies, fostering inclusive governance, and ensuring universal access to health, housing, and essential services. As urbanization continues, building sustainable, people-centered, and adaptable cities is vital to securing a just and thriving urban future for all (Reyes et al., 2020; Hermaputi & Chen, 2017).

Smart buildings are central to future cities, combining technology, sustainable design, and innovative materials to create adaptive and energy-efficient spaces. Textiles—ranging from solar fabrics and thermal-regulating membranes to acoustic panels and responsive shading systems—are redefining building performance (Etukudoh et al., 2024; Li et al., 2019). Integrated into *Architex*, these bright, technical textiles regulate light, temperature, and air quality while enabling real-time monitoring. By enhancing sustainability and occupant comfort, textile-driven innovations position smart buildings as key contributors to resilient and future-ready urban environments (Liu et al., 2021; Juhara et al., 2024).

This chapter explores how smart technologies can work hand in hand with sustainable design to shape cities that are not only intelligent but also more liveable and eco-friendlier. Sustainable design, as defined by organizations such as the GSA, Interaction Design Foundation, and IBM, is the process of designing products, services, and buildings to minimize adverse environmental and human health impacts while maximizing their potential for reuse and recyclability. It is a holistic approach that considers the entire lifecycle of a product, from design to disposal, aiming for long-term solutions that benefit both society and the environment. The integration of energy efficiency in buildings and architecture will always be the better choice for moving toward a smarter, more sustainable city (Lionar et al., 2019). However, only structural elements or architectural design cannot fulfill this dream of a smart city; the process of lighting materials is the key. A conscious selection of materials will always be a top priority when planning a smart city, whether it is underway or in the planning stage. Sustainable materials have the potential not only to enhance building performance by contributing to thermal insulation, light diffusion, and acoustic regulation but also to improve overall indoor environmental quality by absorbing VOCs, thereby enhancing human and ultimately society's well-being (Tiwari, 2021; Liu et al., 2023). In this context, innovative and sustainable materials have great potential. Selection of the right material, in terms of durability, functionality, and biodegradability, can play a critical role in reducing energy consumption and carbon footprint. Renewable and recycled materials not only support sustainable development but also circular design (Tadesse et al., 2025; Filho et al., 2022). For large-scale applications, energy-generating building facades are revolutionizing urban infrastructure (Wu et al., 2022). These facades not only enhance energy efficiency but also contribute to sustainable building certifications, such as LEED (Leadership in Energy and Environmental Design).

In contrast, GRIHA (Green Rating for Integrated Habitat Assessment), an Indian certification for a City of the future, would not only evolve with technology but also with different certifications and laws within sustainable design principles such as site and context sensitivity, energy efficiency and passive design, materials, ensuring resource efficiency, water conservation and management, indoor environmental quality (IEQ), lifecycle and resilience planning, etc. (Oliveira et al., 2024). This chapter explores the potential of materials and the integration of technology to navigate toward the city of the future. Another innovative application is solar parking canopies, which generate electricity for EV charging stations. Similarly, biodegradable and nature-inspired solutions provide innovative pathways for sustainable design, such as Self-Cleaning Fabrics inspired by lotus leaves, reducing the need for water-intensive washing. Biodegradable polymers are designed to break down naturally, mimicking decomposition processes found in nature.

In this era of climate change, temperatures cross 50°C, and it has become normal (Rosen, 2021). Innovation to cool down the building is also required. Different mechanisms to minimize heat, and the utilization of innovative materials to do so, are also becoming key parameters for researchers. Some innovations are already in the commercial stage, such as ultra-white nanophotonic panels that reflect 98% of sunlight and radiate heat into space, and a cooling paint made with crushed scallop shells and aerogel with embedded air pockets that keep surfaces cool (Peng and Cui, 2024). Researchers are also working on phase-change materials (PCMs) in fabrics that absorb heat when warm and release it when cool and on shape-memory polymers that flex with temperature, enabling furniture and blinds to adapt passively. Photonic pigments that change reflectivity based on sunlight are also being researched for adaptation and diversification. Similarly, salt-infused clay mimics human skin by absorbing and releasing moisture, cooling the surrounding environment. Blending traditional materials with passive climate-control measures can be a viable option for this research area.

5.2 Principles of Sustainable Design

Sustainable design in architecture focuses on integrating energy efficiency with ecological sensitivity (Ayu et al., 2021). This ensures how structures can harmonize with their surroundings while reducing negative environmental impacts (Odji and Oni, 2024). Architects can design buildings that are not only functional and aesthetically pleasing but also sustainable, circular, and resilient, using the following principles (Nasrullah and Syafri, 2024; Mageed et al., 2024).

Climate, topography, and urban density are critical factors for optimizing site and context sensitivity. Designing with these elements in mind helps reduce the building's overall footprint and enhance sustainability (Baper and Ismael, 2024; Dénes et al., 2022; Ayoobi et al., 2024). Materials can play a crucial role in site and context-sensitive sustainable design. For example, traditionally, natural Namda is used in India for thermal insulation. However, from a sustainability perspective, it can also be a better option today; instead of natural wool, artificial wool-like properties can be incorporated to provide that insulation while retaining its lightweight properties. Smart coatings and electronic textiles (e-textiles) are advanced technologies applied to textiles to enhance their functionality, adaptability, and performance in various environments (Ghosh et al., 2020). Smart coatings can offer properties such as fire and UV resistance, durability, self-cleaning, self-healing, antimicrobial protection, and temperature regulation. Self-healing coatings use microcapsules filled with healing agents to repair damage, while antimicrobial coatings often incorporate metal-based nanomaterials like silver, copper, zinc, or titanium. Self-cleaning coatings mimic the lotus leaf's surface to repel water and remove dirt, while anti-fouling and UV/IR protective coatings shield textiles from environmental damage (Zou et al., 2024). Thermoregulating coatings utilize phase-change materials (PCMs) that respond to temperature changes, while optical coatings integrate thermochromic polymers and LEDs to produce interactive color and light effects. They can be integrated with photovoltaic cells to generate energy and are used in applications ranging from smart wearables to interior design elements like carpets, walls, and furniture (Hetimy et al., 2024; Patrucco et al., 2024; Procaccini et al. 2023). Bio-fabricated textiles, such as mycelium-based partitions, support thermally insulated modular architecture while maintaining a minimal environmental impact. Mycelium composites are bio-based materials. This is made from fungal mycelium and organic substrates. This kind of bio-based material is emission-free, non-toxic, low-cost, and recyclable. Basidiomycota species such as *Pleurotus ostreatus* and *Ganoderma lucidum* can develop thick, high-growth mycelium, conferring thermal stability, hydrophobicity, and mechanical strength. Biofabricated textiles are superior to conventional textiles (Alemu et al., 2022).

Ultra-white panels are made from multilayer nanophotonic materials. These layers are carefully structured to implement passive strategies such as solar heating, natural ventilation, and daylighting, helping reduce reliance on artificial energy sources. Incorporating high-performance insulation, energy-efficient glazing, and renewable energy systems, such as solar panels, further enhances the building's energy performance (Bonjar et al., 2021; Hilal et al., 2023).

It is also reported that smart materials in kinetic facades adapt to external climate factors, such as solar radiation, light, and humidity, to reduce energy consumption and enhance indoor comfort. Simulations compared a conventional static facade (ECBC base case) with two adaptive scenarios using ethylene tetrafluoroethylene (ETFE) and an innovative auto Heal self-healing material. Material choice significantly impacts facade performance (Tharim et al., 2018). Beyond energy efficiency, smart facades improve thermal comfort, lower costs, and reduce carbon footprints (Bashir et al., 2023; Kawasaki and Kawai, 2006; Kuda et al., 2025). Additionally, phase-change material (PCM) textiles in wall panels passively regulate indoor temperatures; phase-change materials (PCMs) enable energy-efficient cooling by storing and releasing latent heat. They enhance passive cooling methods like solar control, ventilation, and radiative cooling (Beevi et al., 2025). PCM walls delay heat transfer by 2 hours, PCM windows reduce heat by 66%, and nocturnal radiative cooling lowers interior temperatures by 13 °C. Natural ventilation with PCMs can save up to 90% of energy in hot climates. Despite scalability challenges, PCMs improve thermal performance, with experimental studies showing 14–90% energy reductions, proving their potential for sustainable building design (Reddy et al., 2024; Ghamari et al., 2024). Photovoltaic fabric canopies harness solar energy to generate electricity, offering an innovative solution for integrating renewable energy into architectural design. It is reported that MIT developed new ultra-thin solar cells of only 15 microns with electrodes printed on top. They use special inks and printable materials to make them strong; they are adhered to a tough fabric called Dyneema with a UV-curing glue. These solar cells give high power—370 W/kg. That is better than normal solar cells. They are very light and environmentally friendly. Even after being rolled 500 times, they still work well, keeping over 90% of their power. Now, MIT is also working on eco-friendly packaging (Valsalakumar et al., 2022).

Aerogel-infused textile insulation, used in modern green buildings, provides high thermal resistance with minimal thickness, further reducing energy consumption. These textile innovations integrate with passive strategies like solar heating, ventilation, and daylighting, significantly enhancing energy performance while minimizing environmental impact (Olatunde et al., 2024; Azkorra-Larrinaga et al., 2023).

Given their exceptional properties, aerogels are viewed as materials with immense potential. As a natural polymer, cellulose offers the advantage of being both renewable and biodegradable (Sözcü et al., 2023). Cellulose-derived aerogels possess the replenishability, biocompatibility, and inherent natural degradation of cellulose, along with additional benefits such as low weight, extensive porosity, and a large specific surface area. Even with increasing appreciation and acceptance, the undiscovered potential of aerogels in the textiles sphere remains predominantly uninvestigated (Sözcü et al. 2023; Long et al., 2018).

Selecting the right materials and ensuring resource efficiency are also significant for sustainability. Using recycled, locally sourced, and low-impact materials reduces the building's carbon footprint. Additionally, construction waste can be minimized through modular design and efficient material management, ensuring that resources are used responsibly throughout the building's lifecycle (Liang et al., 2023). Globally, buildings like The Edge in Amsterdam use recycled PET-based acoustic panels, demonstrating how textile waste can be repurposed for high-performance architecture. Modular textile partitions and 3D-knitted fabric walls enable efficient material management, reducing construction waste (Narayan et al., 2025). By integrating recycled textiles, natural fibers, and modular design, buildings optimize material efficiency, reduce carbon footprints, and extend the lifecycle of resources, aligning with sustainable construction principles (Ogwu and Kosoe, 2025). A modular room partition made entirely from birch wood composites offers a lightweight, flexible, and sustainable solution for interior spaces. The birch wood frame provides stability while maintaining a natural aesthetic, and the fabric panels—ranging from felt for acoustic absorption to woven textiles for decorative appeal—allow for customization. Features such as interchangeable, roll-up, or sliding textile panels make it ideal for offices, co-working spaces, homes, and exhibitions (Cheng et al., 2016; Dubois et al., 2020; Wei et al., 2023).

As waste from the ion industry, pre-consumer and post-consumer, environmental concerns, fill use this waste as a solution. The building industry also tried this and came up with a beautiful solution: using this waste to make bricks—Fabrick, made from recycled fabric waste (Kowsalya et al., 2025). It supports waste reduction, lowers carbon emissions, and promotes resource circularity. These bricks not only address environmental problems but also enhance thermal insulation, helping regulate indoor temperatures and reducing energy consumption (Paixão et al., 2023; Abdullatif et al., 2020).

Water conservation and management involve incorporating water-saving features such as rainwater harvesting, greywater recycling, and efficient plumbing systems to reduce water consumption. Landscaping with native and drought-resistant plants further minimizes the need for irrigation, promoting a more sustainable use of water resources. Textile materials can play a vital role in effectively implementing this principle. For example, geotextiles are applied in rainwater harvesting pits. This helps to enhance groundwater recharge (MVSS, R. 2014; Cao, 2024). Textile materials can play a role in greywater treatment by serving as adsorbents, capturing pollutants such as dyes and other contaminants. For example, cotton waste and denim fabric can be used as adsorbents to remove dyes and other pollutants from textile wastewater (Choudhury, Majumdar, et al. 2025). Additionally, green walls incorporating textile-based filling media can be used for greywater treatment. For landscaping, DuPont's Plantex range helps conserve water by reducing evaporation and controlling weeds without the need for chemical treatments (Sebastian 2024; DuPont, 2025).

IEQ includes improving air quality by using non-toxic materials and ensuring proper ventilation. Maximizing natural light, providing access to outdoor views, and designing acoustically comfortable spaces contribute to the well-being and productivity of building occupants (Woo et al., 2021). Textile materials play a crucial role in enhancing indoor environmental quality (IEQ) by improving air quality, natural lighting, ventilation, and acoustics. BWF Feltec has introduced polySONIC® fr, a flame-retardant acoustic material made from synthetic polymers. It effectively absorbs and disperses sound, reducing noise levels in interiors. The product is featured in acoustic pearls' "Fiber Ceiling" series, showcased at the Design Post Köln, a key hub for interior design innovation since 2006 (BWF Group; Rana et al., 2023)—Himalayan wool and yak hair-based acoustic panels in interiors to reduce noise levels while ensuring thermal insulation.

Additionally, hand-woven wool rugs and hemp-based curtains improve indoor air quality by absorbing pollutants and regulating humidity. Hand-woven cotton and banana fiber blinds, which allow for natural ventilation and diffused daylight, reduce the need for artificial lighting while maintaining thermal comfort. Globally, the Bullitt Center in Seattle uses wool-based air filtration panels to enhance indoor air quality by capturing airborne contaminants. Subramanian (2025) discusses the integration of textiles and architecture, mainly in the context of sustainability and biophilic design. It is reported that wool-based panels can absorb VOCs. They are biodegradable, moisture-resistant, and maintain structural integrity, making them a sustainable architectural choice (Kumar et al., 2025; Hegyi et al., 2021; Călătan and Dico, 2022). Regulation of indoor temperature by PCM-infused textile and provision of acoustic comfort using a natural sound-absorbing panel also fall under IEQ (Zhu et al., 2025). This principle of lifecycle and resilience planning ensures that buildings are designed to last and adapt over time. Flexible and modular spaces allow for future modifications, reducing the need for demolition and reconstruction. Additionally, incorporating climate resilience strategies—such as flood-resistant designs and energy backup systems—helps buildings withstand environmental changes and natural disasters (Prihartanto et al., 2023). Textile materials, such as bamboo-reinforced composites, are used in floating schools and homes in the Sundarbans, allowing buildings to adapt to rising water levels. In earthquake-prone regions, integrating textile-based seismic reinforcement, such as flexible fiber-reinforced polymer (FRP) wraps around columns and beams, improves structural strength while allowing flexibility during tremors. Globally, the Eden Project in the UK employs ETFE textile membranes for lightweight, durable, and modular roofing, ensuring energy efficiency and long-term adaptability. Additionally, recyclable textile panels in prefabricated structures allow for future modifications without demolition, reducing waste (Zeng and Long, 2021; Hammerl and Kromoser, 2022).

Notable examples of buildings that use textile materials for seismic reinforcement include the Textile Centre in Auckland, New Zealand, and the FRP application to the parapets, spandrels, and other structural elements to enhance seismic performance while preserving the building's external appearance. This ensures the structure could withstand lateral loads during earthquakes (Derakhshan et al., 2019; Giaretton et al., 2016). Similarly, the Fa-bo Building in Japan employs Cabkoma Strand Rods, a thermoplastic carbon fiber composite. These rods are seven times stronger than iron of the same diameter, but are significantly lighter. This is the first structure globally to use such carbon fiber composites for seismic reinforcement (Boem, 2022; Basim et al., 2020).

ETFE has evolved into a mainstream building material that provides extraordinary tear resistance, long life, and transparency to ultraviolet light. Its application in prominent projects such as the Eden Project in the UK and the Beijing Olympic Aquatics Centre has brought this material to public attention (Ozcanlor, 2015).

Weighing approximately 1% of glass, ETFE membranes, and cushions impose significantly less dead load on supporting structures, enabling a reduction in the structural framework. This lightweight nature contributes to overall building efficiency while maintaining exceptional performance characteristics. The material transmits up to 95% of light, making it ideal for applications requiring high transparency, though it can also be printed or fritted with patterns to reduce solar gain when needed (Monticelli and Zanelli, 2021; Shafa, 2024).

Perhaps most significant for lifecycle resilience is ETFE's exceptional durability. Unaffected by UV light, atmospheric pollution, and environmental weathering, ETFE has a projected lifespan exceeding 40 years according to laboratory and field research (Lamnatou et al., 2018).

Seismic retrofitting using FRP jacketing represents a crucial application of textile-based materials for enhancing building resilience, particularly in earthquake-prone regions. FRP composites consist of a polymer matrix reinforced with fibers that provide discernible reinforcing function in one or more directions. This anisotropic property distinguishes FRPs from traditional construction materials such as steel or aluminum, enabling targeted structural reinforcement (Kim et al., 2025).

Wood and recycled textile composite panels represent an innovative approach to sustainable building materials that enhance adaptability and lifecycle performance. Research has evaluated the potential of recycled cotton textiles as fillers and reinforcements in oriented strandboard (OSB) panels, a standard building material. Studies found that panels incorporating 5% recycled textiles maintained performance comparable to that of conventional panels in bending strength, elasticity, and nail withdrawal strength (Danihelová et al., 2021).

Research is being done to integrate smart technologies into buildings to make them smarter and more responsive to the external environment. Integration of technologies can be done in different ways, such as automating the system or using IoT-based, voice-activated systems to open and close the curtains. However, it should be remembered that smartness cannot be achieved by burdening the earth. Simultaneous monitoring of energy and water usage allows for predictive maintenance, making buildings more responsive to environmental conditions. Features like thermal regulation based on the outer environment can be integrated into the interior, such as building curtains and upholstery, and to some extent into the exterior. Thermal-management textiles are integrated into curtains, upholstery, and energy-efficient facades to optimize indoor climate control. Materials such as PCM (phase-change material) coatings, photothermal absorption layers, or conductive heating elements help reduce reliance on conventional HVAC (heating, ventilation, and air conditioning) systems, promoting sustainable living and energy conservation (Zhu et al., 2025).

Photovoltaic fabric is a fabric that incorporates solar cells. This not only generates energy but also serves as shading and insulation. Dutch designer Marjan van Aubel reported the incorporation of organic photovoltaic (OPV) cells into lightweight fabric (Dezeen, 2022). Heliatek also reported on the development of ultra-thin solar films that generate renewable energy while reducing indoor heat (Heliatek).

An innovative project named Solar Fiber also reported the use of photovoltaic threads. These threads can generate energy from sunlight. Pvilion, a US-based company, also commercializes fabric, canopies, and tents of the same concept (Pvilion n.d.; Moldenhauer, 2022). Flexible solar fabric canopies used in outdoor seating areas have already been commercialized. Companies like Envision Solar (Beam Global) have installed solar-powered carports, allowing businesses, residential complexes, and public spaces to provide on-site renewable energy for electric vehicles while reducing heat buildup in parked cars.

Another way to incorporate PCMs. Phase-change materials (PCMs) are thermal-regulating materials that absorb, store, and release heat, helping maintain stable interior temperatures. One notable example is Outlast® technology, which developed nano-capsules of PCMs (ThermoCells®) that respond to changes in body temperature. This technology can be used in sofa covers and bedding. DuPont and BASF have already developed PCM-based drywall and plasterboards. This type of drywall or plasterboard can store and release heat to maintain a building's internal environment (Kurdi et al., 2021).

Photothermal textiles use nanomaterials such as graphene and metallic nanoparticles to trap and retain solar heat, making them ideal for passive heating applications, such as cold-weather blankets (Hasan et al., 2024; Kim, 2020).

Nature-inspired solutions can help minimize unnecessary activities, such as washing, by mimicking the lotus leaf's water-repelling properties. Nano-Tex (Stain-Resistant Textiles) is based on the same concept. It has a Crypton Fabric, which applies nano-coatings to fabrics, making them resistant to liquids and stains (Crypton, n.d.). It is also found that Schoeller Technologies, a Swiss company, has already developed a self-cleaning, water-repellent finish inspired by the lotus effect. Similarly, Devan Chemicals, a Belgian textile innovation company, has developed self-cleaning curtains and upholstery using titanium dioxide (TiO_2) (Schoeller Textiles; Wei, 2023; Heiq, 2022). The TiO_2 coating helps break down organic pollutants when exposed to light. This breakdown automatically removes stains and odors. Odor and self-cleaning fabric have also been developed using mint-based silver particles (Xie, Tu, et al. 2023).

Biodegradability is an important parameter for sustainable design. Biodegradable Polymers follow natural decomposition processes. PLA (Polylactic Acid), derived from renewable sources like corn starch and sugarcane, is a notable example in this category. PLA is used in packaging, disposable cutlery, and medical sutures, as it decomposes into non-toxic components under industrial composting conditions. Its utilization for building and architecture can also be explored. Another prominent material is PHA (Polyhydroxyalkanoates). PHA is produced by bacteria that store energy as polymers. PHA-based plastics are used in food packaging, agricultural films, and even 3D printing, and they break down naturally in soil and water without harming ecosystems (Gürler and Torğut, 2022; Ahsan et al., 2023; Bamps et al., 2022).

5.3 Conceptual Foundations

5.3.1 Defining Smart Buildings

Smart buildings are structures that use digital, sensor-driven technologies to monitor, control, and optimize their own performance. This includes environments that respond adaptively to occupants, environmental conditions, and energy flows. The definition often encompasses automated systems, connectivity (IoT), data analytics, and integration across building services such as HVAC, lighting, security, and comfort (Lavrinovica et al., 2024; Batra et al., 2014).

5.3.2 *Energy Efficiency in Building Design*

This involves both passive and active strategies: passive design (orientation, thermal mass, daylighting, and natural ventilation), high-performance materials (insulation, glazing, and façades), efficient mechanical systems, and integration of renewables. The goal is to minimize both operational energy (for heating, cooling, and lighting) and embodied energy (in materials and construction) (Chen et al., 2020; Hu et al., 2023).

5.3.3 *Principles of Sustainable Building Design*

These principles include resource efficiency, site sensitivity, indoor environmental quality (IEQ), circularity (reuse, adaptive reuse), lifecycle thinking, resilience to climate impacts, water conservation, and occupant health and well-being (Akadiri et al., 2012; Timm et al., 2023).

5.3.4 *Alignment with Circular Economy*

Sustainable buildings increasingly align with circular economy concepts: minimizing waste, using recycled or bio-based materials, designing for disassembly, reusing components, and thinking beyond the operational phase to the end of life of materials (Majumdar, Goswami, et al. 2025). The circular economy also demands systems that enable reuse, recycling, and regeneration rather than just efficient consumption (Akanbi et al., 2020; Akinade et al., 2023).

5.4 Core Technologies Enabling Smart and Sustainable Buildings

5.4.1 *Building Management Systems (BMS)*

These are centralized or semi-centralized control platforms that integrate HVAC, lighting, security, energy, and environmental monitoring. BMS enables automated, scheduled control, fault detection, and occupant feedback loops, thereby improving energy efficiency and comfort (Jradi et al., 2021; Gunay et al., 2018).

5.4.2 *Internet of Things (IoT) and Sensors*

Sensors (temperature, humidity, occupancy, light, CO_2, etc.) connected via IoT networks provide real-time data. These support dynamic control, predictive maintenance, and allow data-driven decision-making. For example, The Edge in Amsterdam uses tens of thousands of sensors to monitor and respond to occupant presence and environmental conditions (Wired, 2014; Rojek et al., 2025).

5.4.3 *Artificial Intelligence and Machine Learning*

Artificial intelligence and machine learning (AI/ML) algorithms analyze large streams of sensor data to detect patterns, optimize systems, forecast energy needs, automatically adjust operations, and even suggest design improvements. They play roles in predictive maintenance, anomaly detection, occupancy forecasting, and optimizing energy use (Bouabdallaoui et al., 2021; Himeur et al., 2021).

5.4.4 *Smart Meters and Grid Integration*

Smart meters measure electricity, gas, and water in detail and in real time. Integration with the power grid (and possibly microgrids) enables buildings to respond to demand signals, participate in demand response, feed excess generation (from PV or other on-site renewable sources), and manage loads to reduce peak demand (Himeur et al., 2021; Toderean et al., 2025).

5.4.5 *Sustainable Construction Technologies*

This covers advanced materials (recycled or bio-based), modular and prefabricated components, low-embodied-carbon products, methods for reducing construction waste, and technologies for improved water, thermal, and acoustic performance. Also, digital tools (BIM, digital twin) that simulate performance and reduce errors are in construction (Päätalo et al., 2024; Wang and Li, 2025).

5.5 Energy Performance and Environmental Impact

The pursuit of sustainable architecture increasingly relies on a suite of interconnected strategies that span energy, water, and material performance. Energy modeling and simulation tools, such as EnergyPlus, IES, and TRNSYS, allow design teams to anticipate heating and cooling demands, daylight availability, and thermal behavior well before construction begins. Post-occupancy, these tools provide critical insights to align operational performance with design intent and continuously fine-tune building systems. Renewable energy integration—through solar, geothermal, or localized wind systems, combined with storage solutions—further reduces reliance on centralized grids and enhances resilience, as seen in projects like The Edge, which uses solar panels and aquifer thermal energy storage (Gil-Ozoudeh et al., 2024; Pavlović et al., 2018).

Water management is another cornerstone of sustainable design. From rainwater harvesting and greywater reuse to sensor-controlled irrigation and drought-tolerant landscapes, strategies now extend beyond conservation to emphasize resilience in increasingly water-scarce contexts. Parallel to this, the measurement and reduction of carbon footprints through Life-Cycle Assessment (LCA) foreground the importance of both embodied impacts—linked to materials, transport, and construction—and operational emissions over a building's lifespan. By targeting net-zero or net-positive outcomes, sustainable buildings aspire not just to reduce harm but to regenerate their environments actively (Pearlmutter et al., 2021).

5.6 Design Strategies for Sustainability

The design and operation of sustainable buildings increasingly rely on advanced tools and strategies that extend across energy, water, and material systems. Energy modeling and simulation tools such as EnergyPlus, IES, and TRNSYS allow architects and engineers to predict and optimize heating and cooling loads, daylighting, shading effects, and thermal performance (Wang et al., 2025). Post-occupancy, these models provide valuable feedback, enabling building systems to be fine-tuned and aligned with original design goals. Alongside this, integrating renewable energy—whether through solar photovoltaics, solar thermal systems, geothermal solutions, or local wind power—reduces dependence on conventional grids. Coupled with storage technologies, these systems balance intermittency and ensure resilience, as exemplified by The Edge's combination of solar panels and aquifer thermal energy storage (Wang et al., 2023; Tinti et al., 2023).

Equally critical are water management strategies that emphasize efficiency and circularity. Greywater reuse, rainwater harvesting, drought-tolerant landscaping, and sensor-based irrigation demonstrate how buildings can conserve and regenerate resources while ensuring occupant comfort (Majumdar, Choudhury, and Cheshmehzangi 2025). At a broader scale, carbon and environmental performance assessments, particularly Life-Cycle Assessment (LCA), help quantify both embodied and operational emissions. This approach reframes sustainability not only as an operational concern but also as a material one, considering transport, construction, and long-term maintenance. Moving toward net-zero or net-positive outcomes demands reducing carbon intensity across the building lifecycle while simultaneously advancing regenerative design principles (Gil-Ozoudeh et al., 2024).

Together, these strategies point to a shared trajectory across architecture and textiles: one where predictive tools, renewable integration, water stewardship, and lifecycle accountability form the foundation of design. Both sectors are moving beyond incremental efficiency toward systemic transformation, in which materials and technologies not only minimize harm but also actively regenerate environmental and social systems (Baharun et al., 2011; Khan et al., 2022; Abidin et al., 2025; Knutsson & Knutsson, 2021).

5.7 Case Studies

Exemplary projects worldwide demonstrate the transformative potential of sustainable, smart architecture. The Edge in Amsterdam, designed by PLP Architecture for Deloitte, has achieved the highest BREEAM score to date, operating as a net-zero energy workplace that produces more energy than it consumes. Its integration of over 30,000 sensors, combined with intelligent lighting, automated shading, and aquifer-based thermal energy storage, sets a benchmark in digital and energy-efficient design (Kumar, 2025; Gelan, 2022).

Similarly, Milan's Bosco Verticale illustrates how biodiversity can be integrated into architecture. The vertical forests composed of more than 900 trees and 20,000 plants improve air quality, regulate microclimates, and provide natural shading while reducing overall energy loads and climate resilience environment (Choudhury, Majumdar, et al. 2025). In Seattle, the Bullitt Center advances these principles through the Living Building Challenge framework, achieving net-positive energy and water performance while maintaining rigorous standards for material transparency and indoor environmental quality (Magdy, 2025).

On a larger scale, Singapore's Smart Public Housing initiative demonstrates how sustainability can be embedded at the community level through retrofitting, green roofs, integrated waste management, and real-time feedback dashboards that encourage responsible energy use among residents (Rajpal et al., 2024). In the Middle East, Masdar City in the United Arab Emirates stands as a living laboratory for renewable energy and low-carbon urban design. Likewise, One Central Park in Sydney integrates vegetated façades and heliostat mirrors to enhance daylight penetration and urban biodiversity (Tan et al., 2021).

These global examples reveal that innovation extends beyond energy efficiency—it encompasses biodiversity, material circularity, and community engagement.

Together, these projects and parallels with textiles point toward a future where sustainability is not engineered into isolated structures but woven holistically into the systems, materials, and practices that shape everyday life. (Bloomberg, 2015; Edge technologies; AIPH 2020)

5.8 Policy and Market Enablers

Policy frameworks and financial instruments are becoming key drivers of sustainable building transitions. Minimum energy performance standards now mandate improved insulation, advanced glazing, orientation strategies, and consideration of local climate conditions, while energy modeling and performance benchmarks are increasingly embedded in building codes. Alongside regulation, economic mechanisms such as green bonds, subsidies for renewable technologies, tax rebates, and performance-based contracts provide incentives for stakeholders to adopt cleaner and more efficient practices. Government-led pilot projects and exemplar buildings further play a vital role by demonstrating feasibility, reducing perceived risks, and setting industry benchmarks (Moore and Doyon, 2023; Hou, 2024).

However, even with firm policy and financial levers, behavioral change remains central to success. Tools such as occupant dashboards, feedback loops, and user control over lighting and temperature can encourage more sustainable choices, while participatory design can strengthen a sense of ownership and accountability. At the same time, building managers and facility operators require ongoing training and awareness to manage these complex systems effectively (Lee et al., 2024).

The textile sector reflects many of these dynamics. Just as building codes enforce energy standards, sustainability certifications for textiles covering insulation properties, thermal regulation, or lifecycle impacts set new baselines for performance. Financial instruments, from green procurement policies to subsidies for natural fibers, similarly accelerate adoption. Behavioral shifts are equally vital: consumer choices, awareness of care practices, and demand for transparency drive sustainable textile markets. Moreover, textiles can directly improve building efficiency, as seen in high-performance fabrics for insulation, adaptive shading, and textile-based façades that respond to light and temperature. In this sense, textiles and buildings share a parallel path, where regulation, finance, and human behavior intersect to transform sustainability from aspiration to practice (Xie, Tyler, et al. 2023; U.S. Energy Codes and Building Performance Standards: Supporting Energy Use and Emissions Reductions in Buildings. International Code Council 2023; Yitmen et al., 2021).

5.9 AI-Driven Design and Construction Automation

Generative design, robotics such as 3D printing and automated assembly, and adaptive technologies are beginning to redefine how we approach the built environment. These innovations enable real-time responsiveness through features such as adaptive façades. At the same time, integrating artificial intelligence offers new ways to optimize material use, streamline construction sequencing, and significantly reduce waste (Jadhav and Anchliya, 2024). At the same time, automation in monitoring and maintenance creates opportunities for long-term efficiency and resilience. However, the successful integration of these technologies demands more than technical capability—it requires rethinking traditional practices, cultivating interdisciplinary collaboration, and addressing the social, ethical, and environmental implications of automation. Only through such a holistic approach can these emerging tools move beyond experimental applications and contribute meaningfully to a more sustainable and intelligent built future (Zhang and Tan, 2025).

5.10 Future Trends and Innovations

Digital twins—virtual, evolving replicas of buildings—are transforming the way we design, construct, and operate the built environment. By enabling predictive maintenance, energy optimization, and early error detection, they reduce inefficiencies across the building lifecycle. Increasingly, the ambition is not just to minimize environmental impact but to create regenerative buildings: structures that generate more energy than they consume, clean and recycle water, restore soils, and actively support surrounding ecosystems (Bongomin et al., 2025).

This vision expands further when buildings are seen as active participants in larger networks. Through microgrids, district heating and cooling, and shared data and resource systems, they can contribute to coordinated energy, water, and transport infrastructures at the neighborhood or city level. While the potential for resilience and efficiency is immense, achieving this requires supportive governance, interdisciplinary collaboration, and community engagement (Ożadowicz, 2025).

5.11 Challenges and Limitations

The transition toward sustainable and smart technologies is filled with both opportunities and obstacles. High initial investments often pose a barrier, particularly when the timeframe for financial returns is unclear. At the same time, the growing reliance on large-scale sensor networks and occupant tracking raises valid concerns around data privacy and the need for stronger security frameworks. Another pressing issue lies in the fragmented landscape of systems, standards, and vendors, which can limit efficiency and increase costs rather than reduce them. Beyond these technical and financial constraints, there is also a knowledge gap: many designers, engineers, facility managers, and policy makers still lack the interdisciplinary expertise required to navigate and integrate these innovations effectively. Overcoming these challenges is vital if we are to move from promise to practice in building a smarter, more sustainable future (Baharetha et al. 2024; Kelly et al., 2022).

5.12 Conclusion

This chapter discussed the basic principles of incorporating energy-efficient and sustainable design, illustrated with some prominent examples, for the city of the future. From this discussion, it is clear that smart solutions cannot work without sustainable design principles. A smart solution without sustainability is like a human without a soul. Integrating the latest technology with sustainable design principles remains a challenge, but realizing how we have deteriorated Mother Earth is the only way to develop solutions that integrate technology while keeping sustainability in mind. Designing buildings and creating architectural articles is valuable, but careful consideration and selection of the right materials are critical. Materials must provide functional performance, aesthetic appeal, and sustainability. They should be circular, biodegradable, and contribute to long-term environmental responsibility. Achieving the vision of a sustainable smart city cannot be the task of a single stakeholder; it requires collaboration among designers, researchers, technologists, governments, and citizens to bring this dream to life (Choudhury and Roy 2025). Textile materials are essential to this change because they provide resilience and adaptability through creative applications. This chapter explained how sustainable design principles can be achieved through different textile materials—from mycelium and bamboo-based textiles for green construction to FRP jacketing for seismic safety and ETFE membranes for lightweight, long-lasting facades. These solutions help buildings meet environmental demands while using minimal resources over their lifespan. They enable the construction of enduring buildings that are not only efficient and flexible but also serve current and future generations responsibly. Advancements in materials—from phase-change materials (PCM) to self-cleaning fabrics, thermoregulating textiles, and energy-harvesting solutions—can seamlessly integrate with modern architecture, creating self-sustaining environments that promote eco-friendly living and smart energy management. These cutting-edge innovations demonstrate how materials science can deliver sustainable, energy-efficient thermal regulation by harnessing nature-inspired passive cooling strategies across a range of applications across industries.

References

Abdullatif, K., Guirguis, M., & Moussa, R. (2020). Analyzing the structural properties of fire clay bricks after adding cigarette filters. *WSEAS Transactions on Environment and Development, 16*, 671–679. https://doi.org/10.37394/232015.2020.16.69

Abidin, N. H. Z., Fauzi, M. A., & Tamyez, P. F. M. (2025). Solar PV application on office buildings: A state of the art analysis through bibliometric approach. *Journal of Engineering and Applied Science, 72*(1), 84.

Ahsan, W., Hussain, A., Lin, C., & Nguyen, M. (2023). Biodegradation of different types of bioplastics through composting—A recent trend in green recycling. *Catalysts, 13*(2), 294. https://doi.org/10.3390/catal13020294

AIPH. (2020). Bosco Verticale, Milan—Case Study. https://aiph.org/green-city-case-studies/milan-italy-bosco-verticale/

Akadiri, P. O., Chinyio, E. A., & Olomolaiye, P. O. (2012). Design of a sustainable building: A conceptual framework for implementing sustainability in the building sector. *Buildings, 2*(2), 126–152.

Akanbi, et al. (2020). Fundamentals of building deconstruction as a circular economy strategy for the reuse of construction materials. *Buildings, 10*(12), 220.

Akinade, et al. (2023). Integration of BIM and design for deconstruction to improve circular economy of buildings. *Journal of Building Engineering, 75*, 106990.

Alemu, D., Tafesse, M., & Mondal, A. K. (2022). Mycelium-based composite: The future sustainable biomaterial. *International Journal of Biomaterials, 2022*(1), 8401528.

Ayoobi, A. W., Inceoğlu, M., & Inceoğlu, G. (2024). A next-generation holistic building design framework: A focus on integrating sustainable and vernacular design principles. *Smart Construction and Sustainable Cities, 2*(1), 18.

Ayu, L., Maimun, A., & Dewi, C. (2021). Multifunctional village hall with sustainable architecture concept: An integrated land use system by considering building energy efficiency. *Journal Innovation of Civil Engineering (Jice), 2*(2), 69. https://doi.org/10.33474/jice.v2i2.13118

Azkorra-Larrinaga, Z., Romero-Antón, N., Martín-Escudero, K., López-Ruiz, G., & Giraldo-Soto, C. (2023). Evaluation of the thermal performance of two passive facade system solutions for sustainable development. *Sustainability, 15*(24), 16737. https://doi.org/10.3390/su152416737

Baharetha, S., Soliman, A. M., Hassanain, M. A., Alshibani, A., & Ezz, M. S. (2024). Assessment of the challenges influencing the adoption of smart building technologies. *Frontiers in Built Environment, 9*, 1334005. https://doi.org/10.3389/fbuil.2023.1334005

Baharun, et al. (2011). Experimental validation of EnergyPlus® simulation of a single storey building. *Journal of Civil Engineering, Science and Technology, 2*(1), 12–17.

Bamps, B., Guimaraes, R., Duijsters, G., Hermans, D., Vanminsel, J., Vervoort, E., & Peeters, R. (2022). Characterizing mechanical, heat seal, and gas barrier performance of biodegradable films to determine food packaging applications. *Polymers, 14*(13), 2569. https://doi.org/10.3390/polym14132569

Baper, S. Y., & Ismael, Z. K. (2024). The impact of site contexts in increasing the diversity of architectural concepts: Using an interactive architectural approach. *Buildings, 14*(8), 2567.

Bashir, et al. (2023). Thermawrap (TW) architecture in Kashmir. SSM College of Engineering Conference Paper, India.

Basim, S., Hejazi, F., Rashid, R., & Esfahani, M. (2020). Development of finite element model for embedded carbon fiber–reinforced polymer rod in reinforced concrete frame joints. *International Journal of Protective Structures, 11*(4), 533–553. https://doi.org/10.1177/2041419620917689

Batra, N., Singh, A., Singh, P., Dutta, H., Sarangan, V., & Srivastava, M. (2014). Data driven energy efficiency in buildings. arXiv preprint arXiv:1404.7227.

Beevi, S., Mahabuba, A., Gopinath, S., Varshini, S., Shoaib, A., & Munaf, R. (2025). Enhancing solar panel efficiency through cost-effective cooling strategies. *Key Engineering Materials, 1008*, 79–92. https://doi.org/10.4028/p-b5wj0o

Bibri, S. (2019). On the sustainability of smart and smarter cities in the era of big data: An interdisciplinary and transdisciplinary literature review. *Journal of Big Data, 6*(1), 25.

Bloomberg. (2015). The Edge: The world's greenest building.

Boem, I. (2022). Multi-layer modelling of masonry structures strengthened through textile-reinforced mortar. *Open Research Europe, 2*, 132. https://doi.org/10.12688/openreseurope.15233.1

Bongomin, O., Mwape, M. C., Mpofu, N. S., Bahunde, B. K., Kidega, R., Mpungu, I. L., & Ngulube, G. (2025). Digital twin technology advancing industry 4.0 and industry 5.0 across sectors. *Results in Engineering, 26*, 105583.

Bonjar, M., Horváth, K., Baranyai, B., & Kistelegdi, I. (2021). Passive system optimization in office buildings using a reference testbed building. *Pollack Periodica, 16*(2), 124–129. https://doi.org/10.1556/606.2020.00212

Bouabdallaoui, Y., Lafhaj, Z., Yim, P., Ducoulombier, L., & Bennadji, B. (2021). Predictive maintenance in building facilities: A machine learning-based approach. *Sensors, 21*(4), 1044.

Călătan, G. A., & Dico, C. (2022). Environmentally friendly building materials with beneficial potential for indoor air quality. *Athens Journal of Technology & Engineering, 9*(4), 305–320.

Cao, S. (2024). Comparative analysis of greywater recycling and rainwater harvesting as supplementary water sources for conventional urban and tourist resort water supplies. *Applied and Computational Engineering, 58*(1), 94–100. https://doi.org/10.54254/2755-2721/58/20240699
Chen, Z., Hammad, A. W., Kamardeen, I., & Akbarnezhad, A. (2020). Optimising embodied energy and thermal performance of thermal insulation in building envelopes via an automated building information modelling (BIM) tool. *Buildings, 10*(12), 218.
Cheng, Y., Yu, Y., Fu, F., Wang, J., Shang, L., Gu, Z., et al. (2016). Controlled fabrication of bioactive microfibers for creating tissue constructs using microfluidic techniques. *ACS Applied Materials & Interfaces, 8*(2), 1080–1086. https://doi.org/10.1021/acsami.5b11445
Choudhury, M., & Roy, P. (2025). Challenges with microplastic pollution in the regime of UN sustainable development goals. *World Development Sustainability, 2025*, 100216. https://doi.org/10.1016/j.wds.2025.100216
Choudhury, M., Majumdar, S., Goswami, S., & Sillanpää, M. (2025). *Smart wastewater systems and climate change: Innovations through spatial intelligence*. Royal Society of Chemistry. https://doi.org/10.1039/9781837676880
Crypton. (n.d.). Nanotex by Crypton | Commercial Upholstery Fabric. https://crypton.com/nanotex-crypton/
Danihelová, A., Sčensný, P., Gergeľ, T., Ondrejka, V., Němec, M., Igaz, R., & Mitterová, I. (2021). Influence of flame retardant impregnation on acoustic and thermophysical properties of recycled technical textiles with the potential for use in wooden buildings. *Polymers, 13*(16), 2598. https://doi.org/10.3390/polym13162598
Dénes, T. O., Iştoan, R., Tămaş-Gavrea, D. R., Manea, D. L., Hegyi, A., Popa, F., & Vasile, O. (2022). Analysis of sheep wool-based composites for building insulation. *Polymers, 14*(10), 2109.
Derakhshan, H., Walsh, K., Ingham, J., Griffith, M., & Thambiratnam, D. (2019). Seismic fragility assessment of nonstructural components in unreinforced clay brick masonry buildings. *Earthquake Engineering and Structural Dynamics, 49*(3), 285–300. https://doi.org/10.1002/eqe.3238
Dezeen. (2022). Marjan van Aubel tapestry project.
Dubois, H., Verkasalo, E., & Claessens, H. (2020). Potential of birch (betula pendula Roth and b. pubescens ehrh.) for forestry and forest-based industry sector within the changing climatic and socio-economic context of western Europe. *Forests, 11*(3), 336. https://doi.org/10.3390/f11030336
DuPont. (2025). Plantex.
Etukudoh, E., Nwokediegwu, Z., Umoh, A., Ibekwe, K., Ilojianya, V., & Adefemi, A. (2024). Solar power integration in urban areas: A review of design innovations and efficiency enhancements. *World Journal of Advanced Research and Reviews, 21*(1), 1383–1394. https://doi.org/10.30574/wjarr.2024.21.1.0168

Filho, W., Perry, P., Heim, H., Dinis, M., Moda, H., Ebhuoma, E., & Paço, A. (2022). An overview of the contribution of the textiles sector to climate change. *Frontiers in Environmental Science, 10*. https://doi.org/10.3389/fenvs.2022.973102

Gelan, E. (2022). Green building concepts and technologies in Ethiopia: The case of wegagen bank headquarters building. *Technologies, 11*(1), 2. https://doi.org/10.3390/technologies11010002

Ghamari, M., See, C. H., Hughes, D., Mallick, T., Reddy, K. S., Patchigolla, K., & Sundaram, S. (2024). Advancing sustainable building through passive cooling with phase change materials, a comprehensive literature review. *Energy and Buildings, 312*, 114164.

Ghosh, S., Nitin, B., Remanan, S., Bhattacharjee, Y., Ghorai, A., Dey, T., et al. (2020). A multifunctional smart textile derived from merino wool/nylon polymer nanocomposites as next generation microwave absorber and soft touch sensor. *ACS Applied Materials & Interfaces, 12*(15), 17988–18001. https://doi.org/10.1021/acsami.0c02566

Giaretton, M., Dizhur, D., Porto, F., & d., & Ingham, J. (2016). Post-earthquake reconnaissance of unreinforced and retrofitted masonry parapets. *Earthquake Spectra, 32*(4), 2377–2397. https://doi.org/10.1193/121715eqs184m

Gil-Ozoudeh, I., Iwuanyanwu, O., Okwandu, A. C., & Ike, C. S. (2024). Water conservation strategies in green buildings: Innovations and best practices. *Engineering Science & Technology Journal, 4*, 651–671.

Gunay, B., Shen, W., Huchuk, B., Yang, C., Bucking, S., & O'Brien, W. (2018). Energy and comfort performance benefits of early detection of building sensor and actuator faults. *Building Services Engineering Research and Technology, 39*(6), 652–666.

Gürler, N., & Torğut, G. (2022). Dielectric biodegradable biopolymer-based graphene nanocomposites for use in the packaging industry and capacitor application. *ChemistrySelect, 7*(32), e202201975. https://doi.org/10.1002/slct.202201975

Hammerl, M., & Kromoser, B. (2022). Bending behaviour of prestressed t-shaped concrete beams reinforced with frp—Experimental and analytical investigations. *Materials, 15*(11), 3843. https://doi.org/10.3390/ma15113843

Hasan, K. F., Bai, S., Chen, S., Lin, K., Ahmed, T., Chen, J., & Tso, C. Y. (2024). Nanotechnology-empowered radiative cooling and warming textiles. Cell reports physical. *Science, 5*(9), 102108.

Hegyi, A., Bulacu, C., Szilagyi, H., Lăzărescu, A. V., Meiţă, V., Vizureanu, P., & Sandu, M. (2021). Improving indoor air quality by using sheep wool thermal insulation. *Materials, 14*(9), 2443.

HeiQ. (2022). *HeiQ mint | plant-based odor control technology*. Heliatek. HeliaSol Solar Fabric. https://www.heliatek.com/en/products/heliasol/

Hermaputi, R., & Chen, H. (2017). Creating urban water resilience: Review of China's development strategies "sponge city" concept and practices. *The Indonesian Journal of Planning and Development, 2*(1), 1. https://doi.org/10.14710/ijpd.2.1.1-10

Hetimy, S., Megahed, N., Eleinen, O. A., & Elgheznawy, D. (2024). Exploring the potential of sheep wool as an eco-friendly insulation material: A comprehensive review and analytical ranking. *Sustainable Materials and Technologies, 39*, e00812.

Hilal, N., Haggag, M., & Saleh, A. (2023). Optimizing energy efficiency in high-rise residential buildings in Abu Dhabi's hot climate: Exploring the potential of double skin façades. *Buildings, 13*(9), 2148. https://doi.org/10.3390/buildings13092148

Himeur, Y., Ghanem, K., Alsalemi, A., Bensaali, F., & Amira, A. (2021). Artificial intelligence based anomaly detection of energy consumption in buildings: A review, current trends and new perspectives. *Applied Energy, 287*, 116601.

Hou, H. C. (2024). Occupants' willingness to reduce in-room energy consumption– a discrete choice experiment (DCE) approach with occupants in a student apartment. *Energy and Buildings, 320*, 114628.

Hu, M., Zhang, K., Nguyen, Q., & Tasdizen, T. (2023). The effects of passive design on indoor thermal comfort and energy savings for residential buildings in hot climates: A systematic review. *Urban Climate, 49*, 101466.

Jadhav, K., & Anchliya, S. (2024). Adaptive architecture society and its productive economy. *Journal of Emerging Technologies and Innovative Research (JETIR), 11*(3), i412–i417.

Jradi, M., Boel, N., Madsen, B. E., Jacobsen, J., Hooge, J. S., & Kildelund, L. (2021). BuildCOM: Automated auditing and continuous commissioning of next generation building management systems. *Energy Informatics, 4*(1), 2.

Juhara, S., Nokeo, R., Ruskardi, R., & Pasambuna, B. (2024). The role of smart materials, sustainable engineering practices, and iot integration in advancing engineering solutions. *Journal of Academic Science, 1*(2), 96–102. https://doi.org/10.59613/3erdzs77

Kawasaki, T., & Kawai, S. (2006). Thermal insulation properties of wood-based sandwich panel for use as structural insulated walls and floors. *Journal of Wood Science, 52*(1), 75–83.

Kelly, S., Harper, S., Mehrnezhad, M., & Mace, J. (2022). User privacy concerns in commercial smart buildings. *Journal of Cybersecurity, 8*(1). https://doi.org/10.3233/JCS-210035

Khan, S. A., Alam, T., Khan, M. S., Blecich, P., Kamal, M. A., Gupta, N. K., & Yadav, A. S. (2022). Life cycle assessment of embodied carbon in buildings: Background, approaches and advancements. *Buildings, 12*(11), 1944.

Kim. (2020). Thermal insulation property of graphene/polymer coated fabrics. *Scientific Reports, 10*(1), 14392.

Kim, Y., Lee, C., & Jeon, J. (2025). Model parameter estimation for hysteretic behavior simulation of frp-jacketed reinforced concrete columns. *Earthquake Engineering and Structural Dynamics, 54*(6), 1657–1677. https://doi.org/10.1002/eqe.4329

Knutsson, J., & Knutsson, P. (2021). Water and energy savings from greywater reuse: A modelling scheme using disaggregated consumption data. *International Journal of Energy and Water Resources, 5*(1), 13–24.

Kowsalya, P., Jennifer Flora, G., Kousalya, A., Toha, M. A., Khan, P. A., Choudhury, M., & Parthasarathi, T. (2025). Circular systems unveiled: Essential components and their impact on recycling, reuse, and resource recovery. In S. Majumdar, M. Choudhury, & A. Cheshmehzangi (Eds.), *Sustainable urban future. Urban sustainability* (pp. 55–71). Springer. https://doi.org/10.1007/978-981-95-2976-6_3

Kuda, A., Yadav, M., & Ali, S. M. (2025). Performance evaluation of Etfe and auto heal for responsive kinetic facades: Improving building efficiency and urban built environment.

Kumar, A. (2025). The Edge Amsterdam: How AI and UX design created the world's smartest green building - A case study in sustainable digital architecture. Syscore. https://www.sys-core.com/blog/the-edge-amsterdam-how-ai-and-ux-design-created-the-world-s-smartest-green-building-a-case-study-in-sustainable-digital-architecture

Kumar, S., Sakagami, K., & Lee, H. P. (2025). From sustainability to regeneration: Improving indoor environment quality through innovative design.

Kurdi, A., Almoatham, N., Mirza, M., Ballweg, T., & Alkahlan, B. (2021). Potential phase change materials in building wall construction—A review. *Materials, 14*(18), 5328.

Lamnatou, C., Moreno, A., Chemisana, D., Reitsma, F., & Clariá, F. (2018). Ethylene tetrafluoroethylene (ETFE) material: Critical issues and applications with emphasis on buildings. *Renewable and Sustainable Energy Reviews, 82*(3), 2186–2201. https://doi.org/10.1016/j.rser.2017.08.072

Lavrinovica, I., Judvaitis, J., Laksis, D., Skromule, M., & Ozols, K. (2024). A comprehensive review of sensor-based smart building monitoring and data gathering techniques. *Applied Sciences, 14*(21), 10057.

Lee, P. H., Han, Q., & de Vries, B. (2024). Advancing a sustainable built environment: A comprehensive review of stakeholder promotion strategies and dual forces. *Journal of Building Engineering, 95*, 110223.

Li, Y., Commenges, H., Bordignon, F., Bonhomme, C., & Deroubaix, J. (2019). The Tianjin eco-city model in the academic literature on urban sustainability. *Journal of Cleaner Production, 213*, 59–74. https://doi.org/10.1016/j.jclepro.2018.12.018

Liang, R., Zheng, X., Wang, P., Liang, J., & Hu, L. (2023). Research progress of carbon-neutral design for buildings. *Energies, 16*(16), 5929. https://doi.org/10.3390/en16165929

Lionar, M., Arslan, T., & Durak, S. (2019). Architecture as a tool for community building in two different Islamic cultures: Case of masjid pathok negoro and külliye. *European Journal of Sustainable Development, 8*(1). https://doi.org/10.14207/ejsd.2019.v8n1p115

Liu, H., Jay, M., & Chen, X. (2021). The role of nature-based solutions for improving environmental quality, health and well-being. *Sustainability, 13*(19), 10950. https://doi.org/10.3390/su131910950

Liu, H., Σκάνδαλος, N., Brasliņa, L., Kapsalis, V., & Karamanis, D. (2023). Integrating solar energy and nature-based solutions for climate-neutral urban environments. *Solar, 3*(3), 382–415. https://doi.org/10.3390/solar3030022

Long, L. Y., Weng, Y. X., & Wang, Y. Z. (2018). Cellulose aerogels: Synthesis, applications, and prospects. *Polymers, 10*(6), 623.

Magdy. (2025). Bosco Verticale: Milan's innovative vertical forest design. IEREK Blog. https://www.ierek.com/news/bosco-verticale-milans-innovative-vertical-forest-design/

Mageed, N., Alsultani, R., & Abbas, A. (2024). The impact of using advanced technologies in sustainable design to enhance usability and achieve optimal architectural design. *International Journal of Sustainable Development and Planning, 19*(11), 4273–4280. https://doi.org/10.18280/ijsdp.191116

Majumdar, S., Choudhury, M., & Cheshmehzangi, A. (Eds.). (2025). *Sustainable urban future: Embracing circular economy in waste management.* Springer. https://doi.org/10.1007/978-981-95-2976-6

Majumdar, S., Goswami, S., Sillanpää, M., & Choudhury, M. (2025). Revolutionizing wastewater management: The emergence and Impact of Smart Systems. In M. Choudhury, S. Majumdar, S. Goswami, & M. Sillanpää (Eds.), (pp. 1–13). Royal Society of Chemistry. https://doi.org/10.1039/9781837676880-00001

Moldenhauer, A. (2022). Interview: Solar designer Marjan van Aubel. STYLEPARK. https://www.stylepark.com/en/news/marjan-van-aubel-studio-solar-design-sun-energy-alternative-light-stylepark

Monticelli, C., & Zanelli, A. (2021). Material saving and building component efficiency as main eco-design principles for membrane architecture: Case-studies of ETFE enclosures. *Architectural Engineering and Design Management, 17*(3–4), 264–280.

Moore, T., & Doyon, A. (2023). *A transition to sustainable housing: Progress and prospects for a low carbon housing future* (p. 289). Springer Nature.

MVSS, R. (2014). Rooftop rainwater harvesting for recharging shallow groundwater. *Journal of Geology & Geosciences, 03*(06). https://doi.org/10.4172/2329-6755.1000172

Narayan, S., Yadav, D., Jha, S., Khatun, B., Choudhary, R., Verma, U. K., Choudhury, M., & Kumar, D. (2025). From past to present: The evolution of waste management and the rise of circular economy principles. In *Sustainable urban future: Embracing circular economy in waste management* (pp. 23–53). Springer Nature. https://doi.org/10.1007/978-981-95-2976-6_2

Nasrullah, N., & Syafri, S. (2024). Innovative sustainable design approaches in urban architecture: Balancing aesthetics and environmental impact. *Global International Journal of Innovative Research, 2*(9), 2007–2024. https://doi.org/10.59613/global.v2i9.290

Odji, E., & Oni, O. (2024). Advancing circular economy in Nigeria's built environment: The roles of industrial design within multidisciplinary teams. *African Journal of environment and Natural Science Research, 7*(3), 158–172. https://doi.org/10.52589/ajensr-gi4vvc6b

Ogwu, M., & Kosoe, E. (2025). Innovative approaches to recycling, upcycling, and downcycling for sustainable waste management. *CleanMat, 2*(3), 242–261. https://doi.org/10.1002/clem.70013

Olatunde, T., Okwandu, A., Akande, D., & Sikhakhane, Z. (2024). Energy efficiency in architecture: Strategies and technologies. *Open Access Research Journal of Multidisciplinary Studies, 7*(2), 031–041. https://doi.org/10.53022/oarjms.2024.7.2.0024

Oliveira, A., Lanzinha, J., & Kern, A. (2024). Building rehabilitation: A sustainable strategy for the preservation of the built environment. *Sustainability, 16*(2), 553. https://doi.org/10.3390/su16020553

Ożadowicz, A. (2025). Towards sustainable buildings and energy communities: AI-driven Transactive energy, smart local microgrids, and life cycle integration. *Energies, 18*(21), 5668.

Ozcanlor, A. (2015). *ETFE—The trending facade material making the trendiest buildings*. WFM Media. https://wfmmedia.com/etfe-the-trending-facade-material-making-the-trendiest-buildings/

Päätalo, J., Alao, P. F., Rohumaa, A., Kers, J., Liblik, J., & Lylykangas, K. (2024). Prefab light clay-timber elements for net zero whole-life carbon buildings. *Journal of Sustainable Architecture and Civil Engineering, 34*(1), 89–100.

Paixão, G., Santos, G., Marotta, L., Santos, R., Oliveira, C., Oliveira, I., & Soares, P. (2023). Ecological bricks: the revolution in sustainable construction. In *Harmony of knowledge: Exploring interdisciplinary synergie*. Seven Publicações Acadêmicas LTDA. https://doi.org/10.56238/sevened2023.006-010

Patrucco, A., Zoccola, M., & Anceschi, A. (2024). Exploring the potential applications of wool fibers in composite materials: A review. *Polymers, 16*(16), 2360.

Pavlović, A., Đorić-Veljković, S., & Karamarković, J. (2018). The importance of ethylene-tetrafluoroethylene for building daylighting. *Facta Universitatis, Series: Architecture and Civil Engineering, 16*(3), 355–374.

Pearlmutter, D., Pucher, B., Calheiros, C., Hoffmann, K., Aicher, A., Pinho, P., & Nehls, T. (2021). Closing water cycles in the built environment through nature-based solutions: The contribution of vertical greening systems and green roofs. *Water, 13*(16), 2165. https://doi.org/10.3390/w13162165

Peng, Y., & Cui, Y. (2024). Thermal management with innovative fibers and textiles: Manipulating heat transport, storage and conversion. *National Science Review, 11*(10), nwae295. https://doi.org/10.1093/nsr/nwae295

Prihartanto, E., Rohman, M., & Wiguna, I. (2023). Assessment of airport conditions in resilience efforts: A review. *E3s Web of Conferences, 429*, 03001. https://doi.org/10.1051/e3sconf/202342903001

Procaccini, G., Prieto, A., Knaack, U., Monticelli, C., & Konstantinou, T. (2023). Textile membrane for façade retrofitting: Exploring fabric potentialities for the development of innovative strategies. *Buildings, 14*(1), 86.

Pvilion. (n.d.). Solar fabric technology. https://www.pvilion.com/

Rajpal, A., Choudhury, M., Goswami, S., Chakravorty, A., & Raghavan, V. (2024). *Waste management and treatment: Advances and innovations* (1st ed.). CRC Press. https://doi.org/10.1201/9781003258377

Rana, A. K., Mostafavi, E., Alsanie, W. F., Siwal, S. S., & Thakur, V. K. (2023). Cellulose-based materials for air purification: A review. *Industrial Crops and Products, 194*, 116331.

Reddy, V. J., Ghazali, M. F., & Kumarasamy, S. (2024). Innovations in phase change materials for diverse industrial applications: A comprehensive review. *Results in Chemistry, 8*, 101552.

Reyes, A., Reyes, A., & Daigle, C. (2020). Looking back to look forward: Evolution of the habitat agenda and prospects for implementation of the new urban agenda. *Current Urban Studies, 08*(02), 337–363. https://doi.org/10.4236/cus.2020.82019

Rojek, I., Mikołajewski, D., Mroziński, A., Macko, M., Bednarek, T., & Tyburek, K. (2025). Internet of things applications for energy management in buildings using artificial intelligence—A case study. *Energies, 18*(7), 1706.

Rosen, R. (2021). Can panel data methodologies determine the impact of climate change on economic growth? https://doi.org/10.36687/inetwp171

Sebastian, S. (2024). Modification of coir geotextiles for enhanced mechanical properties as a sustainable alternative. *Materials Circular Economy, 6*(1), 36.

Shafa, S. (2024). Smart materials in green architecture: The role of etfe and phase change materials in sustainable building design. *Journal of Design Studio, 6*(2), 383–395.

Sözcü, S., Venkataraman, M., Wiener, J., Tomková, B., Militký, J., & Mahmood, A. (2023). Incorporation of cellulose-based aerogels into textile structures. *Materials, 17*(1), 27. https://doi.org/10.3390/ma17010027

Subramanian, N. (2025). *A green living building: Case of the Bullitt Center*. Parametric Architecture. https://parametric-architecture.com/case-of-the-bullitt-center/

Tadesse, M., Abate, M., Lübben, J., & Rauch, M. (2025). Recycling and sustainable design for smart textiles – a review. *Advanced Sustainable Systems, 9*(8). https://doi.org/10.1002/adsu.202401072

Tan, B. A., Gaw, L. Y. F., Masoudi, M., & Richards, D. R. (2021). Nature-based solutions for urban sustainability: An ecosystem services assessment of plans for Singapore's first "forest town". *Frontiers in Environmental Science, 9*, 610155.

Tharim, A., Munir, F., Samad, M., & Mohd, T. (2018). A field investigation of thermal comfort parameters in green building index (gbi)-rated office buildings in Malaysia. *International Journal of Technology, 9*(8), 1588. https://doi.org/10.14716/ijtech.v9i8.2763

Timm, J. F. G., Maciel, V. G., & Passuello, A. (2023). Towards sustainable construction: A systematic review of circular economy strategies and ecodesign in the built environment. *Buildings, 13*(8), 2059.

Tinti, F., Tassinari, P., Rapti, D., & Benni, S. (2023). Development of a pilot borehole storage system of solar thermal energy: Modeling, design, and installation. *Sustainability, 15*(9), 7432. https://doi.org/10.3390/su15097432

Tiwari, A. (2021). Advanced materials research and innovation priorities for accomplishing the sustainable development goals. *Advanced Materials Letters, 12*(6), 1–6. https://doi.org/10.5185/amlett.2021.061633

Toderean, L., Cioara, T., Anghel, I., Sarmas, E., Michalakopoulos, V., & Marinakis, V. (2025). Demand response optimization for smart grid integrated buildings: Review of technology enablers landscape and innovation challenges. *Energy and Buildings, 326*, 115067.

U.S. Energy Codes and Building Performance Standards: Supporting Energy Use and Emissions Reductions in Buildings. International Code Council. (2023). *Guide to Energy Codes and Building Performance Standards.*

Valsalakumar, S., Roy, A., Mallick, T., Hinshelwood, J., & Sundaram, S. (2022). An overview of current printing technologies for large-scale perovskite solar cell development. *Energies, 16*(1), 190. https://doi.org/10.3390/en16010190

Wang, J., Ma, Y., Li, R., & Zhang, S. (2025). Applications of building information Modeling (BIM) and BIM-related Technologies for Sustainable Risk and Disaster Management in buildings: A meta-analysis (2014–2024). *Buildings, 15*(13), 2289.

Wang, Y., & Li, X. (2025). Energy optimization and low-carbon development of prefabricated buildings enabled by BIM technology. *Science and Technology for Energy Transition, 80*, 37.

Wang, Y., Zhong, K., Gao, Y., Sun, Z., Dong, R., & Wang, X. (2023). Feasibility analysis of storing solar energy in heterogeneous deep aquifer by hot water circulation: Insights from coupled hydro-thermo modeling. *Advances in Geo-Energy Research, 10*(3), 159–173. https://doi.org/10.46690/ager.2023.12.03

Wei. (2023). Recent advances in photocatalytic self-cleaning textiles. *Materials, 16*(1), 31.

Wei, W., Sun, Y., & Shim, E. (2023). *Progress of recycled polyester in rheological performance in molding, and economic analysis of recycled fibers in fashion and textile industry.* IntechOpen. https://doi.org/10.5772/intechopen.103864

Wired. (2014). *Deloitte's Amsterdam HQ senses staff movements to save energy.*

Woo, J., Rajagopalan, P., Francis, M., & Garnawat, P. (2021). An indoor environmental quality assessment of office spaces at an urban Australian university. *Building Research and Information, 49*(8), 842–858. https://doi.org/10.1080/09613218.2021.1944037

Wu, Q., Qin, K., Gan, M., Xu, J., Li, Z., & Li, Z. (2022). Recyclable biomass-derived polyethylene-like materials as functional coatings for commercial fabrics: Toward upcycling of waste textiles. *ACS Sustainable Chemistry & Engineering, 10*(51), 17187–17197. https://doi.org/10.1021/acssuschemeng.2c05080

Xie, Y., Tu, P., Xiao, Y., Li, X., Ren, M., Cai, Z., & Xu, B. (2023). Designing non-fluorinated superhydrophobic fabrics with durable stability and photocatalytic functionality. *ACS Applied Materials & Interfaces, 15*(33), 40011–40021. https://doi.org/10.1021/acsami.3c07352

Xie, Y., Tyler, M. T., Huckett, J. C., Bartlett, R., Chen, Y., Salcido, V. R., & Mendon, V. V. (2023). A review of the evaluation of building energy code compliance in the United States. *Energies, 16*(16), 5874. https://doi.org/10.3390/en16165874

Yitmen, I., Alizadehsalehi, S., Akıner, İ., & Akıner, M. E. (2021). An adapted model of cognitive digital twins for building lifecycle management. *Applied Sciences, 11*(9), 4276. https://doi.org/10.3390/app11094276

Zeng, J., & Long, T. (2021). Compressive behavior of frp grid-reinforced uhpc tubular columns. *Polymers, 14*(1), 125. https://doi.org/10.3390/polym14010125

Zhang, Z., & Tan, T. (2025). Advancing modular design generation: A review of modularization workflows and computational design approaches. *Journal of Building Engineering, 116*, 114646.

Zhu, W., Chow, L., Ye, D., Qiu, Y., Li, J., Zhang, B., & Yu, X. (2025). Advances in smart textiles for personal thermal management. *Med-X, 3*(1), 6.

Zou, Y., Xia, Y., & Yan, X. (2024). Effect of melamine formaldehyde resin encapsulated uv acrylic resin primer microcapsules on the properties of uv primer coating. *Polymers, 16*(16), 2308. https://doi.org/10.3390/polym16162308

CHAPTER 6

Analytics-Driven Urban Planning: Transforming Data into Actionable Insights

Mariana Flores-Garcia, María del Pilar Mercader, Elkin Vargas Lopez, and Sushobhan Majumdar

6.1 Introduction

The growing urbanization in cities in the Global South has triggered an imperative need to develop solutions to improve management processes in the face of realities characterized by environments where population growth is a key factor, as well as the conditions of sustainability in social, economic,

The original version of the chapter has been revised. A correction to this chapter can be found at https://doi.org/10.1007/978-3-032-18026-1_17

M. Flores-Garcia (✉)
Faculty of Architecture Universidad Salle Bajío, León, Mexico

M. d. Pilar Mercader
Research Group: Architecture: Design and Technique, University of Seville, Seville, Spain

E. V. Lopez
Director BuroDap, Bogotá, Colombia

S. Majumdar
Department of Geography, Pandit Raghunath Murmu Smriti Mahavidyalaya, Bankura, West Bengal, India

S. Majumdar, M. Choudhury (eds.), *City of the Future*,
https://doi.org/10.1007/978-3-032-18026-1_6

environmental, and cultural terms, generating conditions of equity and social justice in accordance with the Sustainable Development Goals and the New Urban Agenda (ONU HABITAT, 2021; CEPAL, 2017). The future of cities and the territory must consider going beyond sustainability, that is, recognizing the dynamic cycles of the habitat and its ecosystems, as well as the impacts generated by human activity that will be reflected in the short, medium, and long term, with discouraging scenarios, so it is essential to establish a change of thinking toward a new paradigm of circular and regenerative cities (Lares & Henríquez, 2021; Daniel Christian, 2020). This chapter refers to a documentary and exploratory research project that identified key aspects for the development of data-driven urban design and management tools. It is divided into five parts: the first presents a theoretical model for data-driven urban planning; the second identifies common problems associated with circularity that can be addressed using new data-driven approaches. The third section presents successful practices and lessons learned from 50 success stories in cities that adopt circular approaches. The results will be reflected in the fourth section, which offers "actionable insights" to establish methods for future research. The fifth section presents some conclusions and possible future lines of development. It is hoped that the following contribution will serve as a reference framework for guiding strategies and actions in a glocal context (Fig. 6.1).

From 2007 to 2025, citation patterns in the field of analytics-driven urban planning reflect the rapid evolution of data-intensive methods in city governance and spatial analysis. Early work (2007–2012) showed modest but growing scholarly attention as GIS, remote sensing, and early smart-city concepts began integrating with urban planning theory. Between 2013 and 2018, citations increased substantially, driven by the rise of big data, widespread adoption of sensor networks, and advances in machine learning for spatial modeling. Influential publications during this period focused on mobility analytics, real-time urban sensing, and the digital transformation of planning workflows. From 2019 onward, citation volume accelerated even further as research expanded into AI-driven decision-support systems, digital twins, participatory platforms, and resilience planning—particularly in response to global disruptions such as climate challenges and the COVID-19 pandemic, which highlighted the value of data-driven approaches for emergency response and adaptive governance (Choudhury et al., 2025a, b). By 2025, the field had become highly interdisciplinary, drawing citations from urban studies, computer science, environmental science, public policy, and engineering, demonstrating its maturation and centrality to contemporary urban innovation.

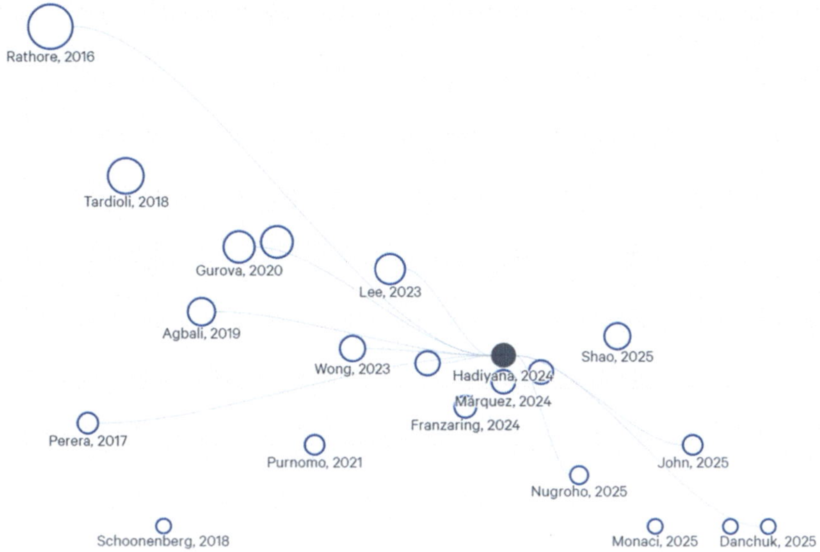

Fig. 6.1 Citation analysis of analytics-driven urban planning

6.1.1 Traditional to Data-Driven Urban Planning

Urban planning has historically relied on qualitative assessments, expert judgment, and periodic surveys to understand how cities function and evolve. While these traditional approaches offered valuable insights, they were often constrained by slow data collection, fragmented information, and an inability to capture real-time urban dynamics (Abdurakhmonov et al., 2024). As cities have grown more complex, characterized by rapid population shifts, expanding infrastructure, and diverse socio-economic patterns, the need for more responsive and comprehensive planning methods has become evident (Alshaikh et al., 2023). This shift has given rise to data-driven urban planning, which leverages digital technologies, sensor networks, geospatial systems, and large-scale datasets to provide a more accurate and dynamic understanding of urban environments. By integrating advanced analytical tools, planners can move beyond intuition-based decisions toward evidence-supported strategies that better reflect the real behaviors and needs of urban populations (Table 6.1).

Table 6.1 Overview of urban data types and analytical methods

Data type	*Primary source*	*Analytical methods used*	*Key insights generated*	*Typical urban applications*
Sensor & IoT data	Traffic sensors, air quality monitors, smart meters	Real-time analytics, anomaly detection, predictive modeling	Traffic flow, pollution levels, energy consumption patterns	Mobility management, environmental monitoring, utility optimization
GIS & Spatial Data	Satellite imagery, cadastral maps, land use records	Spatial analysis, geospatial modeling, heat maps	Land use patterns, accessibility levels, hazard zones	Zoning decisions, infrastructure planning, disaster risk reduction
Mobility data	Public transit systems, GPS devices, ride-sharing platforms	Network analysis, route optimization, demand forecasting	Travel behavior, congestion points, transit demand	Transportation planning, route design, congestion management
Administrative & Census Data	Government records, demographic surveys	Descriptive statistics, trend analysis, socio-economic modeling	Population profiles, housing trends, service coverage gaps	Housing policy, social services planning, resource allocation
Social Media & Crowdsourced Data	Social platforms, citizen apps, surveys	Sentiment analysis, text mining, event detection	Public opinion, activity hotspots, emergent issues	Community engagement, event management, public safety

6.1.2 Analytics in Modern City Management

Analytics now plays a central role in modern city management, enabling governments to make faster, more transparent, and more effective decisions. Through predictive modeling, machine learning, and spatial analytics, city authorities can anticipate future challenges such as congestion, environmental degradation, infrastructure failures, and resource shortages (Hammoumi et al., 2024). Real-time data enables cities to optimize public services across transportation systems, energy distribution, waste management, and emergency response, enhancing operational efficiency and the overall quality of life (Folorunsho et al., 2024). Moreover, analytics supports participatory governance by providing clearer visualizations and accessible insights that help citizens engage more actively in urban decision-making. In essence, analytics empowers cities to be more adaptive, resilient, and equitable in the face of rapid urbanization and global uncertainty.

6.2 The Foundations of Urban Analytics

6.2.1 Key Concepts

Urban analytics refers to the systematic use of data, computational tools, and analytical techniques to understand, model, and improve urban systems. It operates at the intersection of urban planning, data science, and geography, enabling planners to extract meaningful patterns from complex urban environments (Ye et al., 2025). Key concepts include spatial analysis, which examines how geographic factors influence urban processes; predictive analytics, which uses historical data to forecast future trends; and prescriptive analytics, which suggests optimal actions based on modeled scenarios. Together, these concepts form the backbone of evidence-based urban decision-making, helping cities transition from reactive planning to proactive, insight-driven strategies (Raghupathi and Raghupathi, 2021).

6.2.2 Types and Sources of Urban Data

Urban analytics draws on a wide array of data types and sources that capture different dimensions of city life. Sensor networks and IoT devices record real-time information on traffic flows, air quality, energy usage, and environmental conditions. Geographic Information Systems (GIS) provides spatial data layers essential for mapping land use, infrastructure, and demographic patterns (Mammadov et al., 2024). Mobility data from public transit systems, GPS devices, and ride-sharing platforms reveal movement trends and travel behaviors. Administrative records, such as census data, permits, and service logs, offer structured authoritative information on population, housing, and city services (Rahman et al., 2021). Additionally, social media and crowd sourced data provide insights into public sentiment, event detection, and human activity patterns, enriching traditional datasets with real-time human perspectives.

6.2.3 *Data Quality, Interoperability, and Integration Challenges*

Despite the richness of these data sources, cities face significant challenges with data quality, interoperability, and integration. Urban datasets often vary in format, resolution, accuracy, and temporal frequency, making it challenging to combine them meaningfully. Issues such as missing values, inconsistent spatial boundaries, and outdated records can hinder reliable analysis (Hu and Shu, 2023). Interoperability is another hurdle, as data generated by different agencies or proprietary systems may not seamlessly align. Integrating diverse data streams also requires robust governance frameworks, standardized protocols, and scalable technological infrastructures (Gebler et al., 2025). Overcoming these challenges is essential to unlocking the full potential of urban analytics and ensuring that insights are both accurate and actionable.

6.3 Urban Planning and Trends in Data-Driven Technologies for Circular Cities

Data-driven urban planning refers to trends in transforming the processes of managing and planning today's cities and territories, enhanced by the use and application of emerging technologies that concentrate data on social, environmental, economic, and cultural aspects of contemporary cities. From these, problems associated with new ways of living can be identified in conjunction with multidimensional analysis infrastructures for housing, mobility, transportation, energy, services, and the circular economy, among others, to promote actions that facilitate decision-making (Majumdar et al., 2025). Advanced technologies such as AI, algorithms in urban contexts, big data, and others have revolutionized how cities are built, so it is important to delve deeper into how these technologies are linked to one another to establish integrated, more efficient urban and territorial models that meet current planning needs. This leads us to establish new ideas, not only theoretical but also operational, where demographic, geospatial, and environmental data converge alongside increasingly complex but often disjointed information systems, leaving significant gaps in the city's comprehensive strategic planning (Bibri, 2021). Transforming data into actionable data involves processes linked to technologies such as urban sensors, open-access satellite imagery, and crowdsourcing, which can provide tangible examples. Platforms such as

UrbanFootPrint (UrbanFootPrint, 2025) and Oper Urban Planning Toolbox (IDB, 2025) allow the integration of real-time data, which is articulated with innovative methodologies adaptable to diverse contexts and undergoing continuous improvement, supported by case analysis and knowledge transfer across global contexts. The circular economy refers to a paradigm shift in our consumption patterns, toward more conscious and responsible consumption, aiming to recognize the damage done to the planet by anthropocentric activity. This implies, in a more profound sense, recognizing processes from the perspective of the life cycles of products, materials, services, homes, cities, and even people themselves, and the impacts on the planet from the extraction of raw materials to the generation of waste (Narayan et al., 2025). This is why the concept of regenerative is highlighted, focusing on going beyond sustainability (Canada East, 2023; World Habitat, 2022). Thus, concepts such as the 3 Rs are being raised: Reuse, Recycle, Reutilize (CEPAL, 2022), up to models such as the 5Rs within the framework of ICLEI climate neutrality (ICELI, 2025). The main strategies are as follows: Rethink: Redesign urban systems to facilitate circularity; Regenerate: Promote infrastructure and systems that restore ecosystems; Reduce: Minimize resource use and waste generation; Reuse: Extend the useful life of materials and products; and Recover: Maximize the recovery and reintegration of resources into the production system (Kowsalya et al., 2025).

Given the above, it is important to integrate aspects such as material flows, energy efficiency, and available data to optimize processes and technical cycles that are also connected to biological, social, and environmental processes. Given the regional diversity, realities such as those of cities in the Global South are identified, which represent important challenges in diverse scenarios, where realities address, in some cases, contexts of climate vulnerability, informality, and water scarcity, as well as transformations based on population conditions. To conduct an adequate analysis, it is important to identify the production and consumption systems by region, since they differ across countries and, in some cases, in very radical ways. This leads us to establish a starting point for circularity, technology, and resilient cities, which must also keep in mind, at all times, the ethical and equity conditions for the population, the planet, and their ways of life. For this reason, the study has focused on identifying innovative practices, especially in countries in the Global South, as shown in the image.1, shown below (Fig. 6.2).

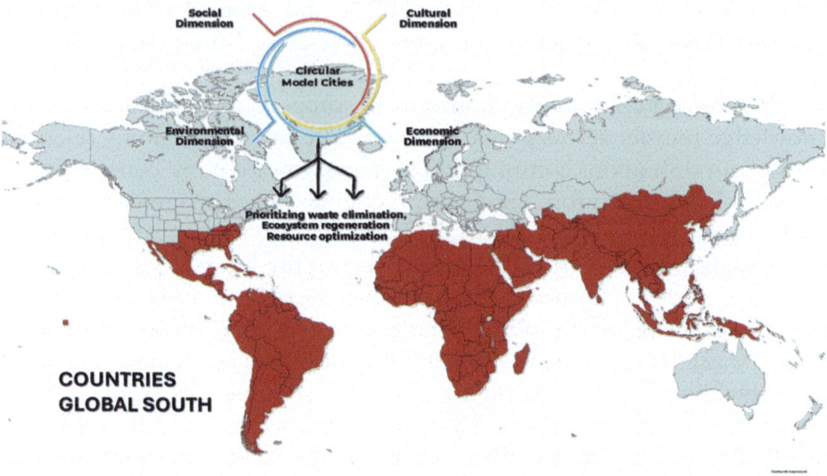

Fig. 6.2 Contextualization of the analysis of cases and problems in the countries of the Global South. (Prepared by authors)

Based on the research carried out, some of the main problems identified are listed:

1. **Institutional fragmentation and poor systemic efficiency.** Infrastructure investments fail to meet their objectives due to the lack of integrated environmental monitoring systems and isolated practices. In Nigeria, specifically Lagos, water is managed by 15 disjointed agencies, which accounts for 80% of undetected failures. In this sense, AI-based platforms using chatbots would centralize information and make management more efficient, not only in the process but also in resource use, even achieving considerable time savings, as in the case of CivicLens, which integrates real-time data on potential system failures in Nairobi (Civiclens, 2022).
2. **Little or no access to disruptive technologies.** A high concentration of startups in Smart Cities has been identified in India, particularly in cities such as Bangalore and Hyderabad, where they account for 80%, in contrast to other peripheral regions, where a digital divide of up to 47% persists. Therefore, it is important to link projects such as Code for South Africa (CfA), which seeks to democra-

tize open-access tools to strengthen the capacities of up to 540 urban planners in 12 countries, with examples such as the use of QGIS and AI for the design of inclusive urban biocorridors (Code, 2025).

3. **Integration approaches and emerging technologies**. Technologies such as AI, the Internet of Things (IoT), and big data offer promising solutions to address these challenges. One example is AI-powered urban planning, which leverages a comprehensive ecosystem of open-access databases to achieve urban sustainability through AI.

6.4 Analytical Frameworks for Urban Planning

6.4.1 *Descriptive, Diagnostic, Predictive, and Prescriptive Analytics*

Analytical frameworks for urban planning provide structured approaches for transforming raw urban data into meaningful insights that guide policy and design decisions. At the core of these frameworks are four complementary analytical types (Busciantella-Ricci et al., 2024). Descriptive analytics summarizes historical and real-time data to reveal existing trends, such as traffic congestion hotspots or patterns of land use change. Diagnostic analytics digs deeper to uncover the underlying causes of these trends, identifying factors that influence issues such as rising energy consumption and housing shortages (Lembhe, 2022). Predictive analytics uses statistical models and machine learning techniques to forecast future urban conditions, helping planners prepare for population growth, mobility demand, or climate-related risks 9 (Zarrin et al., 2024). Finally, prescriptive analytics recommends actionable solutions by evaluating potential interventions and their expected outcomes, supporting evidence-based decision-making across urban sectors.

6.4.2 *Spatial Analysis and Geospatial Modeling*

Analytical frameworks for urban planning provide structured approaches for transforming raw urban data into meaningful insights that guide policy and design decisions. At the core of these frameworks are four complementary analytical types (Harris et al., 2015). Descriptive analytics summarizes historical and real-time data to reveal existing trends, such as traffic congestion hotspots or patterns of land use change. Diagnostic analytics digs deeper to uncover the underlying causes of these trends, identifying factors that influence issues such as rising energy consumption and housing shortages (Cheng et al., 2023). Predictive analytics uses statistical models and machine learning techniques to forecast future urban conditions, helping planners prepare for population growth, mobility demand, or climate-related risks (Zarrin et al., 2024). Finally, prescriptive analytics recommends actionable solutions by evaluating potential interventions and their expected outcomes, supporting evidence-based decision-making across urban sectors.

6.4.3 *Machine Learning and AI Techniques in Urban Contexts*

Machine learning and artificial intelligence (AI) further enhance urban planning by enabling models that learn from large, diverse datasets. Applications range from classifying land cover using satellite imagery to predicting transit demand, detecting anomalies in utility networks, and analyzing social media sentiment for real-time event monitoring (Kazanskiy et al., 2025). AI-driven tools support adaptive planning by continuously updating predictions as new data becomes available.

6.4.4 *Scenario Modeling and Simulation Tools*

Scenario modeling and simulation tools enable planners to test different policy options and visualize their long-term effects. Through agent-based models, system dynamics, and digital twins, cities can simulate how changes in infrastructure, zoning, or transportation systems might affect mobility patterns, energy use, or environmental outcomes (Mehdizadeh et al., 2022). These tools help decision-makers evaluate trade-offs, mitigate risks, and design more resilient and sustainable urban futures.

6.5 Case Studies

In the transition to a new paradigm for information use, geospatial data becomes a structural axis for mapping infrastructure and population flows, leading us to identify enabling technologies powered by AI, cloud computing, 5G, machine learning, and simulations based on digital twins and predictive models. To transform the entire data ecosystem into urban actions, it is important to implement strategic management and planning, as shown in PwC's Smart Urban Planning Framework (PwC, 2025). Moreover, City OS of Barcelona (BarcelonaCat, 2025), these are integrated platforms and systems that manage the city more efficiently with the help of technology, and where various types of analysis are generated, such as cluster analysis, focused predictive models, and complex systems that are subject to trial and error conditions, always leading to a system of continuous improvement.

Based on the case analysis for this study, 50 initiatives in cities of the Global South have been analyzed, which are shown in the Table 6.2.

Based on the analysis, the information has been categorized into various types of technology and their applications, in chronological order. The results are structured into five categories and tools to guide the understanding of a possible integrated system, recognizing the type of technology used, uses, and applications: these are divided into Smart Monitoring, Resource Optimization and Management, Circular Traceability, Predictive Modeling, and Citizen Participation, which are described below.

6.5.1 Smart Monitoring and Resource Optimization and Management

Smart Monitoring includes tools that provide insight into the type of technology used. First, Lot Sensors are identified in waste containers (Choudhury et al., 2025a, b). These are installed devices that can measure fill levels to optimize collection routes. This can reduce operating costs by up to 20% or 30%, and CO_2 emissions by up to 15% annually, as is the case with big data and real-time smart recycling in Spain through SmartWaste (ElMundo, 2023). The use of Lot for waste management improved collection efficiency, reduced costs, sorted waste, and improved citizen participation (Uribe, 2023). Of the 50 cases, 20 used this type of technology. In drone-based aerial mapping of landfills in critical areas and informal dumps, critical points were identified with up to 90% greater accuracy, as demonstrated in India, where drones reduced identification and inspection time by up to 70% (Maganga, 2022). Open data platforms, for their part, enable visualization of waste flows correlated with energy consumption, providing greater transparency and up to 40% citizen participation, as in the case of Austin Materials Marketplace, which has recycled 1400 m^3 of materials since 2014 (BID, 2022; Ellen Macarthur, 2014). RFID tagging optimizes the tracking of goods and products, facilitating their recovery for post-consumption or new uses. It is possible to reduce losses of

Table 6.2 Summary of projects, organized by technology categories, for urban and economic management in developing regions in 2025

Project & description	*Year*	*Country*	*Technology/ Tool used*
1. Biorefinery in Xiangyang to convert sludge and organic waste into bioenergy.95–98% reduction in CO_2 emissions.	2019	China	IoT + big data
2. Biogas buses in Kolkata powered by animal and plant waste.12 times cheaper transport fares and reduced emissions.	2019	India	IoT
3. Online platform in Cape Town for business-to-business waste sharing.Cost savings and reduced pressure on landfills.	2019	Sudáfrica	Big data
4. Air quality sensors and public Wi-Fi in Dhaka.18% improvement in respiratory health.	2021	Bangladesh	IoT
5. Sensors on garbage trucks in Guadalajara to optimize routes.15% reduction in CO_2 emissions from logistics.	2021	México	IoT
6. Water management system in Quezon City with sensors to detect leaks.25% savings in water consumption.	2022	Filipinas	IoT
7. Tracking recycled plastics in Hanoi using blockchain.Supply chain transparency and a 40% increase in recycling.	2022	Vietnam	Blockchain
8. Event management in Kigali through data analysis in public spaces.35% increase in tourism.	2022	Ruanda	Big data
9. Digital certification of recycled materials in Medellín.90% traceability in recycling chains.	2022	Colombia	Blockchain
10. Sustainable mobility platform in Curitiba to integrate public transport.22% increase in non-motorized transport use.	2022	Brazil	Big data
11. Smart irrigation system in Da Nang for urban green spaces.35% reduction in water use.	2022	Vietnam	IoT
12. 3D modeling in Cali for resilient urban planning. Flood mitigation in 12 neighborhoods.	2022	Colombia	Big data
13. Sensors in Surabaya landfills to prevent fires. 90% decrease in garbage burning incidents.	2022	Indonesia	IoT
14. Collaborative economy platform in Pretoria for car sharing.20% reduction in transportation emissions.	2022	Sudáfrica	Big data
15. Analysis of consumption patterns in Ho Chi Minh City to reduce waste.15% reduction in food waste.	2022	Vietnam	Big data

(*continued*)

Table 6.2 (continued)

Project & description	*Year*	*Country*	*Technology/ Tool used*
16. Reverse logistics platform in Mumbai for packaging reuse.25% reduction in packaging costs.	2022	India	Big data
17. Chatbots to report failures in public infrastructure in Cape Town.50% faster response to repairs.	2022	Sudáfrica	IA
18. Solar energy credit registration in Recife.25% increase in solar panel adoption.	2022	Brazil	Blockchain
19. Secondary materials trading platform in Hanoi.35% increase in industrial reuse.	2022	Vietnam	Big data
20. Circular economy network in Buenos Aires to map material flows.Creation of 1200 green jobs and reduction of industrial waste.	2023	Argentina	Big data
21. Processing organic waste in Accra using anaerobic digesters.Biogas production for 10,000 homes.	2023	Ghana	IoT
22. Open traffic platform in Mexico City to optimize vehicle flows.20% reduction in traffic congestion.	2023	México	Big data
23. Predictive models for waste management in Jakarta.30% reduction in untreated waste.	2023	Indonesia	IA
24. Data analysis to optimize waste collection routes in Johannesburg.25% reduction in logistics costs.	2023	Sudáfrica	Big data
25. Chatbots in Buenos Aires for citizen education on the circular economy.60% of the population participates in recycling programs.	2023	Argentina	IA
26. Water quality monitoring in Kumasi using connected sensors.30% reduction in waterborne diseases.	2023	Ghana	IoT
27. Tokenization of carbon credits in reforestation projects in Durban.Capture of 50,000 tons of CO_2 annually.	2023	Sudáfrica	Blockchain
28. Smart e-waste management in Chennai. Recycling of 70% of toxic components.	2023	India	IoT
29. Immutable record of recyclable material transactions in Córdoba.50% increase in investor confidence.	2023	Argentina	Blockchain
30. Diagnostics of urban crops in Tamale using satellite images.40% increase in agricultural productivity.	2023	Ghana	IA

(*continued*)

Table 6.2 (continued)

Project & description	*Year*	*Country*	*Technology/ Tool used*
31. Predictive management of electrical grids in Brasilia.12% reduction in power failures.	2023	Brazil	IA + IoT
32. Optimization of recycling routes in Barranquilla with genetic algorithms.20% fuel savings.	2023	Colombia	IA
33. Mapping of informal recycling areas in Rosario using GIS.Formalization of 35% of recyclers	2023	Argentina	Big data
34. Water demand prediction in León using machine learning models.18% reduction in water waste.	2023	México	IA + big data
35. Road infrastructure diagnosis in Kumasi using computer vision.25% reduction in road accidents.	2023	Ghana	IA
36. Platform to certify circular products in Pereira.40% increase in sales of recycled goods.	2023	Colombia	IA + Blockchain
37. Sensors in Bogotá for monitoring recycling and solid waste management.30% increase in recycling rates.	2024	Colombia	IoT + IA
38. Urban Futurability in São Paulo: Smart grids for renewable energy.15% reduction in electricity consumption.	2024	Brazil	IA + big data
39. Smart public lighting system in Shanghai with automatic adjustment.40% energy savings.	2024	China	IA + IoT
40. Urban agriculture in Bangalore with automated irrigation and soil monitoring.50% increase in local food production.	2024	India	IoT + IA
41. Energy demand prediction in Manila using machine learning.10% savings in electricity consumption	2024	Filipinas	IA + big data
42. Infrastructure diagnosis in Rio de Janeiro using drones and images.Identification of 200 critical points for maintenance.	2024	Brazil	IA
43. Traffic simulations in Monterrey for road redesign.18% reduction in travel times.	2024	México	IA + big data
44. Digital twin of Shenzhen for urban policy simulation.25% optimization in resource allocation.	2024	China	IA + IoT
45. Water quality monitoring in Cebu rivers with real-time sensors.50% improvement in water quality.	2024	Filipinas	IoT
46. Certification of sustainable buildings in Puebla using blockchain.30% increase in eco-friendly construction.	2024	México	Blockchain

(*continued*)

Table 6.2 (continued)

Project & description	*Year*	*Country*	*Technology/ Tool used*
47. Flood early warning system in Jakarta.40% reduction in property damage from floods.	2024	Indonesia	IoT + big data
48. Sensor network in Hangzhou for urban noise monitoring.20% reduction in noise pollution.	2024	China	IoT
49. Smart street lighting management in Hyderabad.30% energy savings.	2024	India	IoT
50. Air quality monitoring in Davao with sensors and predictive analytics.20% improvement in respiratory health indicators.	2024	Filipinas	IoT + big data

Source: Compiled by authors

valuable materials by up to 25%. This has been proven in some European countries with RFID tags, with a recovery rate of at least 30% in textiles, and in some Latin American countries (Ellen Macarthur, 2014). Sewer Sensor Networks has opted to integrate real-time leak and contaminant detection technology, resulting in at least 30% savings in treated water, as demonstrated in Kolkata, where the sensor system reduced water losses by 25% (Fundación Bankinter, 2022). Air quality monitoring systems are integrated into urban settings to measure CO_2 emissions and particulate matter. This can significantly reduce respiratory illnesses by 15% through contingency areas or preventive programs, which can also be enhanced with other infrastructure in specific microenvironments. In São Paulo, pollution decreased by 12% after the implementation of IoT networks (Fundación Bankinter, 2022). Smart weighing systems can automatically record waste from treatment plants, improving classification by up to 20% as in Cape Town, where an additional 500 tons/day are processed (Fundación Bankinter, 2022). Smart weighing systems can automatically record waste from treatment plants, improving classification by up to 20% as in Cape Town, where an additional 500 tons/day are processed (Fundación Bankinter, 2022). Along with the previous one, sensors on collection vehicles help optimize routes using traffic and fueling data, with an 18% reduction in fuel costs and time. An example of this is Smart Waste in Spain, with annual savings of €1.2 million (Fundación Bankinter, 2022). For its part, the socialization of the use of satellite images contributes significantly to detecting changes and deterioration in soils and deforested areas, allowing the development of preventive and action plans, such as the restoration of 100 hectares in Brazil, as Amazonas recorded 50% fewer degraded areas in 2022 (Maganga, 2022).

6.5.2 Resource Optimization and Management

In recent years, strengthening processes through the use of digital twins and infrastructure and services has shown considerable improvements in monitoring systems based on virtual replicas that simulate distributions, leaks, and demands. This can reduce water losses by up to 30%. This is the case in Lushan (China), where 25,000 tons of water per day were saved (Weforum, 2024; CEIT, 2025). Predictive models applied to energy efficiency improvements have enabled dynamic patterns in past solar and wind power generation, enabling energy efficiency gains of up to 25%. In South Africa, solar plants improved their output by 18% (CEIT, 2025). Smart irrigation systems use humidity and weather sensors to boost urban agriculture initiatives, saving up to 30% of water for crops. Pilot projects in India have reduced consumption by 50% (Fundación Bankinter, 2022). Simulations of green infrastructure support better planning for parks, equipment, and vegetated infrastructure, with respect to their impacts on water and CO_2 collection systems. This allows for a 20% increase in green areas per decade, as in Medellín, where 30 green corridors were created by 2023, and has also become a priority strategy in several countries (CEPAL, 2023). Intelligent street lighting management has evolved with adaptive LED lighting, including motion sensors that can reduce electricity consumption by up to 60%. In Santiago, Chile, €3 million was saved annually (Fundación Bankinter, 2022). Predictive maintenance platforms, meanwhile, use AI to anticipate potential failures in infrastructure such as bridges and networks, thereby reducing repair costs by up to 25%. In Rio de Janeiro, 30% of leaks were prevented by 2024 (CEIT, 2025). Water harvesting systems integrated automated rainwater collection and filtration systems, which can provide up to 15% of the water supply in urban areas during droughts. One example is Cape Town, where 10,000 m^3/year were stored (Fundación Bankinter, 2022). Organic waste management with Lot is based on an automated composting system and quality monitoring, which can reduce the amount of waste sent to landfills by 50%. In Brazil, plants with IoT technology process 200 tons per day (Uribe, 2023; Ameijde, 2023; Batty, 2021; Bisello et al., 2021; Elias Bibri

& Krogstie, 2017; Kant Singh et al., 2024; Kumar & Bassill, 2024; PwC Middle East's, 2024).

6.5.3 *Circular Traceability*

Blockchain for textiles creates a record of the origin, composition, and life cycle of garments, increasing fiber reuse by 30% and aligning with TRICK (EU)'s plan to reduce 92 million tons of textile waste (Hinojosa Pérez, 2025; CIT-UPC, 2022). Service tokenization proposes a blockchain-based credit system for recycling activities, increasing citizen participation by 25%. In São Paulo, 2 million tokens will be issued by 2024 (Hinojosa Pérez, 2025). The traceability of metals like gold and silver, using a retrieval mechanism in recycled electronics, allows for the recovery of more gold than in mines. Some projects in Ghana have recovered 1 ton of gold per year (Fundación Bankinter, 2022). Smart contracts for reverse logistics are automated agreements between companies and recyclers that reduce administrative costs by up to 20%. In Mexico, 500,000 shipments per year were optimized (Lnjatrans, 2024). Blockchain in construction enables the recording of recycled materials in buildings, contributing up to 15% to the certification of sustainable buildings. In Chile, 30 projects use blockchain (CEPAL, 2023). The food supply chain, which uses sensors and blockchain, enables waste reduction, reducing crop losses by up to 35%. In Argentina, 1000 tons of food were saved annually (SAP España, 2023). Digital Product Passports, which generate QR codes with repair and reuse history, increase secondhand sales by 50%. In Brazil, 2 million products are sold with a passport (CEPAL, 2023). Automated audits powered by AI enable on-the-spot verification of compliance with circular standards, reducing environmental fraud by up to 30%. In Peru, 100 cases were detected in 2024 (Hinojosa Pérez, 2025).

6.5.4 *Predictive Modeling*

Using AI for collection routes, they develop algorithms that prioritize areas with the highest waste generation, optimizing routes and reducing the kilometers traveled by up to 25%. In Spain, 500 routes were optimized daily (ElMundo, 2023). Climate scenario simulations enable predictive assessments of potential impacts on water and energy infrastructure, which influence mitigation actions for up to 40% of flood damage. In Bangladesh, €10 million in losses were avoided (Weforum, 2024).

6.5.5 *Citizen Participation*

Regarding citizen participation, in all the previous cases it has been an important factor of impact not only in the interaction with the technology itself but in the construction of the data ecosystem for better city management, where applications can be identified to share products and services linked to living value chains, reward systems for recycling, crowdfunding platforms for projects, initiatives based on gamified environmental education games, strengthening of community social networks, citizen reporting platforms, participatory voting systems based on AI, collaborative resource maps, virtual training programs, and corporate volunteering platforms.

In addition to the above, common patterns, methodological tools, and scalable strategies are identified to transform cities into hubs of climate resilience and collective prosperity. Some of the innovative technologies mentioned above include multidimensional circular indicator systems, digital twins for scenario simulation, blockchain and circular material processes, IoT sensors and 5G networks in waste management, smart governance and public policies for circular economies, technology transfer and South-South collaborative networks, community initiatives and citizen participation, co-financing and sustainable business models, and technology-based startup accelerators, among others.

To articulate the information obtained previously, a Comprehensive Strategic Plan for the Urban Circular Economy with Advanced Technology is proposed, from which the various tools, strategies, and technologies for its implementation can be identified. The conceptual approach of the integrated Strategic Plan is shown in the image. 6.3 (Fig. 6.3).

The implementation proposal is described in four phases:

- **Phase 1**. An initial assessment is recommended for the target cities or areas to identify opportunities, limitations, and available resources. Based on this, a conceptual framework should be established to identify the main challenges and available technological infrastructure, to identify potential synergies that can be leveraged by technologies based on AI, the Metaverse, big data, or predictive models, as appropriate or based on specific needs from a circular perspective—phase **2.** The technological context within which strategies and actions will be developed is determined. At this point, possibilities for implementing or improving processes using digital twins, the Metaverse, or predictive AI are identified.

Fig. 6.3 Comprehensive strategic plan for urban circular economy with advanced technology, 2025. (Source: Prepared by authors)

- **Phase 3.** Based on the initial assessments in phases 1 and 2, a roadmap is proposed that outlines actions across three areas. The first refers to practices or pilot projects aimed at unifying existing platforms and tools and integrating compatible or complementary databases. As shown in the cases analyzed, Innovation Labs, Lot technology, and Metaverse enable the exploration of digital environments that facilitate the construction of these convergences, even in hybrid formats. The second category includes successful practices that offer the potential for replicability and even scalability, validated in initial pilot projects. Here, sectoral digital twins can be established alongside instruments such as the blockchain-based materials passport. The third category of this phase contributes to the sustainability of the 2030 roadmap by promoting evaluation, follow-up, and monitoring actions that support the collaborative economy 4.0 and quantum Lot in waste, as well as an operational metaverse.

- **Phase 4.** The above must permeate an integrated governance system that incorporates metrics and management mechanisms, starting with a hybrid governance model, composed of physical infrastructure and a digital ecosystem, based on multi-sector actions between technical consultants and researchers, the public sector, the business sector, and academia, who also form sectoral committees in energy, water, waste, and mobility, among others, always accompanied by citizen observatories where the community participates permanently through gamified applications. The governance ecosystem must be supported by circular systems with key performance indicators (KPIs) to measure and evaluate the performance of all of the above in the practical transition to a circular economy. The roadmap for the transition to circularity could integrate Participatory Diagnosis processes through the Mapping of material flows with citizen sensors and co-creative workshops. Pilot projects could be developed around Rapid Prototyping, leveraging digital twins to test interventions in pilot districts. Enabling policies could be established, including legal frameworks that incentivize circular business models through tax exemptions, and adaptive replication practices based on open knowledge platforms to customize solutions in new contexts. This requires a continuous monitoring system, including dashboards powered by IoT and blockchain to verify impacts in real time.

6.6 Turning Data into Actionable Insights

6.6.1 *Decision-Support Systems and Dashboards*

Turning data into actionable insights is a critical step in ensuring that urban analytics contributes meaningfully to city planning and management. Decision-support systems (DSS) and dashboards play a central role in this process by transforming complex datasets into intuitive visualizations, performance indicators, and interactive maps (Rachansa & Meditya, 2024). These tools help planners, policymakers, and stakeholders quickly interpret trends, assess system performance, and compare alternative scenarios. Dashboards equipped with real-time data streams enable timely responses to emerging issues such as sudden traffic congestion or deteriorating air quality. At the same time, DSS platforms integrate predictive models and simulation outputs to support long-term strategic planning (Stojanović & Stojanović, 2020). By making analytical results accessible and easy to understand, these technologies promote transparency and informed decision-making across government departments.

6.6.2 *Translating Analytics into Urban Policies and Strategic Plans*

Translating analytics into urban policies and strategic plans involves bridging the gap between technical findings and practical interventions. Planners must contextualize data insights within political, social, and economic realities to craft policies that are both evidence-based and feasible (Wickenden et al., 2025). This includes defining priority areas, allocating resources, and setting measurable targets informed by analytical outcomes. Effective translation also requires stakeholder engagement, where visualizations and data narratives help communicate the rationale for proposed changes to communities, decision-makers, and partner organizations (Tomkins & Lange, 2023). By embedding analytics into policy frameworks, cities can create adaptive planning mechanisms that respond to evolving conditions and ensure long-term sustainability.

6.7 Governance, Ethics, and Citizen Engagement

6.7.1 Data Governance and Regulatory Frameworks

Governance, ethics, and citizen engagement form the backbone of responsible and sustainable data-driven urban planning. Effective data governance and regulatory frameworks ensure that data collected from diverse urban systems is managed securely, consistently, and transparently (Lérias et al., 2024). These frameworks define standards for data collection, storage, sharing, and use across government departments and external partners. They establish guidelines for interoperability, ensuring that datasets from transportation, health, environment, and public services can be integrated for comprehensive analysis (Boffetta et al., 2011). Regulatory mechanisms also clarify roles and responsibilities, helping cities manage risks, maintain accountability, and build institutional capacity. Strong governance systems create a stable foundation upon which analytical tools can function effectively and support long-term planning.

6.7.2 Privacy, Transparency, and Ethical Considerations

Privacy, transparency, and ethical considerations are essential to maintaining public trust in data-driven urban initiatives. As urban analytics often relies on personal or location-based data, cities must safeguard individual privacy through anonymization, secure data practices, and compliance with relevant legal frameworks (Lottu et al., 2024). Transparency in how data is collected, analyzed, and applied allows citizens to understand the rationale behind planning decisions and ensures that analytics does not reinforce bias or discrimination. Ethical use of AI and machine learning requires careful attention to fairness, accuracy, and inclusivity, ensuring that models do not disproportionately impact vulnerable communities. By prioritizing ethics, cities can foster equitable and just outcomes (Ramoni et al., 2025).

6.8 Participatory Planning Enhanced by Analytics

Participatory planning enhanced by analytics empowers residents to play an active role in shaping their communities. Digital platforms, interactive dashboards, and geospatial visualization tools make complex data more accessible, enabling citizens to understand local issues better and provide informed feedback (Kock, 2025). Crowdsourcing, mobile apps, and online surveys enable planners to gather real-time community input, which can be integrated with analytical models to yield richer insights. By combining technical analysis with lived experiences, participatory approaches produce more inclusive, context-sensitive urban strategies (Lybeck, 2018). Ultimately, analytics-driven engagement strengthens democratic decision-making and helps build cities that reflect the diverse needs and aspirations of their residents.

6.9 Challenges and Future Directions

6.9.1 Technical, Organizational, and Socio-political Barriers

Despite the growing adoption of urban analytics, cities continue to face significant obstacles in effectively leveraging data-driven approaches. Technical barriers include fragmented data systems, legacy infrastructure, limited computational capacity, and uneven data quality that hinder integration and advanced analysis (Roy et al., 2025). Organizational challenges arise from siloed government departments, insufficient data literacy among staff, and resistance to change, all of which impede the institutionalization of analytics in everyday planning workflows. At a broader level, socio-political barriers such as unequal digital access, concerns over surveillance, and a lack of public trust can limit community acceptance of data-driven initiatives. These issues highlight the need for comprehensive capacity building, ethical governance structures, and strategic coordination across agencies and stakeholders (Jackson et al., 2019).

6.9.2 Emerging Technologies

Rapid technological advancements present new opportunities for transforming urban analytics. Edge computing and IoT ecosystems enable real-time monitoring of infrastructure, mobility, and environmental conditions. Artificial intelligence and deep learning expand capabilities in pattern recognition, predictive modeling, and automated decision-making (Joel & Oguanobi, 2024). Digital twins, virtual replicas of cities, allow planners to simulate complex systems with high precision and test interventions before implementation. Additionally, blockchain technologies provide secure, transparent mechanisms for data sharing, while 5G networks enable high-speed communication between connected devices. These emerging tools promise to enhance accuracy, responsiveness, and scalability in urban planning processes (Joel & Oguanobi, 2021).

6.9.3 Opportunities for more Resilient, Equitable, and Adaptive Cities

By strategically embracing advanced analytics, cities can evolve into more resilient, equitable, and adaptive environments. Data-driven approaches support climate resilience through improved disaster forecasting, risk mapping, and infrastructure stress modeling (Gatla, 2019). Equity can be strengthened by using analytics to identify underserved areas, guide resource allocation, and monitor disparities in service delivery. Adaptive planning becomes more achievable as real-time insights enable cities to adjust policies quickly in response to shifting conditions such as population changes, economic disruptions, or environmental crises (Mutambik, 2025). Ultimately, integrating analytics with inclusive governance and emerging technologies positions cities to build sustainable futures that proactively respond to the needs of all residents.

6.10 Conclusion

Based on the previous analysis, some technological challenges can be identified that relate to fragmented infrastructure and obsolete management systems. Their starting point is a paradigm shift toward circularity and regeneration, which also entails a cross-cutting component in the dilemma of technology, connectivity, and digitalization. Some of the main challenges by sector stem from the lack of sustainable financing systems, as financial resources are limited. Without investment in infrastructure, it is common to encounter barriers when seeking to implement nonexistent hybrid models. Regarding diverse realities, scenarios of exclusion of informal actors in technological solutions are commonly identified, such as the case of the startup Coliba, which includes 2500 informal recyclers through machine learning algorithms. However, there are policies, such as those in Amsterdam or Glasgow, that prioritize competitiveness over inclusion, perpetuating inequalities. This is where gaps are identified in inconsistent metrics and unreliable data. Some of the challenges, by type of impact, relate to the recognition of asymmetries in technical capabilities and governance, such as the lack of technological adaptation to local contexts. Thus, a promising yet challenging outlook for the future is emerging, in which research must be conducted on low-cost technologies and distributed energy, whose impact, from a circularity perspective, must permeate even the level of circular public policy proposals. This also poses a significant challenge for the education and training of professionals, citizens, and communities toward the circular and regenerative transition, fostering multi-stakeholder alliances for multi-level, multi-center circular governance. This path requires adopting a new level of social and environmental awareness and responsibility, where models and programs that integrate holistic well-being metrics must be considered, starting with redefining traditional urban management processes beyond GDP, prioritizing the regeneration of current and future cities and territories.

References

Abdurakhmonov, S., Khamidova, M., Romanyuk, Y., Sattorov, S., Adizov, S., Karimov, E., & Бозоров, M. (2024). Conventional and current approaches of urban mapping and geodetic base formulation for establishing demographic processes database: Tashkent, Uzbekistan. *E3s Web of Conferences, 497*, 02028. https://doi.org/10.1051/e3sconf/202449702028

Alshaikh, R., Abdulmunem, S., & Alkinani, A. (2023). A review on urban planning and its role in managing flood risks. *upc, 1*(1), 1–8. https://doi.org/10.55121/upc.v1i1.104

Ameijde, J. (2023). Data-driven urban design: Conceptual and methodological constructs for people-oriented public spaces. In *Disruptive technologies: The convergence of new paradigms in architecture*. Springer International Publishing. https://doi.org/10.1007/978-3-031-14160-7_5

BarcelonaCat. (2025). City Os Barcelona. Barcelona Ciudad Digital. Obtenido de https://ajuntament.barcelona.cat/digital/es/hagamos-accesible-la-tecnologia/tecnologia-accesible-y-participativa/tecnologia-accesible-y-1

Batty, M. (2021). The digital transformation of planning. *Environment and Planning B: Urban Analytics and City Science*. https://doi.org/10.1177/23998083211016122

Bibri, S. (2021). Data-driven smart sustainable cities of the future: Urban computing and intelligence for strategic, short-term, and joined-up planning. *Computational Urban Science, 1*, 8. https://doi.org/10.1007/s43762-021-00008-9

BID. (2022). Accelerador de Políticas y Proyectos de Economía Circular en América Latina y el Caribe. Proyectos Regionales. Obtenido de https://www.iadb.org/es/proyecto/RG-T4079

Bisello, A., Vettorato, D., Haarstad, H., & Judith, B.-V. (2021). *Smart and sustainable planning for cities and regions. Results of SSPCR 2019*. Springer International Publishing. https://doi.org/10.1007/978-3-030-57764-3

Boffetta, P., Colditz, G., Potter, J., Kolonel, L., Robson, P., Malekzadeh, R., & Kelemen, L. (2011). Cohorts and consortia conference: A summary report (Banff, Canada, June 17–19, 2009). *Cancer Causes & Control, 22*(3), 463–468. https://doi.org/10.1007/s10552-010-9717-0

Busciantella-Ricci, D., Macchi, A., Viviani, S., & Rinaldi, A. (2024). Healthy and inclusive neighbourhoods: A design research toolkit for the promotion of healthy behaviours. *Sustainability, 16*(7), 3059. https://doi.org/10.3390/su16073059

Canada East. (2023). Regenerative design: Moving beyond net zero buildings. Obtenido de https://aeecanadaeast.org/2023/05/16/regenerative-design-moving-beyond-net-zero-buildings/

CEIT. (2025). Gemelos digitales para la economía circular. Obtenido de https://www.ceit.es/gemelos-digitales-para-la-economia-circular

CEPAL. (2017). Nueva Agenda Urbana. Obtenido de Plataforma Urbana y de Ciudades de America LAtina y el Caribe. https://plataformaurbana.cepal.org/es/new-urban-agenda

CEPAL. (2022). Estrategia de economía circular para Chile: Soluciones habitacionales sostenibles. Obtenido de https://www.cepal.org/es/publicaciones/48573-estrategia-economia-circular-chile-soluciones-habitacionales-sostenibles

CEPAL. (2023). The Circularity Gap Report: América Latina y el Caribe. (C. E. (CEPAL), Ed.) *Repositorio digital CEPAL*. Obtenido de https://repositorio.cepal.org/entities/publication/67111cb9-3d5f-4813-931b-b90a95b28149

Cheng, W., Luo, H., Long, Z., Xu, G., & Tian, L. (2023). Statistical modeling of water shortage in water distribution systems in Guangzhou. *Water, 15*(18), 3257. https://doi.org/10.3390/w15183257

Choudhury, M., Dixit, G., & Majumdar, S. (Eds.). (2025a). *Sustainable synergy: Harnessing ecosystems for climate resilience.* Springer Nature Switzerland. https://doi.org/10.1007/978-3-031-77957-2

Choudhury, M., Suhag, M., Jha, P., Harne, K. R., Samanta, P., Choudhury, M., Borgohain, R. B., Majumdar, S., Kravchenko, Y., & Choudhary, R. (2025b). Urban waste management and challenges in emerging smart city systems governance and circular economy perspectives. *Ecological Engineering & Environmental Technology, 26*(10), 243–266. https://doi.org/10.12912/27197050/211014

CIT-UPC. (2022). Tecnología blockchain para impulsar la economía circular en el sector textil. *Instituto de Investigación Textil y Cooperación Industrial de Terrassa de la Universitat Politècnica de Catalunya (UPC).* Obtenido de https://cit.upc.edu/es/portfolio-item/tecnologia-blockchain-per-impulsar-leconomia-circular-en-el-sector-textil/

Civiclens. (2022). Backup utility for ArcGIS online and portal for ArcGIS. *Civilens.* Obtenido de https://www.civiclens.com/backups.

Code. (2025). Code for South Africa. Obtenido de https://code4sa.org/

Daniel Christian, D. (2020). *Design de Culturas Regenerativas* (Bambual Editora LTDA ed.). Obtenido de https://books.google.com.mx/books?hl=es&lr=&id=hwtIEAAAQBAJ&oi=fnd&pg=PT5&dq=%22culturas+regenerativas%22+daniel+wahl&ots=HIQEMt8tNg&sig=0ohW0FffenrqLNOILbbQKxuznLw#v=onepage&q=%22culturas%20regenerativas%22%20daniel%20wahl&f=false

Elias Bibri, S., & Krogstie, J. (2017). Smart sustainable cities of the future: An extensive interdisciplinary literature review. *Sustainable Cities and Society.* https://doi.org/10.1016/j.scs.2017.02.016

Ellen Macarthur. (2014). Austin materials marketplace foundation. Obtenido de https://www.ellenmacarthurfoundation.org/circular-examples/developing-the-materials-marketplace-austin

ElMundo (2023). *Big data y reciclaje inteligente para impulsar la economía circular.* Obtenido de https://www.elmundo.es/ciencia-y-salud/medio-ambiente/2023/03/17/640b40affdddfff38c8b45a5.html

Folorunsho, S., Adenekan, O., Ezeigweneme, C., Somadina, I., & Okeleke, P. (2024). Developing smart cities with telecommunications: Building connected and sustainable urban environments. *Engineering Science & Technology Journal, 5*(8), 2492–2519. https://doi.org/10.51594/estj.v5i8.1441

Fundación Bankinter. (2022). Internet of waste: IoT para reciclar y recuperar desperdicios. Obtenido de https://www.fundacionbankinter.org/noticias/internet-of-waste/?_adin=11551547647

Gatla, T. (2019). A cutting-edge research on AI combating climate change: Innovations and its impacts. *International Journal of Innovations in Engineeri ng Research and Technology, 6*(9), 1–8. https://doi.org/10.26662/ijiert.v11i3.pp1-8

Gebler, R., Reinecke, I., Sedlmayr, M., & Goldammer, M. (2025). Enhancing clinical data infrastructure for AI research: Comparative evaluation of data management architectures. *Journal of Medical Internet Research, 27*, e74976. https://doi.org/10.2196/74976

Hammoumi, L., Maanan, M., & Rhinane, H. (2024). Characterizing smart cities based on artificial intelligence. *Smart Cities, 7*(3), 1330–1345. https://doi.org/10.3390/smartcities7030056

Harris, P., Friel, S., & Wilson, A. (2015). 'Including health in systems responsible for urban planning': A realist policy analysis research programme. *BMJ Open, 5*(7), e008822. https://doi.org/10.1136/bmjopen-2015-008822

Hinojosa Pérez, A. (2025). Blockchain y la Cadena de Suministro con enfoque de Sostenibilidad Económica de las Industrias Textiles de Alpaca, Arequipa, Perú. https://doi.org/10.31637/epsir-2025-1003.

Hu, L., & Shu, Y. (2023). Enhancing decision-making with data science in the internet of things environments. *International Journal of Advanced Computer Science and Applications, 14*(9). https://doi.org/10.14569/ijacsa.2023.01409120

ICELI. (2025). The ICLEI climate neutrality framework. *The ICLEI Climate Neutrality Framework, ities and regions on the path to net zero emissions and climate resilience.* Obtenido de https://iclei.org/climate_neutrality/

IDB. (2025). Digital appliations for urban planning. Oper Urban Planning Toolbox. Obtenido de https://www.iadb.org/en/who-we-are/topics/urban-development-and-housing/digital-applications-urban-planning

Jackson, C., Gardy, J., Shadiloo, H., & Silva, D. (2019). Trust and the ethical challenges in the use of whole genome sequencing for tuberculosis surveillance: A qualitative study of stakeholder perspectives. *BMC Medical Ethics, 20*(1). https://doi.org/10.1186/s12910-019-0380-z

Joel, O., & Oguanobi, V. (2024). Navigating business transformation and strategic decision-making in multinational energy corporations with geodata. *International Journal of Applied Research in Social Sciences, 6*(5), 801–818. https://doi.org/10.51594/ijarss.v6i5.1103

Kant Singh, K., Rho, S., Singh, A., & Sergei, C. (2024). Urban analytics: Integrating big data and machine learning. *Complex & Intelligent Systems, 10*, 1–2. https://doi.org/10.1007/s40747-023-01050-2

Kazanskiy, N., Khabibullin, R., Никоноров, A., & Khonina, S. (2025). A comprehensive review of remote sensing and artificial intelligence integration: Advances, applications, and challenges. *Sensors, 25*(19), 5965. https://doi.org/10.3390/s25195965

Kock, C. (2025). Citizen participation enabled by technology in urban planning. https://doi.org/10.21203/rs.3.rs-7751298/v1

Kowsalya, P., Jennifer Flora, G., Kousalya, A., Toha, M. A., Khan, P. A., Choudhury, M., Saiyyeda, F., & Parthasarathi, T. (2025). Circular systems unveiled: Essential components and their impact on recycling, reuse, and resource recovery. In S. Majumdar, M. Choudhury, & A. Cheshmehzangi (Eds.), *Sustainable urban future. Urban Sustainability* (pp. 55–71). Springer. https://doi.org/10.1007/978-981-95-2976-6_3

Kumar, D., & Bassill, N. P. (2024). Analysing trends of computational urban science and data science approaches for sustainable development. *Computational Urban Science, 4*, 33. https://doi.org/10.1007/s43762-024-00142-0

Lares, L., & Henríquez, A. (2021). Diseño Regenerativo y Economía Circular. *Cuadernos del Centro de Estudios en Diseño y Comunicación. Ensayos. Ejemplar dedicado a: Diseño y territorio social: Pasajeros en Tránsito*. Obtenido de https://dialnet.unirioja.es/servlet/articulo?codigo=8371845

Lembhe, P. (2022). Data analytics and predictive modeling for toll violation prevention. *Design of Single Chip Microcomputer Control System for Stepping Motor*, 1–7. https://doi.org/10.47363/jaicc/2022(1)246

Lérias, E., Guerra, C., & Ferreira, P. (2024). Literacy in artificial intelligence as a challenge for teaching in higher education: A case study at Portalegre polytechnic university. *Information, 15*(4), 205. https://doi.org/10.3390/info15040205

Lnjatrans. (2024). Economía circular en la logística inversa. Obtenido de https://lanjatrans.com/economia-circular-en-la-logistica-inversa/

Lottu, O., Jacks, B., Ajala, O., & Okafor, E. (2024). Enhancing data management and security protocols in financial sector projects. *World Journal of Advanced Research and Reviews, 21*(3), 2655–2666. https://doi.org/10.30574/wjarr.2024.21.3.0735

Lybeck, R. (2018). Mobile participation in urban planning; exploring a typology of engagement. *Planning Practice and Research, 33*(5), 523–539. https://doi.org/10.1080/02697459.2018.1534579

Maganga, M. (2022). https://secmotic.com/casos-uso-iot-gestion-residuos/. Archdaily. Obtenido de https://www.archdaily.mx/mx/986758/que-hacemos-con-los-residuos-la-economia-circular-y-el-sur-global

Majumdar, S., Choudhury, M., & Cheshmehzangi, A. (Eds.). (2025). *Sustainable urban future: Embracing circular economy in waste management*. Springer Singapore. https://doi.org/10.1007/978-981-95-2976-6

Mammadov, E., Asgarov, A., & Mammadova, A. (2024). Applications of IoT in civil engineering: From smart cities to smart infrastructure. *LUMIN, 1*(1), 13–28. https://doi.org/10.69760/lumin.202400003

Mehdizadeh, M., Nordfjærn, T., & Klöckner, C. (2022). A systematic review of the agent-based modelling/simulation paradigm in mobility transition. *SSRN Electronic Journal*. https://doi.org/10.2139/ssrn.4006249

Mutambik, I. (2025). IoT-enabled adaptive traffic management: A multiagent framework for urban mobility optimisation. *Sensors, 25*(13), 4126. https://doi.org/10.3390/s25134126

Narayan, S., Yadav, D., Jha, S., Khatun, B., Choudhary, R., Verma, U. K., Choudhury, M., & Kumar, D. (2025). From past to present: The evolution of waste management and the rise of circular economy principles. In *Sustainable urban future: Embracing circular economy in waste management* (pp. 23–53). Springer Nature Singapore. https://doi.org/10.1007/978-981-95-2976-6_2

ONU HABITAT. (2021). La Nueva Agenda Urbana en español NAU. HABITAT III. Obtenido de onuhabitat.org.mx/index.php/la-nueva-agenda-urbana-en-espanol

PwC. (2025). Smart urban planning framework de PwC. Obtenido de https://www.pwc.com/gx/en/industries/government-public-sector/smart-cities.html

PwC Middle East's. (2024). Data and AI-driven urban data and AI-driven urban-Data and AI-driven urban. *PwC Middle East's Smart Urban Planning Framework*. Obtenido de https://www.pwc.com/m1/en/publications/documents/2024/data-and-ai-driven-urban-planning.pdf

Rachansa, K., & Meditya, W. (2024). Integrated multi-income stream performance dashboard: A Japanese corporate banking case. *International Journal of Advances in data and Information systems, 5*(1), 12–28. https://doi.org/10.59395/ijadis.v5i1.1313

Raghupathi, W., & Raghupathi, V. (2021). Contemporary business analytics: An overview. *Data, 6*(8), 86. https://doi.org/10.3390/data6080086

Rahman, M., Paul, K., Hossain, M., Ali, G., Rahman, M., & Thill, J. (2021). Machine learning on the covid-19 pandemic, human mobility, and air quality: A review. https://doi.org/10.20944/preprints202103.0396.v1

Ramoni, D., Scuricini, A., Carbone, F., Liberale, L., & Montecucco, F. (2025). Artificial intelligence in gastroenterology: Ethical and diagnostic challenges in clinical practice. *World Journal of Gastroenterology, 31*(10). https://doi.org/10.3748/wjg.v31.i10.102725

Roy, S., Pal, P., Saha, S., Behera, N., & Lahiri, S. (2025). Bridging academia and industry: A comprehensive review of advances, gaps, and future directions of fault detection and diagnosis (fdd) systems in the chemical industry. *The Canadian Journal of Chemical Engineering, 103*(10), 4718–4750. https://doi.org/10.1002/cjce.25701

SAP España. (2023). Logística inversa y economía circular: redefiniendo la sostenibilidad en la cadena de suministro. *SAP Spain Centro de Noticias*. Obtenido de https://news.sap.com/spain/2023/10/logistica-inversa-y-economia-circular-redefiniendo-la-sostenibilidad-en-la-cadena-de-suministro/

Stojanović, N., & Stojanović, D. (2020). Big mobility data analytics for traffic monitoring and control. *Facta Universitatis Series Automatic Control and Robotics, 19*(2), 087. https://doi.org/10.22190/fuacr2002087s

Tomkins, A., & Lange, E. (2023). Planning and designing natural and urban environments with an adaptive visualization framework: The case of Pazhou Island, Guangzhou, Pearl river delta. *Land, 12*(2), 377. https://doi.org/10.3390/land12020377

UrbanFootPrint. (2025). Resilience insights. Obtenido de https://urban-footprint.com/

Uribe, I. (2023). Casos de uso IoT en la gestión de residuos y reciclaje. *Secmotic Smart Inside.* Obtenido de https://secmotic.com/casos-uso-iot-gestion-residuos/

Weforum. (2024). Cómo los gemelos digitales están transformando la gestión del agua. Obtenido de https://es.weforum.org/stories/2024/11/como-los-gemelos-digitales-estan-transformando-la-gestion-del-agua/

Wickenden, A., Dean, L., Molyneux, S., Godwin-Akpan, T., Kollie, K., Phillip, M., Zaizay, ZF, Rogers, E., van der Grinten, E., Nana-Kwadwo, B, Theobald, S. (2025). 'We can not just keep it in our palm': A policy analysis of the integration of the case management of neglected tropical diseases into the health system of liberia. https://doi.org/10.1101/2025.02.06.25321795.

World Habitat. (2022). Urban regeneration: Social value creation. Obtenido de https://urbanoctober.unhabitat.org/sites/default/files/2022-10/whd-2022-roundtable-4-concept-note-2_0.pdf

Ye, X., Yiğitcanlar, T., Goodchild, M., Huang, X., Li, W., Shaw, S., Fu, Y., Gong, W., & Newman, G. (2025). Artificial intelligence in urban science: Why does it matter? *Annals of GIS, 31*(2), 181–189. https://doi.org/10.1080/19475683.2025.2469110

Zarrin, Z., Hamidi, O., Amini, P., & Maryanaji, Z. (2024). Predicting the pulse of urban water demand: A machine learning approach to deciphering meteorological influences. *BMC Research Notes, 17*(1). https://doi.org/10.1186/s13104-024-06878-6

CHAPTER 7

Innovations in Public Service: Leveraging Smart Technologies for Community Well-Being

Xinge Zhang, Yuanru Gao, and Sushobhan Majumdar

7.1 Introduction

7.1.1 Smart Technologies in Public Service

The rapid advancement of smart technologies has initiated a paradigm shift in the way public services are designed, managed, and delivered. Traditionally, public service delivery has been characterized by bureaucratic processes, centralized decision-making, and limited citizen interaction (Latupeirissa et al., 2024; Ghosh et al., 2024; Majumdar et al., 2024). However, the integration of digital innovations such as artificial intelligence (AI), the Internet of Things (IoT), big data analytics, blockchain,

X. Zhang (✉)
Weitzman School of Design, University of Pennsylvania, Philadelphia, PA, USA

Y. Gao
Faculty of the Built Environment, University College London, London, UK

S. Majumdar
Department of Geography, Pandit Raghunath Murmu Smriti Mahavidyalaya, Bankura, West Bengal, India

S. Majumdar, M. Choudhury (eds.), *City of the Future*,
https://doi.org/10.1007/978-3-032-18026-1_7

and mobile applications has transformed these conventional systems into more agile, transparent, and citizen-centric models (Acharya et al., 2024). Smart technologies enable governments to collect, analyze, and utilize real-time data to make informed decisions and anticipate community needs more effectively. For example, IoT-based sensors can monitor traffic, waste management, and environmental conditions, enabling timely interventions and efficient resource use (Hadian, 2023). Similarly, AI-powered platforms can streamline administrative processes, automate routine tasks, and enhance the accuracy of service delivery. Furthermore, digital platforms and mobile applications have improved accessibility and engagement, empowering citizens to participate directly in governance processes through feedback systems, e-services, and online consultations (Manita et al., 2023). This shift from reactive to proactive governance promotes accountability and strengthens public trust. Despite these advancements, challenges such as digital inequality, data privacy, and the need for regulatory frameworks remain critical considerations (Pratomo et al., 2024; Santibanez et al., 2023). Therefore, harnessing the potential of smart technologies in public service requires a balanced approach, one that integrates innovation with inclusivity, ethical governance, and sustainable development principles. Ultimately, smart technologies are not just tools for efficiency but catalysts for transforming public institutions into adaptive, responsive, and citizen-focused entities.

7.1.2 Community Well-Being Through Innovation

Innovation in public service plays a crucial role in enhancing community well-being by addressing social challenges with creative, technology-driven, and inclusive solutions (Nijnik et al., 2023). By adopting smart technologies, public institutions can move beyond traditional service delivery models to create systems that are more responsive, personalized, and sustainable (Wu et al., 2025). These innovations enable communities to experience improvements in health, safety, education, mobility, and environmental quality, core dimensions of overall well-being (Rasweswe et al., 2025; Majumdar & Sivaramakrishnan, 2025). For instance, digital health platforms and telemedicine services have expanded access to healthcare, particularly in remote or underserved areas. At the same time, data analytics helps detect health trends early and allocate medical resources efficiently. Similarly, smart transportation systems contribute to safer and more efficient mobility, reducing congestion and pollution. In the context of education, e-learning and digital literacy programs empower individuals with the skills required to thrive in an increasingly digital society (Maharani & Rahmawati, 2023). Moreover, innovations that foster citizen participation, such as open data initiatives and digital feedback mechanisms, enhance transparency and trust between governments and communities. They also create opportunities for co-creation, where citizens contribute directly to policy development and service design (Colovic et al., 2025). Ultimately, the transformative potential of innovation lies in its ability to integrate technology with human-centered values, promoting equity, inclusivity, and resilience. By aligning technological progress with social objectives, public service innovation becomes a driving force for holistic community well-being and sustainable societal advancement (Afzal et al., 2024).

7.1.3 *Smart Cities and Digital Governance*

The concept of "Smart Cities" represents a forward-thinking approach to urban development that integrates technology, data, and innovation to enhance the quality of life for citizens and improve the efficiency of urban management (Mohamed, 2023). At the core of this transformation lies "digital governance," the strategic use of digital tools, and data-driven systems to facilitate transparent, participatory, and efficient public administration. Together, smart cities and digital governance redefine how governments interact with citizens, manage resources, and address complex urban challenges. Smart city initiatives rely on interconnected technologies such as the Internet of Things (IoT), artificial intelligence (AI), cloud computing, and big data analytics to collect and analyze real-time data across sectors such as transportation, energy, waste management, and public safety (Adenekan et al., 2024). These technologies enable predictive decision-making, optimized service delivery, and rapid responses to emerging issues. For example, smart traffic management systems can reduce congestion and pollution, while intelligent energy grids enhance sustainability and cost-efficiency (Almeida et al., 2024). Digital governance complements these innovations by promoting open data policies, e-governance platforms, and digital participation tools that empower citizens to engage in policymaking and community improvement. Through digital governance, governments can foster greater accountability, inclusiveness, and responsiveness (Ologunebi & Taiwo, 2025). However, the development of smart cities also raises challenges related to data privacy, cyber security, and digital inequality. Addressing these concerns requires robust legal frameworks, ethical standards, and inclusive design strategies that ensure technology benefits all citizens equally.

7.1.4 Citizen-Centered Services

In the digital era, public expectations of government performance have evolved significantly, driven by rapid technological advancement, increased access to information, and the growing influence of private-sector service models (Lui & Avelino, 2025). Citizens now demand public services that are not only efficient and reliable but also transparent, accessible, and tailored to their individual needs. This shift represents a move away from traditional bureaucratic structures toward more agile, data-driven, and participatory governance (Nnenna et al., 2024). Efficiency has become a central priority, with citizens expecting faster service delivery, simplified procedures, and reduced administrative burdens. Smart technologies such as automation, AI-driven decision-making, and online service platforms are helping governments streamline processes and improve responsiveness (Al-Ansi et al., 2024). For instance, digital portals for tax payments, licensing, and welfare programs allow citizens to access services seamlessly, reducing time and operational costs. Transparency is equally critical in building trust and accountability. Open data initiatives and digital reporting tools provide citizens with real-time insights into government operations, expenditures, and policy outcomes (Lindquist & Huse, 2017). By making information accessible, governments can foster public confidence and encourage civic oversight. At the same time, the focus on citizen-centered services underscores the importance of inclusivity and engagement. Modern governance models increasingly incorporate citizen feedback, participatory budgeting, and service co-creation, ensuring that policies reflect real community needs. Ultimately, the shift toward efficient, transparent, and citizen-centered services redefines the relationship between governments and the public, transforming citizens from passive recipients into active partners in governance and community development (Guo & Zhang, 2024).

Citation analysis of research on "public service" from "1991 to 2024" (Fig. 7.1) reveals a significant evolution in both thematic focus and scholarly influence. Early studies in the 1990s primarily emphasized administrative efficiency, public management reforms, and the shift from traditional bureaucratic models to New Public Management (NPM). During the 2000s, citation patterns reflected growing academic interest in governance, accountability, and citizen-centric service delivery, with

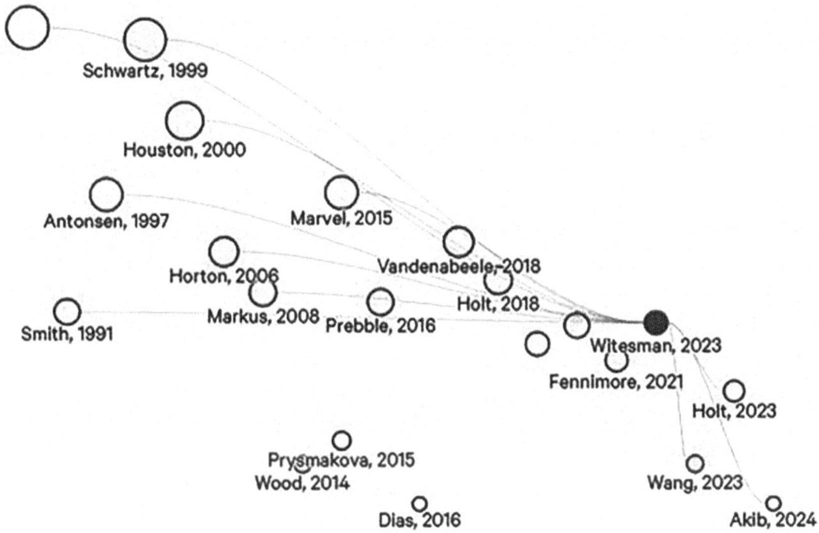

Fig. 7.1 Citation analysis of public service

cross-disciplinary inputs from economics, sociology, and political science. From 2010 onward, the surge in citations corresponded to the expansion of digital governance, e-government, and sustainable public service innovation, indicating a paradigm shift toward technology-enabled and participatory models. The period from 2020 to 2024 shows heightened research activity addressing public service resilience, inclusivity, and ethical governance, especially in response to global challenges such as the COVID-19 pandemic and climate change. Overall, citation trends suggest that the field of public service has transitioned from a managerial to a multidimensional and sustainability-oriented domain, with increasing global collaboration and impact across academic and policy circles.

7.2 Defining Smart Technologies in Public Service

7.2.1 Overview of Smart Technologies

"Smart technologies" describe a connected ecosystem of digital tools that combine sensing, automation, and data-driven intelligence to transform service delivery and public-sector governance. They anchor the broader shift toward smart cities and digital government emphasizing efficiency, transparency, and citizen-centricity. Table 7.1 showing smart technologies with their applications in public service.

Networks of sensors, actuators, and connected devices instrument physical infrastructure for real-time monitoring of systems such as energy, transport, and utilities. Connectivity, interoperability, and continuous sensing enable fine-grained situational awareness and faster operational decisions. Beyond automation, AI supports predictive and prescriptive analytics across functions like resource forecasting, fraud detection, and natural-language interfaces that expand access to services. Adaptability, pattern recognition, and predictive accuracy help governments move from reactive to anticipatory operations. The explosion of administrative, sensor, and earth-observation data supports evidence-based policymaking, predictive governance, and early warning mechanisms. Hallmarks include high volume, velocity, variety, and veracity the "four Vs" which allow analysis at scale and discovery of latent correlations. Distributed ledgers can enhance trust, auditability, and tamper-evidence in specific public records and transactions. Attributes include immutability, transparency, and decentralization useful where integrity and non-repudiation are paramount. Cloud infrastructure provides elastic, scalable, and cost-efficient capacity especially valuable for resource-constrained administrations while enabling shared services and interdepartmental data integration. Together,

Table 7.1 Key smart technologies and their applications in public service

Technology	*Public service applications*	*Benefits to community well-being*
Internet of Things (IoT)	Smart traffic management, waste monitoring, environmental sensing	Reduced congestion, cleaner environments, real-time services
Artificial intelligence (AI)	Predictive policing, resource optimization, chatbots	Efficiency, personalized services, enhanced safety
Big data analytics	Policy planning, health surveillance, disaster response	Informed decisions, early warnings, better allocation
Cloud computing	e-governance portals, digital records, remote services	Improved access, scalability, reduced costs
Blockchain	Digital identity, land registry, transparent procurement	Trust, transparency, reduced fraud
Robotics & automation	Administrative automation, public facility maintenance	Speed, accuracy, minimized human error

Source: Compiled by the authors

Fig. 7.2 Medical education complex as a smart learning hub for advancing community health services

these technologies push governance away from siloed hierarchies toward flexible, networked systems capable of adaptive management in complex environments. Figure 7.2 shows medical education complex in China.

7.2.2 Public Sector Applications

Digital health ecosystems combine IoT devices for continuous monitoring, AI-assisted triage and diagnostics, and big-data epidemiology for outbreak modeling and resource allocation. India's eSanjeevani telemedicine platform scaled affordable remote consultations nationwide, expanding access in underserved areas. Rwanda's national drone logistics shorten emergency blood-delivery times and reduce waste, improving time-critical care in rural settings. Cloud-based learning management and AI-driven adaptive tutoring expand access and personalization. IoT-enabled "smart campus" systems optimize learning environments and energy use. In low-connectivity contexts, platforms such as Kenya's M-Shule leverage SMS and AI to deliver personalized learning without Internet access. Intelligent

Transport Systems (ITS) integrate IoT sensing, AI optimization, and cloud platforms to manage congestion, emissions, and safety. Researchers and pilot projects have also explored blockchain for trusted ticketing and settlement in multimodal systems. Hangzhou's City Brain applies AI to signal control and network management, with reports of congestion reductions of around 15%. Agencies increasingly rely on geospatial analytics, satellite imagery, and AI-based prediction to anticipate hazards, allocate resources, and communicate with the public in near real time. Japan's earthquake early warning system integrates dense seismic sensing with rapid analytics to issue life-saving alerts within seconds. Automated eligibility checks using administrative and alternative data, digital-ID-enabled transfers, and AI-assisted case management streamline processes and improve equity. Estonia's e-governance stack secures sensitive data and enables real-time benefits delivery; X-Road provides secure data exchange. Separate deployments of KSI blockchain add tamper-evidence for critical records.

7.2.3 Key Characteristics

Smart technologies in public service are defined by several key characteristics that collectively enhance efficiency, transparency, and citizen-centered governance. These technologies are data-driven, leveraging real-time data collection, analytics, and artificial intelligence to inform decision-making and improve service delivery (Chatterjee et al. 2023). They are interconnected through digital platforms and the Internet of Things (IoT), enabling seamless communication across systems and stakeholders. Smart technologies also emphasize adaptability and responsiveness, allowing public services to anticipate and address community needs dynamically. Citizen engagement is central, with participatory tools such as mobile apps and digital platforms fostering co-creation and feedback (Majumdar, 2020).

Furthermore, they prioritize sustainability and inclusivity, using digital innovations to optimize resource use and ensure equitable access to services. Altogether, these characteristics make smart technologies pivotal in transforming public administration into a more proactive, transparent, and people-oriented system that advances community well-being.

Table 7.2 shows smart public service innovations with their outcomes.

Table 7.2 Smart public service innovations and their outcomes

Innovation area	*Public-sector impact*	*Community outcomes*
Smart health services	Greater service reach, reduced hospital burden	Improved access to care, lower costs
Smart governance	Faster service delivery, efficient bureaucracy	Convenience, reduced wait times
Smart safety & security	Faster response times, proactive policing	Increased trust, safer communities
Smart education	Expanded learning options, real-time monitoring	Equity in education, improved outcomes
Smart mobility	Reduced congestion, better transit reliability	Cleaner air, time saved
Smart environment	Data-driven sustainability policies	Healthier living conditions

Source: Compiled by the authors

7.3 Drivers of Innovation in Public Service

Innovation emerges from a convergence of technological, institutional, social, and global forces that collectively shape public-sector transformation.

7.3.1 Technological Advancements

Rapid advances have lowered adoption barriers by improving accessibility, interoperability, and affordability. Cloud services enable even small municipalities to deliver at scale without major capital outlays. Estonia's X-Road secures data exchange across agencies and sectors separately, Estonia has deployed KSI blockchain to provide integrity guarantees for critical records. In Singapore's Smart Nation program, AI-driven analytics and IoT systems such as smart lighting illustrate how mature, off-the-shelf capabilities can optimize urban systems. Public–private innovation labs and open-source collaborations further reduce costs and de-risk experimentation, helping smaller jurisdictions benefit from frontier technologies. Figure 7.3 shows the Community self-service EV charging station in China.

Fig. 7.3 Community self-service EV charging station: a smart mobility solution for sustainable public service

7.3.2 Policy and Governance Reforms

Institutional frameworks enable technology to scale. India's Digital India program, anchored by Aadhaar, extends secure digital identities to over a billion residents, supporting direct benefit transfers and reducing leakage in welfare delivery. The European Union's Digital Europe Programme (2021–2027) dedicates multiyear funding to AI, cybersecurity, and advanced digital skills, signaling durable commitment to capability-building. Brazil's Digital Governance Strategy consolidated services into a single gov.br portal, reducing fragmentation and improving access. Open-data policies in the UK and EU continue to catalyze civic innovation via dashboards and third-party applications built on public datasets. Overall, reforms are shifting administrations from procedural rigidity toward adaptive, innovation-oriented governance.

7.3.3 *Community Expectations and Demands*

Residents increasingly expect public services to match the usability and responsiveness of private digital platforms. New York City's 311 enables real-time reporting, status tracking, and analysis of service requests supporting transparency and operational responsiveness. Barcelona's *Decidim* institutionalizes digital participation (proposals, deliberation, and voting) with strong auditability; blockchain has been explored in pilots for additional traceability. In Seoul, *mVoting* lets residents weigh in on local issues from transport to environmental policy via mobile participation at city scale. These examples underscore communities as co-creators of public value, driving co-design, crowdsourcing, and co-production models.

7.3.4 *Global Challenges*

Systemic shocks and megatrends accelerate innovation. Climate change has prompted predictive modeling and IoT-based water monitoring, for example, the Netherlands' *Room for the River* program to manage flood risk. Pandemics highlighted the need for scalable digital health: Taiwan's integrated datasets enabled rapid contact tracing; Germany operationalized real-time ICU dashboards during COVID-19. Rapid urbanization is fueling large-scale smart-city pilots; Hangzhou's *City Brain* applies AI to network traffic management, with reports of ~15% congestion reductions. Rwanda's *Kigali Innovation City* illustrates how digital infrastructure and smart mobility are being integrated into long-term urban development strategies. These pressures expose vulnerabilities but also generate political will for resilient, inclusive, tech-enabled governance. Table 7.3 shows challenges and opportunities in public services.

Table 7.3 Challenges and opportunities in implementing smart public services

Category	*Challenges*	*Opportunities*	*Implications for policymakers*
Technology	Legacy systems, integration issues	Modernization, interoperability	Need for standards and digital infrastructure
Governance	Limited technical capacity, slow bureaucracy	Data-driven governance models	Invest in skills, encourage innovation culture
Funding	High initial costs	Long-term savings, partnerships	Explore PPPs, phased investment
Social & ethical	Privacy concerns, digital divide	Inclusive design, community engagement	Develop safeguards, promote digital literacy
Legal & regulatory	Lack of updated policies	Regulatory sandboxes, flexible frameworks	Update laws to support tech adoption
Implementation	Resistance to change, training needs	Workforce transformation	Continuous training and change management

Source: Compiled by the authors

7.4 Smart Technologies and Community Well-Being

Smart technologies not only increase operational efficiency but also reshape equity, participation, and sustainability outcomes. They redefine how citizens interact with public institutions and how collective well-being is maintained in digital societies.

7.4.1 Improving Access and Equity

Smart platforms can reduce geographic and socioeconomic barriers to essential services. India's *eSanjeevani* surpassed 100 million remote consultations by 2023, expanding affordable access for rural communities. Rwanda's Zipline drones have dramatically shortened delivery times for blood and vaccines and reduced wastage, improving the reliability of emergency care in remote clinics. In education, Kenya's *M-Shule* pairs SMS with AI to deliver personalized learning without Internet access, supporting students in low-resource settings. Big-data tools further enable targeting (e.g., poverty mapping), aligning delivery with Sustainable

Fig. 7.4 Smart agriculture base: integrating digital technologies to enhance food security and community well-being

Development Goal objectives. Persistent challenges affordability, connectivity, and digital literacy underscore the need to close the digital divide. Figure 7.4 shows smart agriculture base in China.

7.4.2 Enhancing Service Delivery

Embedding intelligence into workflows makes services faster, more adaptive, and proactive. AI chatbots such as the UK's *Ask NHS* help triage inquiries and schedule care around the clock, easing provider workloads. *In* social protection, Estonia's e-welfare system secures data and supports timely benefits disbursement; X-Road underpins secure data exchange, and KSI blockchain deployments provide tamper-evidence where required. Brazil's *Bolsa Família* leverages digital ID and analytics to improve targeting and reduce fraud. These shifts illustrate a transition from reactive to anticipatory models that better predict demand and allocate resources while raising important expectations for transparency and accountability in algorithmic decision-making. Table 7.4 shows metrics for the smart public service initiatives.

Table 7.4 Metrics for evaluating smart public service initiatives

Evaluation dimension	*Key metrics*	*Purpose*
Efficiency	Service processing time, cost savings, automation rate	Assess improvements in operational performance
Accessibility & inclusion	Digital access rates, user satisfaction, equity indicators	Evaluate reach across diverse populations
Service quality	Response time, accuracy of services, service uptime	Determine reliability and quality of digital public services
Community well-being	Health indicators, safety rates, environmental quality	Measure direct social outcomes of smart initiatives
Engagement & trust	Citizen participation rates, transparency scores	Assess civic involvement and institutional trust
Sustainability	Energy consumption, carbon reduction, resource efficiency	Track environmental impacts of smart services
Economic impact	Job creation, public–private partnerships, innovation index	Evaluate contribution to local economic growth

Source: Compiled by the authors

7.4.3 *Strengthening Public Participation*

Digital participation platforms are moving governance toward co-creation. Barcelona's *Decidim* has engaged large resident cohorts in budgeting and policy debates, with open auditing and data export to foster trust; blockchain has been piloted for traceability in some contexts. Seoul's *mVoting* enables mobile input on transportation, environmental, and safety issues. Taiwan's *vTaiwan* applies AI-assisted clustering of public input to help officials identify consensus on complex technology policies. Such tools enhance responsiveness and inclusivity while reinforcing the personalization and integration capacities highlighted earlier. Figure 7.5 shows a vehicle charging station in China.

7.4.4 *Safety and Sustainability*

Smart technologies contribute to immediate safety and long-term sustainability. Japan's Earthquake Early Warning integrates dense IoT seismic

Fig. 7.5 Smart vehicle charging station: advancing sustainable mobility in public service infrastructure

sensing with rapid analytics to issue alerts within seconds. The Netherlands' water management programs use real-time sensors and predictive models to reduce flood risk. In transport, Hangzhou's *City Brain* has been reported to reduce congestion by ~15% while lowering emissions. *Dubai's* Smart Grid Strategy couples AI-based demand forecasting with distributed solar integration to improve energy efficiency and resilience. Collectively, these approaches advance resilient, low-carbon development aligned with community well-being and the Sustainable Development Goals. Figure 7.6 shows transportation facilities in China.

Fig. 7.6 Real-time bus stop and route facilities: enhancing public mobility through smart transit technologies

7.5 Innovative Case Studies

The following four case studies demonstrate how smart technologies are being applied in practice across four key public domains: healthcare, education, transport, and governance. These examples illustrate how technological innovation impacts efficiency, accessibility, and citizen engagement.

7.5.1 Smart Health Initiatives

The COVID-19 lockdowns accelerated the digital transformation of healthcare systems worldwide. Technologies such as the Internet of Things (IoT), big data, cloud computing, and artificial intelligence (AI) collectively form the "smart healthcare technology system," enabling real-time data exchange between patients and clinicians while enhancing system resilience (Majumdar, 2021a, Majumdar, 2021b). The Telemedicine Practice Guidelines (March 2020) established a legal framework for virtual consultations conducted by registered physicians, facilitating the integrated application of AI-assisted diagnostic tools. Figure 7.7 shows medical facilities in China.

Fig. 7.7 Hospital self-service machine: streamlining appointments, inquiries, and billing through smart automation

India's most significant challenge during the pandemic was the uneven distribution of resources between urban and rural areas. With nearly 75% of healthcare professionals concentrated in urban centers, rural regions faced significant service gaps. Embedding AI in remote diagnostic workflows mitigated this imbalance, allowing geographically isolated communities to access specialist care. During the pandemic, AI-supported telediagnosis programs benefited over 11.4 million individuals, expanding access to quality healthcare in underdeveloped regions. Figure 7.8 shows healthcare facilities in China.

An example from Rwanda in 2019 demonstrated a complementary model integrating logistics with last-mile delivery. To overcome the dual challenges of complex terrain and underdeveloped infrastructure, the government partnered with the California-based firm Zipline to deploy

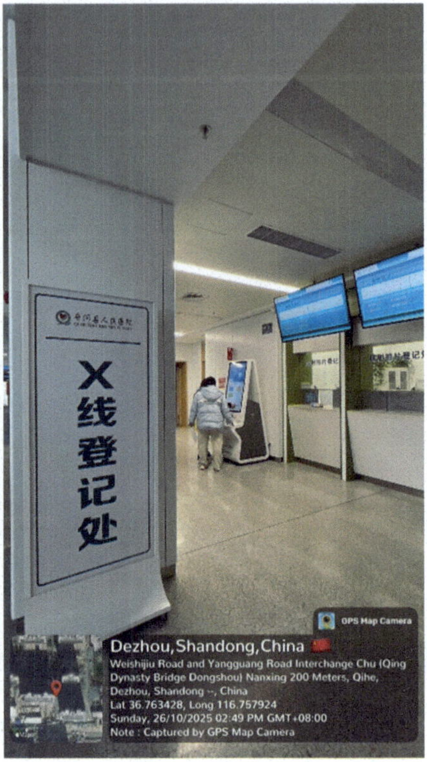

Fig. 7.8 Smart medical facilities enhancing community health through technology-driven services

fixed-wing drones for the rapid delivery of blood, vaccines, and emergency medicines. This reduced journeys that previously required over an hour's drive to less than 15 minutes. Orders were placed via digital tools such as WhatsApp, ensuring service accessibility even in areas with weak network coverage. This case demonstrates how intelligent systems enhance the resilience and equity of healthcare systems. Practices in both India and Rwanda prove that smart technologies are effectively bridging geographical and infrastructural divides, advancing the universalization of public health services.

7.5.2 *Digital Education Platforms*

Digital tools in education enhance teaching effectiveness by fostering higher-order cognitive and communication skills through interactive scenarios (Majumdar & Mandal, 2021). Cardiff University's Master's program in Urban Design vividly demonstrates the practical application of digital education platforms during home-based learning. Pre-recorded videos cover theoretical course content, enabling students to preview material and review lessons at any time, conveniently. Discussions can also be conducted via Microsoft Teams, facilitating real-time supervision and personalized guidance from tutors (Peimani & Kamalipour, 2021). This dual-mode design implements artificial intelligence adaptive learning, integrating personalized learning concepts into online education and providing an actionable solution for digital teaching practices during public health crises (Kaswan et al., 2024).

7.5.3 Smart Transportation Systems

Information technology and shared mobility platforms are key drivers in achieving sustainable urban transport. Research indicates that these technologies play a vital role in optimizing efficiency, providing technical and operational guidance for urban planning (Majumdar et al., 2025).

Copenhagen's multimodal public transport network serves 800,000 journeys daily. Paulsen's team analyzed three months of Automatic Vehicle Location (AVL) data and tested five adaptive passenger route selection models under varying conditions of real-time information availability. Their research quantified citywide door-to-door journey delay scenarios, revealing significant benefits of real-time information in reducing travel times.

Within shared micro-mobility, smartphone applications effectively alleviate congestion and enhance operational stability by coordinating vehicle availability with station utilization. A case study in Delft, the Netherlands, demonstrated that shared bicycles, owing to their low cost and minimal risk of theft or damage, substantially reduced reliance on walking, public transport, and private vehicles. Collectively, these findings provide empirical guidance for municipal transport planning, highlighting the transformative potential of intelligent transport systems.

7.5.4 Civic Tech and E-Government

The paradigm shift toward digital government relies on systematically integrating citizen experience into iterative technological design. Australian research indicates that citizen satisfaction with e-government services necessitates simultaneously meeting three key elements: precision, simplicity, and system stability. User-generated application reviews enhance feedback accessibility and cost-effectiveness. Citizens spontaneously sharing their usage experiences provide government agencies with signals for continuous improvement.

Porto Alegre's participatory budgeting scheme offers a complementary case study, where partial municipal budgetary authority is transferred to citizens through legally binding referendums. Empirical evidence indicates that granting citizens substantive control throughout the full proposal–debate–oversight cycle effectively enhances trust and civic engagement. Therefore, institutionalizing real-time feedback within policy formulation and resource allocation cycles, supported by digital technologies, can facilitate the realization of adaptive, inclusive, and citizen-centered governance models.

7.6 Challenges and Ethical Considerations

While the preceding case studies demonstrate the transformative potential of intelligent technologies in enhancing cross-sectoral efficiency, accessibility, and civic engagement, their implementation also presents complex ethical, societal, and institutional challenges. To ensure these innovations maintain fairness, transparency, and accountability, critical scrutiny is needed of issues such as the digital divide, data privacy, algorithmic bias, and organizational resistance.

7.6.1 Digital Divide

Despite widespread adoption of smart technologies in public services, significant inequalities persist in device access, network connectivity, and digital literacy, collectively creating the digital divide.

Meanwhile, usage disparities are equally pronounced. Rural populations in low- and middle-income countries (LMICs) are 37% less likely to use mobile Internet than their urban counterparts, reflecting persistent geographical inequalities. Furthermore, the economic burden of acquiring smart devices further constrains adoption in LMICs, where households often face higher relative costs for digital services.

The digital divide not only exacerbates existing socioeconomic inequalities but also undermines the potential efficiency, inclusivity, and impact of smart public services. Addressing these disparities requires targeted infrastructure investment, subsidized device access, and community-centered digital literacy programs. Without intervention, technological innovation risks over-benefiting urban, affluent, and digitally literate populations, thereby undermining equitable governance models.

7.6.2 *Privacy and Data Security*

The widespread deployment of intelligent technologies generates vast quantities of personal and behavioral data, raising significant privacy and security concerns. AI-driven analytical tools and continuous surveillance via closed-circuit television networks readily provoke ethical debates concerning consent mechanisms and oversight frameworks.

Whether through unauthorized access, commercial exploitation, or inadequate anonymization, such practices constitute misuse of personal data. These improper methods introduce additional risks, particularly in sectors that handle sensitive information, such as healthcare, education, and social services. Regulatory frameworks in many nations lag behind technological advancements, creating vulnerabilities in data protection, accountability mechanisms, and enforcement capabilities. For instance, the 2018 SingHealth data breach in Singapore resulted in the unlawful theft of approximately 1.5 million patients' personal information. This incident not only eroded public trust but also incurred substantial remediation costs.

Such risks undermine public confidence and diminish engagement with digital services. Effective mitigation requires the integrated application of robust privacy legislation, transparent data governance, and citizen-centered consent mechanisms to ensure that intelligent public services remain efficient while upholding ethical responsibilities.

7.6.3 *Algorithmic Bias and Accountability*

The adoption of artificial intelligence-driven tools and platforms within public services holds significant potential to enhance efficiency, deliver personalized services, and foster civic engagement. However, this simultaneously raises critical ethical and accountability challenges. A systematic evaluation of artificial intelligence and telemedicine in rural communities underscores the urgent need for national-level regulations, protocols, and guidelines to govern telemedicine practices, ensure ethical compliance, and mitigate inherent biases in AI processing of medical data. The absence of such frameworks risks increasing disparities in service provision and potentially reinforcing existing social inequalities.

In education, AI-based personalized learning tools offer both opportunities and risks. While these tools can enhance learning outcomes by tailoring content to students' abilities and skills, it is imperative to prioritize privacy protection, ethical considerations, and potential algorithmic biases to ensure their responsible and equitable use. Effective implementation further requires robust infrastructure, access to high-quality datasets, and comprehensive teacher training to realize the educational benefits of AI interventions fully. In this context, selecting tools appropriate to learners' age and developmental stage is essential to ensure digital solutions align with students' cognitive levels and skill tiers.

Beyond healthcare and education, AI-driven civic technology platforms demonstrate how system configurations influence accountability mechanisms and democratic participation. Analysis by Palacin and colleagues of 31 European *Decidim* participatory budgeting cases indicates that identical technological infrastructure yields markedly divergent levels of civic engagement when configured differently. Consequently, alongside developing these intelligent interfaces, we must concurrently establish minimum democratic standards to ensure digital engagement translates meaningfully into substantive civic influence.

7.6.4 Resistance to Change

Institutional and human barriers frequently constrain the adoption of intelligent technologies in public services. Outdated systems, hierarchical workflows, and staff uncertainty regarding new tools can all delay implementation and diminish effectiveness. When healthcare practitioners and educators lack sufficient involvement in planning or training, they may adopt a wait-and-see approach toward AI-driven tools.

Resistance can be mitigated through participatory planning, targeted training, and user-centered implementation strategies. This ensures technologies integrate with existing workflows, enhance acceptance, and enable digital innovations to achieve their intended efficiency, inclusivity, and engagement objectives.

7.7 Best Practices and Success Factors

7.7.1 *Citizen-Centered Design*

Citizen-centered design lies at the heart of innovative and effective public service delivery (Baek & Kim, 2018). It emphasizes understanding and addressing the real needs, experiences, and expectations of citizens rather than focusing solely on administrative efficiency or technological capability. This approach ensures that digital transformation in the public sector remains human-centric, inclusive, and responsive to diverse communities (Ewim et al., 2024). By adopting citizen-centered design principles, governments can co-create solutions with the people they serve. Techniques such as user research, participatory design workshops, and continuous feedback mechanisms allow policymakers and service designers to gain deep insights into citizen behaviors and challenges. This participatory approach enhances the relevance, usability, and accessibility of public services, especially for marginalized or vulnerable groups who are often excluded from conventional decision-making processes (Gwaleba et al., 2023). Moreover, citizen-centered design promotes empathy and transparency in governance. By involving citizens early in the design process, governments can build trust and legitimacy while ensuring that services are intuitive, equitable, and adaptable to changing social conditions. Successful implementation also depends on cross-sector collaboration, bringing together public institutions, private innovators, academia, and civil society to design integrated and sustainable solutions.

7.7.2 *Cross-Sector Collaboration*

Cross-sector collaboration is a vital success factor in driving innovation and achieving sustainable outcomes in public service delivery (Han & Yu, 2024). It involves partnerships among government agencies, private enterprises, academic institutions, and civil society organizations to address complex social, economic, and environmental challenges collectively. By pooling diverse expertise, resources, and perspectives, such collaborations enable the development of holistic, scalable, and practical solutions that a single sector alone could not achieve (Absor & Ratriana, 2024). In the context of smart governance and public service innovation, cross-sector collaboration facilitates the integration of advanced technologies and data-driven practices into public systems. For example, partnerships with technology firms can support the implementation of smart city infrastructure, while collaborations with universities and research institutions can foster evidence-based policymaking and innovation labs (Esashika et al., 2023). Similarly, engagement with non-profit organizations ensures that the needs of marginalized or vulnerable communities are represented in policy design and service delivery. Effective collaboration requires shared goals, transparent communication, and clearly defined roles among partners. Establishing governance frameworks that promote accountability, trust, and data sharing is essential to maintaining long-term cooperation. Additionally, flexible and adaptive management approaches help accommodate the evolving nature of technology and community needs.

7.7.3 Agile Policy and Regulation

Agile policy and regulation are essential components of modern governance, particularly in an era defined by rapid technological advancement and constant societal change. Traditional regulatory frameworks, often rigid, hierarchical, and slow to adapt, can hinder innovation and limit the responsiveness of public institutions (Adeyinka & Adewumi, 2023). In contrast, agile governance embraces flexibility, experimentation, and continuous learning to ensure that policies remain relevant, effective, and aligned with emerging realities. Agile policymaking emphasizes iterative development, where regulations evolve based on evidence, stakeholder feedback, and real-time data (Simpson et al., 2024). This approach allows governments to test new ideas on a smaller scale, evaluate outcomes, and refine frameworks before implementing them broadly. For example, regulatory sandboxes in sectors such as fintech, health tech, and smart mobility provide controlled environments for innovation, enabling policymakers to balance technological progress with risk management and public safety.

Furthermore, agile regulation promotes collaboration between government, industry, academia, and civil society. By engaging stakeholders early and consistently, policymakers can co-create solutions that address diverse interests while fostering transparency and accountability. Digital tools and data analytics also enhance this process by providing insights that support timely and informed decision-making (Elugbaju et al., 2024).

7.7.4 Continuous Learning and Evaluation

Continuous learning and evaluation are critical to ensuring that public service innovations remain effective, relevant, and aligned with evolving community needs (McDonnell et al., 2025). In a rapidly changing technological and social landscape, static policies and one-time interventions are insufficient. Instead, governments must adopt a culture of ongoing assessment, feedback, and adaptation to improve service delivery and maximize community impact (Ines, 2013). Continuous learning involves systematically gathering insights from service outcomes, citizen feedback, and operational data to identify successes, gaps, and emerging challenges. Evaluation processes, both formative (during implementation) and summative (after implementation), enable policymakers to measure performance, assess impact, and make evidence-based adjustments (Isramirawati, 2021). For example, analyzing data from smart city systems, e-governance platforms, or digital health initiatives can highlight inefficiencies, inequities, or opportunities for innovation. Moreover, fostering organizational learning encourages public institutions to remain flexible, responsive, and innovative. By institutionalizing knowledge-sharing practices, training programs, and adaptive management techniques, governments can rapidly incorporate lessons learned into policy revisions, program design, and technological upgrades (Persaud, 2014). Continuous evaluation also strengthens accountability and transparency, enabling citizens and stakeholders to track progress, provide feedback, and co-create solutions (Bouw et al., 2023). Ultimately, embedding continuous learning and evaluation within public service frameworks ensures that innovation is sustainable, inclusive, and impactful, driving ongoing improvements in efficiency, citizen satisfaction, and overall community well-being.

7.8 Future Directions

7.8.1 Toward Inclusive Smart Communities

The future of public service innovation lies in building inclusive smart communities in urban and rural environments, leveraging technological advancements to enhance the quality of life for all citizens, regardless of socioeconomic status, age, or ability (Perboli & Rosano, 2020). Inclusivity ensures that smart technologies do not exacerbate existing inequalities but instead empower marginalized and underserved populations to access essential services, participate in governance, and benefit from sustainable development initiatives (Hassebo et al., 2024). Inclusive smart communities integrate digital infrastructure, data-driven decision-making, and citizen-centered approaches to address social, economic, and environmental challenges (Qi et al., 2023). Examples include accessible e-governance platforms, AI-assisted healthcare for remote regions, smart transportation systems that serve diverse mobility needs, and community-based digital literacy programs. Beyond technology deployment, inclusivity requires proactive community engagement in planning, design, and evaluation to ensure that innovations reflect local contexts, priorities, and cultural sensitivities. Policy frameworks and regulatory mechanisms must also support equity, data privacy, and transparency, while fostering cross-sector collaboration and capacity-building (Sujithra et al., 2025). By combining human-centered design with sustainable technology strategies, inclusive smart communities can promote resilience, participation, and social cohesion.

7.8.2 AI and Machine Learning in Public Services

Artificial Intelligence (AI) and Machine Learning (ML) are rapidly reshaping public service delivery by enabling governments to analyze large volumes of data, automate complex processes, and make evidence-based decisions (Sobrino-García, 2021). These technologies offer unprecedented opportunities to enhance efficiency, accuracy, and responsiveness across a wide range of sectors, including healthcare, transportation, education, public safety, and social services. In public healthcare, AI algorithms can predict disease outbreaks, optimize resource allocation, and support diagnostic decision-making, improving both outcomes and accessibility (Adenyi et al., 2024). In urban management, ML-driven analytics help optimize traffic flow, energy usage, and waste management, fostering smarter, more sustainable cities (Choudhary et al., 2025). AI-powered chatbots and virtual assistants enhance citizen engagement by providing real-time responses to queries and facilitating seamless access to government services (Yiğitcanlar et al., 2024). However, leveraging AI and ML in public services requires careful attention to ethical, legal, and social implications. Concerns such as algorithmic bias, data privacy, and transparency must be addressed to maintain public trust. Governments must implement robust governance frameworks, regulatory oversight, and inclusive design practices to ensure AI benefits all citizens equitably.

7.8.3 Decentralized and Open Technologies

Decentralized and open technologies are transforming public service delivery by promoting transparency, trust, and citizen empowerment. These technologies, including blockchain, open-source software, and distributed ledger systems, enable secure, verifiable, and accessible management of public data and resources, reducing reliance on centralized authorities and fostering greater accountability in governance (Judijanto, 2023). Blockchain, for example, can streamline administrative processes such as land registries, identity management, and financial transactions by providing tamper-proof, transparent records for all stakeholders. Open-source platforms facilitate collaboration between government agencies, developers, and civil society, allowing services to be co-designed, continuously improved, and tailored to local needs (Costopoulou et al., 2021). The use of decentralized systems also enhances data security and resilience, minimizing the risks posed by single points of failure or cyberattacks. Beyond technical advantages, decentralized and open technologies empower citizens by giving them greater control over their data and enabling participatory governance. They support initiatives such as community-led decision-making, transparent budgeting, and real-time public accountability mechanisms. However, adopting these technologies requires careful attention to regulatory, ethical, and infrastructural considerations. Governments must ensure accessibility, interoperability, and inclusivity while safeguarding privacy and mitigating misuse (Adeusi et al., 2024).

7.8.4 Resilient and Adaptive Systems

Resilient and adaptive systems are essential for modern public service delivery, particularly in the face of complex, unpredictable challenges such as pandemics, climate change, natural disasters, and rapid technological disruption (Huang et al., 2024). These systems are designed to anticipate, absorb, and recover from shocks while continuously learning and evolving to meet changing community needs. Resilience in public services involves building redundancy, flexibility, and responsiveness into infrastructure, processes, and governance structures (Agum & Adofu, 2024). For instance, smart grid energy systems can quickly adjust to demand fluctuations, while digital health platforms can scale rapidly to respond to sudden increases in patient demand. Adaptive systems leverage real-time data, predictive analytics, and AI-driven insights to monitor conditions, identify emerging risks, and guide timely interventions. Such systems also emphasize human-centered design and stakeholder collaboration (Alkhatib et al., 2025). By integrating feedback from citizens, civil society, and cross-sector partners, governments can refine services to remain relevant, equitable, and effective under diverse conditions. Continuous evaluation and iterative improvement further enhance adaptability, ensuring that lessons learned from past challenges inform future strategies (Nnenna et al., 2024).

7.9 Conclusion

The integration of smart technologies and innovative practices in public service delivery represents a transformative opportunity to enhance community well-being, foster inclusive governance, and create resilient, adaptive societies. From AI and machine learning to decentralized and open technologies, these tools enable governments to improve efficiency, transparency, and responsiveness, while empowering citizens to participate actively in decision-making. The success of these innovations depends not only on technological adoption but also on human-centered design, cross-sector collaboration, agile policy frameworks, and continuous learning. By prioritizing equity, accessibility, and ethical governance, public institutions can ensure that technological advancements benefit all members of society, including marginalized and underserved populations. Looking forward, the vision of inclusive smart communities underscores the importance of balancing innovation with social responsibility. Resilient and adaptive systems, informed by real-time data and citizen feedback, provide the flexibility necessary to respond to emerging challenges and evolving community needs.

References

Absor, U., & Ratriana, K. (2024). Enhancing stunting reduction through collaborative governance: Concepts and implementation in Brebes regency. *Jurnal Keluarga Berencana, 9*(1), 24–37. https://doi.org/10.37306/w6f8xj35

Acharya, S., Saldana, S., & Markus, H. (2024). Developing supply chain as a service through standardized digital integration. *Supply Chain Conferences, 1*(1), 1–20. https://doi.org/10.33422/supplychainconf.v1i1.463

Adenekan, O., Ezeigweneme, C., & Chukwurah, E. (2024). The evolution of smart cities: Integrating technology, governance, and sustainable development.

International Journal of Applied Research in Social Sciences, 6(5), 891–902. https://doi.org/10.51594/ijarss.v6i5.1131

Adenyi, A., Okolo, C., Olorunsogo, T., & Babawarun, O. (2024). Leveraging big data and analytics for enhanced public health decision-making: A global review. *GSC Advanced Research and Reviews, 18*(2), 450–456. https://doi.org/10.30574/gscarr.2024.18.2.0078

Adeusi, K., Jejeniwa, T., & Jejeniwa, T. (2024). Advancing financial transparency and ethical governance: Innovative cost management and accountability in higher education and industry. *International Journal of Management & Entrepreneurship Research, 6*(5), 1533–1546. https://doi.org/10.51594/ijmer.v6i5.1099

Adeyinka, F., & Adewumi, A. (2023). Enhancing public service delivery in Nigeria through agile practice. *Jurnal Administrasi Publik public Administration Journal, 13*(2), 192–206. https://doi.org/10.31289/jap.v13i2.10647

Afzal, B., Li, X., & Hernández-Lara, A. (2024). The innovation journey and crossroads of sustainability, resilience and human-centeredness: A systematic literature review. *Transforming Government People Process and Policy, 18*(3), 368–383. https://doi.org/10.1108/tg-10-2023-0151

Agum, C., & Adofu, I. (2024). Assessing the impact of Nigerian government policies on building resilience and facilitating transformational recovery: A comparative analysis with other African countries. https://doi.org/10.21203/rs.3.rs-3966573/v1

Al-Ansi, A., Garad, A., Jaboob, M., & Al-Ansi, A. (2024). Elevating e-government: Unleashing the power of AI and IoT for enhanced public services. *Heliyon, 10*(23), e40591. https://doi.org/10.1016/j.heliyon.2024.e40591

Alkhatib, S., Katmah, R., Kosaji, D., Afzal, S., Tariq, M., Simsekler, M., & Ellahham, S. (2025). AI-driven decision support framework for preventing medical equipment failure and enhancing patient safety: A new perspective. *Journal of Multidisciplinary Healthcare, 18*, 6299–6313. https://doi.org/10.2147/jmdh.s528612

Almeida, F., Guimarães, C., & Amorim, V. (2024). Exploring the differences and similarities between smart cities and sustainable cities through an integrative review. *Sustainability, 16*(20), 8890. https://doi.org/10.3390/su16208890

Baek, S., & Kim, S. (2018). Participatory public service design by gov.3.0 design group. *Sustainability, 10*(1), 245. https://doi.org/10.3390/su10010245

Bouw, K., Wiekens, C., Tigchelaar, C., & Faaij, A. (2023). Involving citizens in heat planning: A participatory process design for informed decision-making. *Sustainability, 15*(3), 1937. https://doi.org/10.3390/su15031937

Chatterjee, U., Antipova, A., Ghosh, S., Majumdar, S., & Setiawati, M. (2023). *Urban environment and smart cities in Asian countries.* https://doi.org/10.1007/978-3-031-25914-2

Choudhary, R., Kumar, A., Narayan, S., Mishchenko, I., Choudhury, M., Bhardwaj, L. K., & Boiko, O. (2025). Integrating fuzzy logic into life cycle assessment for sustainable municipal solid waste management in metro cities. *Ecological Engineering & Environmental Technology, 26*(7), 316–331. https://doi.org/10.12912/27197050/205500

Colovic, A., Bagherzadeh, M., & Liévin, J. (2025). Choosing the right crowdsourcing strategy: Implications for governments' crowdsourcing initiatives. *Public Administration Review, 85*(5), 1302–1315. https://doi.org/10.1111/puar.13917

Costopoulou, C., Ntaliani, M., & Ntalianis, F. (2021). Evolution of e-participation in Greek local government. *Information Polity, 26*(3), 311–325. https://doi.org/10.3233/ip-190174

Elugbaju, W., Okeke, N., & Alabi, O. (2024). Saas-based reporting systems in higher education: A digital transition framework for operational resilience. *International Journal of Applied Research in Social Sciences, 6*(10), 2512–2532. https://doi.org/10.51594/ijarss.v6i10.1663

Esashika, D., Masiero, G., & Mauger, Y. (2023). Living labs contributions to smart cities from a quadruple-helix perspective. *Journal of Science Communication, 22*(03). https://doi.org/10.22323/2.22030202

Ewim, C., Okeke, N., Alabi, O., Igwe, A., & Ofodile, O. (2024). Customer-centric digital transformation framework: Enhancing service delivery in SMES for underserved populations. *International Journal of Management & Entrepreneurship Research, 6*(10), 3493–3516. https://doi.org/10.51594/ijmer.v6i10.1658

Ghosh, S., Majumdar, S., & Cheshmehzangi, A. (2024). Cities of tomorrow: Urban resilience and climate change preparedness. *Springer Singapore*. https://doi.org/10.1007/978-981-97-9658-8

Guo, J., & Zhang, H. (2024). Digital age: The path choice of government-citizen value co-creation. *Heliyon, 10*(15), e35482. https://doi.org/10.1016/j.heliyon.2024.e35482

Gwaleba, M., Kongela, S., & Kombe, W. (2023). Governing land use planning in pursuit of customary tenure security: A case of Kilombero district in rural Tanzania. *Journal of Property Planning and Environmental Law, 15*(3), 109–129. https://doi.org/10.1108/jppel-11-2022-0035

Hadian, N. (2023). Mapping information technology adoption in smart cities: A literature review and research themes. *Jiko (Jurnal Informatika Dan Komputer), 6*(3), 196–202. https://doi.org/10.33387/jiko.v6i3.6967

Han, L., & Yu, T. (2024). Enhancing healthcare delivery through integrated management strategies: A multi-sector approach. *IJSSPA, 4*(1), 219–226. https://doi.org/10.62051/ijsspa.v4n1.23

Hassebo, A., Tealab, M., & Hamouda, M. (2024). From a traditional city to a smart city: The measurement of cities' readiness for transition, Egypt as a case study. *Urban Science, 8*(4), 212. https://doi.org/10.3390/urbansci8040212

Huang, A., Cooke, S., Garsden, C., Behne, C., & Borkoles, E. (2024). Transitioning to sustainable, climate-resilient healthcare: Insights from a health service staff survey in Australia. *BMC Health Services Research, 24*(1). https://doi.org/10.1186/s12913-024-10882-8

Ines, M. (2013). *The social media innovation challenge in the public sector.* https://doi.org/10.3233/978-1-61499-244-8-71

Isramirawati, I. (2021). *The impact of formative assessment on students' academic achievement.* https://doi.org/10.2991/assehr.k.210202.024

Judijanto, L. (2023). Implementation of blockchain technology in improving transparency of public services: A case study on government service delivery in Indonesia. *WSIST, 1*(02), 63–71. https://doi.org/10.58812/wsist.v1i02.477

Latupeirissa, J., Dewi, N., Prayana, I., Srikandi, M., Ramadiansyah, S., & Pramana, I. (2024). Transforming public service delivery: A comprehensive review of digitization initiatives. *Sustainability, 16*(7), 2818. https://doi.org/10.3390/su16072818

Lindquist, E., & Huse, I. (2017). Accountability and monitoring government in the digital era: Promise, realism and research for digital-era governance. *Canadian Public Administration, 60*(4), 627–656. https://doi.org/10.1111/capa.12243

Lui, L., & Avelino, C. (2025). Policy design spaces in Brazil's and Argentina's digital government. *Journal of Public Affairs, 25*(1). https://doi.org/10.1002/pa.70010

Maharani, A., & Rahmawati, F. (2023). The development of digital literacy smart e-book (ceri) as a resource for learning digital literacy skills in elementary schools. *Journal JPSD (Jurnal Pendidikan Sekolah Dasar), 10*(2), 61. https://doi.org/10.26555/jpsd.v10i2.a27982

Majumdar, S. (2020). 3. Sustainability in urban expansion of a metropolitan city: Impacts of urban growth towards the outer fringe of Kolkata, 59-72. https://doi.org/10.1515/9783110628593-003

Majumdar, S. (2021a). Spatiotemporal pattern and hotspot detection of malaria using spatial analysis and GIS in West Bengal: An approach to medical GIS. In V. E. Balas & S. Pal (Eds.), *Healthcare paradigms in the internet of things ecosystem* (pp. 83–100). Academic Press. https://doi.org/10.1016/B978-0-12-819664-9.00005-3

Majumdar, S. (2021b). A study of dengue disease data by GIS in Kolkata City: An approach to healthcare informatics. In A. Suresh & S. Paiva (Eds.), *Deep learning and edge computing solutions for high performance computing. EAI/springer innovations in communication and computing.* Springer. https://doi.org/10.1007/978-3-030-60265-9_3

Majumdar, S., & Sivaramakrishnan, L. (2025). Impacts of vegetation and impervious surfaces on urban sustainability: A case study of Kolkata. In L. Sivaramakrishnan & S. Bandyopadhyay (Eds.), *Sustainable urban forms and communities: Urban geographies of eastern India (chapter 11)*. Routledge. https://doi.org/10.4324/9781003509400-11

Majumdar, S., Choudhury, M., & Cheshmehzangi, A. (Eds.). (2025). *Sustainable urban future: Embracing circular economy in waste management*. Springer Singapore. https://doi.org/10.1007/978-981-95-2976-6

Majumdar, S., Kandpal, V., & Anthopoulos, L. G. (2024). *Smart cities: Innovations, challenges and future perspectives*. Springer Nature Switzerland AG. https://doi.org/10.1007/978-3-031-59846-3

Majumdar, S., & Mandal, M. (2021). Transforming the environment of education by internet of things: A review. In M. Al-Emran & K. Shaalan (Eds.), *Recent advances in technology acceptance models and theories. Studies in systems, decision and control* (Vol. 335). Springer. https://doi.org/10.1007/978-3-030-64987-6_23

Manita, S., Khaghaany, H., & Mensah, I. (2023). Digital technologies and transformation of governance processes: A study of Syddanmark regional council in Denmark. *Journal of Public Policy & governance, 7*(3), 13–22. https://doi.org/10.53819/81018102t5224

McDonnell, K., Percy, J., Lassen, A., Brown, M., Richman, A., Deardorff, J., Ruiz, M. S., Liu, J., Russell, K., Snyder, A., & Marshall, C. (2025). Investing in the development of the next generation of MCH leaders. *Frontiers in Public Health, 13*. https://doi.org/10.3389/fpubh.2025.1606108

Mohamed, A. (2023). Nurturing sustainable urban space: Integrating smart city innovations and earthship design principles for eco-friendly futures. *International Journal of Architectural Engineering and urban Research, 6*(2), 376–400. https://doi.org/10.21608/ijaeur.2024.265873.1066

Nijnik, M., Sarkki, S., Melnykovych, M., & Miller, D. (2023). Social innovation to address sustainability challenges and enlarge the opportunities provided by forests for a green and just transition of marginalised mountain areas. https://doi.org/10.20315/silvaslovenica.0022.02

Nnenna, U., Perpetua, U., Nchaga, A., Hadijah, T., Mabonga, E., & Nyamboga, T. (2024). Impact of digital governance on public service delivery in East Africa. *IAA JAH, 11*(1), 18–29. https://doi.org/10.59298/iaajah/2024/11.11829.11

Ologunebi, J., & Taiwo, E. (2025). *The impact of digital communication on governance, political dynamics, and leadership; a case study of the Nigerian people and process*. https://doi.org/10.2139/ssrn.5094521

Perboli, G., & Rosano, M. (2020). A taxonomic analysis of smart city projects in North America and Europe. *Sustainability, 12*(18), 7813. https://doi.org/10.3390/su12187813

Persaud, D. (2014). Enhancing learning, innovation, adaptation, and sustainability in health care organizations. *The Health Care Manager, 33*(3), 183–204. https://doi.org/10.1097/hcm.0000000000000014

Pratomo, A., Santoso, J., Nugroho, A., Fildansyah, R., & Vandika, A. (2024). Analysis of data privacy policy, data processing ethics, and technology ethics awareness on user privacy protection in west java. *WSSHS, 2*(03), 412–422. https://doi.org/10.58812/wsshs.v2i03.715

Qi, J., Mazumdar, S., & Vasconcelos, A. (2023). Datafication in smart cities: Understanding how the public experience urban environments. *ECRM, 22*(1), 260–263. https://doi.org/10.34190/ecrm.22.1.1576

Rasweswe, M., Ntho, T., Hlahla, L., & Ntshayintshayi, P. (2025). Using photovoice to transform undergraduate nursing students into critical thinkers during environmental health assessment in South Africa. https://doi.org/10.21203/rs.3.rs-7010662/v1

Santibanez Gonzalez, E. D. R., Kandpal, V., Machado, M., Martens, M. L., & Majumdar, S. (2023). A bibliometric analysis of circular economies through sustainable smart cities. *Sustainability, 15*(22), 15892. https://doi.org/10.3390/su152215892

Simpson, B., Johnson, E., Adeleke, G., Amajuoyi, C., & Lande, O. (2024). Leveraging big data for agile transformation in technology firms: Implementation and best practices. *Engineering Science & Technology Journal, 5*(6), 1952–1968. https://doi.org/10.51594/estj.v5i6.1216

Sobrino-García, I. (2021). Artificial intelligence risks and challenges in the Spanish public administration: An exploratory analysis through expert judgements. *Administrative Sciences, 11*(3), 102. https://doi.org/10.3390/admsci11030102

Sujithra, R., Marin, V., & Sreya, B. (2025). Commentary: Leveraging big data in health care and public health for AI driven talent development in rural areas. *Frontiers in Public Health, 13.* https://doi.org/10.3389/fpubh.2025.1685569

Wu, D., Liao, W., Yu, H., Majumdar, S., Xu, J., & Choudhury, M. (2025). Assessing traditional waste management practices in urban environments and recent trends. In S. Majumdar, M. Choudhury, & A. Cheshmehzangi (Eds.), *Sustainable urban future. Urban sustainability.* Springer. https://doi.org/10.1007/978-981-95-2976-6_5

Yiğitcanlar, T., David, A., Li, W., Fookes, C., Bibri, S., & Ye, X. (2024). Unlocking AI adoption in local governments: Best practice lessons from smart cities. https://doi.org/10.20944/preprints202406.0159.v1

CHAPTER 8

Infrastructure Insights Evaluating and Upgrading for a Smarter Future

Reza Mirzaei, Elnaz Lakizadeh, and Basudeb Pyne

8.1 Introduction

The industrial heritage as an integral part of the historical and cultural identity of societies plays a crucial role in the formation and development of cities. This heritage not only reflects the technological and economic progress of a community but also serves as a testament to its industrial history and social transformations, thereby holding particular significance (Oevermann et al., 2024). However, many industrial heritage sites worldwide, particularly in urban areas, face serious challenges that threaten their survival and sustainability (Oevermann, 2015).

One of the principal challenges confronting industrial heritage is its vulnerability to urban changes and new industrial developments. Many of these sites are in urban areas that have been affected by urban expansion and land-use changes. For instance, in the cities of Ahvaz and Abadan,

R. Mirzaei (✉) • E. Lakizadeh
Faculty of Architecture and Urban Planning, Iran University of Art, Tehran, Iran

B. Pyne
Department of Education, Ramakrishna Mission Sikshanamandira, Belur Math, Howrah, India

S. Majumdar, M. Choudhury (eds.), *City of the Future*,
https://doi.org/10.1007/978-3-032-18026-1_8

industrial heritage sites related to the oil industry have been marginalized and eroded by demographic changes and urban development.

Furthermore, the lack of attention to the historical and architectural value of these sites by residents and responsible institutions has led to inappropriate alterations to their structures and the loss of their authenticity. This not only diminishes the cultural value of these sites but also eliminates the associated economic and tourism opportunities.

In this context, the need for innovative, intelligent approaches to the preservation and revitalization of industrial heritage is increasingly evident. Digital technologies and sustainable approaches can serve as powerful tools for assessing, reconstructing, waste management, and managing these sites (Choudhury et al., 2025). For example, the use of innovative technologies, such as digital monitoring systems, public awareness programs, and green solutions, can contribute to reducing pollution and erosion affecting these sites (Bibri & Krogstie, 2017).

This chapter aims to examine the challenges facing industrial heritage in the cities of Ahvaz and Abadan and to propose innovative, sustainable solutions to protect and revitalize these sites. Leveraging the latest research and related theories, this study seeks to determine how innovative technologies and sustainable approaches can be employed to conserve industrial heritage effectively.

8.2 Background Knowledge

Industrial heritage, as part of cultural heritage, serves not only as a testament to the technological and economic advancements of societies but also as a historical and social asset that plays a crucial role in shaping local and national identities (Alfrey & Putnam, 2003). However, conserving this heritage faces numerous challenges, necessitating novel and intelligent approaches (Ashworth & Tunbridge, 2000). In this section, by examining the theories and perspectives of experts, the key components of industrial heritage conservation and the use of innovative technologies will be analysed (Fig. 8.1).

8.2.1 Industrial Heritage: Definitions and Concepts

Industrial heritage refers to a collection of structures, machinery, and sites associated with industrial activities that are historically, culturally, and architecturally valuable. This heritage includes factories, mines, railways, and other industrial facilities recognized as evidence of industrial and

Fig. 8.1 Symbiotic development heritage with smart infrastructure. (Source: Drawn by the authors)

social transformations (TICCIH, 2024; ICOMOS, 2024). Some experts believe that industrial heritage is not only part of a community's industrial history but also a resource for sustainable development and cultural tourism (Oevermann, 2015).

8.2.2 Theories Related to the Conservation of Industrial Heritage

8.2.2.1 Integrated Conservation Theory

This theory, proposed by experts such as Jokilehto, is based on the idea that heritage conservation should be carried out in an integrated manner and in relation to its surrounding environment. This approach involves considering the historical, cultural, and social values of industrial heritage, as well as the needs of the local community (Jokilehto, 2017).

8.2.2.2 Sustainable Development Theory

The sustainable development theory is based on the principle that development should be carried out in a way that meets the needs of the present generation without compromising the ability of future generations to meet their own needs. In the context of industrial heritage, this theory emphasizes the use of green technologies and sustainable approaches for the conservation and revitalization of industrial sites (Li, 2021).

8.2.2.3 Smart City Theory

This theory focuses on the use of digital and data-driven technologies to improve urban management and the conservation of cultural heritage. In the context of industrial heritage, this theory can serve as a framework for the utilization of smart technologies, such as digital monitoring systems and public awareness programs (Bibri & Krogstie, 2017).

8.2.3 Key Components in the Conservation of Industrial Heritage

8.2.3.1 Historical and Cultural Values

Industrial heritage, as a part of the industrial history of a community, possesses historical and cultural values that should be considered in the conservation process (Nikolić et al., 2024). These values include industrial architecture, traditional technologies, and historical events related to industrial sites.

8.2.3.2 Environment and Sustainability

The conservation of industrial heritage must be carried out with consideration for environmental issues and sustainability. The use of green technologies and sustainable approaches can contribute to the reduction of pollution and erosion of industrial sites (Li, 2021).

8.2.3.3 Local Community and Public Participation

The participation of the local community in the conservation process of industrial heritage is of particular importance. Increasing public awareness and the active involvement of residents can contribute to the effective and sustainable conservation of these sites (Cimadomo, 2015).

The theoretical foundations of this research indicate that the conservation of industrial heritage requires integrated, sustainable, and smart approaches. By utilizing relevant theories and considering key components such as historical values, environmental sustainability, and public participation, effective and sustainable conservation of industrial heritage can be achieved (Fig. 8.2).

Fig. 8.2 Smart infrastructure ecosystem with public and private participation. (Source: Drawn by the authors)

8.3 Research Context and Procedure

One of the fundamental requirements regarding decision-making, planning, and policymaking is the evaluation of the experiences of countries in different parts of the world. To this end, in the present research, official government documents, laws and regulations, research reports, and strategic programs were taken into consideration. The criteria for reviewing and analysing these documents are the objectives, stakeholders, focus on smart infrastructure programs, legal framework, scale of implementation, and level of accessibility.

8.4 Evaluating Practical Experiences in the Field of Smart Infrastructure

One of the fundamental pillars that plays a crucial role in improving decision-making, planning, and policymaking is the evaluation of documents, reports, laws, and regulations utilized by other countries (Benjaafar et al., 2024). Before proposing a solution in the domain of smart infrastructure, assessing successful and unsuccessful models can be highly effective. This evaluation will assist managers and policymakers in developing a better understanding of the barriers and challenges, enabling them to put forth more informed strategies. In this manner, the assessment of countries' results and achievements can delineate the optimal pathways (Gohil & Srivastava, 2023). Consequently, urban managers can better understand the relationship between policymaking and implementation and engage in evidence-based decision-making by examining documents and laws rather than relying on conjecture and speculation. This process will elucidate the legal, technical, and managerial challenges. The following sections will examine the practical experiences of Singapore, Brazil, the Netherlands, Finland, Ukraine, Greece, Germany, and Iran.

8.4.1 Practical Experience in Singapore

Singapore can be introduced as a global pioneer in smart infrastructure and collaborative innovative governance. The Smart Nation initiative, launched in 2014, is one of the most significant programs of this city-state, aimed at transforming Singapore into an innovative and sustainable city.

Through this initiative, Singapore's urban managers have adopted innovative technologies, actively engaged citizens, improved the quality of life, and attracted investment. This country has considered the local, national, and international levels. According to available documents and reports on the websites, this initiative was in place from 2014 to 2023 (Government of Singapore, 2021; World Economic Forum, 2023; Smart Nation Singapore, 2023).

Key Objectives of the Smart Nation Initiative:

- Improving the quality of life for citizens
- Enhancing Singapore's competitiveness and attracting foreign investment

Specific Objectives of the Smart Nation Initiative:

- Improving the quality of urban services
- Increasing citizen participation in policymaking and city management
- Enhancing transparency

To advance in the field of smart infrastructure, Singapore's urban managers have established legal frameworks such as the Electronic Transactions Act (enacted in 1998, amended in 2004), the Personal Data Protection Act (enacted in 2012), and the Cyber Security Act (enacted in 2018), as well as developing six comprehensive national ICT master plans (Ministry of Law Singapore, 2024; Cyber Security Agency of Singapore, 2024; PDPC, 2024). In 2023, the International Institute for Management Development (IMD) recognized Singapore as one of the world's smartest cities for its smart infrastructure.

The vision of Singapore's Smart Nation is to continuously leverage information and communications technology, networks, and data to support its communities better. The Smart Nation framework, published in 2018 by the Smart Nation and Digital Government Office of Singapore, covers areas such as transportation (focusing on smoother traffic flow and increased public transportation), urban solutions (using sensors and smart systems for safer homes), finance (greater efficiency in financial transactions and strengthened supervision), healthcare (smart healthcare systems), and education (collaborative learning, relationships between students, teachers, and parents) (Smart Nation Singapore, 2023).

In this country, the government, citizens, the private sector, and non-governmental organizations work together. Singapore's citizens have public access to most collaborative innovative governance platforms. However, in some cases, this access is limited, such as the Government Information Management System (GIMS), the management sections of the Singapore eGovernment portal (eCitizen), and the Integrated Collaboration System (ICS) platform. To facilitate greater citizen engagement with the government and the successful implementation of innovative policies, the following platforms have been defined:

- *Smart Nation Platform:* Allows citizens to report urban issues and challenges, request services, and follow up on their requests online (Smart Nation Singapore, 2023).
- *Data.gov.sg:* Provides citizens with defined access to urban data (Data.gov.sg - Singapore's open data platform, 2024).
- *Singpass:* Enables citizens to share their data with the government (Singpass - your improved digital ID, 2024).

The design and deployment of these platforms demonstrate that Singapore's urban managers have a strong focus on the infrastructure of networks (high-speed broadband, high-speed Internet, and the Internet of Things), security, data (processing large volumes of information, cloud computing), user interfaces, cloud computing, and organizational collaboration as the foundations of smart infrastructure that form the basis of a platform.

8.4.2 Practical Experience of Rio de Janeiro/Brazil

In recent years, Rio de Janeiro has faced numerous challenges regarding urban development. To address these challenges, the city has focused on utilizing digital technologies, emphasizing smart infrastructure, and implementing smart governance. The focus on smart city initiatives has gained momentum in Rio de Janeiro since 2016. The city's urban managers have stated that the objective of this effort is to increase citizen participation in solving problems (Prefeitura do Rio de Janeiro, 2023; Rede Cidade Digital, 2023).

Key Objectives of Rio de Janeiro:

- Increasing citizen participation
- Transforming Rio de Janeiro into a smart city
- Developing a sustainable and equitable society

Specific Objectives of Rio de Janeiro:

- Improving the efficiency and effectiveness of public management through citizen collaboration
- Promoting innovation and digital transformation
- Addressing social inclusion and enhancing social justice

To achieve these objectives at the local, regional, national, and international levels, Rio de Janeiro's urban managers have established legal frameworks, such as the Access to Information Act (enacted in 2011), the Electronic Government Act (enacted in 2007), the Digital Signature Act (enacted in 2006), the General Data Protection Law (enacted in 2018), the Civil Code (enacted in 2002), the Cybercrime Act (enacted in 2012), and the National Cybersecurity Act (enacted in 2019) (Presidência da República do Brasi, 2024).The local government (city hall and city council), citizens, private sector, non-governmental organizations, and universities and research centres have been involved in the process of Rio de Janeiro's smart city transformation, with each stakeholder assigned a specific role. Platforms such as the Urban Management Information System (SIM), the management sections of the Rio de Janeiro City Hall portal, the Integrated Organizational Communication System (SIC), the Crisis Management Platform, and the Video Surveillance Platform have limited access (Prefeitura do Rio de Janeiro, 2023). One of the most important platforms that has been effective in the two-way interaction between the government and citizens is the Government Feedback Portal and the SGSecure mobile application. Additionally, local governments have increased transparency in Rio de Janeiro's management by providing and receiving data from citizens. The smart platforms in this city include:

- *DataRio:* Allows citizens to access urban data (DATA RIO, 2024).
- *Participa Rio:* Assists citizens in participating in government decision-making (Participa RIO, 2024).
- *Conecta Rio:* Enables citizens to access healthcare services online (Conecta RIO, 2024).

In Rio de Janeiro, smart infrastructure has focused on healthcare, transportation, and citizen welfare. The foundation of smartness in Rio de Janeiro lies in citizen participation, achieved through the design and deployment of digital platforms. Aspects such as high-speed broadband, high-speed Internet, the Internet of Things, security, processing large volumes of data, cloud computing, user interfaces, and organizational collaboration have been prioritized.

According to a 2023 report by the Agência Brasileira de Promoção de Exportações e Investimentos, Brazil is becoming a significant hub for the digital economy.

8.4.3 *The Practical Experience of Amsterdam, Netherlands*

Amsterdam, a leading pioneer of smart cities and smart infrastructure worldwide, has been the focus of a digital development and smart city platform since 2014. This program is implemented at the local (specific neighbourhoods or urban areas), regional (different cities in the Amsterdam metropolitan region), national (local, regional, and national governments), and international levels.

The primary factor behind the success of the Amsterdam Smart City project is the citizens. Mobile applications and public meetings with citizens were developed for the success of smart cities (Amsterdam Data Strategy, 2022; Amsterdam Privacy Policy, 2022; Amsterdam Smart City Strategy, 2022). These smart infrastructures in the Netherlands have been developed across transportation, particularly in the environment and climate (Mirzaei & Lakizadeh, 2024), municipal services, health, and education.

The overarching goals of Amsterdam are to enhance smart services, encourage citizen collaboration and participation through the development of smart infrastructures, and improve the quality of life. The specific objectives are to increase access to municipal services, help reduce traffic, protect the environment, and improve air and water quality.

The urban managers of Amsterdam, the Netherlands, have established legal frameworks, including the Open Data Act (2011), the Personal Data Protection Act (2001), the Cybersecurity Act (2015), the E-Commerce Act (2000), and the Intersectoral Cooperation Act (2013), to optimize smart infrastructure. These laws refer to the roles of the local government (municipality and city council), citizens, the private sector, civil society organizations (collaboration with local government and the private sector), universities and research centres (providing expertise and knowledge), technology companies (developing smart solutions), infrastructure companies (developing smart infrastructures), consulting firms (assisting the government and private sector), and international development organizations (financial and technical assistance) (Overheid.nl., 2024).

The Amsterdam government has implemented practical measures to organize in-person events across the city to ensure the participation of all citizens and social justice, provide online platforms in multiple languages, and offer technical and educational support to citizens, all of which align with the goal of universal access.

According to the report by the International Institute for Management Development, Amsterdam is one of the top smart cities in the area of smart governance. Reports have shown that the Amsterdam Citizen Participation Platform has more than 100,000 registered users (Amsterdam Economic Board, 2023). The platforms and programs used in the field of smart infrastructures are as follows:

- *Amsterdam Participation:* Allowing citizens to express their opinions on important issues and participate in decision-making since 2014 (Gemeente Amsterdam, 2024a, b).
- *Amsterdam Smart City Platform:* Providing information on citizen participation initiatives since 2016 (Amsterdam Smart City, 2024).
- *Amsterdam Data Portal:* Publishing public data openly and accessible for all (Gemeente Amsterdam, 2024a, b).
- *Citizens as Sensors Project:* Launching a project in 2025 for citizens to collaborate in providing data on the environment and climate (Amsterdam Smart City, 2015).
- *Citizens as Planners Project:* Launching a project in 2016 for citizen participation in urban development projects (Amsterdam Smart City, 2016).
- *Citizens as Innovators Project:* Launching a project in 2017 for citizens to provide new ideas to solve urban challenges (Amsterdam Smart City, 2017).

8.4.4 The Practical Experience of Helsinki, Finland

The city of Helsinki is moving forward with the motto "Smart City + Smart Village = Smart Region." This approach was initiated in 2016, aiming to increase people's willingness to participate in decision-making.

Investigations have shown that in this city, increased citizen and public participation in urban decision-making processes at the local, regional, national, and international levels has led to greater decision reliability, transparency, and public trust in local governments (the municipality and city council). Mobile applications, websites, and public meetings have been considered to achieve this goal (My Helsinki, 2024).

Helsinki's overarching goals are sustainability, improving citizens' quality of life, and increasing citizens' roles in urban and rural management. The specific objectives are to use smart technologies to improve municipal services, reduce environmental problems, and improve collaboration between the government, the public, and the private sector.

The urban managers of Helsinki have established legal frameworks, such as the Finnish Freedom of Information Act (1999), the Finnish Personal Data Protection Act (2018), and the Helsinki Smart City Strategy (2019), to support the stratification of Helsinki. According to these laws and regulations, the local government (municipality and city council), the private sector, civil society organizations, citizens, and universities and research centres are defined as actors and stakeholders (Finlex, 2024). In 2019, this city was recognized as one of the top smart cities in the world by the IESE Cities in Motion Research Institute for improving the efficiency and effectiveness of public services. The programs and platforms include:

- *My Helsinki:* This platform provides the opportunity for participation in urban management and decision-making processes, sharing ideas, and discussion and polling on urban issues (My Helsinki, 2024).
- *Helsingin Kanava:* This platform enables people to communicate with city officials, ask questions, and submit feedback and complaints (City of Helsinki, 2024).
- *Open Data Helsinki:* This platform makes open data available to the public for the development of new programs and services (Helsinki Region Infoshare, 2024).
- *Helsinki Smart City Platform:* This platform enables citizens, businesses, and organizations to connect with Helsinki's smart city data and services (Helsinki Smart, 2024).
- *Aalto University Design Factory:* This innovation centre develops new solutions for urban challenges through collaboration with citizens, businesses, and organizations to develop and test new ideas (Aalto University, 2024).

- *Helsinki Think Company:* This organization helps organizations develop innovative solutions for urban challenges to implement new ideas (Think Company, 2024).
- *Forum Virium Helsinki:* This organization collaborates with citizens, businesses, and organizations to evaluate and monitor new ideas (Forum Virium Helsinki, 2024).
- *Helsinki Urban Lab:* This laboratory develops and tests new solutions for urban challenges, with the help of citizens, businesses, and organizations, and conducts needs assessments and feasibility studies for new ideas (University of Helsinki, 2024).
- *Helsinki Design Week:* This annual event involves citizens, businesses, and organizations in design and innovation in Helsinki (Helsinki Design Week, 2024).
- *Slush Helsinki:* This startup event provides a platform for presenting new ideas to investors (Slush, 2024).

8.4.5 The Practical Experience of Ukraine

In Ukraine, the innovative infrastructure project focused on protecting heritage. Therefore, this country can be considered a pioneer in the application of intelligent infrastructure for heritage protection. On January 1, 2013, a fully automated electronic national cadastre system was launched for the first time in Ukraine. The Ministry of Culture has decided to develop and implement an electronic registration system for tangible and intangible cultural heritage. The areas of buildings, protected areas, reserves, and archaeological protected areas belonging to lands with historical and cultural purposes have been included in the land cadastre, and access to the public cadastre is possible through the KadastrUA website and application (Hiraskar, 1992; Ukraine, 2000, 2011a, 2011b).

Forty-one years after the adoption of the International Covenant on Economic, Social and Cultural Rights and 5 months after the publication of the Fribourg Declaration, in November 2007, Ukraine published its fifth report on cultural rights. During this period, significant positive changes have been made to laws directly related to economic, social, and cultural rights. The Cabinet of Ministers has tried to complete parts of the report in written responses. Ukraine has taken measures to protect the cultural heritage of ethnic minorities (Hiraskar, 1992; Ukraine, 2000, 2011a, 2011b).

Ukraine has placed strong emphasis on cultural heritage and its preservation, especially in times of war and crisis, in its attention to cultural rights. There are numerous international initiatives for the protection of cultural heritage in Ukraine, focusing more on: (1) monitoring of damages and risks, (2) emergency relief measures such as the delivery of protective materials or storage through cultural institutions, (3) training of cultural heritage professionals and authorities, (4) digitization of inventories and archives, and (5) support for the cultural and educational sectors, both in Ukraine and abroad (UNESCO, 2023).

In Ukraine, any legal standard related to the rights and duties of citizens can be enforced only if it has been published in specific forms, including the mass media. Every judge must have an electronic reference system in their work that contains the text of these laws.

An examination of the legal-administrative structure governing heritage protection through e-governance of the cadastre shows that the Ministry of Culture and the State Service for Geodesy, Cartography, and Cadastre are responsible for heritage protection through the application of e-governance of the cadastre. These two are subordinate to the executive branch, and the effective actors in this matter include local governments, local communities, and the state. This is mentioned in the 2021 amended Law of Ukraine on the Regulation of Urban Planning Documents. Both the judiciary and the executive branch play a role in preparing the cadastre, with the judiciary responsible for land registration and the executive branch for preparing the necessary information layers (Hiraskar, 1992; Ukraine, 2000, 2011a, 2011b).

8.4.6 The Practical Experience of Greece

Greece can be identified as another country that has utilized smart infrastructure to protect its heritage. The preservation of cultural heritage encompasses the identification, assessment, registration, documentation/study, protection, and restoration, as well as the prevention of illegal exportation and the facilitation of public access and information regarding cultural heritage. This law also includes the enhancement and integration of heritage into contemporary social life, the education and aesthetic enjoyment of cultural heritage, and public awareness. The protection of cultural heritage is closely linked to land-use policies, residential growth, and oversight of construction activities and infrastructure development in sensitive areas of historical significance. The principal legislation governing cultural heritage includes a comprehensive and detailed system for protecting both movable and immovable monuments and intangible heritage. Consequently, the archaeological cadastre and digital collections have emerged as two primary digital tools of the Ministry of Culture and Sports of Greece for the implementation of the convention for the protection of archaeological heritage, serving as instruments for the integrated protection and management of archaeological heritage, which encompasses immovable monuments, digital collections of movable monuments, and historical archives of the archaeological organization. The electronic cadastre and heritage are accessible on the website (Greece, 2002; Hellenic Ministry of Culture, 2014; Maria Dimelli & Associates, 2019).

Reports by the Committee on Economic, Social and Cultural Rights of UNESCO indicate that the Greek government has adopted educational policies and cultural heritage protection measures that have led to a national program and stringent laws concerning the protection of archaeological heritage, which continued throughout much of the nineteenth and twentieth centuries, prioritizing actions in the field of cultural rights. Notably, this area has focused on cultural heritage and its impacts on climate change (Report of the Committee on Economic, Social and Cultural Rights of UNESCO, 2015). The legal-administrative structure for heritage protection reveals that the Ministry of Culture and Sports and the Ministry of Environment, as subunits of the executive branch, along with the Ministry of Energy and Assets, are responsible for heritage protection, with assistance from the cadastre. Additionally, at the local level, there are regional services for historical monuments and specialized services, as well as the Department of Antiquities, a significant authority in protecting the heritage fabric. The archaeological cadastre operates as an executive body in the realm of heritage fabric protection (Greece, 2002). Both the judiciary and the executive branch play roles in preparing the cadastre, with the judiciary responsible for land registration and the executive branch for preparing the necessary information layers. An examination of existing documents and laws indicates the presence of a distinct document titled "The Archaeological Cadastre in Greece."

The primary tools for archaeology and planning throughout the planning process, during implementation, and post-implementation monitoring involve continuous oversight of the condition of monuments and sites, control of construction activities, and issuance of building permits. These tools serve as instruments for documentation, facilitating researchers' access to materials, disseminating scientific information, and thereby enhancing the exchange and aggregation of information and promoting access and public awareness. They are essential for both archaeologists and developers, ensuring the reliability, generality, and availability of spatial and legal data about properties, thus guaranteeing public trust and the security of transactions. Furthermore, they establish an integrated, continuously updated information system encompassing legal, technical, and other rights-related information, creating a modern, fully automated registration system that enhances transparency and security, reduces bureaucracy, and serves as a critical tool for rational organization and development. This system ensures the general availability of legal information, with the cadastre information technology system including spatial and legal data, providing a reliable means to meet the need for secure, up-to-date, and validated information for the protection of property rights of Greek and foreign citizens, as well as public assets (Hellenic Ministry of Culture, 2014; Maria Dimelli & Associates, 2019).

8.4.7 The Practical Experience of Germany

Germany can be recognized as another pioneer in the use of smart infrastructure to protect heritage. By judicial and legal requirements, both the federal government and the governments of the sixteen states of Germany are responsible for formulating, developing, and implementing, to the greatest extent possible, a policy aimed primarily at coordinating and utilizing all available scientific, technical, cultural, and other resources to ensure the adequate protection, preservation, and presentation of cultural heritage. The German Law on the Protection of Cultural Property came into force on August 6, 2016, and protects not only cultural objects of outstanding importance to the country's cultural identity but also modern facilities that meet technical standards. Furthermore, the internal use of the cadastre imposes additional requirements for the information about the cadastre concerning temporal aspects. Consequently, a cultural landscape cadastre was introduced. The cultural landscape cadastre comprises three components: historic buildings, historic grounds, and elements of cultural and historical landscapes.

The Basic Law of the Federal Republic of Germany assigns the responsibility for land registration laws to the federal republic, while the states enact laws about property cadastres. Inter-state bodies (the working group of government auditing agencies) ensure the uniformity of property cadastres across all states. Land registration offices are part of the administration of justice in the sixteen states of Germany. The jurisdiction of land registration authorities, the manner of establishing and maintaining land registers, and additional directives from the Federal Minister of Justice are delineated; the state ministries of justice issue regulations governing the registration of documents and properties. Land registration offices are part of the local courts. Cadastre maps are digitally processed in automated property maps. This information is available on the website of the Federal Government Commissioner for Culture and the Media (2016; Seifert, 2018).

The Federal Republic of Germany wishes to emphasize its commitment to the International Covenant on Economic, Social, and Cultural Rights, particularly Article 15. The questionnaire from the United Nations Special Rapporteur on Cultural Rights indicates that policymakers must promote the arts and culture, care for cultural heritage, and create a conducive environment for a rich and diverse cultural life. Germany owes its cultural wealth to the diversity of its historical cultural landscape. Compared to other countries, Germany possesses a dense network of cultural institutions. Cultural centres and prominent cultural attractions exist not only in major urban cities but also in smaller towns and suburban areas. This well-developed structure reflects Germany's cultural landscape within today's cultural federalism (International Monetary Fund, 2015). An examination of the legal-administrative structure related to heritage protection through the cadastre is conducted by regional associations. Both the judiciary and the executive branch play roles in preparing the cadastre, with the judiciary responsible for land registration and the executive branch for preparing the necessary information layers. A review of laws and documents in Germany indicates that the cultural landscape cadastre has been developed to protect heritage.

It should be noted that the Charter on Digital Heritage (2003) (UNESCO, 2003), the Moscow Declaration on the Preservation of Digital Information (2011) (UNESCO, 2011), the Documentary Heritage at Risk: Policy Gaps in Digital Preservation (2015) (UNESCO, 2021), the Recommendation on the Preservation and Access to Documentary Heritage, including in Digital Form (2016) (UNESCO, 2015), and the UNESCO document Protecting and Preserving Cultural Diversity in the Digital Era (2023) (UNESCO, 2023) contain provisions for documenting heritage through the use of digital technologies and smart infrastructures. Publications and web-based portals, as well as electronic and analogue content, are among the technical infrastructures emphasized in these documents. Digital materials encompass texts, databases, still and moving images, sound, graphics, software, and web pages, among a vast and growing array of formats. Strategies and policies for preserving digital heritage must be formulated with consideration of urgency levels, local conditions, available tools, and future projections. The continuity of digital heritage is fundamental. To preserve digital heritage, actions must be taken throughout the lifecycle of digital information, from creation to access. Long-term preservation of digital heritage begins with the design of reliable systems and procedures that produce authentic and sustainable digital objects. Proposed strategies can be classified into legal and managerial domains concerning the identification, access, and protection of heritage. The legal domain often addresses issues such as the formulation of policies and guiding principles, data sovereignty and Indigenous peoples' rights, the inclusion of digital preservation in national laws and regulations, and legal access support. The managerial domain frequently focuses on topics such as capacity-building for the identification, preservation, and access to heritage; ensuring participation and collaboration; data exchange; training for professionals and Indigenous communities; promoting the development and use of open-source and specialized software; and disseminating information. Another important aspect in this regard is the attention to cultural rights, which emphasizes the necessity of smartening up, whereby the incorporation of recognized rights in this declaration into national laws and practices, the implementation of rights enshrined in the declaration, ensuring access for all, and strengthening international cooperation tools are crucial elements that legitimize the necessity of creating smart infrastructures and smartening up (Mirzaei et al., 2024).

8.4.8 Practical Experience of Tehran/Iran

The city of Tehran has designed initiatives aimed at smartening up. An examination of documents revealed that, for the first time in the country's legislation, particularly in the Electronic Services Development Regulations for Executive Bodies ratified in 2014 (Supreme Administrative Council (Iran), 2014) and the Regulations for the Protection of Owners, Legal Occupants, and Private and Non-Governmental Users of Registered Immovable Properties and Historical-Cultural Contexts ratified in 2021(Cabinet of Ministers (Iran), 2021), there is an emphasis on the necessity of establishing databases in the field of the country's heritage and history. In this regard, the need for a cadastre database, website, and platform was felt.

Moreover, the Electronic Services Development Regulations for Executive Bodies ratified in 2014 (Cabinet of Ministers (Iran), 2021), the resolution on the requirements for executive bodies concerning the Single Window of Government Services to the Public ratified in 2022 (Executive Council of Information Technology (Iran), 2022), and the document on the Technical-Executive Standards for the Development of E-Government ratified in 2013 (Supreme Council of Information Technology (Iran), 2014) highlight the necessity of free flow of information, information dissemination, the creation of informational portals, online access for citizens, access to minimum information, and citizen education in alignment with the acceptance of electronic services. The focus of Iranian documents and regulations is primarily on the necessity of citizen access to information and their participation.

It is noteworthy that these documents address the technical necessity of establishing required infrastructure, such as web services, databases, and other electronic platforms. In the managerial domain, these documents address executive and managerial actions, including the need to prepare electronic development plans, create databases, identify services, and promote education and culture.

8.4.9 Practical Experience in India

India is moving forward with the Smart City Mission (SCM), an approach initiated in 2015, aiming to have rural and urban areas evolve together through inclusive digital transformation (Praharaj, 2021; Kesar & Ache, 2024). The initiative emphasizes people-centric government, aiming to increase citizens' participation in decision-making processes at the local, regional, and national levels, thereby enhancing decision reliability, transparency, and public trust in local governments (Kahyaoğlu et al., 2024). Cities like Pune, Kolkata, Surat, and Bhubaneswar indicate heightened public involvement in the planning and monitoring processes and have strengthened trust between citizens and local government (Mullick & Patnaik, 2024).

To involve local people, city governments have implemented digital platforms, smartphone applications, and in-person discussions. For example, the MyGov India portal, the SCM portal, and city-level applications such as the NKDA portal, Pune Connect, and the Surat Smart City Portal allow residents to share their opinions, monitor projects, and participate in the planning process (National Institute of Urban Affairs, 2023).

India's smart infrastructure initiative fosters sustainability, enhances the quality of life, and promotes inclusive governance, extending digital benefits beyond the metropolitan areas. The mission's motto has included innovative technologies to enhance delivery services, reduce environmental impacts, increase mobility, and promote cooperation with public and private sectors and communities (Das, 2024).

Modernizing farmwork is supported by numerous legal and strategic tools, including the Digital India Initiative (2015), the National Data Sharing and Accessibility Policy (2012), the Personal Data Protection Act (2023), and the Smart Cities Mission (2015) (National Institute of Urban Affairs, 2023). Together, these policies define the roles of municipal bodies, private enterprises, research institutes, civil society, and citizens as active stakeholders in India's urban transformation. Overall, India's smart region approach demonstrates how technological modernization and participatory governance can together create a more inclusive and sustainable model of development.

8.4.10 Practical Experience in China

China has emerged as a global pioneer in smart infrastructure and digital governance, driven by the Digital China strategy launched in 2012 and reinforced by the New-Type Smart City Development Plan introduced in February 2023, which has a comprehensive plan to promote the overall layout of "Digital China" development goals from 2025 to 2035 (Shan et al., 2021). This initiative aims to transform China into an intelligent, sustainable ecosystem that integrates data, technology, and citizen participation at the local, regional, and national levels, and to become the global leader in smart infrastructure (Wei et al., 2024). The key goals of China's smart city vision are to enhance citizens' quality of life, increase national competitiveness, and support innovation-driven economic growth. Specifically, it improves urban services through automation and data sharing, integrates public participation into governance, and enhances efficiency in city management (Li, 2025).

Legal and institutional frameworks such as the Cyber Security Law (2017), Data Security Law (2021), and Personal Information Protection Law (2021) provide a strong foundation for innovative city development. Most prominent programs include Hangzhou's City Brain, Shanghai's

Citizen Cloud, and Beijing's Digital Twin City, which integrate AI and IoT for traffic and emergency management, public-service platform, as well as integrate 3D modelling for urban planning (The State Council Information Office of the People's Republic of China, 2023; Li, 2025).

8.5 Planning and Prioritizing Upgrades

Effective smart infrastructure planning balances the historical legacy with contemporary technologies, social, environmental, and economic goals—while preserving the cultural and architectural identity of cities. To ensure modern development, incorporating innovative roadmaps that prioritize sustainability, inclusivity, and heritage conservation should be the government's and planners' primary priority. They should have designed the priority areas such as transportation, energy, water, and digital connectivity. These frameworks can help align investments in smart infrastructure with national goals for growth, resilience, and cultural continuity. Cost-benefit analysis is essential not only in financial terms but also in protecting and maintaining historical assets and community character to evaluate long-term viability and preservation. Policymakers can include more responsive plans for stakeholders by balancing economic benefits with environmental protection and heritage preservation. Public-private partnerships further enhance this balance by including privatization, technological advancement, and financial assistance with public oversight that promotes both modernization and heritage preservation. Collaboration with governments, the commercial sector, academia, and citizens ensures smart upgrades and protects cultural heritage.

However, strategic planning, thorough evaluation, and collaborative governance can pave the way for the development of smart infrastructure that enhances efficiency while preserving the cultural essence which defines a local identity.

8.6 Implications

8.6.1 The Conceptual Framework for Designing Smart Infrastructures

The innovative infrastructure framework, with a focus on objectives, domains of smartness, mechanisms, and roadmap planning, is a conceptual, multilayered construct that captures all necessary elements. The

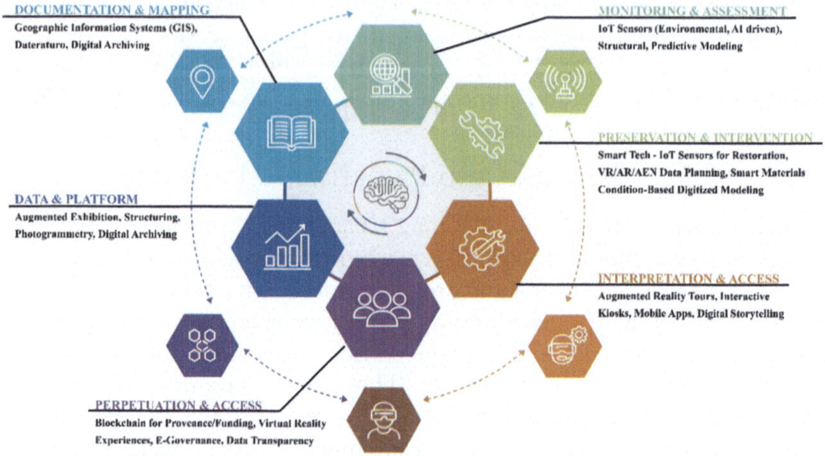

Fig. 8.3 Framework for designing smart infrastructures with integration of heritage. (Source: Drawn by the authors)

elements of this framework are continuous, follow one another, and have a reciprocal relationship. The first part of this framework outlines the objectives, which define the purpose of evaluating and developing smart infrastructure. Improving the quality of life, enhancing citizens' roles in city and rural management, improving air quality, enhancing the transportation system, improving municipal services, and enhancing public welfare and health are among the objectives of this endeavour (Dixit et al., 2025). The second part is the mechanism of operation, which includes the characteristics of goods and services, institutional characteristics, stakeholder characteristics, the managerial-institutional dimension, infrastructures, capabilities and facilities, and the domains of innovative infrastructure application. The final part is the application process and its output (Fig. 8.3, Table 8.1).

8.7 Future Technology Enabled Smarter Infrastructure

The development of more innovative infrastructure depends on integrating technologies that enhance efficiency, safety, and sustainability in urban management without harming the existing cultural heritages. Sensors and

Table 8.1 The conceptual framework for designing smart infrastructures

Dimensions			*Description*
Objective	Improve quality of life		Reduce pollution, access to services, optimize energy
	Citizen participation		Participation in startups, events, urban labs to solve urban problems, development of digital tools, provision of e-services
	Digital preservation of heritage		Development of digital tools
	Environmental sustainability		Smart management of water resources and waste
	Improving transportation		Optimizing the public transportation system and reducing traffic
	Social justice		Equal access to technologies
Application areas			Transportation, urban services, health, education, health and health services, heritage
Mechanism	Institutional feature	Legal framework	Regulation of partnership contracts, regulation of laws and regulations
		Governance model	Developing a program for the participation of citizens, rural people, local government and ministries, and in general, horizontal and vertical relationships between actors
		Financing	Budgeting and investment
	Product and service feature	Data-centric	Using various urban data for smart decision-making
		Service integration	Connecting all infrastructure to a central hub
	Characteristics of activists	Government and local government	Local government (municipality and city council)
		People	Citizens and villagers
		Other sectors	Private sector, NGOs, and universities and research centres
	Infrastructure and smart facilities	Database	Cadastre database
		Technical infrastructure	Network (high-speed broadband, high-speed Internet, and Internet of Things), security, data (large-scale data processing, cloud computing), user interfaces, cloud computing and enterprise collaboration

(*continued*)

Table 8.1 (continued)

Dimensions		*Description*
Process	Planning	Preparation and arrangement of the participation program of activists and actors, legal support of the government and judicial authorities for this program, formation of a roundtable of activists, discussion and debate on issues, examination of the best infrastructure for greater efficiency and effectiveness
	Data collection	Determining the data collection method, assessing challenges
	Review and evaluation	Review and evaluation as well as feasibility of creating the required infrastructure, feasibility of providing digital platforms
	Policymaking	Designing and formulating policies
Output		Formation of smart infrastructures with usability by the general public in cities and villages and organizations

Source: Authors' studies

IoT networks form the foundation of real-time monitoring systems, enabling continuous data collection from roads, buildings, bridges, and utilities. These tools can detect issues such as traffic jams, fluctuations in air quality, pipe leaks, and structural damage, allowing authorities to respond immediately. Based on this data, AI technology can derive actionable insights from raw data on equipment failures and maintenance traffic systems, and improve resource allocation—leading to cost savings and greater operational reliability. Furthermore, digital twins and simulation technologies can create virtual replicas of physical assets and systems that accurately reflect their real-time performance. Through these digital tools, urban planners, engineers, and the government can develop predictive models, test ideas before implementation, and evaluate the impact of infrastructure changes or environmental stresses. Collaboration with these technologies—IoT connectivity, AI-driven intelligence, and digital simulation—forms a robust infrastructure that enables governance and planners to advance a data-driven, futuristic approach, ensuring more innovative, safer, and more sustainable infrastructure (Fig. 8.4).

Fig. 8.4 Smart infrastructural benefits and challenges (Source: Drawn by the authors)

8.8 Conclusion

In recent years, the existing issues and problems in cities have consistently driven urban managers to create smart infrastructure, and successful examples have also led urban and rural populations to demand more from their governments, but this must be without harming the existing heritage. The use of technology and smart infrastructure has transformed the city's architecture, so it must be protected. An examination of the experiences of developing smart infrastructure in the cities of Brazil, Finland, Singapore, Ukraine, Germany, Greece, Iran, India, and China has revealed that these countries are primarily focused on creating smart infrastructure to enhance citizens' roles in solving urban issues. Platforms such as My Helsinki, Open Data Helsinki, Smart Nation Platform, Singpass, Participa Rio, and Amsterdam participation have been designed and deployed in these countries.

This chapter has presented a framework for developing smart infrastructure. According to this framework, the domains of innovative infrastructure application include transportation, municipal services, healthcare, education, health and medical services, and historical heritages. In the implementation mechanism, the institutional characteristics (legal framework, governance model, and financing), the characteristics of goods and services (data-centricity and service integration), the characteristics of stakeholders (government, people, and other actors), and the intelligent infrastructure and capabilities (information base and technical infrastructure) must be considered.

Acknowledgements In this section, we would like to express our gratitude to Professor Majumdar and all the editors for providing us with this opportunity.

References

Aalto University. (2024). Design factory. https://designfactory.aalto.fi/

Alfrey, J., & Putnam, T. (2003). *The industrial heritage: Managing resources and uses.* Routledge. https://doi.org/10.4324/9780203392911/

Amsterdam Data Strategy. (2022). City of Amsterdam. https://www.amsterdam.nl/en/policy/data-strategy/

Amsterdam Economic Board. (2023). Amsterdam smart city strategy. https://www.amsterdameconomicboard.com/smart-city-strategy/

Amsterdam Privacy Policy. (2022). City of Amsterdam. https://www.amsterdam.nl/en/policy/privacy-policy/

Amsterdam Smart City Strategy. (2022). City of Amsterdam. https://www.amsterdam.nl/en/policy/smart-city/

Amsterdam Smart City. (2015, October 27). Citizens as sensors. https://smartcity.amsterdam.nl/en/projecten/citizens-as-sensors/

Amsterdam Smart City. (2016, March 8). Citizens as planners. https://smartcity.amsterdam.nl/en/projecten/citizens-as-planners/

Amsterdam Smart City. (2017, May 15). Citizens as innovators. https://smartcity.amsterdam.nl/en/projecten/citizens-as-innovators/

Amsterdam Smart City. (2024). https://amsterdamsmartcity.com/

Ashworth, G. J., & Tunbridge, J. E. (2000). *The tourist-historic city.* Routledge. https://doi.org/10.4324/9780080519470/

Benjaafar, H., Berrichi, A., Jai, F., & Arabi, M. (2024). Review: Green Morocco plan and Moroccan legislation for the socioeconomic and sustainable development of agricultural cooperatives: Challenges and prospects. *E3s Web of Conferences, 527*, 03003. https://doi.org/10.1051/e3sconf/202452703003

Bibri, S. E., & Krogstie, J. (2017). Smart sustainable cities of the future: An extensive interdisciplinary literature review. *Sustainable Cities and Society, 31*, 183–212. https://doi.org/10.1016/j.scs.2017.02.016/

Cabinet of Ministers (Iran). (2021). *Regulations for the protection of owners, legal occupants, and private and non-governmental users of immovable works and historical-cultural contexts registered in the list of national, historical, c ultural, and valuable works.* https://qavanin.ir/Law/TreeText/?IDS=139730361032559761

Choudhury, M., Suhag, M., Jha, P., Harne, K. R., Samanta, P., Choudhury, M., Borgohain, R. B., Majumdar, S., Kravchenko, Y., & Choudhary, R. (2025). Urban waste management and challenges in emerging smart city systems governance and circular economy perspectives. *Ecological Engineering & Environmental Technology, 26*(10), 243–266. https://doi.org/10.12912/27197050/211014

Cimadomo, G. (2015). Community participation for heritage conservation. *Sustainability in Heritage Protected Areas, Poland: AESOP*, 88–95. https://www.researchgate.net/publication/281684109_Community_Participation_for_Heritage_Conservation/

City of Helsinki. (2024). Contact the city of Helsinki. https://kopio.hel.fi/en/decision-making/contact-the-city-of-helsinki.html/

Conecta RIO. (2024). https://conectario.com/

Cyber Security Agency of Singapore. (2024). https://www.csa.gov.sg/

Das, D. (2024). Exploring the symbiotic relationship between digital transformation, infrastructure, service delivery, and governance for smart sustainable cities. *Smart Cities, 7*(2), 806–835. https://doi.org/10.3390/smartcities7020034

DATA RIO. (2024). https://data.rio.gov.br/

Data.gov.sg - Singapore's open data platform. (2024). https://data.gov.sg/

Dixit, G., Majumdar, S., & Choudhury, M. (2025). Introduction to ecosystem services for climate resilience: A theoretical framework. In M. Choudhury, G. Dixit, & S. Majumdar (Eds.), *Sustainable synergy: Harnessing ecosystems for climate resilience. Climate change management.* Springer. https://doi.org/10.1007/978-3-031-77957-2_1

Executive Council of Information Technology (Iran). (2022). *Resolution No. 1 of the 24th Session (Executive instruction and technical requirements for paragraph "w," Note 7 of the 1401 Budget Law).* https://qavanin.ir/Law/TreeText/?IDS=8597782347407598284

Federal Government Commissioner for Culture and the Media. (2016, August 6). *The German Act on the Protection of Cultural Property (Kulturgutschutzgesetz, KGSG)*. https://www.kulturgutschutzdeutschland.de/EN/EverythingAboutTheProtectionOfCulturalProperty/TheActOnTheProtectionOfCulturalProperty/theactontheprotectionofculturalproperty_node.html

Finlex. (2024). Finlex data bank. https://www.finlex.fi/en

Forum Virium Helsinki. (2024). https://forumvirium.fi/

Gemeente Amsterdam. (2024a). Data en informatie. https://data.amsterdam.nl/

Gemeente Amsterdam. (2024b). https://www.amsterdam.nl/

Gohil, K., & Srivastava, S. (2023). Analysing the drivers of smart city implementation using dematel method. *Eco Cities, 4*(2), 2674. https://doi.org/10.54517/ec.v4i2.2674

Government of Singapore. (2021). Smart nation Singapore: A progress report. https://www.smartnation.gov.sg/about-smart-nation/smart-nation-progress-report/

Greece. (2002). *Law 3028/2002: On the protection of antiquities and cultural heritage in general* (Government Gazette 153/A/28-6-2002). https://www.law-archaeology.gr/index.php/en/greek-cultural-herutage-law

Hellenic Ministry of Culture. (2014). *About the Archaeological Cadastre* (The Greek Archaeological Cadastre Program). https://www.arxaiologikoktimatologio.gov.gr/en/content/about-archaeological-cadastre/

Helsinki Design Week. (2024). Home. https://helsinkidesignweek.com/

Helsinki Region Infoshare. (2024). Traffic and tourism. https://hri.fi/data/en_GB/group/liikenne-ja-matkailu/

Helsinki Smart. (2024). https://helsinkismart.fi/

Hiraskar, G. K. (1992). *Fundamentals of town planning*. Dhanpat Rai. https://cis-legislation.com/document.fwx?rgn=16052/

ICOMOS. (2024). International council on monuments and sites: Industrial heritage. https://www.icomos.org/en/

International Monetary Fund. (2015, May 11). *Germany: Concluding statement of the 2015 Article IV Mission*. https://www.imf.org/en/News/Articles/2015/09/28/04/52/mcs051115

Jokilehto, J. (2017). *A history of architectural conservation*. Routledge. https://doi.org/10.1088/1755-1315/631/1/012033

Kahyaoğlu, S., Abdieva, R., & Baigonushova, D. (2024). Citizen perception and participation in local government in post-soviet countries: Case of Kyrgyzstan. *Journal of Eurasian Studies, 16*(2), 261–274. https://doi.org/10.1177/18793665241283476

Kesar, P., & Ache, P. M. (2024). Past, present and future of the smart city in India: An institutional perspective. *Cities, 146*, 104775.

Li, Q. (2021). Study on the sustainable development model of industrial heritage. In IOP conference series: Earth and environmental science (Vol. 631, No. 1,

p. 012033). IOP Publishing. https://doi.org/10.1088/1755-1315/631/1/012033/

Li, Y. (2025, May 30). *Smart city and sustainable urban development in China: A comprehensive overview*. Swissnex in China. https://swissnex.org/china/news/smart-city-and-sustainable-urban-development-in-china-a-comprehensive-overview/

Maria Dimelli & Associates. (2019). *Greek cadastre*. http://dimelli-law.gr/services/greek-cadastre/?lang=en

Ministry of Law Singapore. (2024). https://www.mlaw.gov.sg/

Mirzaei, R., & Lakizadeh, E. (2024). Developing a climate resilience framework for cities with an emphasis on innovative technology applications. In *Cities of tomorrow: Urban resilience and climate change preparedness* (pp. 269–289). Springer Nature Singapore. https://doi.org/10.1007/978-981-97-9658-8_14/

Mirzaei, R., Purohit, S., & Lakizadeh, E. (2024). Development of the framework for the application of cultural rights in smart urban governance. *Journal of Global Resources, 11*, 01. https://doi.org/10.46587/JGR.2025.v11i01.001/

Mullick, M., & Patnaik, A. (2024). Urban governance, technologically equipped citizens, and their (non) participation in the Indian smart cities of Bhubaneswar and Bhopal. *GeoJournal, 89*(4), 177.

My Helsinki. (2024). https://www.myhelsinki.fi/

National Institute of Urban Affairs. (2023, March). *75+ case studies of innovative projects of Smart Cities Mission: Smart cities and academia towards action & research: Part C: Urban infrastructure* [Compendium]. Ministry of Housing & Urban Affairs, Government of India. https://niua.in/intranet/sites/default/files/2778.pdf

Nikolić, M., Šćekić, J., Drobnjak, B., & Takač, E. (2024). Examined in theory—Applicable in practice: Potentials of sustainable industrial heritage conservation in a contemporary context—The case of Belgrade. *Sustainability, 16*(7), 2820. https://doi.org/10.3390/su16072820/

Oevermann, H. (2015). Industrial heritage management in the context of urban planning. https://www.researchgate.net/publication/299751694_Industrial_heritage_management_in_the_context_of_urban_planning/

Oevermann, H., Wergeland, E. S., & Hanika, S. (2024). Industrial heritage and pathways for cultural-creative development in Bamberg, Germany. *Urban Planning, 9*, 10.17645/up.8072/.

Overheid.nl. (2024). Wettenbank. https://wetten.overheid.nl/

Participa RIO. (2024). https://participa.rio/

Personal Data Protection Commission Singapore (PDPC). (2024). https://www.pdpc.gov.sg/

Praharaj, S. (2021). Area-based urban renewal approach for smart cities development in India: Challenges of inclusion and sustainability Urban Planning, 6(4), 202–215. https://doi.org/10.17645/up.v6i4.4484.

Prefeitura do Rio de Janeiro. (2023). CidadeInteligente. https://www.rio.rj.gov.br/web/riodigital/cidade-inteligente/

Presidência da República do Brasil. (2024). https://www.planalto.gov.br/

Rede Cidade Digital. (2023). Cidades Inteligentes no Brasil. https://www.redecidadedigital.org.br/cidades-inteligentes-no-brasil/

Seifert, M. (2018). *Germany* (Cadastral Template 2.0). CSDILA. https://cadastraltemplate.org/germany.php

Shan, Z., Zhang, Y., Zhang, Y., Tang, S., & Wang, W. (2021). A review of recent progress and developments in China smart cities. *IET Smart Cities, 3*(4), 189–200. https://doi.org/10.1049/smc2.12020

Singpass - your improved digital ID. (2024). https://www.myinfo.gov.sg/

Slush. (2024). https://slush.org/

Smart Nation Singapore. (2023). About smart nation. https://www.smartnation.gov.sg/about-smart-nation/

Supreme Administrative Council (Iran). (2014). *Regulations for the development of electronic services for executive agencies* (آييننامهتوسعهخدماتالكترونيكيدستگاههاياجرايي). https://rc.majlis.ir/fa/law/show/902007/

Supreme Council of Information Technology (Iran). (2014). *Technical-executive criteria for e-government development.* https://qavanin.ir/Law/TreeText/?IDS=17000274089048541169/

The State Council Information Office of the People's Republic of China. (2023, March 16). *China's law-based cyberspace governance in the new era* [White paper]. https://dig.watch/wp-content/uploads/2022/10/Chinas-Law-Based-Cyberspace-Governance-in-the-New-Era-16-March-2023.pdf

Think Company. (2024). https://www.thinkcompany.fi/

TICCIH. (2024). The international committee for the conservation of the industrial heritage. https://ticcih.org/

Ukraine. (2000). *Law of Ukraine No. 1805-III: About protection of cultural heritage* (Amended May 30, 2023). https://cis-legislation.com/document.fwx?rgn=17469

Ukraine. (2011a). *Law of Ukraine No. 3038-VI: On regulation of town-planning activities* (Amended Apr. 9, 2024). https://cis-legislation.com/document.fwx?rgn=32960

Ukraine. (2011b). *Law of Ukraine No. 3613-VI: About the State land cadastre* (Amended Apr. 9, 2024). https://cis-legislation.com/document.fwx?rgn=46230

UNESCO. (2003, October 17). *Charter on the preservation of the digital heritage.* https://dig.watch/resource/charter-preservation-digital-heritage/

UNESCO. (2011, October). *The Moscow declaration on digital information preservation: Final document* (Programme and meeting document). https://unesdoc.unesco.org/ark:/48223/pf0000383397/

UNESCO. (2015, November 17). *Recommendation concerning the preservation of, and access to, documentary heritage including in digital form.* https://unesdoc.unesco.org/ark:/48223/pf0000244675

UNESCO. (2021). *Documentary heritage at risk: Policy gaps in digital preservation.* UNESCO PERSIST Programme. https://unescopersist.org/wp-content/uploads/2021/10/documentary_heritage_at_risk_policy_gaps_in_digital_preservation_en.pdf

UNESCO. (2023, April 20). *Cutting edge | Protecting and preserving cultural diversity in the digital era.* https://www.unesco.org/en/articles/cutting-edge-protecting-and-preserving-cultural-diversity-digital-era/

United Nations Educational, Scientific and Cultural Organization (UNESCO). (2023, October 31). *Report on the protection of cultural property in Ukraine* (Document C54/23/18.COM/6). https://unesdoc.unesco.org/ark:/48223/pf0000387970/

University of Helsinki. (2024). Research units. https://www.helsinki.fi/en/research/research-units/

Wei, Y., Yuan, H., & Li, H. (2024). Exploring the contribution of advanced systems in smart city development for the regeneration of urban industrial heritage. *Buildings, 14*(3), 583.

World Economic Forum. (2023). Smart cities: Singapore. https://www.weforum.org/agenda/2023/01/smart-cities-singapore/

PART III

Designing for the Future: Implementing Circular Economy Solutions

CHAPTER 9

IoT and Beyond: Smart Technologies in Waste Collection and Management

Yfantidou Anastasia, Hazi Mohammad Azamathulla, Shilpi Yadav, Temple Probyne Abali, and Sushobhan Majumdar

9.1 Introduction

In the twenty-first century, cities around the world are facing mounting pressure to manage waste sustainably amid rapid urbanization, industrial growth, and population expansion. According to the United Nations (2019), more than 68% of the global population is projected to live in urban areas by 2050, leading to an exponential rise in municipal solid waste generation. This trend not only strains local governments and waste

Y. Anastasia
Department of Civil Engineering and Geomatics, Cyprus University of Technology, Limassol, Cyprus

H. M. Azamathulla
University of the West Indies, St Augustine, Trinidad

S. Yadav
Department of Geography, School of Humanities and Social Science, Shri Guru Ram Ram Rai University, Dehradun, India

S. Majumdar, M. Choudhury (eds.), *City of the Future*,
https://doi.org/10.1007/978-3-032-18026-1_9

management infrastructures but also poses significant threats to environmental sustainability and public health. The traditional "collect–transport–dispose" model of waste handling is no longer sufficient in the context of increasingly complex waste streams, limited landfill capacity, and growing public demand for sustainable solutions (Majumdar et al., 2025).

Sustainable waste management has therefore become a cornerstone of modern urban planning and environmental governance. It is closely linked to several United Nations Sustainable Development Goals (SDGs), including SDG 11 (Sustainable Cities and Communities) and SDG 12 (Responsible Consumption and Production). Effective waste systems help reduce pollution, minimize greenhouse gas emissions, conserve natural resources, and promote recycling and reuse practices (UNEP, 2020). In this context, the emergence of digital technologies, particularly the Internet of Things (IoT), artificial intelligence (AI), and blockchain, represents a transformative opportunity for cities to optimize waste operations, enhance transparency, and move toward circular-economy models (Choudhury et al., 2025).

Despite these opportunities, the path toward innovative waste management is not without obstacles. The integration of advanced technologies requires significant financial investment, institutional coordination, and data infrastructure. Developing countries, in particular, face challenges in scaling digital waste solutions due to funding constraints, lack of skilled labor, and limited policy support. Therefore, this chapter begins by establishing the broader context and rationale for improving waste management systems before delving into how IoT and related technologies are reshaping the sector.

9.1.1 Importance of Effective Waste Collection and Management

Effective waste collection and management systems are vital components of sustainable urban and rural development. As global urbanization

---

T. P. Abali
Department of Geography and Environment, Rivers State University, Port Harcourt, Nigeria

S. Majumdar (✉)
Department of Geography, Pandit Raghunath Murmu Smriti Mahavidyalaya, Bankura, West Bengal, India

accelerates, cities are producing unprecedented volumes of municipal solid waste, estimated to exceed 3.4 billion tons annually by 2050. Poorly managed waste leads to severe environmental, health, and economic consequences, including pollution of air, soil, and water resources, greenhouse gas emissions, and the proliferation of disease vectors (Guerrero et al., 2013). In urban areas, the concentration of population and commercial activity intensifies these challenges, requiring efficient systems to ensure public hygiene and maintain livability.

Citation analysis of smart waste management research (Fig. 9.1) from 2007 to 2024 shows a steady, significant rise in scholarly attention, reflecting the growing global push toward sustainable urban infrastructure and digital transformation (Majumdar et al. 2024). Early studies (2007–2012) concentrated on foundational concepts, such as sensor-based waste monitoring and optimization algorithms, but citations were relatively low as the field was still emerging. From 2013 to 2017, citations increased notably as the Internet of Things (IoT) expanded and wireless sensor networks were integrated into municipal waste systems, leading to more influential papers on real-time bin monitoring, route optimization, and data-driven

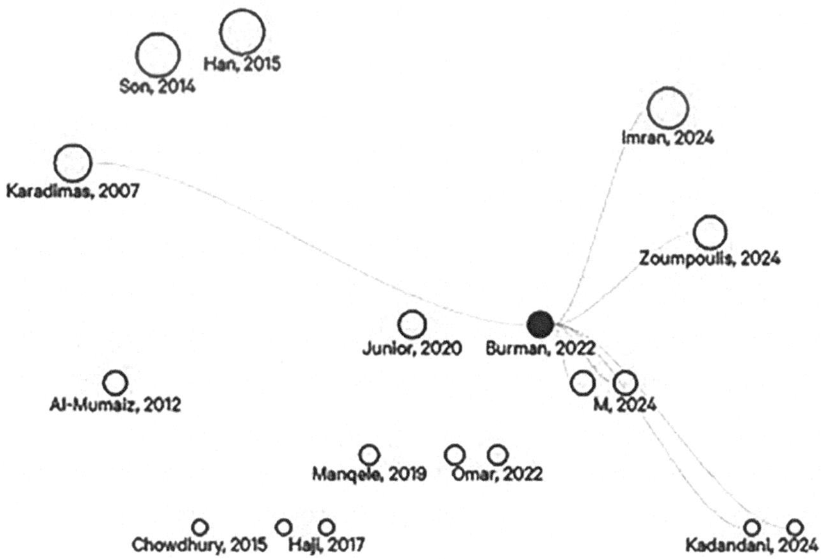

Fig. 9.1 Citation analysis of smart waste management

decision-making. The period from 2018 to 2020 saw accelerated citation growth driven by smart city initiatives, low-power communication technologies, and machine-learning-based waste prediction models. After 2020, citations continued to grow as research shifted toward AI-enhanced systems, circular-economy frameworks, blockchain-enabled traceability, and sustainability assessment. By 2024, the literature displayed a mature and interdisciplinary citation network, indicating that smart waste management had evolved into a core component of broader smart city, sustainability, and environmental informatics research (Kowsalya et al., 2025).

In rural settings, waste management often receives less attention due to dispersed populations and limited municipal capacity. However, inadequate disposal practices, such as open dumping or burning, pose comparable risks to ecosystems and human health (Marshall & Farahbakhsh, 2013). The integration of sustainable waste practices in both urban and rural contexts supports broader environmental objectives, including reducing carbon footprints, conserving natural resources, and promoting circular-economy principles. Effective waste management also contributes to social equity, as reliable services improve the quality of life across all communities, especially in developing regions where informal waste handling remains prevalent (Table 9.1).

9.1.2 *Challenges of Traditional Waste Management Systems*

Traditional waste management systems are characterized by several inefficiencies and limitations that undermine sustainability goals (Wu et al., 2025). Many municipalities still rely on static collection schedules and manually operated processes that fail to account for variable waste generation rates (Zhu et al., 2008). As a result, collection vehicles often travel inefficient routes, collecting half-empty bins or overflowing containers, leading to wasted fuel, higher operational costs, and unnecessary emissions (Mokalled et al., 2019). Moreover, the absence of real-time data and monitoring hampers effective planning and decision-making. In many cities, waste data are collected infrequently and lack accuracy, preventing timely responses to fluctuations in waste generation patterns (Nour et al., 2020). The situation is further complicated by inadequate segregation at the source, limited recycling infrastructure, and weak enforcement of waste regulations. In developing countries, institutional and financial constraints exacerbate these issues, often leading to uncollected waste that accumulates in public spaces and informal settlements.

Table 9.1 Overview of smart technologies in waste management

Technology	*Key components*	*Benefits*	*Challenges*
IoT sensors	Ultrasonic sensors, RFID, GSM/LoRa modules, microcontrollers	Reduces collection frequency, lowers costs, data-driven routing	Connectivity issues, power supply
AI & machine learning	Predictive models, neural networks, image recognition	Improved accuracy, reduced contamination	Requires large datasets, training complexity
Robotics	Robotic arms, conveyor systems, AGVs	Faster, safer sorting; labor cost reduction	High initial cost, maintenance
Blockchain	Smart contracts, distributed storage	Transparency, accountability	Scalability and interoperability issues
Drones	GPS, cameras, remote sensing	Faster monitoring, safety	Regulatory restrictions, weather dependence
Big data analytics	Cloud platforms, analytics engines	Better planning and resource allocation	Data privacy and integration complexity
Smart vehicles	GPS, telematics, on-board diagnostics	Fuel savings, improved efficiency	Technical integration issues

Source: Compiled by the authors

Traditional systems also struggle to address the growing complexity of modern waste streams, such as electronic waste, plastics, and hazardous materials, which require specialized treatment (Kaza et al., 2018). Additionally, public awareness and participation in waste reduction and recycling remain low in many communities, limiting the effectiveness of existing programs. Collectively, these challenges underscore the urgent need for innovative approaches that combine technology, policy, and behavioral change to create more adaptive and sustainable systems.

9.1.3 Role of Technology in Transforming Waste Management

Technological innovation, particularly the advent of the Internet of Things (IoT), artificial intelligence (AI), and data analytics, is revolutionizing waste management practices. IoT-enabled smart bins equipped with sensors can monitor fill levels, temperature, and weight, transmitting data in real time to centralized platforms (Pazhani et al., 2020). This data allows for dynamic route optimization of collection vehicles, minimizing fuel use and reducing greenhouse gas emissions (Fatimah et al., 2021). Through predictive analytics, municipalities can forecast waste generation trends, allocate resources efficiently, and anticipate maintenance needs for waste infrastructure.

Artificial intelligence and machine learning play increasingly significant roles in automating waste sorting and recycling processes. For instance, computer-vision technologies can identify and categorize recyclable materials, improving recovery rates and reducing contamination (Xiao et al., 2019). Blockchain technology is also emerging as a tool for transparent tracking of waste streams, enhancing accountability, and supporting regulatory compliance (Singh & Ordoñez de Pablos, 2021). Furthermore, robotics and automation are being applied in material recovery facilities, enabling safer and more efficient operations, especially for hazardous or mixed waste. Recent advances in computational tools, including optimization algorithms and digital twins, are further enhancing the efficiency of smart waste management systems by enabling real-time simulation and decision support (Nesmachnow et al., 2025). These technological advancements align with broader smart city initiatives, where waste management systems are integrated with urban data networks to support sustainable planning and circular-economy models. Such integration fosters collaboration between municipalities, private operators, and citizens, paving the way for participatory and data-driven environmental governance (Mora et al., 2019).

9.1.4 *Purpose and Scope of the Chapter*

The purpose of this chapter is to explore how IoT and related smart technologies are transforming the waste management landscape, addressing long-standing inefficiencies and environmental challenges. It provides an analytical overview of how connected devices, intelligent systems, and data-driven decision-making are reshaping the collection, transportation, treatment, and disposal of municipal solid waste. The chapter situates these technological developments within the broader framework of sustainable urban management and circular-economy strategies.

Specifically, the discussion encompasses both the technical and socio-institutional dimensions of smart waste management. It examines case studies from developed and developing countries to illustrate practical implementations, measurable outcomes, and context-specific barriers to adoption. Moreover, the chapter acknowledges that technological innovation alone cannot resolve the systemic challenges of waste management. It emphasizes the need for enabling policies, financial mechanisms, capacity-building, and community engagement to ensure the equitable and sustainable deployment of smart waste systems.

Continuous research into computational modeling and intelligent systems highlights the rapid evolution of smart waste technologies and their integration into broader smart city infrastructures (Nesmachnow et al., 2025). The scope also extends beyond IoT-based solutions to include emerging frontiers such as robotics, AI-driven analytics, and blockchain applications, highlighting how these technologies collectively contribute to the transition toward resilient, low-carbon, and resource-efficient urban systems. Ultimately, the chapter argues that smart waste management is not merely a technological upgrade but a strategic transition that can redefine the relationship between cities, citizens, and the environment.

9.2 Overview of IoT in Waste Management

The integration of the Internet of Things (IoT) into waste management systems represents a fundamental transformation in how societies conceptualize and operationalize material flows. Moving beyond the linear "collect–transport–dispose" paradigm, IoT-based frameworks enable continuous monitoring, data-driven decision-making, and predictive system optimization (Al-Mashaqbeh & Al Nawayseh, 2021; UNEP, 2024). This digital evolution aligns with broader global trends toward smart urbanism, environmental informatics, and circular-economy governance, where technology functions as an intermediary between environmental processes and human decision systems (Elijah et al., 2018).

9.2.1 Definition of the Internet of Things (IoT)

The Internet of Things refers to a distributed network of interconnected physical devices equipped with sensors, communication modules, and computational intelligence that enable autonomous interaction and information exchange over the Internet (Atzori et al., 2017). The conceptual foundation of IoT lies in cyber-physical systems hybrid configurations where digital and physical elements operate in synchronized feedback loops (ITU, 2022). Within waste management, IoT encompasses the digital augmentation of waste-handling infrastructure, creating an intelligent ecosystem capable of self-diagnosis, adaptive scheduling, and real-time analytics (Mekki et al., 2019).

This technological paradigm is distinguished by its scalability and interoperability. Devices embedded within bins, collection vehicles, and transfer stations can communicate via standardized protocols, enabling municipalities to monitor operational performance continuously (Al-Mashaqbeh & Al Nawayseh, 2021). The resulting datasets transform waste management from a static logistical service into a responsive environmental information system that supports evidence-based governance.

9.2.2 IoT in Waste Management

IoT operates through the triad of sensing, communication, and data intelligence. Sensors installed in collection points measure waste fill levels, mass, and environmental variables such as temperature and humidity (Anagnostopoulos et al., 2017). These data are transmitted via wireless networks commonly LPWAN, LoRaWAN, or NB-IoT to centralized data repositories or cloud platforms (Mekki et al., 2019). Machine learning and artificial intelligence algorithms subsequently process the data to identify patterns, predict waste generation rates, and optimize collection routes (Elijah et al., 2018; Raut et al., 2021).

In practice, this digital orchestration enables municipal agencies to replace fixed collection timetables with dynamic scheduling. Vehicles can be deployed only when bins reach predefined thresholds, reducing unnecessary trips, fuel consumption, and carbon emissions (Sharma & Rajoriya, 2021). Additionally, the integration of IoT with geographic information systems (GIS) and enterprise resource planning (ERP) software enhances spatial planning and financial tracking, thereby improving both operational efficiency and accountability (Table 9.2) (ITU, 2022).

9.2.3 Core Components: Smart Bins, Sensors, Connectivity, Data Analytics

IoT-based waste management systems are composed of several interdependent components that together enable real-time control and strategic planning. Smart bins are the most recognizable feature, incorporating ultrasonic or infrared sensors to determine fill levels and load cells to measure weight (Anagnostopoulos et al., 2017). Some designs also

Table 9.2 Comparison of traditional versus smart waste collection

Aspect	*Traditional system*	*Smart system (IoT-enabled)*
Collection method	Fixed schedules	Dynamic, data-driven routing
Monitoring	Manual inspections	Real-time sensor feedback
Operational cost	High	Reduced through optimization
Environmental impact	Higher fuel use & emissions	Lower emissions due to fewer routes
Workforce requirements	High manpower	Lower due to automation
Transparency	Limited	High, with digital dashboards
Waste sorting	Manual & error-prone	AI-based and automated

Source: Compiled by the authors

include odor and gas sensors to detect decomposition or hazardous emissions.

The connectivity infrastructure forms the backbone of IoT functionality. Depending on spatial scale and bandwidth requirements, networks may utilize short-range communication (e.g., ZigBee, Bluetooth) or long-range protocols (e.g., LoRa, Sigfox, NB-IoT) to ensure continuous data flow (Mekki et al., 2019). Data are transmitted to cloud platforms where data analytics applications aggregate, clean, and interpret information in real time. Advanced systems integrate predictive modeling, anomaly detection, and performance dashboards, enabling managers to assess the state of the system and forecast future demands (Elijah et al., 2018).

Crucially, these components rely on interoperability standards and cybersecurity frameworks to ensure data integrity and user trust (ITU, 2022). The fusion of IoT with artificial intelligence, blockchain, and big-data analytics is increasingly cited as a key enabler of "next-generation" smart waste ecosystems (UNEP, 2024).

9.2.4 *Case Examples of IoT Integration*

Globally, the deployment of IoT in waste management has accelerated, particularly in smart city programs. Barcelona, Spain, implemented one of Europe's earliest sensor-based waste systems, achieving up to a 25% reduction in collection costs and significant emission savings (European Commission, 2019). Seoul, South Korea, applies IoT-enabled RFID food-waste bins linked to a pay-as-you-throw model, resulting in a 47% reduction in food waste over a decade (Korea Environment Corporation,

2021). In Singapore, real-time data from IoT-equipped bins feed into centralized analytics dashboards under the Smart Nation initiative, guiding policy and infrastructure investment (NEA, 2024).

In the Global South, IoT adoption demonstrates both opportunities and challenges. Pune, India, introduced GPS-tracked vehicles and sensorized bins to enhance route optimization and compliance monitoring, yielding measurable cost and labor efficiency improvements (Raut et al., 2021). Cape Town, South Africa, has piloted IoT-based systems to monitor landfill leachate and air quality, integrating environmental data into its sustainability framework (UNEP, 2024). Meanwhile, Dubai, UAE, employs IoT-driven platforms that link public and private waste contractors through performance dashboards, improving service transparency and regulatory oversight (Sharma & Rajoriya, 2021).

Collectively, these examples illustrate that IoT serves not merely as a technological enhancement but as a catalyst for institutional transformation. The integration of intelligent sensing and data analytics supports systemic efficiency, while also reinforcing transparency and accountability core dimensions of sustainable urban governance. As global cities continue to digitalize environmental infrastructure, IoT will likely remain pivotal in aligning operational efficiency with sustainability objectives (Fig. 9.2).

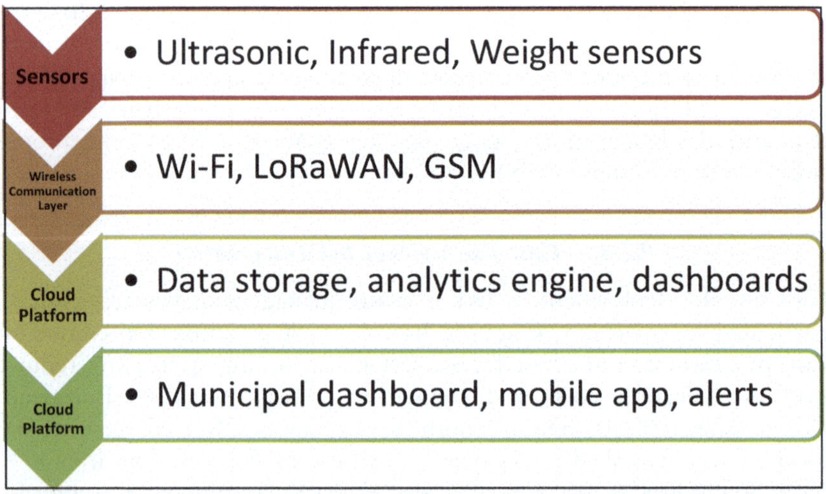

Fig. 9.2 IoT-enabled smart waste collection system architecture

9.3 Smart Waste Collection Systems

9.3.1 Sensor-Based Monitoring of Bin Fill Levels

Smart bins assess fill levels and waste characteristics in real time or at predetermined intervals using technologies including ultrasonic, infrared, time-of-flight (ToF), capacitive sensors, or camera-based vision systems. Municipalities can prioritize collections, identify almost full bins (and prevent overflow), and feed data into analytics and optimization modules later on thanks to sensorized monitoring (Ferrer & Alba, 2019). While camera/vision techniques can add material-type categorization when higher sophistication is desirable, recent comparative and deployment studies show that ultrasonic/ToF sensors are more affordable for normal municipal and biomedical bins (Fig. 9.3).

9.3.2 Real-Time Data Transmission and Cloud Integration

For storage, dashboarding, and analytics, sensor data is sent to cloud or hybrid edge-cloud systems over wireless connectivity (Wi-Fi, GSM/3G/4G/5G, LoRaWAN, NB-IoT, Sigfox). According to Kumar Singu (2024), cloud integration offers centralized dashboards, multi-stakeholder access, automatic alarms, scalable storage, and API connectors with routing and ERP systems. To address sporadic connection and privacy/security issues in urban deployments, architectures that employ secure data encoding/encryption or edge pre-processing (to minimize bandwidth) have been proposed (Popa et al., 2017).

9.3.3 Route Optimization for Waste Collection Vehicles Using GPS and Data Analytics

Real-time fill-level data, along with vehicle GPS and street network information, enables dynamic route optimization, which minimizes trip distance, fuel consumption, and labor costs when compared to fixed-route schedules. Among the strategies are hybrid dynamic routing, multi-criteria

Fig. 9.3 Real-time data flow for intelligent waste collection

decision methods (like TOPSIS), integer programming, and metaheuristics (like ACO and GA) (Adel et al., 2025; de Morais et al., 2019). While dynamic routing formulations like the Dynamic Reverse Inventory Routing Problem (D-DRIRP) use predicted fill levels to plan routes that change in near-real time, recent applied studies demonstrate that IoT-aided, data-driven routing may minimize both the total distance and the number of stops (Adel et al., 2025).

9.3.4 Predictive Scheduling and Automated Alerts

Regression, tree-based ensembles, time-series models, and deep learning are examples of predictive analytics that forecast trash creation trends and bin fill trajectories, allowing for proactive scheduling as opposed to only reactive collection. Automated threshold warnings alert operators when observed or predicted fill levels surpass critical levels, and predictive scheduling combined with routing minimizes missed collections and avoids overflow situations (Garapati, 2025). In order to enhance privacy and data integrity for dispersed sensor networks, emerging deployments integrate blockchain technology with federated or edge learning.

9.4 Data-Driven Decision-Making

9.4.1 Role of Big Data and AI in Analyzing Waste Patterns

The combination of big data and artificial intelligence (AI) has changed how we analyze waste generation and management systems. By processing large datasets from sensors in smart bins, GPS-equipped waste collection vehicles, and IoT monitoring systems, AI algorithms can recognize patterns in waste generation and disposal over time and space (Dandong & Yifan, 2020). These technologies help create models that predict waste amounts, improve collection routes, and find problems in current systems. With machine learning and data analytics, we can classify waste composition and anticipate future trends. This allows policymakers to create better and more sustainable management strategies that support smart city goals (Ferronato & Torretta, 2019). AI-based image recognition and robotic sorting systems improve the recycling process by increasing the accuracy and speed of waste sorting (Zambrano-Monserrate et al., 2021). When combined with cloud computing and real-time data visualization tools, these systems allow for continuous monitoring and better decision-making for local authorities.

This combination of AI and big data helps with efficient waste management, resource recovery, and the shift toward circular-economy practices (Dandong & Yifan, 2020; Zambrano-Monserrate et al., 2021).

9.4.2 Machine Learning for Predictive Analysis

Machine learning (ML) has become a strong analytical tool in modern waste management systems. It enables predictive analysis of waste generation trends, including the identification of peak waste generation times. By using large datasets collected from sensors, smart bins, and historical waste records, ML algorithms can find connections between waste volumes and factors such as population density, socioeconomic conditions, festivals, weather changes, and tourism activities (Mishra et al., 2021). Supervised learning models, such as regression and decision trees, are often used to forecast daily or seasonal patterns of waste generation. This helps authorities improve collection frequency, allocate resources, and plan routes (Nanda & Berruti, 2021).

Moreover, unsupervised learning methods, such as clustering, can group areas based on their waste-generation habits. This improves the design of management strategies tailored to specific areas. Predictive analytics from ML not only boosts operational efficiency but also supports proactive decision-making to prevent system overload during busy periods (Dey & Chatterjee, 2023). As cities continue to adopt data-driven waste management strategies, integrating ML models into innovative waste systems is proving vital for improving sustainability, cost-effectiveness, and environmental performance (Choudhary et al., 2025).

9.4.3 Dashboards and Decision-Support Systems for Municipalities

Dashboards and decision-support systems (DSS) play a crucial role in improving the efficiency, transparency, and responsiveness of municipal solid waste management operations. These digital platforms combine data from various sources, such as IoT-enabled bins, GPS-equipped collection vehicles, and citizen reporting apps, to provide real-time visualization of waste collection status, bin fill levels, and route efficiency (Yadav et al., 2020). By showing key performance indicators (KPIs) through interactive dashboards, municipal authorities can monitor operations, spot service delays, and allocate resources more effectively. Such systems allow for data-driven decisions, ensuring timely actions during peak waste

Table 9.3 Key metrics in smart waste management systems

Aspect	*Traditional system*	*Smart system (IoT-enabled)*
Collection method	Fixed schedules	Dynamic, data-driven routing
Monitoring	Manual inspections	Real-time sensor feedback
Operational cost	High	Reduced through optimization
Environmental impact	Higher fuel use & emissions	Lower emissions due to fewer routes
Workforce requirements	High manpower	Lower due to automation
Transparency	Limited	High, with digital dashboards
Waste sorting	Manual & error-prone	AI-based and automated

Source: Compiled by the authors

generation periods and enhancing overall service delivery (Curtis & Sarc, 2021).

Decision-support systems use predictive analytics, GIS-based mapping, and AI algorithms to help with urban planning and policy creation. For example, these tools can simulate different waste collection scenarios to find the best bin placements, assess environmental impacts, and predict future waste generation trends (Strand et al., 2020). Additionally, integrating dashboards with citizen feedback mechanisms encourages public participation, transparency, and accountability in waste management governance. Overall, these smart tools are vital to modern municipal management systems, supporting sustainable, evidence-based decision-making (Table 9.3).

9.5 Integration with Smart City Infrastructure

9.5.1 Linking Waste Management with Traffic Control, Public Health, and Energy Systems

Integrating waste management with broader smart city infrastructure enables a more holistic, efficient, and sustainable urban ecosystem by leveraging synergies across traffic control, public health, and energy systems (Bauermann et al., 2024; Majumdar, 2023; Majumdar, 2021). IoT-enabled waste collection vehicles can be coordinated with smart traffic management systems to optimize routes, reduce congestion, and minimize fuel consumption, ensuring timely collection without exacerbating urban traffic (Jaafreh and Allouzi, 2023). Similarly, sensors monitoring waste accumulation, decomposition, and contamination can provide early warning signals for public health risks, such as pathogen outbreaks or exposure to hazardous materials, allowing authorities to take preventive actions. Waste-to-energy systems can also be integrated, converting organic and combustible waste into biogas or electricity, feeding renewable energy into the city grid, and supporting circular-economy initiatives (Megha, 2023). By linking these systems, municipalities can create a data-driven, interconnected urban environment in which waste management contributes not only to cleanliness and recycling but also to traffic efficiency, public health monitoring, and sustainable energy production, exemplifying the multi-dimensional benefits of smart city integration.

9.5.2 Interoperability with Other Smart City Components

Interoperability between waste management systems and other smart city components is essential for creating a cohesive, data-driven urban ecosystem where information flows seamlessly across sectors (Majumdar et al. 2024, Majumdar et al. 2025, Kuai et al., 2024). IoT-enabled waste infrastructures, such as smart bins, sensor-equipped collection vehicles, and AI-powered sorting facilities, can be integrated with traffic management, water and energy grids, environmental monitoring systems, and public safety networks to enhance operational efficiency and decision-making (Bibri et al., 2024). For example, real-time waste data can inform dynamic traffic routing, energy recovery scheduling, and pollution monitoring, while insights from environmental sensors can optimize waste collection frequency and resource allocation. Achieving interoperability requires standardized communication protocols, open data platforms, and cross-sector collaboration, enabling diverse technologies and stakeholders to share information effectively (Elendu et al., 2024). This interconnected approach not only improves efficiency and reduces costs but also supports broader smart city objectives, including sustainability, resilience, and citizen engagement, by enabling a holistic view of urban infrastructure and facilitating coordinated, intelligent responses to emerging challenges (Ugochukwu et al., 2024).

9.5.3 Urban Sustainability and Circular-Economy Models

Integrating waste management within urban sustainability and circular-economy models is critical for minimizing resource consumption, reducing environmental impact, and promoting long-term resilience in cities

(Saeed et al., 2023). Smart technologies, including IoT-enabled monitoring, AI-driven sorting, and blockchain-based traceability, enable efficient collection, segregation, and recovery of materials, transforming waste from a disposal problem into a valuable resource stream (Devi et al., 2025). By linking waste flows to recycling, composting, and energy recovery systems, cities can close material loops, reduce reliance on landfills, and support the principles of a circular economy. Furthermore, data-driven insights allow policymakers to design targeted interventions, incentivize responsible consumer behavior, and optimize resource allocation across sectors (Saeed et al., 2023). Through such integration, urban waste management becomes not only a mechanism for maintaining cleanliness but also a strategic tool for achieving sustainability goals, enhancing resource efficiency, and fostering regenerative urban environments that align with broader environmental, economic, and social objectives (Sharma et al., 2023).

9.6 Beyond IoT: Emerging and Future Technologies

9.6.1 AI and Robotics in Sorting and Recycling

Beyond the capabilities of IoT-based monitoring and data acquisition, emerging technologies such as artificial intelligence (AI) and robotics are increasingly transforming sorting and recycling processes within smart waste management systems (Khan, 2023). Driven by computer vision, machine learning, and deep learning, this technology enables highly accurate identification and classification of diverse waste materials, reducing contamination, increasing recovery rates, and optimizing operational efficiency (Sivaranjani et al., 2024). When integrated with advanced robotic systems, these technologies enable automated material handling, with robotic arms equipped with sensors and vision systems rapidly picking, sorting, and separating recyclables with precision surpassing that of manual labor (Sodiya et al., 2024). This synergy of IoT, AI, and robotics creates a closed-loop intelligent ecosystem in which IoT devices collect real-time data, AI algorithms interpret and make decisions, and robots physically execute sorting tasks, resulting in efficient, continuous, and cost-effective recycling operations (Gamberini and Pluchino, 2024). Future developments such as self-learning robots, edge AI for on-site decision-making, swarm robotics, and AI-based material fingerprinting further extend the possibilities for fully autonomous recycling infrastructure (Singh and Singh, 2023). Collectively, these innovations represent a pivotal shift toward intelligent circular-economy systems, where waste is not only monitored and collected but also autonomously identified, processed, and reintegrated into sustainable resource cycles.

9.7 Blockchain for Traceability and Accountability in Waste Logistics

Blockchain technology is increasingly recognized as a powerful tool for enhancing traceability, transparency, and accountability across waste logistics systems, offering capabilities that extend far beyond those of traditional record-keeping or IoT-based tracking alone (Faridi et al., 2023). By providing a decentralized, immutable ledger, blockchain ensures that every transaction from waste generation and collection to transportation, sorting, and final disposal is securely recorded and cannot be altered, thereby reducing opportunities for illegal dumping, data manipulation, or misreporting (Faiz et al., 2024). When integrated with IoT sensors and smart bins, blockchain enables real-time verification of waste volumes, routes, and handling practices, creating a trusted chain of custody those stakeholders, including municipalities, recycling companies, and regulators, can access with confidence (Addas et al., 2024). Smart contracts further enhance operational efficiency by automating compliance checks, service payments, and performance-based incentives, ensuring contractual obligations are met without human intervention (Lin, 2024). In addition, blockchain facilitates improved auditing, supports extended producer responsibility (EPR) schemes, and enhances consumer trust by providing verifiable information on how waste is processed and where recovered materials ultimately end up (Ta, 2025). Collectively, these capabilities make blockchain a critical enabler of transparent, secure, and accountable waste logistics within the broader framework of smart and sustainable waste management systems (Choudhury et al., 2025).

9.7.1 Drones and Autonomous Vehicles in Large-Area Waste Monitoring

Drones and autonomous vehicles are emerging as powerful technologies for large-area waste monitoring, offering rapid, high-resolution, and cost-effective surveillance capabilities that significantly enhance traditional inspection and waste management practices (Askarzadeh et al., 2024). Equipped with advanced sensors, cameras, LiDAR, and thermal imaging, drones can cover expansive or hard-to-reach areas such as landfills, industrial zones, waterways, and illegal dumping hotspots, allowing real-time detection of waste accumulation, environmental hazards, and operational inefficiencies (Muto and Tonooka, 2025). Autonomous ground vehicles, meanwhile, can traverse predefined routes in landfills, transfer stations, or urban environments to collect continuous data on waste distribution, bin fill levels, and site conditions without human intervention (Aziz et al., 2018). By integrating these platforms with IoT networks and AI-driven analytics, waste managers gain access to accurate, time-stamped geospatial information that supports predictive decision-making, improves regulatory compliance, and accelerates response to improper disposal activities. Furthermore, the automated and remote nature of these systems reduces labor requirements and enhances worker safety by limiting exposure to hazardous or unstable environments. Together, drones and autonomous vehicles represent a critical advancement in large-scale waste monitoring, enabling smarter, more proactive, and data-driven waste management strategies.

9.7.2 Smart Packaging and Materials Innovation

Smart packaging and materials innovation are increasingly central to the evolution of sustainable waste management systems, offering new ways to reduce waste generation, enhance recyclability, and support circular-economy objectives (Solomon et al., 2024). Smart packaging incorporates embedded sensors, QR codes, RFID tags, or chemical indicators that provide real-time information on product freshness, usage, and disposal pathways, enabling more accurate sorting and reducing contamination in recycling streams (Stewart, 2024). Innovations in materials science, such as biodegradable polymers, compostable composites, edible films, and easily separable multilayer packaging, further help minimize environmental impact by promoting materials that degrade safely or can be efficiently recovered (Ciawi et al., 2024). Additionally, advanced design strategies like "design for disassembly," mono-material packaging, and digital product passports help streamline recycling processes and extend product lifecycles. When combined with IoT-enabled waste systems, innovative packaging enhances traceability and supports intelligent sorting technologies that rely on digital identifiers. Overall, these innovations play a crucial role in shifting from linear consumption patterns toward more regenerative and resource-efficient waste management models.

9.8 Case Studies and Real-World Implementations

9.8.1 Examples from Smart Cities

Smart cities around the world provide compelling examples of how advanced technologies are being integrated into waste collection and management to improve efficiency, sustainability, and citizen engagement (Bauermann et al., 2024). Cities such as Singapore, Seoul, and Amsterdam have leveraged IoT-enabled smart bins, sensor-equipped waste collection vehicles, and data-driven route optimization to reduce operational costs and minimize environmental impact (Lloret et al., 2025). In Barcelona and Copenhagen, smart sensors and analytics platforms monitor bin fill levels in real time, enabling dynamic collection schedules that reduce fuel consumption and traffic congestion (Salih et al., 2025). Some cities, like San Francisco and Stockholm, have integrated AI and robotics into material recovery facilities to automate sorting, improve recycling rates, and lower contamination, while also employing blockchain to ensure transparent tracking of waste flows and accountability across stakeholders. These real-world implementations demonstrate the potential of combining digital technologies, automation, and data analytics to create more responsive, sustainable, and citizen-centric waste management systems, providing valuable lessons for other urban centers aiming to transition toward smart, circular, and resilient waste infrastructure (Apanavičienė and Shahrabani, 2023).

9.8.2 Developing World Innovations Using Low-Cost IoT Solutions

In developing countries, resource constraints and rapid urbanization have driven the development of innovative waste management approaches using low-cost IoT solutions, demonstrating that smart technologies can be adapted to local contexts without requiring high investment (Huang et al., 2024). Cities such as Pune and Bangalore in India, and Accra in Ghana, have implemented sensor-equipped bins, GPS-tracked collection vehicles, and mobile-based monitoring platforms to optimize collection routes, reduce operational costs, and improve service coverage in densely populated areas (Srikantha, 2017). These low-cost IoT systems often rely on open-source hardware, affordable sensors, and cloud-based data platforms, making them accessible to municipal authorities and small-scale waste service providers (Aghenta and Iqbal, 2019). In addition, community engagement is enhanced through mobile apps and SMS notifications, enabling residents to report overflowing bins or illegal dumping, thereby creating participatory and responsive waste management systems. Such initiatives not only increase collection efficiency but also support environmental monitoring, data-driven planning, and gradual adoption of smart city practices in regions with limited infrastructure, highlighting the potential of scalable, cost-effective IoT solutions to transform waste management in the developing world (Ugochukwu et al., 2024).

9.8.3 Public-Private Partnerships and Innovation Hubs

Public-private partnerships (PPPs) and innovation hubs are playing a pivotal role in advancing smart waste management by combining the expertise, resources, and technological capabilities of both sectors (Katah and Ozturk, 2024). Through PPPs, municipalities collaborate with technology companies, recycling firms, and startups to deploy IoT-enabled waste collection systems, AI-driven sorting facilities, and data analytics platforms, often sharing investment costs and operational responsibilities (Sami et al., 2023). Innovation hubs and incubators further accelerate the development and testing of new solutions, providing platforms for startups to pilot smart bins, autonomous collection vehicles, blockchain-based tracking systems, and smart packaging technologies in real-world urban environments (Liao, 2022). Examples include Singapore's Urban Solutions & Sustainability (USS) Lab and the Netherlands' Circular Innovation Hubs, where cross-sector collaboration has led to scalable, data-driven, and sustainable waste management solutions. By fostering knowledge exchange, encouraging experimentation, and aligning public policy with technological innovation, these partnerships help create resilient, efficient, and citizen-focused waste management ecosystems while promoting circular-economy practices and environmental sustainability (Lopes et al., 2025).

9.9 Challenges and Limitations in IoT-Enabled Smart Waste Management

Even with its great potential, implementing IoT in waste management is quite challenging (Yevle and Mann, 2025). These challenges touch on many areas, from financial constraints and infrastructure to technical issues, society's readiness, knowledge management, environmental effects, and moral considerations—all interwoven (Fuqaha and Nursetiawan, 2025).

9.9.1 *Financial Concerns*

The primary challenge comes from the high upfront expenditure of creating IoT systems. The cost of intelligent bins, communication hubs, cloud storage, and analytic software adds up. For cities working with tight purse strings, this can be a deal breaker. Beyond the initial setup, keeping these systems running requires maintenance, battery changes, sensor tuning, and data subscriptions. These constant costs need to be covered down the road, which might put extra strain on the city's already stretched funds. Savings from smart systems, such as improved collection rounds or reduced fuel use, may take time to materialize, making it harder for those at the highest levels to accept investment in projects that scale widely from ground zero.

9.9.2 *Structural and Network Issues*

The success of IoT hinges on network performance, which involves reliability and availability. Rural, remote sections that are highly populated might experience consistent failures across coverage, leading to losses and slowing response times (Lin et al., 2024). Overall, device batteries as a power source pose a significant limitation due to their limited capacity; therefore, their management is a critical element from the design phase onward, as otherwise frequent swapping soon disrupts workflows that depend on the data for continuous cycles (Sahoo et al., 2024).

Current setup methods are usually outdated because most do not support automated forms. Integration of backward-facing poses presents a complicated technical hurdle, plus extra budgets, considering there are older methods still in use for cities using data documentation based on paper (Liu et al., 2020). Data from connected devices puts strain on connectivity pathways from the cloud to servers, slowing processing and drastically increasing response times, depending on how efficiently traffic is streamlined.

9.9.3 *Technical Aspects to Device Installation*

Practical installation, along with continuous, smooth task performance, is part of everyday care and requires oversight from specialists. Degradations arise directly from nature, due to dust and heat exposure, ruining the quality of the signals emitted and impacting the levels initially provided by gadgets, which compromises the precision of collected feeds (Dey et al. 2023). Impedance introduced by objects can cause data interference, muddying measurements. Incompatible speaking language-wise from distinct firms that designed creates challenges for a smooth flow of actions when connecting.

Reacting speeds critically affect process actions because latency reduces efficiency; optimizing for real time is vital to avoid problems with critical responsiveness. Risks related to hacking breaches become significant because system designs provide entry points.

However, implementation, including programmed routine maintenance that addresses how precisely tools operate, should include regular evaluations. Hence, all detectors operate smoothly, delivering solid feedback while keeping environmental harm at a distance through sturdy encasements. Communication must occur through universal standards like MQTT, enabling systems to integrate operations without hitches caused by coding differences across diverse brands. Safety guidelines must be practiced, including password encryption, to prevent attacks and maintain the secure integrity and operational levels that devices provide. Finally, setting up contingency networks ensures activities do not close in the event of tech failures, preventing downtime.

9.9.4 *People, Organizational, and Social Issues*

Efficient waste management requires individuals within organizations to demonstrate readiness to adopt modern digital forms. Training is needed to avoid under-employment or improper execution despite innovative benefits (Dey et al. 2023). Resistance remains common, ranging from fear of job losses to shifts throughout the work cycle. Society vandalizes devices out of ignorance, destroying public property, while insufficient guidelines exacerbate these concerns, making project implementation harder.

To address this trend, investing in courses and employee training enhances knowledge of proper use and professional skill development. Modification plans must be implemented to limit negativity by informing about the evolution of capabilities, rather than removing placements. Education targeted at raising public awareness and engagement in community trash initiatives must be prioritized. Structures with precise control and connections among institutions strengthen collaboration, enabling the effective use of the IoT project's action plans for greater efficiency.

9.9.5 *Information and Analytical Issues*

Functionality counts heavily on the quality of collected data; therefore, there come struggles due to false feedback, technical disturbances, and environmental distortions (Nikouei et al. 2025). Building forecasting models is highly challenging due to significant changes in population, seasonal behavior, and regional influences. Analytics are compromised by cross-border computing, even via cloud solutions, compared to location-based analytics and centralized data processing. Latency versus strength throughout processing creates dilemmas, while concerns about protecting sensitive user data versus personal data remain chief challenges.

Significantly, automated confirmations for validation/cleaning improve the quality of the data collected. Employing refined study tools improves calculations for projections with debris generation. A hybrid approach that blends both clouds and data hubs maintains high speed while balancing efficiency. Defenses designed to protect data and maintain user privacy will consist of encryption and strict access controls that improve societal engagement, including openness, which encourages trust among groups and enhances teamwork within the scientific community, also boosting public belief.

9.9.6 *Environmental Consequences*

IoT's impact is significant because the fabrication of equipment and power components consumes energy, leading to emissions and waste. Changing parts frequently creates issues and raises concerns about smart containers in areas prone to damage and vandalism, resulting in shorter lifespans and increased financial concerns.

Adopting concepts that boost recycling alongside eco-materials from manufacturers must be a goal. Policy involving collection must ensure accountability for materials after use, ensuring proper recycling of products at the end of their life span. Using sectioned construction eases the removal of parts for swaps, preventing removals throughout pieces. Protective coverings and cameras throughout the location reduce malicious damage, thus extending functionality across devices.

9.9.7 Ethical Issues

The use of modern devices raises ethical questions because unequal access creates societal divides, undermining trust. After all, tracking could resemble government monitoring of certain groups. Job-cut potential also arises from the automation of job duties, negatively impacting economic conditions in parts-worker sectors (Dhinakaran et al. 2025).

Administration should employ approaches that promote equity, so that everyone shares the benefits of smart waste actions in the area. Discussions must focus on why data gathering is necessary to enhance reliability across societal groups. Ethical measurement practices that review prospective community involvement are essential. Job retraining programs help employees adapt to rapidly changing technology.

9.10 Policy and Regulatory Considerations to IoT-Enabled Smart Waste Management

The successful merging of technology and waste management depends on a policy that ensures accountability, enhances compatibility, safeguards civil rights, and protects the sustainability of the surrounding ecosystem (Onifade et al. 2024). Below are some of the policy and regulatory considerations:

9.10.1 Securing Data and Keeping it Private

The tools in question generate volume information about location, linked to personal data points, which poses ethical and legal risks without tight controls. Administration must implement a robust information defense that specifies which data stores exist, who owns them, and how they are stored. Governments must come from universal practices while requiring assessments. Punishments are strictly enforced for breaches of privacy to prevent misuse of information.

9.10.2 Making Systems Work Together (Interoperability)

Missing standards across devices increases incompatibilities and costs, including inefficiencies when upgrades are included in trash programs (Ankit et al., 2021). Regulatory agencies encourage shared formats plus standards. Certification schemes ensure adherence levels meet performance goals for safe operations. Open sourcing enhances adaptation and discourages locked-in vendors, thereby improving product compatibility.

9.10.3 Fair and Sustainable Funding

Investment-intensive often, requires stable finances that is long term because changeable budgets, processes, while contracts reduces sustainability levels (Pérez-Sánchez et al. 2022). Approaches must show apparent competing approaches that support IoT programs. Administration establishes a changeable agreement that approves updates with time alongside incentives encourage from private sectors into environmentally IoT types. Assigning funds specific that is innovative sustains continuity finances, enhancing actions among city substructures.

9.10.4 Accountability and Liability

Determining liabilities becomes complex when many individuals work together; device failure can result in mismanaged rubbish, leading to liabilities and problems (Bassan and Rabitti, 2024). Structures from legal systems point to parts plus responsibility from organizations or manufacturers/service providers. Contracts that specify responsibilities include procedures and standards, such as insurance, to ensure the successful handling of operational threats and offer remedies, including conflict-resolution protocols.

9.10.5 Environmental and Ethical Responsibility

Equipment produces waste that raises questions, plus protecting ecological sustainability matters when guarantees benefits reach groups similarly (Hajam et al., 2023). Policies must promote the removal of waste, the efficient use of recyclable materials, and the fair sharing of benefits throughout the area, while maintaining standards. Analyzing environments must precede expanding IoT to assess risks.

9.11 Conclusion

In conclusion, this study highlights how digital innovation is reshaping modern waste systems, moving them from reactive, labor-intensive operations to proactive, data-driven, and highly efficient urban services. The chapter demonstrates that IoT-enabled sensing, real-time monitoring, AI-driven analytics, and emerging technologies such as blockchain and robotics not only streamline collection processes but also reduce environmental impacts, operational costs, and resource consumption. As cities continue to expand and sustainability pressures increase, the integration of advanced technologies becomes not just beneficial but essential for achieving long-term circular-economy goals. Ultimately, the chapter reinforces that the future of waste management lies in adaptable, interconnected, and intelligent systems that support cleaner, smarter, and more resilient communities.

References

Addas, A., Khan, M. N., & Naseer, F. (2024). Waste management 2.0 leveraging internet of things for an efficient and eco-friendly smart city solution. *PLoS One, 19*(7). https://doi.org/10.1371/journal.pone.0307608

Adel, R., Hatem, A., & Sherif, S. (2025). Route optimization for school buses using real-time GPS and geospatial data. *International Telecommunications Conference (ITC-Egypt)*, 465–470. https://doi.org/10.1109/itc-egypt66095.2025.11186642

Aghenta, L., & Iqbal, M. (2019). Low-cost, open source IOT-based SCADA system design using Thinger.IO and esp32 thing. *Electronics, 8*(8), 822. https://doi.org/10.3390/electronics8080822

Al-Mashaqbeh, I., & Al Nawayseh, M. (2021). Smart waste management system based on IoT technology. *Environmental Monitoring and Assessment, 193*(6), 377. https://doi.org/10.1007/s10661-021-09042-3

Anagnostopoulos, T., Zaslavsky, A., Kolomvatsos, K., Medvedev, A., Khoruzhnikov, S., & Hadjiefthymiades, S. (2017). Challenges and opportunities of waste management in IoT-enabled smart cities: A survey. *IEEE Transactions on Sustainable Computing, 2*(3), 275–289. https://doi.org/10.1109/TSUSC.2017.2691049

Ankit, S., Saha, L., Kumar, V., Tiwari, J. S., Rawat, S., Singh, J., & Bauddh, K. (2021). Electronic waste and their leachates impact on human health and environment: Global ecological threat and management. *Environmental Technology & Innovation, 24*, 102049. https://doi.org/10.1016/j.eti.2021.102049

Apanavičienė, R., & Shahrabani, M. (2023). Key factors affecting smart building integration into smart city: Technological aspects. *Smart Cities, 6*(4), 1832–1857. https://doi.org/10.3390/smartcities6040085

Askarzadeh, T., Bridgelall, R., & Tolliver, D. (2024). Monitoring nodal transportation assets with uncrewed aerial vehicles: A comprehensive review. *drones, 8*(6), 233. https://doi.org/10.3390/drones8060233

Atzori, L., Iera, A., & Morabito, G. (2017). The internet of things: A survey. *Computer Networks, 56*(15), 2787–2805. https://doi.org/10.1016/j.comnet.2012.05.020

Aziz, F., Arof, H., Mokhtar, N., Shah, N., Khairuddin, A., Hanafi, E., & Talip, M. (2018). Waste level detection and hmm based collection scheduling of multiple bins. *PLoS One, 13*(8), e0202092. https://doi.org/10.1371/journal.pone.0202092

Bassan, F., & Rabitti, M. (2024). From smart legal contracts to contracts on blockchain: An empirical investigation. *Computer Law & Security Review, 55*, 106035. https://doi.org/10.1016/j.clsr.2024.106035

Bauermann, B., Bussador, A., Bauermann, H., & Matrakas, M. (2024). Connecting the green to the digital: Integrating eco cities and smart regions. *Eco Cities, 5*(1), 2755. https://doi.org/10.54517/ec.v5i1.2755

Bibri, S., Huang, J., Jagatheesaperumal, S., & Krogstie, J. (2024). The synergistic interplay of artificial intelligence and digital twin in environmentally planning sustainable smart cities: A comprehensive systematic review. *Environmental Science and Ecotechnology, 20*, 100433. https://doi.org/10.1016/j.ese.2024.100433

Choudhary, R., Kumar, A., Narayan, S., Mishchenko, I., Choudhury, M., Bhardwaj, L. K., Aqueel, A., Mykola, A., Petro, G., & Boiko, O. (2025). Integrating fuzzy logic into life cycle assessment for sustainable municipal solid waste management in metro cities. *Ecological Engineering & Environmental Technology, 26*(7), 316–331. https://doi.org/10.12912/27197050/205500

Choudhury, M., Suhag, M., Jha, P., Harne, K. R., Samanta, P., Choudhury, M., Borgohain, R. B., Majumdar, S., Kravchenko, Y., & Choudhary, R. (2025). Urban waste management and challenges in emerging smart city systems governance and circular economy perspectives. *Ecological Engineering & Environmental Technology, 26*(10), 243–266. https://doi.org/10.12912/27197050/211014

Ciawi, Y., Tonyes, S., & Dwipayanti, N. (2024). Redefining packaging solutions: The advantages of split-layer packaging for waste reduction and climate action. *Journal of Modern Industry and Manufacturing, 3*, 12. 10.53964/jmim.2024012.

Curtis, A., & Sarc, R. (2021). Real-time monitoring of volume flow, mass flow and shredder power consumption in mixed solid waste processing. *Waste Management, 131*, 41–49. https://doi.org/10.1016/j.wasman.2021.05.024

Dandong, W., & Yifan, Y. (2020). Design of municipal solid waste intelligent supervision platform based on big data. In *2020 2nd international conference on machine learning, big data and business intelligence (MLBDBI)* (pp. 261–265). https://doi.org/10.1109/mlbdbi51377.2020.00055

de Morais, C. S., Ramos, T. R., & Barbosa-Póvoa, A. P. (2019). Dynamic approaches to solve the smart waste collection routing problem. In *Springer proceedings in mathematics & statistics* (pp. 173–188). https://doi.org/10.1007/978-3-030-10731-4_12

Devi, A., Singh, N., Thandra, N., Gupta, N., Salman, Z., & Karuppiah, N. (2025). AI and IoT-enabled smart urban waste management system for efficient collection, segregation, and disposal. *E3s Web of Conferences, 619*, 03002. https://doi.org/10.1051/e3sconf/202561903002

Dey, P., & Chatterjee, K. (2023). Application of machine learning for optimization. *Advances in Business Information Systems and Analytics*, 113–127. https://doi.org/10.4018/978-1-6684-7105-0.ch007

Dey, S., Ganugula, T. V., Padavala, A. B., & Manoj, A. (2023). Recycling of e-waste materials for controlling the environmental and human health degradation in India. *Green Analytical Chemistry, 7*, 100085. https://doi.org/10.1016/j.greeac.2023.100085

Dhinakaran, D., Edwin Raja, S., Ramathilagam, A., Vennila, G., & Alagulakshmi, A. (2025). Ethical and legal challenges with IoT in home digital twins. *MethodsX, 14*, 103409. https://doi.org/10.1016/j.mex.2025.103409

Elendu, C., Elendu, T., & Elendu, I. (2024). 5g-enabled smart hospitals: Innovations in patient care and facility management. *Medicine, 103*(20), e38239. https://doi.org/10.1097/md.0000000000038239

Elijah, O., Rahman, T., Orikumhi, I., Leow, C. Y., & Hindia, M. N. (2018). An overview of internet of things (IoT) and data analytics in smart waste management. *IEEE Access, 6*, 12380–12395. https://doi.org/10.1109/ACCESS.2018.2804539

European Commission. (2019). *Smart waste collection in Barcelona*. Smart Cities Marketplace. https://smart-cities-marketplace.ec.europa.eu

Faiz, F., Ninduwezuor-Ehiobu, N., Adanma, U., & Solomon, N. (2024). Blockchain for sustainable waste management: Enhancing transparency and accountability in waste disposal. *Comprehensive Research and Reviews in Science and Technology, 2*(1), 045–069. https://doi.org/10.57219/crrst.2024.2.1.0032

Faridi, M., Ali, S., Wang, G., Awan, S., & Iqbal, M. (2023). Chainapparel: A trustworthy blockchain and IoT-based traceability framework for apparel industry 4.0. *Computers Materials & Continua, 77*(2), 1837–1854. https://doi.org/10.32604/cmc.2023.041929

Fatimah, Y. A., Govindan, K., Murniningsih, R., & Setiawan, A. (2021). Smart waste management system: A review. *Journal of Cleaner Production, 294*, 126254. https://doi.org/10.1016/j.jclepro.2021.126254

Ferrer, J., & Alba, E. (2019). Bin-CT: Urban waste collection based on predicting the container fill level. *Biosystems, 186*, 103962. https://doi.org/10.1016/j.biosystems.2019.04.006

Ferronato, N., & Torretta, V. (2019). Waste mismanagement in developing countries: A review of global issues. *International Journal of Environmental Research and Public Health, 16*(6), 1060. https://doi.org/10.3390/ijerph16061060

Fuqaha, S., & Nursetiawan, N. (2025). Artificial intelligence and IoT for smart waste management: Challenges, opportunities, and future directions. *Journal of Future Artificial Intelligence and Technologies, 2*, 24–46. https://doi.org/10.62411/faith.3048-3719-85

Gamberini, L., & Pluchino, P. (2024). Industry 5.0: A comprehensive insight into the future of work, social sustainability, sustainable development, and career. *Australian Journal of Career Development, 33*(1), 5–14. https://doi.org/10.1177/10384162241231118

Garapati, R. S. (2025). Real-time monitoring systems: From sensor networks to predictive analytics. *Artificial Intelligence-Based Systems, Cloud Computing, Web Interfaces, IoT/Connected Devices, Smart Automation, Real-Time Monitoring*. https://doi.org/10.70593/978-93-7185-625-6_6

Guerrero, L. A., Maas, G., & Hogland, W. (2013). Solid waste management challenges for cities in developing countries. *Waste Management, 33*(1), 220–232. https://doi.org/10.1016/j.wasman.2012.09.008

Hajam, Y. A., Kumar, R., & Kumar, A. (2023). Environmental waste management strategies and vermi transformation for sustainable development. *Environmental Challenges, 13*, 100747. https://doi.org/10.1016/j.envc.2023.100747

Huang, C., Chen, Q., Lin, Z., Su, W., Zhang, Z., & Liu, S. (2024). The economic impact and application challenges of IoT technology in smart home and infrastructure. *Journal of Innovation and Development, 6*(1), 35–39. https://doi.org/10.54097/y54jkp60

International Telecommunication Union (ITU). (2022). *IoT for smart sustainable cities*. ITU Publications. https://www.itu.int

Jaafreh, M., & Allouzi, M. (2023). Smart cities adoption in Saudi Arabia: A comprehensive review and future drivers. *British Journal of Multidisciplinary and Advanced Studies, 4*(5), 20–39. https://doi.org/10.37745/bjmas.2022.0300

Katah, I. and Ozturk, O. (2024). Smart cities and sustainability: Comparative analysis and strategic insights for Qatar. https://doi.org/10.20944/preprints202410.1106.v1

Kaza, S., Yao, L., Bhada-Tata, P., & Van Woerden, F. (2018). *What a waste 2.0: A global snapshot of solid waste management to 2050*. World Bank. https://doi.org/10.1596/978-1-4648-1329-0

Khan, M. (2023). Manufacturing waste for sustainable energy generation: A comprehensive review of current methods and future trends. https://doi.org/10.20944/preprints202311.0935.v1

Korea Environment Corporation. (2021). Smart food waste management in South Korea. https://www.keco.or.kr

Kowsalya, P., Jennifer Flora, G., Kousalya, A., Toha, M. A., Khan, P. A., Choudhury, M., & Parthasarathi, T. (2025). Circular systems unveiled: Essential components and their impact on recycling, reuse, and resource recovery. In S. Majumdar, M. Choudhury, & A. Cheshmehzangi (Eds.), *Sustainable urban future. Urban sustainability* (pp. 55–71). Springer. https://doi.org/10.1007/978-981-95-2976-6_3

Kuai, X., He, X., He, B., Liu, Y., Zhigang, Z., & Guo, R. (2024). Smart city ontology framework for urban data integration and governance applications. https://doi.org/10.20944/preprints202410.2577.v1

Kumar Singu, S. (2024). Real-time data integration in the cloud with informatica powercenter: A practical approach. *International Journal of Science and Research (IJSR), 13*(2), 1886–1891. https://doi.org/10.21275/sr240210091406

Liao, J. (2022). Self-employment in the construction machinery industry with blockchain technology in mind. *Kinetic Mechanical Engineering, 3*(4), 10.38007/kme.2022.030403.

Lin, H. (2024). A secure transaction and smart contract architecture for internet of vehicles. *Science Progress, 107*(4). https://doi.org/10.1177/00368504241296039

Lin, H., Kishk, M., & Alouini, M.-S. (2024). Performance evaluation of RF-powered IoT in rural areas: The wireless power digital divide. In *IEEE transactions on green communications and networking* (pp. 1–1). https://doi.org/10.1109/TGCN.2024.3350787

Liu, W., Huang, G., Zheng, A., & Liu, J. (2020). Research on the optimization of IIoT data processing latency. *Computer Communications, 151*. https://doi.org/10.1016/j.comcom.2020.01.007

Lloret, A., Peral, J., Ferrández, A., Auladell, M., & Muñoz, R. (2025). A data-driven framework for digital transformation in smart cities: Integrating AI, dashboards, and IoT readiness. *Sensors, 25*(16), 5179. https://doi.org/10.3390/s25165179

Lopes, J., Gomes, S., & Nogueira, E. (2025). Pathways to circularity: Engagement patterns of European SMEs in the circular economy. *Business Strategy and the Environment, 34*(3), 3848–3864. https://doi.org/10.1002/bse.4192

Majumdar, S. (2021). A study of dengue disease data by GIS in Kolkata city: An approach to healthcare informatics. In A. Suresh & S. Paiva (Eds.), *Deep learning and edge computing solutions for high performance computing. EAI/springer innovations in communication and computing*. Springer. https://doi.org/10.1007/978-3-030-60265-9_3

Majumdar, S. (2023). Uses of artificial intelligence and GIS technologies in healthcare services. In R. S. Tomar, R. K. Chauhan, & S. K. Nayak (Eds.), *Artificial intelligence and data science in contemporary healthcare (chapter 18)*. CRC Press. https://doi.org/10.1201/9781003363606-18

Majumdar, S., Kandpal, V., & Anthopoulos, L. G. (2024). *Smart cities: Innovations, challenges and future perspectives*. Springer Nature Switzerland AG. https://doi.org/10.1007/978-3-031-59846-3

Majumdar, S., Choudhury, M., & Cheshmehzangi, A. (Eds.). (2025). *Sustainable urban future: Embracing circular economy in waste management*. Springer. https://doi.org/10.1007/978-981-95-2976-6

Marshall, R. E., & Farahbakhsh, K. (2013). Systems approaches to integrated solid waste management in developing countries. *Waste Management, 33*(4), 988–1003. https://doi.org/10.1016/j.wasman.2012.12.023

Megha, A. (2023). Cross-efficiency evaluation framework for assessing the sustainability of agriculture farms with biogas plants: Mathematical modeling and solution approach. *International Research Journal of Modernization in Engineering Technology and Science*. https://doi.org/10.56726/irjmets43684

Mekki, K., Bajic, E., Chaxel, F., & Meyer, F. (2019). A comparative study of LPWAN technologies for large-scale IoT deployment. *ICT Express, 5*(1), 1–7. https://doi.org/10.1016/j.icte.2017.12.005

Mishra, R., Singh, E., Kumar, A., & Kumar, S. (2021). Artificial intelligence models for forecasting of municipal solid waste generation. *Soft Computing Techniques in Solid Waste and Wastewater Management*, 289–304. https://doi.org/10.1016/b978-0-12-824463-0.00019-7

Mokalled, T., Ahmad, M., & Fawaz, W. (2019). Smart waste collection system based on IoT for smart cities. *International Journal of Advanced Computer Science and Applications, 10*(4), 480–485. https://doi.org/10.14569/IJACSA.2019.0100463

Mora, L., Bolici, R., & Deakin, M. (2019). The first two decades of smart-city research: A bibliometric analysis. *Journal of Urban Technology, 26*(2), 3–27. https://doi.org/10.1080/10630732.2019.1578837

Muto, M., & Tonooka, H. (2025). Monitoring long-term waste volume changes in landfills in developing countries using aster time-series digital surface model data. *Sensors, 25*(10), 3173. https://doi.org/10.3390/s25103173

Nanda, S., & Berruti, F. (2021). A technical review of bioenergy and resource recovery from municipal solid waste. *Journal of Hazardous Materials, 403*, 123970. https://doi.org/10.1016/j.jhazmat.2020.123970

National Environment Agency (NEA). (2024). Smart waste management initiatives. Government of Singapore. https://www.nea.gov.sg

Nesmachnow, S., Rossit, D., & Moreno-Bernal, P. (2025). A literature review of recent advances on innovative computational tools for waste management in smart cities. *Urban Science, 9*(1), 16. https://doi.org/10.3390/urbansci9010016

Nikouei, M., Baroutian, B., Nabavi, S., Taraghi, F., Aghaei, A., Sajedi, A., & Ebrahimi Moghaddam, M. (2025). Small object detection: A comprehensive survey on challenges, techniques and real-world applications. *Intelligent Systems with Applications, 27*, 200561. https://doi.org/10.1016/j.iswa.2025.200561

Nour, M., Ismail, M., & Quddus, N. (2020). Waste management data gaps and challenges: A review of municipal waste management practices. *Waste Management & Research, 38*(7), 708–719. https://doi.org/10.1177/0734242X20916592

Onifade, M., Zvarivadza, T., Adebisi, J. A., Said, K. O., Dayo-Olupona, O., Lawal, A. I., & Khandelwal, M. (2024). Advancing toward sustainability: The emergence of green mining technologies and practices. *Green and Smart Mining Engineering, 1*(2), 157–174. https://doi.org/10.1016/j.gsme.2024.05.005

Pazhani, A., Khursheed, S., & Kausar, F. (2020). IoT-based smart waste management system for smart cities. *Procedia Computer Science, 172*, 448–453. https://doi.org/10.1016/j.procs.2020.05.072

Pérez-Sánchez, L. À., Velasco-Fernández, R., & Giampietro, M. (2022). Factors and actions for the sustainability of the residential sector: The nexus of energy, materials, space, and time use. *Renewable and Sustainable Energy Reviews, 161*, 112388. https://doi.org/10.1016/j.rser.2022.112388

Popa, C., Carutasu, G., Cotet, C., Carutasu, N., & Dobrescu, T. (2017). Smart city platform development for an automated waste collection system. *Sustainability, 9*(11), 2064. https://doi.org/10.3390/su9112064

Raut, R., Gardas, B., & Narkhede, B. (2021). Framework for sustainable smart waste management using IoT: A case of India. *Journal of Cleaner Production, 307*, 127133. https://doi.org/10.1016/j.jclepro.2021.127133

Saeed, S., Arshad, M., & Ahmed, A. (2023). Advancing circular economy in industrial chemistry and environmental engineering: Principles, alignment with United Nations sustainable development goals, and pathways to implementation. *European Journal of Chemistry, 14*(3), 414–428. https://doi.org/10.5155/eurjchem.14.3.414-428.2452

Sahoo, S., Kumar, S., & Sivarajah, U. (2024). Blockchain for sustainable supply chain management: Trends and ways forward. *Electronic Commerce Research, 24*, 1563–1618. https://doi.org/10.1007/s10660-022-09569-1

Salih, S., Abdelmaboud, A., Husain, O., Motwakel, A., Elshafie, H., Sharif, M., & Hamdan, M. (2025). IoT in urban development: Insight into smart city applications, case studies, challenges, and future prospects. *PeerJ Computer Science, 11*, e2816. https://doi.org/10.7717/peerj-cs.2816

Sami, A., Manzoor, U., Irfan, A., & Sarwar, F. (2023). Characterizing circular supply chain practices in industry 5.0 with respect to sustainable manufacturing operations. *Journal of Management and Research, 10*(1), 10.29145/jmr.101.04.

Sharma, A., Choudhury, M., Agarwal, S., Sharma, R., & Sharma, R. (2023). Corporate social responsibility and roles of developers for sustainability in companies. In *The route towards global sustainability: Challenges and management practices* (pp. 313–332). Springer International Publishing. https://doi.org/10.1007/978-3-031-10437-4_16

Sharma, R., & Rajoriya, S. (2021). Role of IoT in smart waste management: A comprehensive review and case evidence from India. *Sustainable Cities and Society, 68*, 102772. https://doi.org/10.1016/j.scs.2021.102772

Singh, K., & Singh, P. (2023). Fog-based edge AI for robotics: Cutting-edge research and future directions. *EAI Endorsed Transactions on AI and robotics, 2.* https://doi.org/10.4108/airo.3619

Singh, S. K., & Ordoñez de Pablos, P. (2021). Blockchain technology for waste management: A review. *Environmental Science and Pollution Research, 28*(22), 28126–28141. https://doi.org/10.1007/s11356-021-13206-1

Sivaranjani, S. S. P., Deepaharshini, G., & Eunice, J. (2024). Green symphony: A brief review of waste segregation techniques. *International Research Journal on Advanced Engineering Hub, 2*(03), 381–386. https://doi.org/10.47392/irjaeh.2024.0056

Sodiya, E., Umoga, U., Amoo, O., & Atadoga, A. (2024). Ai-driven warehouse automation: A comprehensive review of systems. *GSC advanced Research and Reviews, 18*(2), 272–282. https://doi.org/10.30574/gscarr.2024.18.2.0063

Solomon, N., Simpa, P., Adenekan, O., & Obasi, S. (2024). Sustainable nanomaterials' role in green supply chains and environmental sustainability. *Engineering Science & Technology Journal, 5*(5), 1678–1694. https://doi.org/10.51594/estj.v5i5.1136

Srikantha, N. (2017). Waste management in IoT-enabled smart cities: A survey. *International Journal of Engineering and Computer Science.* https://doi.org/10.18535/ijecs/v6i5.53

Stewart, C. (2024). A day in the life of a food packaging engineer. *Nutrition Today, 60*(1), 40–45. https://doi.org/10.1097/nt.0000000000000716

Strand, M., Syberfeldt, A., & Geertsen, A. (2020). A decision support system for sustainable waste collection. *Waste Management*, 347–365. https://doi.org/10.4018/978-1-7998-1210-4.ch016

Ta, V. (2025). Investigating the role of blockchain-enabled drivers on consumers' organic food consumption: A stimulus-organism-response perspective. *Environmental Research Communications, 7*(3), 035003. https://doi.org/10.1088/2515-7620/adb9be

Ugochukwu, C., Ofodile, O., Okoye, C., & Akinrinola, O. (2024). Sustainable smart cities: The role of fintech in promoting environmental sustainability. *Engineering Science & Technology Journal, 5*(3), 821–835. https://doi.org/10.51594/estj.v5i3.906

United Nations Environment Programme (UNEP). (2020). *Waste management outlook for Latin America and the Caribbean.* https://www.unep.org

United Nations Environment Programme (UNEP). (2024). *World must move beyond waste era and turn rubbish into resources UN report.* https://www.unep.org/news-and-stories/press-release/world-must-move-beyond-waste-era-and-turn-rubbish-resource-un-report

United Nations. (2019). *World urbanization prospects: The 2018 revision.* Department of Economic and Social Affairs. https://population.un.org/wup

Wu, D., Liao, W., Yu, H., Majumdar, S., Xu, J., & Choudhury, M. (2025). Assessing traditional waste management practices in urban environments and recent trends. In S. Majumdar, M. Choudhury, & A. Cheshmehzangi (Eds.), *Sustainable urban future. Urban sustainability.* Springer. https://doi.org/10.1007/978-981-95-2976-6_5

Xiao, Y., Zhao, Q., & Wang, S. (2019). Artificial intelligence in waste sorting: A review. *Resources, Conservation & Recycling, 150*, 104433. https://doi.org/10.1016/j.resconrec.2019.104433

Yadav, S. K., Rajagopal, K., Priya, A. K., & Sharma, G. D. (2020). Smart waste management and energy extraction from waste in Indian smart cities—A review. *Contaminants and Clean Technologies*, 321–330. https://doi.org/10.1201/9780429275852-19

Yevle, D. V., & Mann, P. S. (2025). Artificial intelligence based classification for waste management: A survey based on taxonomy, classification & future direction. *Computer Science Review, 56*, 100723. https://doi.org/10.1016/j.cosrev.2024.100723

Zambrano-Monserrate, M. A., Ruano, M. A., & Ormeño-Candelario, V. (2021). Determinants of municipal solid waste: A global analysis by countries' income level. *Environmental Science and Pollution Research, 28*(44), 62421–62430. https://doi.org/10.1007/s11356-021-15167-9

Zhu, D., Asnani, P. U., Zurbrugg, C., Anapolsky, S., & Mani, S. (2008). *Improving solid waste management in India: A sourcebook for policymakers and practitioners.* World Bank. https://doi.org/10.1596/978-0-8213-7361-3

CHAPTER 10

Circular Solutions: Designing Waste Processing Facilities for Resource Recovery

Hongbin Zhang, Linyu Wang, Zhaochen Zhang, Mingxuan Zhong, Jiajun Xu, Wanying Liao, and Sushobhan Majumdar

H. Zhang
School of Management, Zhejiang University, Hangzhou, China

L. Wang
School of Drama and Film, Jilin University of Arts, Changchun, China

Z. Zhang
School of Geosciences, China University of Petroleum, Qingdao, China

M. Zhong
Department of Chemical and Biological Engineering, University of British Columbia, Vancouver, BC, Canada

J. Xu (✉)
Institute for Advanced Studies, Universiti Malaya, Kuala Lumpur, Malaysia

W. Liao (✉)
Faculty of Arts and Social Sciences, Universiti Malaya, Kuala Lumpur, Malaysia

S. Majumdar
Department of Geography, Pandit Raghunath Murmu Smriti Mahavidyalaya, Bankura, West Bengal, India

S. Majumdar, M. Choudhury (eds.), *City of the Future*,
https://doi.org/10.1007/978-3-032-18026-1_10

10.1 Introduction

The intensification of urbanization has reshaped both the material and ecological dynamics of cities, generating profound governance challenges (Wu et al., 2025; Xu et al., 2023). Rapid demographic concentration in metropolitan regions (Liao et al., 2024a), accelerated spatial expansion (Angel, 2023), and mounting infrastructural stress (Savela et al., 2020) have amplified systemic vulnerabilities. Meanwhile, ecological degradation—evident in escalating pollution (Bhardwaj et al., 2025; Xu et al., 2026), wastewater contamination (Ilyas et al., 2024; Ren et al., 2025), and persistent reliance on fossil-based energy (Khan et al., 2024; Zhong et al., 2024; Zhang et al., 2025)—undermines the resilience of urban systems (Thompson et al., 2024; Ogwu et al., 2025; Wu et al., 2025). Although technological innovations have improved the precision of interventions, the systemic and interdependent nature of these crises resists rapid resolution (Xu et al., 2024; Liao et al., 2024b; Choudhury et al., 2025; Liao et al., 2025). Among these interlinked crises, the accelerating growth of urban waste stands out as a clear and pressing manifestation of unsustainable urbanization, highlighting the contradictions between material consumption, ecological limits, and governance capacity (Almusaed & Almssad, 2023; Das et al., 2024; Ahmad et al., 2025).

Within this context, the accelerating volume of urban waste has become one of the most acute expressions of unsustainable urbanization (Chen, 2018). The traditional "take–use–dispose" model (Fig. 10.1), defined by linearity, inefficiency, and resource loss, is increasingly incompatible with the imperatives of sustainable urban development (Han, 2023; Zhang et al., 2025). By contrast, advances in smart technologies and the growing adoption of circular economy principles provide new opportunities to restructure waste management systems (Luttenberger, 2020). In response, many scholars advocate for integrating intelligent technologies—including the Internet of Things (IoT) (Wang et al., 2021), big data analytics (BDA) (Lu et al., 2018), and artificial intelligence (AI) (Abdallah et al., 2020)—throughout the waste management cycle under the paradigm of circular economy and innovative waste management systems (Zhang et al., 2019; Luttenberger, 2020; Han, 2023).

Therefore, this chapter examines the management framework and technological optimization of waste disposal and resource recycling. Specifically, the chapter examines how intelligent technologies can function synergistically across the three hierarchical levels of waste management—source

Fig. 10.1 Traditional facilities for urban operations: waste collection points and local transfer stations (photographed from Wuhan, China)

avoidance in the production process, comprehensive waste classification and recycling, and final terminal disposal—to deliver systematic, integrated solutions. By embedding digital processes from micro-scale collection to macro-level infrastructure, such systems aim to achieve holistic optimization and circular operation, shifting waste management from end-of-pipe burden to resource regeneration. Ultimately, this chapter responds to its focus by proposing an integrated, driver-based architectural framework that aligns technological innovation with circular economy principles across all operational tiers.

10.2 Case Studies: Integrating Smart Waste Management Perspective

Waste management and recycling can be structured using a wide range of classification frameworks that reflect different disciplinary and practical perspectives (Fig. 10.2). Accordingly, this chapter adopts a case-oriented approach that integrates smart waste management as one of the key analytical perspectives, providing a representative entry point into the broader field of waste management research; using it as an applied lens to examine the processes of source avoidance, classification; and recycling across different operational levels and final disposal.

10.2.1 Source Avoidance in the Production Process

Source avoidance represents the highest priority in the waste management hierarchy, emphasizing proactive reduction of waste generation (Axil, 2024). Intelligent technologies—particularly AI, machine learning (ML), and IoT—are central to domains. On the one hand, AI and ML optimize product design by advancing the concept of "circularity by design" (Ghoreishi, 2019). Algorithms analyzing product lifecycle data enhance durability, reparability, disassembly, and recyclability, thereby lowering material consumption and improving recycling efficiency (Han et al., 2023). They also predict emissions, recycling costs, and reuse values across alternative scenarios, enabling firms to integrate environmental considerations into design (Yu et al., 2021).On the other hand, IoT devices monitor production processes in real time. Sensors and edge computing systems track water, energy, and material use, enabling dynamic operational adjustments to prevent overconsumption and minimize industrial solid

Fig. 10.2 Waste management and recycling classification: a mainstream overview

waste (Joshi et al., 2022). A notable example is Philips' SmartPath model, which embeds circular economy principles into the design of medical equipment. By extending product lifecycles and reducing waste generation, this approach has redefined sustainable design practices (Bakker & Hollander, 2013; Smith-Gillespie et al., 2018). In addition to sustainable product design, SmartSpeed's AI-driven acceleration enhances energy and resource efficiency: shorter scan times and fewer repeat examinations reduce power consumption, maintenance demand, and medical waste, reinforcing the environmental sustainability of Magnetic Resonance Imaging practice (Philips, 2025).

10.2.2 Comprehensive Waste Classification and Recycling

Resource recovery is the cornerstone of smart waste management, aiming to maximize the extraction of usable materials while reducing dependence on virgin resources (Axil, 2024). Intelligent systems that integrate comprehensive waste classification and recycling can enhance precision, automation, and transparency across three operational levels as follows (Fig. 10.3).

10.2.2.1 Front-End: Smart Collection Points

IoT-enabled collection points equipped with sensors, imaging, and weight recognition automate waste identification and reporting (Thürer et al., 2019; Badve et al., 2020). In India, RFID (Radio Frequency Identification)-based tracking systems reward compliant residents with points, improving participation and supporting optimized route planning (Ranjana et al., 2021). It demonstrates how technological integration can simultaneously enhance operational efficiency, improve environmental quality, and foster citizen engagement (Fig. 10.4).

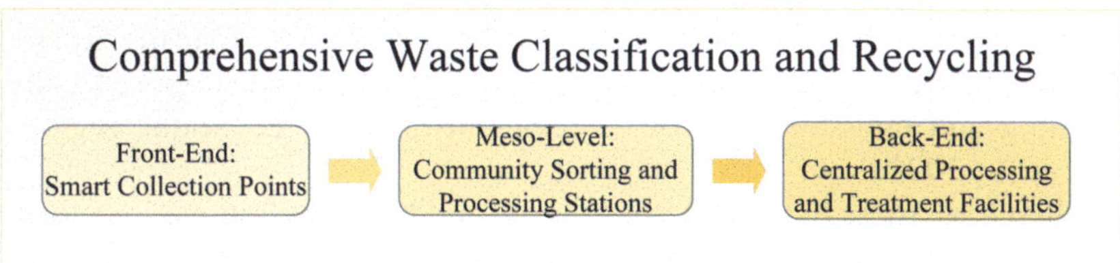

Fig. 10.3 Process chain of comprehensive waste classification and recycling

Fig. 10.4 ATRenew (formerly AiHuiShou): an example of smart recycling points operating for paper, metal, plastic, and fabric (photographed from Wuhan, China)

10.2.2.2 Meso-Level: Community Sorting and Processing Stations

Community stations link household disposal with centralized treatment. By employing AI-driven recognition and automated weighing systems, they improve classification accuracy while fostering behavioral change through feedback mechanisms (Sutikno et al., 2024). Shanghai's 2024 pilot program illustrates this approach, combining mobile app notifications, smart collection routes, and AI-enabled monitoring to enhance waste segregation and reduce visible street waste (Shanghai Municipal People's Government, 2024).

10.2.2.3 Back-End: Centralized Processing and Treatment Facilities

AI, robotics, and BDA support automated sorting, energy recovery, and emission control. In waste-to-energy plants, these technologies improve conversion efficiency through real-time monitoring and predictive analytics (Melinda et al., 2024). Artificial neural network models have been applied to predict waste-to-energy capacity and flue gas pollutant emissions at the city level, drawing on 322,926 monitoring values from 481 plants across China, which achieved high accuracy (Mean Square Errors = 0.003–0.190), illustrating the potential of AI to support facility planning and emission control strategies in different economic regions (Ma et al., 2023).

Therefore, these front-end, meso-level, and back-end innovations form a digital closed-loop system. At the micro scale, smart devices guide individual behavior; at the meso scale, community systems facilitate resource flows; and at the macro scale, AI-enhanced facilities maximize recovery while reducing risks.

10.2.3 Final Terminal Disposal

Although terminal disposal is the lowest priority in the hierarchy, it remains unavoidable (Axil, 2024). Intelligent landfills equipped with sensor networks and automated monitoring enable real-time tracking of gas emissions, leachate, and biological activity (Swapna et al., 2019). AI models further predict capacity, recovery potential, and risks, providing evidence for post-closure planning (Lakhouit, 2025).

For example, the Sudokwon Landfill in South Korea represents one of the world's largest and most advanced waste treatment complexes, integrating a 50-MW (Megawatt) landfill gas power plant and an environmental energy town that produces 2.61 million Gcal of energy annually; with a daily capacity of 18,000 tons and reclaimed areas converted into an eco-park, it exemplifies environmentally sound regional landfill management (Asian Development Bank, 2017). This model highlights the role of intelligent technologies in transforming terminal disposal into a component of broader ecological resilience.

10.3 The Smart Waste Management Hierarchy System (SWMHS)

Building on the preceding case studies, which demonstrated how digital tools and circular practices are being integrated into waste management systems (Fig. 10.5), this chapter proposes and advances the concept of SWMHS. The framework is designed not as an entirely novel paradigm but as a synthesis that consolidates empirical insights into a structured model, aligning smart technologies, circular economy principles, and the waste management hierarchy (Fig. 10.6) (Zhang et al., 2022). These three dimensions—technological capacity, value orientation, and structural logic—respectively provide a coherent foundation for both conceptual development and practical application. The distinct contribution of SWMHS lies in its deliberate integration of these domains into an adaptive system characterized by technological dynamism, resource circularity, and hierarchical coordination.

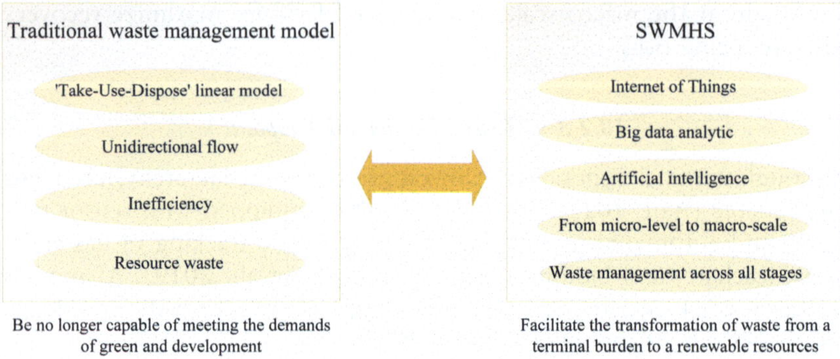

Fig. 10.5 The comparison between the traditional waste management model and SWMHS

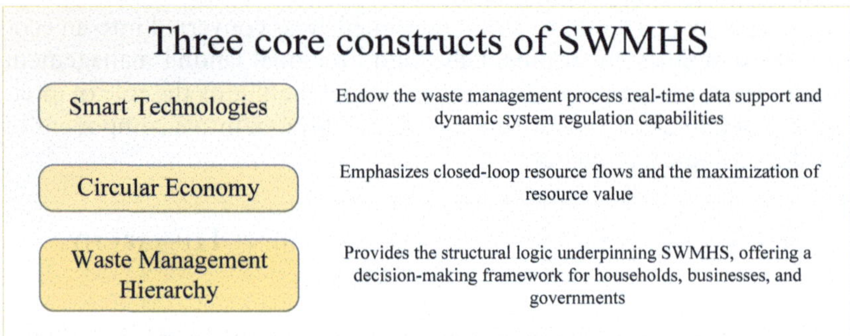

Fig. 10.6 The central core components of SWMHS

10.3.1 Smart Technologies

Digital innovations underpin SWMHS's operational infrastructure, enabling real-time sensing, adaptive regulation, and evidence-based optimization. IoT devices record waste generation frequency, bin overflow, and classification accuracy, while cloud computing and AI-driven analytics enhance the precision of sorting, collection, and resource recovery (Addas et al., 2024). Thus, these tools establish a "neural network" of perception, analysis, and execution that sustains closed-loop system management.

- Perception: Smart bins equipped with sensors transmit data to central platforms, improving collection efficiency and reducing manual interventions (Vishnu et al., 2021).
- Analysis: AI and BDA techniques detect patterns and forecast trends, supporting adaptive, data-informed decision-making (Lu et al., 2018).
- Execution: Centralized platforms adjust routes, classification standards, and treatment strategies in response to analytical outputs, ensuring responsive and integrated system control (Fang et al., 2023).

Emerging technologies such as edge computing and 5G (Fifth-Generation Wireless Technology) enhance processing speed and responsiveness (Longo et al., 2021), while digital platforms extend the scope of public participation through apps, tracking tools, and online information portals, thereby strengthening community engagement in waste governance (Liu, 2024).

10.3.2 Circular Economy

The circular economy perspective situates SWMHS within a broader urban transition from resource consumption to regeneration. By embedding the principles of "reduce, reuse, and recycle" across all stages of the waste cycle, SWMHS fosters systematic closed-loop flows, maximizes resource value, and supports the transformation of waste governance into resource-regenerative systems (Goyal et al., 2018).

10.3.3 Waste Management Hierarchy

The hierarchy provides the structural logic of SWMHS, guiding prioritization among households, industries, and governments. Preventive strategies to reduce waste generation are emphasized first, followed by recovery through reuse, recycling, and energy conversion, with environmentally responsible disposal reserved as the last resort (Axil, 2024; EPA, 2025). Within the SWMHS framework, this hierarchical logic is reinforced by intelligent technologies that eliminate traditional bottlenecks and discontinuities across stages, enabling an integrated, highly efficient system that spans from source reduction to final treatment.

10.4 Challenges and Barriers in SWMHS

While the proposed SWMHS demonstrates how data sharing, platform collaboration, and standardized interfaces can create interconnectivity across various scales, its operationalization remains hindered by structural constraints. For instance, behavioral data generated at smart recycling points can, in principle, be synchronized with dispatch operations at transfer stations and load forecasting at treatment facilities, thereby establishing a chain-responsive system. This reverse data flow—from individual behavior at the micro-level to city-wide regulation at the macro-level—suggests the possibility of dynamic governance and closed-loop control. Nevertheless, despite its conceptual advantages, the pathway to implementation faces persistent barriers that extend across social, technological, and institutional dimensions.

10.4.1 Behavioral Challenges

Public acceptance and participation remain uneven, undermining the system's capacity to function as intended. Particularly in less developed regions, limited environmental awareness and inconsistent participation rates compromise the efficiency of smart recycling initiatives (Zhang et al., 2019). While IoT-enabled devices can monitor, nudge, and regulate user behavior, their effectiveness diminishes significantly without the active cooperation of residents (Fig. 10.7), illustrating the limits of technological determinism in behavior-dependent systems (Zhang et al., 2019).

Fig. 10.7 Local guidance of free and paid used items and waste recycling practices on the public (photographed from Wuhan, China)

10.4.2 Technological and Infrastructural Barriers

Issues of interoperability, data standardization, and platform integration remain unresolved across administrative boundaries, leading to resource fragmentation and inefficiencies (Chang & Chang, 2003). In contexts where digital infrastructure is underdeveloped or technical maturity is lacking, system deployment is further constrained. These technical discontinuities create systemic vulnerabilities that undermine the very interconnectivity SWMHS seeks to achieve. In addition, most smart systems rely heavily on sensors, cloud platforms, and algorithmic models, requiring substantial upfront investment, which remains a considerable barrier, particularly in underdeveloped or rural areas (Fang et al., 2023).

10.4.3 Policy and Institutional Gaps

Existing municipal frameworks often fail to align the advancement of intelligent technologies with the objectives of a circular economy. Outdated regulations, limited subsidy schemes, and fragmented supervisory standards fail to provide sustained institutional support for the large-scale rollout of SWMHS (Fidje et al., 2023). Without adaptive and forward-looking governance structures, the system risks remaining at the pilot-project stage rather than achieving systemic integration. Meanwhile, given that these systems depend extensively on residents' behavioral, location, and consumption data, balancing the need for management efficiency with the protection of individual privacy has become an increasingly salient ethical concern to government (Schafer, 2014). Furthermore, regional disparities in infrastructure, policy support, and public awareness have resulted in varying levels of adaptability and effectiveness in system deployment and promotion (Chang & Chang, 2003). Collectively, the promise of SWMHS is tempered by multi-dimensional barriers that cannot be resolved solely through technological innovation. Its successful realization requires coordinated responses across behavioral engagement, technical standardization, and institutional reform. Only through such integrated governance can the system evolve from a conceptual framework into an operational cornerstone of sustainable urban waste management.

10.5 Recommendations for Advancing SWMHS

Building on the preceding case studies and conceptual framework, the effective advancement of SWMHS requires more than the deployment of isolated technologies. It demands systemic strategies that integrate technical, institutional, and behavioral dimensions into a coherent whole. The following recommendations provide a roadmap for realizing this integration.

10.5.1 Multi-level Deployment of Smart Technologies

To ensure systemic coherence, innovative technologies should be embedded across all hierarchical levels of waste management—from intelligent bins at the source to community-based transfer stations to AI-assisted treatment facilities, and ultimately to smart landfills (Fang et al., 2023). This full-spectrum deployment establishes an end-to-end digital infrastructure that enhances monitoring accuracy, operational efficiency, and predictive management capabilities (Fang et al., 2023). Importantly, the deployment strategy should be adaptive to local contexts, recognizing that infrastructural maturity varies across regions (Awino & Apitz, 2024). Without such contextual tailoring, the diffusion of technology risks reproducing spatial inequalities in system performance (Awino & Apitz, 2024).

10.5.2 Policy Frameworks for Seamless Waste Flows and Circular Economy Alignment

Institutional reforms are required to synchronize waste flows across the hierarchy. Governments, organizations, and academia should develop comprehensive policies and regulatory frameworks that establish uniform standards for sorting, transportation, recovery, and disposal, ensuring interoperability across jurisdictions and platforms (Anuardo et al., 2022). Crucially, these policies must be designed not merely to regulate existing practices, but to actively steer waste management systems toward circular economy objectives. By embedding requirements for reuse, recycling, and energy recovery into regulatory instruments, governments can transform SWMHS from a technological experiment into a structural enabler of urban sustainability.

10.5.3 Incentivizing Behavioral Participation Through Digital Engagement

As public participation remains the linchpin of system effectiveness, strategies must move beyond technical monitoring to cultivate sustained behavioral change (Konstantinidou et al., 2024). Incentive mechanisms that leverage gamification and digital platforms—such as reward points, discounts, or community recognition—can increase participation rates and normalize sustainable practices (Lim et al., 2025). However, such strategies must be accompanied by long-term environmental education and trust-building measures to avoid reducing civic engagement to short-term compliance driven solely by extrinsic rewards.

10.5.4 Cross-Sectoral Collaboration and Knowledge Sharing

The success of SWMHS hinges on collaboration between governments, enterprises, and civil society. Joint platforms for data sharing, innovation pilots, and knowledge exchange can reduce institutional fragmentation and promote scalability (Tutak & Brodny, 2025). Moreover, aligning municipal initiatives with international best practices creates opportunities for benchmarking and continuous improvement, preventing system stagnation and policy myopia.

In sum, advancing SWMHS requires coordinated interventions across technological, institutional, and social domains. Its effectiveness will not be determined solely by technological sophistication, but by the degree to which it is embedded in inclusive governance, adaptive regulation, and civic participation. Only through this multi-dimensional alignment can SWMHS evolve from a conceptual model into an operational paradigm for sustainable urban waste management.

10.6 Conclusion

This chapter, based on a review of conventional waste management systems, has proposed a more forward-looking and systematic SWMHS. This framework transcends the traditional end-of-treatment-centric model by integrating digital technologies and circular economy principles, comprehensively restructuring each phase of waste collection, classification, processing, and reuse. By incorporating technologies such as IoT, BDA, and AI, the system not only enables real-time visualization and dynamic management of waste flows but also promotes a strategic shift from "quantity-based control" to "quality-driven efficiency."

The proposed SWMHS retains the classical structure of "avoidance, recovery, and disposal," but embeds digital and intelligent technologies at each stage to enhance efficiency and sustainability.

(i) At the avoidance stage, AI- and IoT-enabled product design supports optimized lifecycle management, reducing resource use and extending service life to minimize waste generation at the source.
(ii) At the recovery stage, technology-driven recognition and automated sorting improve the precision of material classification, increase recycling rates, and expand waste-to-energy conversion, while algorithmic dispatch and optimized routing enhance collection efficiency and reduce environmental burdens.
(iii) At the disposal stage, intelligent technologies enable more effective landfill operation, increase resource recovery, and mitigate ecological risks, advancing sustainable treatment and energy recovery pathways.

Overall, the system's flexibility and scalability enable adaptation to diverse regional contexts, linking waste governance to broader smart city agendas in which the circular economy plays a central role in reconciling economic growth with environmental sustainability. By operationalizing the sequence of "avoidance, recovery, and disposal," SWMHS is positioned as a strategic lever for building closed-loop resource systems. Nevertheless, implementing such intelligent hierarchies remains constrained by trade-offs between technological advancement and financial costs.

Looking forward, SWMHS is expected to evolve toward greater precision, efficiency, and coordination, gradually forming an intelligent circular network that aligns with the objectives of smart city governance. This network requires not only technical upgrades but also inter-system integration and multi-stakeholder collaboration, linking households, enterprises, sanitation departments, and recycling operators through unified data platforms. In doing so, waste governance can shift emphasis from reactive end-treatment to preventive, process-oriented, and full-cycle optimization.

Future iterations will rely on more advanced sensing and analytics of behavioral patterns, waste composition, and facility performance, facilitating predictive and adaptive management. As smart waste systems strengthen their integration with other urban subsystems such as energy and mobility, they will accelerate the transition from conventional back-end management to a model of comprehensive smart governance—anchored in front-end guidance, process optimization, and resource reintegration—thereby advancing the co-evolution of smart and sustainable cities.

Acknowledgments The authors sincerely thank Ms. Zhao Chunping (赵春平) of New Frontiers International Planning & Design (Hubei Province) Co., Ltd. [新前沿国际规划设计(湖北省)有限责任公司], China, for kindly providing the on-site photographs (photographed from Wuhan, Hubei, China, on 8 October 2025) used in this chapter and for their valuable assistance and support throughout this research.

References

Abdallah, M., Talib, M. A., Feroz, S., Nasir, Q., Abdalla, H., & Mahfood, B. (2020). Artificial intelligence applications in solid waste management: A systematic research review. *Waste Management, 109*, 231–246. https://doi.org/10.1016/j.wasman.2020.04.057

Addas, A., Khan, M. N., & Naseer, F. (2024). Waste management 2.0 leveraging Internet of Things for an efficient and eco-friendly smart city solution. *PLOS One, 19*(7), e0307608. https://doi.org/10.1371/journal.pone.0307608

Ahmad, G., Aleem, F. M., Alyas, T., Abbas, Q., Nawaz, W., Ghazal, T. M., Aziz, A., Aleem, S., Tabassum, N., & Ibrahim, A. M. (2025). Intelligent waste sorting for urban sustainability using deep learning. *Scientific Reports, 15*(1), 27078. https://doi.org/10.1038/s41598-025-08461-w

Almusaed, A., & Almssad, A. (2023). Sustainable regional planning. *IntechOpen*. https://doi.org/10.5772/intechopen.105225

Angel, S. (2023). Urban expansion: Theory, evidence and practice. *Buildings and Cities, 4*(1), 124–138. https://doi.org/10.5334/bc.348

Anuardo, R. G., Espuny, M., Costa, A. C. F., & Oliveira, O. J. (2022). Toward a cleaner and more sustainable world: A framework to develop and improve waste management through organizations, governments and academia. *Heliyon, 8*(4), e09225. https://doi.org/10.1016/j.heliyon.2022.e09225

Asian Development Bank. (2017). Retrieved 15/04/2025, from https://events.development.asia/materials/20170516/sudokwon-landfill-site-management

Awino, F. B., & Apitz, S. E. (2024). Solid waste management in the context of the waste hierarchy and circular economy frameworks: An international critical review. *Integrated Environmental Assessment and Management, 20*(1), 9–35. https://doi.org/10.1002/ieam.4774

Axil. (2024). *What is a waste management hierarchy?* Retrieved 14/04/2025, from https://axil-is.com/blogs-articles/waste-management-hierarchy/

Badve, M., Chaudhari, A., Davda, P., Bagaria, V., & Kalbande, D. (2020). Garbage collection system using IoT for Smart City. In 2020 Fourth International Conference on I-SMAC (IoT in Social, Mobile, Analytics and Cloud) (I-SMAC) (pp. 138–143). IEEE. https://doi.org/10.1109/I-SMAC49090.2020.9243387

Bakker, C., & Hollander, M. D. (2013). Six design strategies for longer lasting products in circular economy. *The Guardian*. Retrieved from https://research.tudelft.nl/en/publications/six-design-strategies-for-longer-lasting-products-in-circular-eco

Bhardwaj, L. K., Choudhury, M., Goswami, L., Rahman, F., & Xu, J. (2025). Micro- and nanoplastics: Its impact on Antarctica and similar ecosystems. In S. Goswami, M. Choudhury, & P. Samanta (Eds.), *Micro-nano plastics exposure, environmental degradation and public health crisis: Perspectives and concerns for sustainable development* (pp. 127–135). Springer Nature. https://doi.org/10.1007/978-981-96-5047-7_11

Chang, Y. H., & Chang, N. B. (2003). Compatibility analysis of material and energy recovery in a regional solid waste management system. *Journal of the Air & Waste Management Association, 53*(1), 32–40. https://doi.org/10.1080/10473289.2003.10466113

Chen, Y. (2018). Effects of urbanization on municipal solid waste composition. *Waste Management, 79*, 828–836. https://doi.org/10.1016/j.wasman.2018.04.017

Choudhury, M., Adithya, A., Xu, J., Liao, Y., Herlekar, M., & Srivastava, V. (2025). Transforming wastewater management: Machine learning and artificial intelligence in decision-making processes. In M. Choudhury, S. Majumdar, S. Goswami, & M. Sillanpää (Eds.), (Vol. 15, pp. 246–256). Royal Society of Chemistry. https://doi.org/10.1039/9781837676880-00246

Das, S., Choudhury, M. R., Chatterjee, B., Das, P., Bagri, S., Paul, D., Bera, M., & Dutta, S. (2024). Unraveling the urban climate crisis: Exploring the nexus of urbanization, climate change, and their impacts on the environment and human well-being – A global perspective. *AIMS Public Health, 11*(3), 963–1001. https://doi.org/10.3934/publichealth.2024050

EPA. (2025). *The waste hierarchy*. Retrieved from https://www.epa.nsw.gov.au/Your-environment/Recycling-and-reuse/warr-strategy/The-waste-hierarchy

Fang, B., Yu, J., Chen, Z., Osman, A. I., Farghali, M., Ihara, I., Hamza, E. H., Rooney, D. W., & Yap, P. (2023). Artificial intelligence for waste management in smart cities: A review. *Environmental Chemistry Letters, 21*(4), 1959–1989. https://doi.org/10.1007/s10311-023-01604-3

Fidje, E., Haddara, M., & Langseth, M. (2023). Sustainable smart waste management adoption challenges in developing countries. *Procedia Computer Science, 219*, 905–914. https://doi.org/10.1016/j.procs.2023.01.366

Ghoreishi, S. (2019). *Assessing the role of artificial intelligence in product design towards circularity* [Master's Thesis, Lappeenranta-Lahti University of Technology LUT]. Retrieved from https://urn.fi/URN:NBN:fi-fe2019072223160

Goyal, S., Esposito, M., & Kapoor, A. (2018). Circular economy business models in developing economies: Lessons from India on reduce, recycle, and reuse paradigms. *Thunderbird International Business Review, 60*(5), 729–740. https://doi.org/10.1002/tie.21883

Han, Y. (2023). Advancing rural waste management program in China: Enlightenment from the case study. *Molodyi Vchenyi*, 27–30. Retrieved from https://molodyivchenyi.ua/omp/index.php/conference/catalog/download/41/829/1724-1

Han, Y., Shevchenko, T., Yannou, B., Ranjbari, M., Shams Esfandabadi, Z., Saidani, M., Bouillass, G., Bliumska-Danko, K., & Li, G. (2023). Exploring how digital technologies enable a circular economy of products. *Sustainability, 15*(3), 2067. https://doi.org/10.3390/su15032067

Ilyas, M., Liao, Y., Xu, J., Wu, S., Liao, W., & Zhao, X. (2024). Removal of anthracene from vehicle-wash wastewater through adsorption using eucalyptus wood waste-derived biochar. *Desalination and Water Treatment, 317*, 100115. https://doi.org/10.1016/j.dwt.2024.100115

Joshi, L. M., Bharti, R. K., Singh, R., & Malik, P. K. (2022). Real time monitoring of solid waste with customized hardware and Internet of Things. *Computers and Electrical Engineering, 102*, 108262. https://doi.org/10.1016/j.compeleceng.2022.108262

Khan, R., Haram, Z., Ahmad, W., Sohni, S., Xu, J., & Ilyas, M. (2024). Removal of hydrocarbon pollutants from refinery wastewater using N-hexadecylchitosan as an efficient adsorptive platform. *Scientific Reports, 14*(1), 17236. https://doi.org/10.1038/s41598-024-66429-8

Konstantinidou, A., Ioannou, K., Tsantopoulos, G., & Arabatzis, G. (2024). Citizens' attitudes and practices towards waste reduction, separation, and recycling: A systematic review. *Sustainability, 16*(22), 9969. https://doi.org/10.3390/su16229969

Lakhouit, A. (2025). Revolutionizing urban solid waste management with AI and IoT: A review of smart solutions for waste collection, sorting, and recycling. *Results in Engineering, 25*, 104018. https://doi.org/10.1016/j.rineng.2025.104018

Liao, W., Wu, D., Guo, Z., Zhao, C., Xu, J., & Xu, Y. (2024a). Variation analysis in historical demographic distribution within urban agglomeration: Experimental evidence from the Wuhan "1+8" City Circle. *Sustainable Social Development, 2*(4), 2778. https://doi.org/10.54517/ssd.v2i4.2778

Liao, Y., Liao, W., Wu, D., Zhang, C., Xu, J., & Majumdar, S. (2024b). Inspired applications of artificial intelligence in disaster-resilient regional construction for climate risk. In S. Ghosh, S. Majumdar, & A. Cheshmehzangi (Eds.), *Cities of tomorrow: Urban resilience and climate change preparedness* (pp. 305–315). Springer Nature. https://doi.org/10.1007/978-981-97-9658-8_16

Liao, W., Tan, X., Xu, J., Kumar, P., & Choudhury, M. (2025). Navigating the wastewater landscape: Defining spatial intelligence for effective management. In M. Choudhury, S. Majumdar, S. Goswami, & M. Sillanpää (Eds.), (15, pp. 58–71). Royal Society of Chemistry. https://doi.org/10.1039/9781837676880-00058

Lim, W. M., Das, M., Sharma, W., Verma, A., & Kumra, R. (2025). Gamification for sustainable consumption: A state-of-the-art overview and future agenda. *Business Strategy and the Environment, 34*(1), 1510–1549. https://doi.org/10.1002/bse.4021

Liu, N. (2024). Influence and mechanisms of perceived information disclosure in shaping public attitudes towards waste disposal facilities: Evidence from China. *Ecological Indicators, 166*, 112514. https://doi.org/10.1016/j.ecolind.2024.112514

Longo, E., Sahin, F. A., Redondi, A. E. C., Bolzan, P., Bianchini, M., & Maffei, S. (2021). Take the trash out... to the edge. Creating a Smart Waste Bin based on 5G Multi-access Edge Computing. In *GoodIT '21: Proceedings of the Conference on Information Technology for Social Good* (pp. 55–60). ACM. https://doi.org/10.1145/3462203.3475921

Lu, W., Webster, C., Peng, Y., Chen, X., & Chen, K. (2018). Big data in construction waste management: Prospects and challenges. *Detritus, 4*, 129–139. https://doi.org/10.31025/2611-4135/2018.13737

Luttenberger, L. R. (2020). Waste management challenges in transition to circular economy–case of Croatia. *Journal of Cleaner Production, 256*, 120495. https://doi.org/10.1016/j.jclepro.2020.120495

Ma, W., Cui, J., Abdoulaye, B., Wang, Y., Du, H., Bourtsalas, A. C., & Chen, G. (2023). Air pollutant emission inventory of waste-to-energy plants in China and prediction by the artificial neural network approach. *Environmental Science & Technology, 57*(2), 874–883. https://doi.org/10.1021/acs.est.2c01087

Melinda, V., Williams, T., Anderson, J., Davies, J. G., & Davis, C. (2024). Enhancing waste-to-energy conversion efficiency and sustainability through advanced artificial intelligence integration. *International Transactions on Education Technology, 2*(2), 183–192. https://doi.org/10.33050/itee.v2i2.597

Ogwu, M. C., Kosoe, E. A., Alaka, A. C., & Xu, J. (2025). Green infrastructure and environmental health security: The role of nature-based solutions in improving urban health. In S. C. Izah & M. C. Ogwu (Eds.), *Innovative approaches in environmental health management: Processes, technologies, and strategies for a sustainable future* (pp. 225–255). Springer Nature. https://doi.org/10.1007/978-3-031-81966-7_9

Philips. (2025). Retrieved 13/04/2025, from https://www.philips.co.uk/healthcare/resources/landing/smartspeed

Ranjana, P., Varsha, S., & Eliyas, S. (2021). IoT based smart garbage collection using RFID and sensors. *Journal of Physics: Conference Series, 1818*(1), 12225. https://doi.org/10.1088/1742-6596/1818/1/012225

Ren, Y. S., He, Q., Liao, W., Liao, Y., Zhan, Z., Xu, J., Acioly, T. M. D. S., Viana, D. C., & Ilyas, M. (2025). Sustainable adsorbent: Activated carbon from waste Styrofoam for efficient aluminum removal. *Microscopy Research and Technique, 88*(4), 1044–1054. https://doi.org/10.1002/jemt.24770

Savela, N., Levänen, J., Lindeman, S., Kgabi, N., Koivisto, H., Olenius, M., John, S., Mashauri, D., & Keinänen-Toivola, M. M. (2020). Rapid urbanization and infrastructure pressure: Comparing the sustainability transition potential of water and energy regimes in Namibia. *WORLD, 1*(2), 49–66. https://doi.org/10.3390/world1020006

Schafer, B. (2014). D-waste: Data disposal as challenge for waste management in the Internet of Things. *The international review of information ethics, 22*, 101–107. https://doi.org/10.29173/irie122

Shanghai Municipal People's Government. (2024). Accessed 12/04/2025, from https://english.shanghai.gov.cn/en-Latest-WhatsNew/20240826/6936861e41cc4dedb30a6146bbe99017.html

Smith-Gillespie, A., Morwood, D., & Aries, T. (2018). *PHILIPS SMARTPATH: A circular economy business model case.* Retrieved from http://hdl.handle.net/10347/20427

Sutikno, T., Ardiansyah, A., Pujiyanta, A., & Handayani, L. (2024). Artificial intelligence technologies in urban smart waste management. *High-Tech and Innovative Series, 1*(1), 55–66. Retrieved from http://journal2.uad.ac.id/index.php/tech/article/view/10941

Swapna, B., Kamalahsan, M., Sowmiya, S., Konda, S., & SaiZignasa, T. (2019). Design of smart garbage landfill monitoring system using Internet of Things. *IOP Conference Series: Materials Science and Engineering, 561*(1), 12084. https://doi.org/10.1088/1757-899X/561/1/012084

Thompson, O. P., Kosoe, E. A., & Xu, J. (2024). Green infrastructure and urban planning for sustainable clean air. In M. C. Ogwu & S. C. Izah (Eds.), *Sustainable strategies for air pollution mitigation: Development, economics, and technologies* (pp. 343–375). Springer Nature. https://doi.org/10.1007/698_2024_1122

Thürer, M., Pan, Y. H., Qu, T., Luo, H., Li, C. D., & Huang, G. Q. (2019). Internet of Things (IoT) driven Kanban system for reverse logistics: Solid waste collection. *Journal of Intelligent Manufacturing, 30*(7), 2621–2630. https://doi.org/10.1007/s10845-016-1278-y

Tutak, M., & Brodny, J. (2025). Stakeholder collaboration and open innovation in smart cities: A Quattro helix model for technological and social transformation. *Journal of Open Innovation: Technology, Market, and Complexity, 11*(3), 100594. https://doi.org/10.1016/j.joitmc.2025.100594

Vishnu, S., Ramson, S. R. J., Senith, S., Anagnostopoulos, T., Abu-Mahfouz, A. M., Fan, X., Srinivasan, S., & Kirubaraj, A. A. (2021). IoT-enabled solid waste management in smart cities. *Smart Cities, 4*(3), 1004–1017. https://doi.org/10.3390/smartcities4030053

Wang, C., Qin, J., Qu, C., Ran, X., Liu, C., & Chen, B. (2021). A smart municipal waste management system based on deep-learning and Internet of Things. *Waste Management, 135*, 20–29. https://doi.org/10.1016/j.wasman.2021.08.028

Wu, D., Liao, W., Yu, H., Majumdar, S., Xu, J., & Choudhury, M. (2025). Assessing traditional waste management practices in urban environments and recent trends. In S. Majumdar, M. Choudhury, & A. Cheshmehzangi (Eds.), *Sustainable urban future: Embracing circular economy in waste management*. Springer Nature. https://doi.org/10.1007/978-981-95-2976-6_5

Xu, J., Wu, D., Wang, H., & Fong, C. S. (2023). Introducing sustainable intervention strategies of operation planning and environment design to activate void areas on the urban fringe. *E3S Web of Conferences, 437*, 03002. https://doi.org/10.1051/e3sconf/202343703002

Xu, J., Liao, W., Lim, N. H. H., Fong, C. S., Wang, H., & Xu, Y. (2024). Operational and environmental assessment for night markets under the informal economy impacts: Insights from the case study of Selangor, Malaysia. *C TIC, 13*(1), 189–219. https://doi.org/10.17993/3ctic.2024.131.189-219

Xu, J., Liao, W., Lim, N. H. H., & Fong, C. S. (2026). Strategic solutions: Water resource assessment and planning for sustainable futures. In S. Majumdar, M. Choudhury, & M. Sillanpää (Eds.), *Advances in hydrology: Spatial intelligence, climate change, and sustainable water resource management*. Elsevier. https://doi.org/10.1016/B978-0-443-26777-2.00002-8

Yu, K. H., Zhang, Y., Li, D., Montenegro-Marin, C. E., & Kumar, P. M. (2021). Environmental planning based on reduce, reuse, recycle and recover using artificial intelligence. *Environmental Impact Assessment Review, 86*, 106492. https://doi.org/10.1016/j.eiar.2020.106492

Zhang, A., Venkatesh, V. G., Liu, Y., Wan, M., Qu, T., & Huisingh, D. (2019). Barriers to smart waste management for a circular economy in China. *Journal of Cleaner Production, 240*, 118198. https://doi.org/10.1016/j.jclepro.2019.118198

Zhang, C., Hu, M., Di Maio, F., Sprecher, B., Yang, X., & Tukker, A. (2022). An overview of the waste hierarchy framework for analyzing the circularity in construction and demolition waste management in Europe. *Science of the Total Environment, 803*, 149892. https://doi.org/10.1016/j.scitotenv.2021.149892

Zhang, H., Wang, L., Zhang, Z., Li, J., Liao, W., Xu, J., & Majumdar, S. (2025). Overcoming hurdles: Analyzing challenges and solutions in traditional waste disposal methods. In M. Choudhury, S. Majumdar, & M. Sillanpää (Eds.), *The circular path: Rethinking waste for a sustainable future*. Springer Nature.

Zhong, M., Wang, C., Xu, J., Cai, Y., Xiao, B., Xu, T., & Wang, X. (2024). Composite $xNiFe_2O_4/(1\text{-}x)SrFe_{12}O_{19}$ oxygen carriers for chemical looping reforming of bioethanol coupled with water splitting to coproduce syngas and hydrogen. *Journal of the Energy Institute, 117*, 101780. https://doi.org/10.1016/j.joei.2024.101780

CHAPTER 11

Innovative Mobility: Integrating Electric and Shared Options for Sustainable Cities

Scaria Alex, Aswathy K. Cherian, M. Vaidhehi, B. Sowmiya, M. K. Vidhyalakshmi, N. A. S. Vinoth Kumar, Arun Varghese Kolanjiyil, and Sundaravadivazhagn Balasubaramanian

S. Alex (✉)
IT Department, Amal Jyothi College of Engineering, Affiliated to APJ Abdul Kalam Technological University, Kottayam, Kerala, India
e-mail: scariaalex@amaljyothi.ac.in

A. K. Cherian • M. Vaidhehi • B. Sowmiya • M. K. Vidhyalakshmi • N. A. S. Vinoth Kumar
Department of Computing Technologies, SRM Institute of Science and Technologies, Chennai, India
e-mail: aswathyc1@srmist.edu.in; vaidhehm@srmist.edu.in; sowmiyab@srmist.edu.in; vidhyalm1@srmist.edu.in; vinoths@srmist.edu.in

A. V. Kolanjiyil
Department of Mechanical and Nuclear Engineering, Virginia Commonwealth University, Richmond, VA, USA

S. Balasubaramanian
College of Computing and Information Sciences, University of Technology and Applied Sciences, Al Mussanah, Oman
e-mail: sundaravadi.balasubaramanian@utas.edu.com

S. Majumdar, M. Choudhury (eds.), *City of the Future*,
https://doi.org/10.1007/978-3-032-18026-1_11

11.1 Introduction

Over the past few decades, urban transport has become a key technological area in modern cities, with significant social, economic, and environmental implications. Along with rapid urban growth come its negative aspects, such as pollution, congestion, and climate change. The demand for increased mobility in the congested city results in an overloaded transportation system, fueled by rising energy use, emissions, and extreme productivity losses, while people suffer declining health and economic well-being. The world's cities suffer from traffic congestion due to urbanization for a variety of reasons, including inefficient resource use, increased stress levels, and CO_2 emissions (Majumdar et al., 2025). Urban transport systems have been and remain a primary target for sustainability efforts because fossil fuels account for most vehicle pollution. As the rate of climate change increases, so do the measures and methods implemented to tackle it. In this example, urban pollution must be tackled head-on through its most prominent form—urban transportation (Dixit & Majumdar, 2025). Ride-sharing, bike-sharing, and public transport are mobility-on-demand services used to mitigate congestion, as they reduce the number of cars on roads, leading to less pollution and less traffic. Addressing unmet public transport needs by building the right shared mobility framework fosters economic savings and accessibility while enhancing land development policies (Mogire et al., 2025). To achieve this goal, a combination of electric shared vehicles, renewable energy, and improved public transport must be utilized to enhance urban mobility. For it to be effective, all efforts must aim toward a clean, integrated solution to urban mobility. There is a pressing urgency to transition to smart, clean cities to serve future generations better while mitigating climate change and the improper development of city spaces (Kumaraswamy et al., 2025).

The backbone of urban transport is a paradigm shift in mobility from car-centricaecar-centricntriappeople-centrico the environment, with a view to accessibility, safety, protection, aprotectionusionnd service service-responsives cohesiveness politically and economically is also contained in the emerging smart-information technology constructs, which focus on maximizing the collection and use of transit resources, the transit eco-balance, and reducing the overall emission performance of the mobility system through the integration of electrification of transit resources. This is underpinned by the development of intelligent mobility systems, which harness the power of ICTs, real-time data, and user interface-based service model to provide seamless high-level, adaptive, and integrated mobility solutions. Collectively, they form the pillars upon which urban transport systems can be built to be adaptive, sustainable, and responsive to the challenges of the future climate (Mogire et al., 2025).

11.2 The Rise of Electric and Shared Mobility Services

The introduction of electric shared mobility services is critical for addressing serious concerns about urban infrastructure, such as traffic congestion, pollution, and climate change (Majumdar, 2020; Majumdar & Sarkar, 2020). This new configuration of incorporated shared mobility with electric cars (EVs) is changing how cities are moved in a more environmentally friendly way. Technology, changing consumer preferences, and government intervention are driving the transition from conventional to electric transportation systems.

11.2.1 Evolution of Electric and Shared Mobility

In just a couple of years, electric mobility has advanced remarkably. Most people do not know that the first electric cars were designed more than a century ago. Even after that, and up to recently, adoption rates remained significantly low due to insufficient infrastructure and the high cost of electric cars. However, now, electric vehicles are widely used around the world due to modern charging systems, improved battery technology, and government subsidies (Marzouk, 2025). As many cities around the world seek to reduce vehicle fossil fuel consumption, tax benefits and EV subsidies are on the rise to improve air quality in dense urban areas (Marzouk, 2025). Public transport has taken another leap forward with the development of ride- and car-sharing, along with e-scooters, a new form of micromobility, all enabled by digital platforms. Pollution in the environment can be reduced through electric vehicles and shared mobility, thereby enabling sustainable mobility (Zhong et al., 2023).

11.2.2 Growth of Shared Mobility Services

In recent years, mobile applications have played a vital role in shared mobility booking, such as ride-sharing (Uber, Lyft), car-sharing (Zipcar,

Share Now), and bike-sharing (Lime, Citi Bike). This allows citizens to book a vehicle by phone, thereby reducing traffic. A new program, where a car is rented for a few hours from another person for a fee, also reduces the cost for an individual to buy a new car (Litman, 2022). The introduction of e-scooters and e-bikes transforms urban commuting by eliminating the need for short car trips.

11.2.3 Technological Innovations Driving Shared Electric Mobility

Emerging technologies are playing a significant role in electric shared mobility. The range of electric vehicles, as well as other micro-mobility devices like e-scooters, has become more affordable due to innovations in battery technology, such as lithium-ion and solid-state batteries (Marzouk, 2025). The broad deployment of AI-powered fleet management systems and supercharger networks is also making it easier to adopt shared electric fleet vehicles by ensuring efficient operations. The emergence of electric and shared mobility services is reshaping transport in megacities (Zhong et al., 2023). These technologies collectively create car-free spaces by transforming the city into a beautiful microenvironment, thereby achieving zero emissions, reduced congestion, and improved access to densely populated areas. Continuous acceleration of technology and enabling policies are key enablers that will ensure electric shared mobility becomes a significant component of urban transport.

11.2.4 Electric Mobility (e-Mobility)

Electric mobility (e-Mobility) broadens the mobility landscape by introducing emission- and fuel-independent options such as EVs, e-buses, e-bikes, and e-scooters, increasing the system's energy efficiency and expanding the mobility ecosystem. These e-mobility solutions have been attributed to advancements in innovative charging technology, strategic integration with renewables, and innovative solutions to range anxiety and grid-interconnected challenges (Buttigieg, 2023).

11.2.4.1 Principal Elements

Electric mobility comprises a range of transport options, including electric vehicles, e-buses, e-bikes, and e-scooters. These options provide flexible, effective, and sustainable solutions to their fuel-based counterparts.

11.2.4.2 Social and Environmental Value

The range of e-mobility benefits includes reductions in greenhouse gas emissions and improvements in air quality, as well as fuel savings and enhanced energy efficiency. There is also conservation of fossil fuels and an overall economic value to people and cities in the long run (Spanos et al., 2022).

11.2.4.3 Energy Integration and Charging Infrastructure

Integration of renewable energy and smart grids is critical to the e-mobility infrastructure (Spanos et al., 2022). The use of fast chargers, battery swapping, and other readily accessible chargers, coupled with e-mobility infrastructure, is critical within the energy transition ecosystem to renewable, smart grids, and conservation.

11.2.4.4 Barriers to Adoption

The e-mobility infrastructure is currently dominated by and hindered by the limited number of charging points. High-range anxiety, rivalry over power grids, the e-mobility infrastructure, and the overall shift to e-mobility ports also constrain the e-mobility contenders (Wu et al., 2025). The resolution of these issues is vital to facilitating access to e-mobility in urban centers.

11.3 Benefits of Electric and Shared Mobility in Urban Areas

Cities continue to experience severe traffic congestion, pollution, and inefficient transport systems. The situation has reached a critical point where electric and shared mobility solutions need to be implemented to reduce dependency on private cars (Wu et al., 2025). Integrating mobility platforms with EVs fosters cities with usable and reliable transport facilities, which in turn creates a sustainable environment. Figure 11.1 shows the pictorial representation of the benefits of electric and shared mobility in urban areas.

11.3.1 Reducing Private Car Ownership and Usage

The most significant impact of shared and electric mobility is, without a doubt, a reduction in the adoption and utilization of private vehicles. The

Fig. 11.1 Pictorial representation of benefits of electric and shared mobility in urban areas

cost of owning a car is often very high, including maintenance, insurance, parking, and fuel. Shared mobility services such as ride-hailing apps and bike and e-scooter rentals make transport more affordable, thereby helping reduce vehicle ownership. This is especially true for micro-mobility devices (Zhong et al., 2023). A single shared vehicle may eliminate countless cars within city limits. Ortega et al. (2023) emphasize that cities that have initiated car-sharing services have seen a drop in vehicle ownership per capita from 25% to 40%. By contrast, the International Transport Forum (2021) argues that mobile services deployed in cities can halve the number of cars, thereby increasing greenery, pavements, and cycling lanes.

Moreover, the lower number of personal cars on the streets results in mild traffic congestion, shorter travel times, and higher overall productivity in the transportation sector. As a result of increased investment in electric shared mobility, cities can further reduce car traffic, improving traffic and urban life as a whole.

11.3.2 Lower Carbon Emissions and Improved Air Quality

The emission of gas from transport contributes to environmental pollution and greenhouse gas emissions. This can be reduced by using shared and electric mobility. There are no tailpipe emissions from EVs, which means reduced CO_2, nitrogen oxides (NO_x), and PM—all of which harm health and the environment (Marzouk, 2025). Cities that adopted so-called electric shared mobility offered services have reported lower emissions. Zhang et al. (2024) state that after the launch of EVs, the amount of carbon dioxide in the atmosphere has decreased by 30%.

11.3.3 Economic Benefits of Shared Mobility

The growing adoption of electric vehicles and shared mobility is boosting the country's economy. Users of mobility services do not bear the charges for parking, gas, and maintenance, which significantly increase the fighting power. Martinez et al. (2024) note that people who have stopped using privately owned cars and adopted shared mobility systems save around 40% annually on transportation. These EV-powered rides are also cheaper to maintain, since they are easier to maintain than traditional vehicles. Besides benefiting users as mentioned above, these services spawn new economic activities, such as ride-sharing, charging, EV maintenance, and even the construction of charging stations, all of which contribute significantly to the city's economy. Also, passengers are more flexible in transitioning between different modes of transport with the new Mobility-as-a-Service (MaaS) systems, which make modern cities more efficient, economical, and innovative. Electrification and the adoption of shared mobility services remain the most important levers for improving urban transport.

11.3.4 Shared Mobility Services

11.3.4.1 Forms of Shared Mobility

There are several forms of shared mobility including vehicle car-sharing, e-scooters, ride-hailing, and microtransit. All these alternatives are beneficial for diminishing the gap of access to untapped transportation resources and infrastructure while simultaneously reducing the demand for personal car ownership (Bösehans et al., 2024).

11.3.4.2 Benefits

Numerous urban mobility challenges are alleviated by shared mobility systems, such as reduced personal vehicle ownership, less spatial congestion, and lower negative environmental impacts, including lower emissions and fuel use. In addition, shared systems enhance access to the city and encourage the use of multiple modes of transportation (Tönnies et al., 2025).

11.3.4.3 Enabling Technologies

Technology is central to the provision of shared mobility services. User-friendly apps, GPS systems, mobile payments, and complex algorithms all work to seamlessly assist users with vehicle booking, route customization, and service availability (Tönnies et al., 2025).

11.3.4.4 Challenges

Some of the most notable challenges to shared mobility include complexities in the legal framework, organizational conflicts in gig economies, and public littering and obstruction. Additionally, ensuring equitable access across different socio-economic groups remains a critical issue for policymakers and service providers.

11.3.5 Shared Mobility

11.3.5.1 Forms of Shared Mobility

Shared mobility provides flexible access to vehicles without ownership. Popular forms include car-sharing for longer trips, a variety of ride services for convenient point-to-point travel, and bike and e-scooter rentals for short- and mid-distance trips. Shared mobility also includes microtransit services with flexible routes to and from areas underserved by public transport. All these services together reduce reliance on personal vehicles and expand travel options simultaneously (Zhang et al., 2024).

11.3.5.2 Benefits

Shared mobility plays a significant role in reducing private vehicle ownership and, in turn, congestion and parking woes in densely populated areas. It also improves space efficiency, as fewer vehicles are needed to serve a larger population (Burger, 2025). Overall, it also supports reducing greenhouse gas emissions and fuel consumption. Together with other modes of shared mobility, it promotes the adoption of cleaner, healthier, and more pleasant cities.

11.3.5.3 Enabling Technologies

Unlike other forms of mobility, shared mobility retains significant value for demonstrating technology. Users can easily access near-real-time services on mobile phones, which also allow them to pay and secure bookings. GPS monitoring provides users with real-time location information for vehicle services. The processing of cashless payments has eased transaction completion, while sophisticated algorithms on the platform enhance the effectiveness of ride matching, fleet redistribution, and dynamic pricing (Marsden, 2022). Technologies of this type provide a smooth, easy experience, enabling reliable, accessible shared mobility.

Despite the prospects shared mobility offers, there are still important issues to tackle. Regulatory barriers often lag and evolve more slowly than the technological backbone, leading to conflicts in integrating services across safety, licensing, and automation. Moreover, there are disputes over the assistive services provided through gig-work-based ride-hailing, where drivers' rights to earn wages are challenged (Mogire et al., 2025). Misuse of public spaces for shared bicycles and scooters, including clogging pedestrian pavements and inequitable distribution across different income groups and neighborhoods, remains a common problem.

11.3.6 Integration of Electric and Shared Mobility

The integration of electric mobility with shared mobility harnesses the advantages of both to provide cleaner, more efficient, and more inclusive urban transport. Fleet electrification with shared EV platforms for taxis, logistics, and public services demonstrates synergy. In contrast, the emerging Mobility as a Service (MaaS) enables the integration of multiple travel modes into a single digital platform, delivering seamless user experiences. Intermodality builds on public transport's last-mile connectivity by alleviating the need for private transport on first-mile connections to transit terminals, using micro-mobility devices such as e-bikes and scooters. Also, as shown by smart charging hubs that charge and dock EVs, integrated parking, and designed-for-access connections, Infrastructure and Urban Design greatly help with this shift (International Transport Forum, 2020). Reallocation of streetspace in favor of pedestrians, cyclists, and shared mobility eases the transition to these modes. It enhances safety while facilitating resource allocation through data-driven planning for route optimization, station placement, and modal choice. All these factors contribute to the development of integrated urban mobility systems that are resilient, user-friendly, and ecologically sound (Buttigieg, 2023).

Sustainable mobility transitions require investment in policies, governance, and financing to ensure that appropriate supporting systems complement innovation. The implementation of strong regulation, such as the establishment of Low Emission Zones, licensing of shared-use fleets, and emission-based incentives for electric vehicles, directs cities toward greener ways of functioning (Coppola et al., 2024). Collaborative, public-private partnerships allow for shared improvement of bridges left by government intervention or private investment through infrastructure, operations, and management.

The subsidization of green bonds, congestion pricing, and electrification all help scale transport systems and automobiles to reduce dependence. Subsidization policies also help promote relative social equity (Hsieh, 2025). Shared and electric mobility must also be affordable and accessible to low-income populations, reducing mobility poverty through sustainable and equitable urban development.

11.4 Case Studies: Cities Leading the Transition

11.4.1 Helsinki, Finland

With the development of Mobility as a Service, Helsinki has also been a global leader in combining shared electric and mobility systems. The launch of Whim and MaaS Global was one of the first projects worldwide to attempt to consolidate various transport modes. This included commuting, taxis, car-sharing, and bike terminals through a single, subscription-based software platform. The MaaS software simplified the transport system for users and private car owners alike, leading to a shift toward flexible, multimodal transport. Helsinki has been a model for the world, effectively offering sustainable urban transport and improving mobility through advanced software systems (Zhang et al., 2024).

11.4.2 Paris, France

With its unparalleled visionary strides toward sustainable, low-emission urban policies, Paris was among the first to embrace shared mobility. Programs like the world-famous Vélib′ public bike-sharing system and the groundbreaking Autolib′ electric car-sharing program made mobility extremely affordable. The city also implemented low-emission corridors and car-use policies, which, along with shared services, improved the transport ecosystem and air quality (Burger, 2025). Autolib′ was discontinued due to some operational challenges. However, it was instrumental in developing new Electric Vehicle (EV) services, while Vélib′ bike-sharing systems still dominate globally. Paris is a good example of the advantages and challenges of shared and electric mobility in urban areas.

Complementing this, ride-sharing platforms are being adapted to local needs, integrating digital payment solutions and mobile-based booking systems that suit Rwanda's rapidly growing digital economy. These initiatives demonstrate how electric and shared mobility can be tailored to resource-constrained contexts, addressing both sustainability and accessibility. Kigali's approach highlights the potential for African cities to leapfrog toward greener mobility models while also supporting economic development and inclusive urban transport (Burger, 2025). To improve air quality and promote sustainable public transportation, cities around the world are incorporating mobility solutions that use shared electric vehicles into their public transport systems. This is also done to combat climate change, which is becoming a global threat. Three case studies are presented here to illustrate successful transitions to electric and shared mobility, along with lessons learned.

11.4.3 Amsterdam, Netherlands: A Leader in Electric and Shared Mobility

For years now, Amsterdam has been an international leader in green urban transport by integrating EVs, bike-sharing schemes, and public transportation into a single transportation ecosystem (Tönnies et al., 2025). Subsidies for EV ownership provided by the city administration, along with a vast network of charging stations, support this claim. Furthermore, OV-fiets—the city's bicycle-sharing system—and e-scooter services are alternatives to public transport that reduce reliance on private cars.

11.4.4 Shenzhen, China: The World's First Fully Electrified Bus Fleet

Shenzhen has become a global leader in electrifying public transport by fully transitioning to electric buses and taxis (Marsden, 2022). The city is home to more than 16,200 electric buses and 22,000 electric taxis, which reduce transport emissions while supporting improved infrastructure development. The city also invested in advanced battery swapping and charging facilities. All of these steps enabled Shenzhen to be a pacesetter in reducing electric vehicle emissions by reducing the carbon footprints of electric buses.

11.4.5 Oslo, Norway: The EV Capital of the World

In Oslo, 80% of the newly sold cars are electric vehicles. The city has put in place unique perks, such as no tolls, free parking, and 1st-class buses, which help entice locals to own electric cars. Furthermore, Oslo provides electric car leasing and public electric taxi services, which help reduce traffic jams (Bindhya et al., 2025).

11.5 Challenges in Implementing Sustainable Electric Mobility

11.5.1 Infrastructure Limitations

A key obstacle to EV adoption is the insufficient availability of charging infrastructure. Many urban areas lack a reliable, widespread network of charging stations, making it difficult for EV owners to find convenient charging options and leading to range anxiety—the fear of running out of battery before reaching a charging point. The unavailability of charging stations, the variety of connector types, and the voltage levels set by manufacturers affect the use of EVs in the country. Establishing charging stations in urban areas is a tedious process due to limited space. In addition, the growing popularity of electric vehicles is putting further strain on the electric grid, which incurs additional costs for required upgrades (Lutsey & Nicholas, 2019). Up until now, these infrastructure concerns have impeded the full potential of electric mobility from being realized.

11.5.2 Policy and Regulatory Hurdles

Changes and initiatives related to electric mobility rely on government charging management. In this regard, fragmented and non-optimal policies in many countries create problems for both consumers and service providers. When policies exist, tax rebates, subsidies, or lower tolls are provided, but other governments do not support such initiatives (Hall & Lutsey, 2017). Often, zoning ordinances and other policy measures are not drafted to require EV charging stations in new homes or business premises, making it inefficient for people in urban areas. The restrictions on establishing EV charging stations are significant; this requires specific policies and liability to support them (M.S. 1.,25).

11.5.3 Public Perception and Behavioral Resistance

The acceptance of electric vehicles (EVs) is low even though the awareness about it is huge among the citizens. Public perception is still a challenge that should be tackled. One concern is "range anxiety," the fear that EVs do not provide sufficient battery life for daily travel. Battery technology has improved, which makes this less of a problem, but skepticism regarding the reliability of EVs and accessibility of charging remains (Li et al., 2021). To increase the adoption rate of electric mobility, allowing for the overcoming of infrastructural, policy, and perception barriers is important. Charging infrastructure should be installed, alongside logical regulatory market policies, and grants should be paid out to both consumers and businesses. Behavioral change and accepting the greater public trust that comes through addressing the problems are made easier by shared mobility service improvement and education (Gogoi et al., 2025). With these problems solved, cities can provide non-polluting and effective transport systems that greatly benefit the public in the long term, both economically and environmentally.

11.6 Mobility as a Service (MaaS): The Future of Urban Transportation

Mobility as a Service (MaaS) influences transportation planning in a city. In this application, various transport modes are combined into a single and very easy-to-use service. This enables users to book and pay for all types of transportation, such as buses, trains, cars, and even two-wheelers (Mola et al., 2020; Baldassa et al., 2022). This makes travel easier and less dependent on private vehicles. By integrating several transport services into a single system, MaaS enables users to choose the most suitable route based on time, cost, or other factors. The idea is to effectively replace vehicle ownership and its excessive use in modern metropolitan areas to alleviate traffic jams, pollution, and other problems associated with the inefficiency of large cities. Figure 11.2 shows the different modes of transport combined in a MaaS application.

The following are the main features of the MaaS model:

- Smooth transit: On MaaS, all methods of transportation can be encompassed into one system which enables users to move between the services with relative ease. These could be a combination of public transport, shared vehicles, or private vehicles, depending on what the users want.

- Custom services: A user's situation drives the development of services offered by MaaS platforms, whether it is speed, cost, or environmental friendliness; the platform enables personalized and special trips (Ogunkunbi et al., 2022).
- Digital transparency and effortless access: On a digital marketplace, transportation is effortlessly accessible and transparent, allowing real-time access to information, ride booking, payment processing, and route planning all in a single application.

11.6.1 Integration of Multiple Transport Modes into a Seamless Mobility Service

Mobility-as-a-Service (MaaS) combines different modes of transportation into a single application, thus eliminating the need for multiple apps and payment methods. This application will reduce the number of individual cars, traffic, and air pollution. MaaS thus encourages public transportation, which optimizes cities' transportation systems (Faghih-Imani & Eluru, 2025). By reducing reliance on private vehicles, MaaS frees up urban space and promotes a more efficient, sustainable, and interconnected mobility ecosystem.

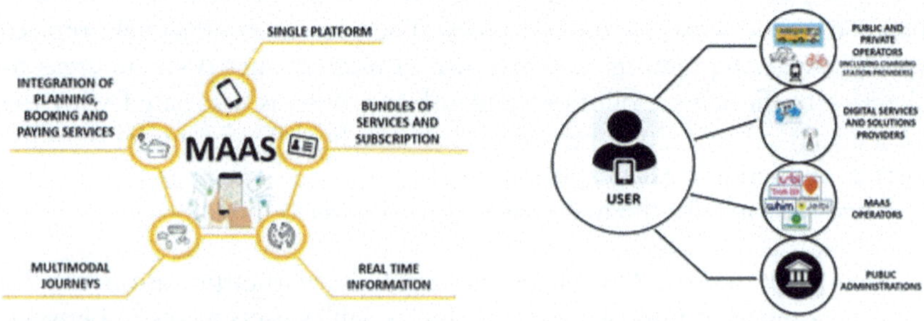

Fig. 11.2 The application of MaaS

11.6.2 Digital Platforms and the Role of Data-Driven Decision-Making

At the core of the MaaS approach is a powerful digital platform that integrates real-time data, forecasts, and machine learning (Li et al., 2021). These platforms gather information from multiple transit agencies—such as traffic, vehicle checkouts, and public transportation schedules—to help users make the best travel choices. To implement these changes in MaaS, data-focused approaches are highly significant not only for enhancing user experience but also for improving the effectiveness of transport systems. With insights from mobility behavioral data, infrastructure development, service delivery, and policies can be more accurately set by understanding travel behavior, demand trends, and congestion hotspots (Ditmore & Miller, 2021; Kant et al., 2023).

In addition, the empty miles (unoccupied distance a vehicle traverses) can be reduced, and the efficiency of the public transport system can be improved using MaaS platforms by adjusting to real-time demand. For example, during peak times, the platform can optimize ride-sharing services by filling uncovered seats or changing public transport routes to accommodate more passengers (Ogunkunbi et al., 2022).

11.7 Challenges and Risks

11.7.1 Technological Fragmentation

One of the main factors in the development of shared and electric mobility is the decentralization of technology. Most systems and platforms work in silos which makes it difficult to offer integrated services like ride-hailing, bike-sharing, and public transit. This is inconvenient for the user, wastes resources, and sets back the evolution of fully integrated Mobility as a Service (MaaS) ecosystems (Li et al., 2021).

11.7.2 Environmental Trade-offs

E-mobility and shared services reduce emissions during operation, but such services are not free of environmental issues. The production of lithium-ion batteries requires significant energy and resources, which are detrimental to the environment and society. Likewise, shared micro-mobility devices like e-scooters are used and discarded after a short period, posing problems of waste and material use (Rajpal et al., 2024).

11.7.3 Data Privacy and Surveillance Concerns

The digitalization of shared mobility systems also results in the collection and storage of large volumes of user-generated data, including travel and usage history, Personally Identifiable Information (PII), and geographical location. Service systems are optimized using such data, but the privacy implications, ethical issues around data surveillance and ownership, and data misuse are alarming. Poor governance of user mobility and data collection systems may erode user confidence and trust in these systems.

11.7.4 Greenwashing and Tokenism

A further critique is about the risk of green washing. It is when a company or state advertises some form of electric or shared mobility as an ecomobility initiative yet makes no other transformative changes. For example, promoting and implementing small, superficial pilot projects, or giving vacuous claims of "sustainability" satisfaction for marketing goals, fails to capture the more profound challenges of renewable over-reliance, inequitable access, and feeble political will (Choudhury et al., 2025).

11.8 Future Directions

11.8.1 Vehicle-to-Grid (V2G) Integration

The integration of electric vehicles (EVs) and the power grid is a transformative development for sites. Vehicles-to-Grid (V2G) technologies enable EVs (Electric Vehicles) and distributed energy resources (DERs) to store and discharge "excess" renewable energy to the grid during peak production times. This balance, in turn, stabilizes the energy supply and fosters grid resiliency, lowering fossil fuel dependency.

11.8.2 Autonomous Electric Mobility

The combination of self-driving technology and electrification is a game-changer for shared mobility systems. Autonomous electric vehicles can enhance safety, reduce congestion through sophisticated routing, and drive down costs. However, there are ethical and governance issues related to liability, data privacy, job loss, and equitable access that need to be addressed through appropriate policies.

11.8.3 Circular Design of Vehicles

Future mobility systems must adopt circular-economy principles by ensuring that vehicles are designed for longevity, reuse, and recycling. Modular EVs, which can be upgraded rather than discarded, shared vehicle fleets are of longer-lasting modules, and closed-loop battery recycling systems will reduce resource consumption and waste, resulting in sustainable mobility throughout their life cycles (Creutzig, 2021).

11.8.4 Climate-Resilient Mobility Systems

As climate change intensifies, mobility systems must be built to withstand extreme weather events and other disruptions. This includes charging stations with flood-resilient architecture, flexible urban transport design, and networks that support emergency evacuation and services during disasters. Building systems that make mobility more resilient ensures sustainable, safe, and effective mobility in the face of climate change (Choudhury et al., 2025).

11.9 Policy Recommendations for Sustainable Urban Mobility

Cities will need to develop policies, foster partnerships, and encourage citizen participation to make a successful shift toward electric mobility and smart urban transport systems. The below-stated policies support sustainable urban mobility.

11.9.1 Government Incentives and Regulations Supporting Electric Mobility

Governors should encourage the use of electric vehicles (EVs) by relaxing taxes, rebates, and subsidies for citizens. This adds to the economic support for citizens adopting EVs. Reduced toll and vehicle registration fees should also be granted to owners of electric vehicles. There has to be a stronger focus on charging infrastructure: clearly defined goals for the spatial coverage of charging stations need to be set, and public grants should be issued to support private investment. The integration of overarching legislation to alleviate regional differences is vital (Hall & Lutsey, 2017).

11.9.2 Public-Private Partnerships for Smart Urban Transport Solutions

The public and private sectors need to work together in order to roll out smart transport solutions. Public-private partnerships can help develop smart mobility ecosystems that include electric vehicles, shared mobility, and data intelligence platforms. Governments need to incentivize infrastructure development and ease regulations to enable private companies to invest in shared and electric transport services. With PPPs, cities can put in place an integrated transport system that enables multimodal transport through a single platform—featuring ride- and bike-sharing, electric scooters, and more (Creutzig, 2021).

11.9.3 Citizen Engagement and Awareness Initiatives

For awareness to be effective regarding participation in the transition to sustainable mobility, citizens ought to be engaged as well. A wide outreach needs to be organized by the Government, highlighting the benefits of electric mobility. Public participation in the project's planning phase could help resolve problems and stimulate acceptance and compliance. This also validates whether the community's requirements are met (Zhong et al., 2023).

11.10 Future Prospects

The future of urban mobility envisions AI-integrated, flexible, and resilient transportation networks that reduce reliance on private vehicles while offering efficient, sustainable alternatives. In such systems, individuals will be empowered to make environmentally responsible travel choices through innovative city technologies that promote social equity, support economic growth, and safeguard the environment. These intelligent mobility platforms will enable seamless multimodal travel, optimize traffic flow, and provide real-time information to enhance user convenience and accessibility.

By fostering continuous collaboration among governments, industry leaders, and communities, cities can accelerate innovation and build sustainable, inclusive transportation ecosystems. This collective effort will help develop mobility systems that not only address current urban challenges but also adapt to future needs. Ultimately, the integration of advanced technologies, shared mobility solutions, and strong governance will shape urban environments that are cleaner, smarter, and more resilient, ensuring a higher quality of life for all.

11.11 Conclusion

Sustainable urban mobility can drastically reduce environmental pollution and enhance the overall efficiency of city traffic systems. As urban populations continue to grow, traditional transportation models centered around private vehicle ownership are no longer sufficient or environmentally viable. To address these challenges, governments must make strategic and sustained investments in business innovation, advanced infrastructure development, and the careful selection of transportation modes that support long-term sustainability. A key component of this transition is the promotion and popularization of electric vehicles (EVs), which offer a cleaner alternative to conventional fuel-based transportation. When supported by an extensive charging network and favorable policy incentives, EVs can significantly reduce greenhouse gas emissions and urban air pollution. Equally important is the expansion of shared mobility services, such as car-sharing, bike-sharing, and ride-sharing, which help reduce traffic congestion and optimize the use of existing transport resources.

Furthermore, integrating artificial intelligence and smart city technologies can make urban mobility systems more intelligent, responsive, and efficient. AI-driven solutions enable real-time traffic management, predictive maintenance of transport infrastructure, and enhanced safety features. Smart mobility platforms can also streamline multimodal travel, making public and shared transport more convenient and appealing to users. Through coordinated investments in these areas, governments can create cleaner, safer, and more efficient urban mobility ecosystems that contribute to sustainable development and improved quality of life for all residents.

References

Baldassa, A., Ceccato, R., Orsini, F., Rossi, R., & Gastaldi, M. (2022). MaaS bundling and acceptance in the pandemic era: Evidence from Padua, Italy. *Journal of Advanced Transportation, 2022*, 9833689. https://doi.org/10.1155/2022/9833689

Bindhya, M. S., Kartha, N. M., Jacob, G., Lukose, A., & Jeena, J. (2025). Electrifying the road: A comprehensive analysis of factors influencing consumer adoption of electric vehicles. *International Journal of Energy Economics and Policy, 15*(1), 552. https://www.econjournals.com/index.php/ijeep/issue/view/242

Bösehans, G., Bell, M. C., & Dissanayake, D. (2024). Shared mobility-novel insights on mode substitution patterns, trip and user characteristics. *Journal of Cycling and Micromobility Research, 2*, 100029. https://doi.org/10.1016/j.tra.2024.104036

Burger, K. (2025). Towards equitable, smart, and sustainable urban mobility: Governance archetypes and their relations. *Transportation Research Part D: Transport and Environment, 145*, 104797. https://doi.org/10.1016/j.trd.2025.104489

Buttigieg, E. (2023). Towards a zero-carbon future: Investigating the barriers that are limiting the adoption of electric vehicles in the Maltese islands (Master's thesis, University of Malta). https://www.um.edu.mt/library/oar/handle/123456789/119013

Choudhury, M., Dixit, G., & Majumdar, S. (2025). *Sustainable synergy: Harnessing ecosystems for climate resilience*. Springer Nature. https://doi.org/10.1007/978-3-031-77957-2

Coppola, P., Costa, F., Luè, A., Mazzeschi, V., Tanelli, M., Milić, V. J., & Stanković, J. (2024, April). Design and assessment of shared and electric mobility hubs for an integrated transport system. In *Transport research arena conference* (pp. 210–215). Springer Nature. https://re.public.polimi.it/retrieve/46f0a793-d8ea-400a-8e88-309444fdd7ce/Coppola%20et%20al._TRA24.pdf

Creutzig, F. (2021). From smart city to digital urban commons: Institutional considerations for governing shared mobility data. *Environmental Research: Infrastructure and Sustainability, 1*(2), 025004. https://doi.org/10.1088/2634-4505/ac0a4e

Ditmore, C. J., & Miller, A. K. (2021). Mobility as a service operating model to enable public policy. *Transportation Research Record, 2675*(11), 141–149, 177–188. https://doi.org/10.1177/03611981211026664

Dixit, G., & Majumdar, S. (Eds.). (2025). *Sustainable synergy: Harnessing ecosystems for climate resilience. Climate change management*. Springer. https://doi.org/10.1007/978-3-031-77957-2_13

Faghih-Imani, A., & Eluru, N. (2025). Sustainable optimization strategies for on-demand transportation systems: Enhancing efficiency and reducing energy use. *Transportation Research Part D: Transport and Environment*. Advance online publication. https://doi.org/10.1080/27658511.2025.2464388 Taylor & Francis Online

Gogoi, N., Borah, D., & Choudhury, M. (2025). Electronic wastes: The man-made crisis. In S. Goswami, M. Choudhury, & S. Agarwal (Eds.), *Electronic waste: Impact on health, animals, and the environment* (pp. 16–24). CRC Press. https://doi.org/10.1201/9781003582311-2

Hall, D., & Lutsey, N. (2017). Emerging best practices for electric vehicle charging infrastructure. *The international council on clean transportation (ICCT): Washington, DC, USA*, 54. https://theicct.org/wp-content/uploads/2021/06/EV-charging-best-practices_ICCT-white-paper_04102017_vF.pd

Hsieh, F.-S. (2025). Emerging research issues and directions on MaaS, sustainability and shared mobility in smart cities with multi-modal transport systems. *Applied Sciences, 15*(10), 5709. https://doi.org/10.3390/app15105709

Kant, P., Machavarapu, P. K., & Gajjar, H. (2023). Assessment of drivers and barriers in the adoption of mobility as a service (MaaS): A case study of Noida, India. *Urban, Planning and Transport Research, 11*(1), 2247044. https://doi.org/10.1080/21650020.2023.2247044

Kumaraswamy, T. R., Choudhury, M., Ponappa, S., & Majumdar, S. (2025). A significance of climate resilience: Strategies for local, regional, and national levels. In M. Choudhury, G. Dixit, & S. Majumdar (Eds.), *Sustainable synergy: Harnessing ecosystems for climate resilience. Climate change management*. Springer. https://doi.org/10.1007/978-3-031-77957-2_13

Li, W., Long, R., & Chen, H. (2021). Consumers' perceptions of barriers to electric vehicle adoption: A systematic review. *Energy Reports, 7*, 5762–5775. https://doi.org/10.1016/j.egyr.2021.09.059

Litman, T. (2022). *Evaluating car-sharing benefits and costs: Transportation research board analysis*. Victoria Transport Policy Institute. https://doi.org/10.3141/1702-04

Lutsey, N., & Nicholas, M. (2019). *Update on electric vehicle costs in the United States through 2030*. Working Paper, (2019-06). https://theicct.org/wp-content/uploads/2021/06/EV_cost_2020_2030_20190401.pdf

Majumdar, S. (2020). Prospects of sustainable transport of automobile vehicles in Kolkata City. In G. R. Kanagachidambaresan, R. Maheswar, V. Manikandan, & K. Ramakrishnan (Eds.), *Internet of Things in smart technologies for sustainable urban development. EAI/Springer Innovations in Communication and Computing*. Springer. https://doi.org/10.1007/978-3-030-34328-6_5

Majumdar, S., & Sarkar, B. (2020). Developing a smart and sustainable transportation plan for a large-sized city: An approach to smart city modeling, 145–152. https://doi.org/10.1007/978-981-15-2780-7_18

Majumdar, S., Choudhury, M., & Cheshmehzangi, A. (Eds.). (2025). *Sustainable urban future: Embracing circular economy in waste management.* Springer Singapore. https://doi.org/10.1007/978-981-95-2976-6

Marsden, G. (2022). *Social benefits of shared mobility: Metrics and methodologies.* https://www.acea.auto/files/Social_benefits_of_shared_mobility-26th_ACEA_SAG-report.pdf

Martinez, L. M., Pritchard, J. P., & Crist, P. (2024). Shared mobility's role in sustainable mobility: Past, present, and future. *Annual Review of Environment and Resources, 49.* https://doi.org/10.1146/annurev-environ-121522-041916

Marzouk, O. A.(2025). Summary of the 2023 report of TCEP (tracking clean energy progress) by the International Energy Agency (IEA), and proposed process for computing a single aggregate rating. In *E3S Web of Conferences* (Vol. 601, p. 00048). EDP Sciences. https://doi.org/10.1051/e3sconf/202560100048

Mogire, E., Kilbourn, P., & Luke, R. (2025). Smart charging for E-mobility in urban areas: A bibliometric review. *Energies, 18*(17), 4655. https://doi.org/10.3390/en18174655

Mola, L., Berger, Q., Haavisto, K., & Soscia, I. (2020). Mobility as a service: An exploratory study of consumer mobility behaviour. *Sustainability, 12*(19), 8210. https://doi.org/10.3390/su12198210

Ogunkunbi, G., Al-Zibaree, H. K. Y., & Mészáros, F. (2022). Modeling and evaluation of market incentives for battery electric vehicles. *Sustainability, 14*(7), 4234. https://doi.org/10.3390/su14074234

Ortega, A., Haq, G., & Tsakalidis, A. (2023). Carsharing in Europe: A critical review of policy, research, innovation, and practice. *Transportation Planning and Technology, 46*(4), 381–406. https://doi.org/10.1080/03081060.2023.2194717

Rajpal, A., Choudhury, M., Goswami, S., Chakravorty, A., & Raghavan, V. (Eds.). (2024). Waste management and treatment: Advances and innovations (1st ed.). CRC Press. https://doi.org/10.1201/9781003258377.

Spanos, A., Angelis, A., et al. (2022). Integration of EVs into the smart grid: A systematic literature review. *Energy Informatics, 5,* 65. https://doi.org/10.1186/s42162-022-00251-2

Tönnies, J., Ahrens, J., Hasselder, P., van Houten, P., van Eerten, J. J., White, M., Bolte, G., et al. (2025). Tackling transport poverty with offers for shared mobility–a qualitative analysis on mobility needs and barriers in the context of a new mobility hub in a deprived neighbourhood. *Cities & Health,* 1–15. https://doi.org/10.1080/23748834.2025.2539608

Wu, S.-M., Wang, Q., Chung, S.-H., Hu, L., Lau, Y.-y., & Liu, S. Q. (2025). Improving green urban mobility: A study on shared electric vehicles versus taxis. *Transportation Research Part E: Logistics and Transportation Review*. Advance online publication. https://doi.org/10.1016/j.transe.2025.1358

Zhang, Y., Fu, W., Chao, H., Mi, Z., & Kong, H. (2024). A comparative analysis of the potential of carbon emission reductions from shared micro-mobility. *Sustainable Energy Technologies and Assessments, 72*, 104088. https://doi.org/10.1016/j.seta.2024.104088

Zhong, S., Liu, A., Jiang, Y., Hu, S., Xiao, F., Huang, H. J., & Song, Y. (2023). Energy and environmental impacts of shared autonomous vehicles under different pricing strategies. *npj Urban Sustainability, 3*(1), 8. https://doi.org/10.1038/s42949-023-00092-2

CHAPTER 12

Collective Action: Enhancing Circular Economy Efforts Through Multi-sector Collaboration

Mariana Flores-Garcia, Sushobhan Majumdar, and Elkin Vargas Lopez

12.1 Introduction

Circularity refers to a paradigm shift in our way of living, reconfiguring our linear production and consumption systems based on recognizing the life cycles of things, products, materials, services, and everything around us, to minimize waste, a permanent pollutant of the planet (Kowsalya

The original version of the chapter has been revised. A correction to this chapter can be found at https://doi.org/10.1007/978-3-032-18026-1_17

M. Flores-Garcia (✉)
ITC Researcher, Research Directorate and Faculty of Architecture, Universidad Salle Bajío, León, Mexico
e-mail: mfloresg@lasallebajio.edu.mx

S. Majumdar
Department of Geography, Pandit Raghunath Murmu Smriti Mahavidyalaya, Bankura, West Bengal, India

E. V. Lopez
Director BuroDap, Bogotá, Colombia

S. Majumdar, M. Choudhury (eds.), *City of the Future*,
https://doi.org/10.1007/978-3-032-18026-1_12

et al., 2025). The objective of circularity seeks to maintain maximum use of products and materials under the premise of efficiency and responsibility, with minimal impact on the earth and available resources (Majumdar et al., 2025). From this emerge positions such as recycling, reuse, repair, renovation, and similar practices. Some strategies that have been socialized in recent years within the disciplines of Architecture, Design, and Urban Planning, and among the community, include circular design, rental models, the 5Rs, and opting for low-energy products, but these are not enough. All of the above suggest benefits such as reduced pollution impacts, greater industry competitiveness, and greater impact on social well-being. However, in most cases, these are isolated, disjointed actions, and the impacts are usually not as expected. An example of these advances in this field is the manufacturing sector, which will account for 35.7% of the global circular economy market in 2023, followed by the electronics and construction industries, which are identified as the sectors with the highest levels of pollution and waste (Goswami et al., 2025). Therefore, strategies toward new waste management systems and new modular systems are beginning to be developed (King Research, 2024).

12.2 Theoretical Approach: Multi-sector Collaboration Trends Toward Circularity

Circular ecosystems require new models of collaboration from their design, conceptualization, and implementation, and this involves industry, businesses, governments, academia, and civil society in an ongoing engagement focused on developing transition strategies from linear models to those that, based on circularity, can innovate in the face of an emerging market. Innovation through technology, based on the use of digital platforms, is an example derived from the exchange of resources or the promotion of the design and formation of consortia between national and international public–private partnerships focused on the management and planning of circular infrastructure and processes, where figures such as open urban innovation laboratories are also identified.

In addition to the above, trends can be identified regarding collaborative environments ranging from public–private partnerships, linking academia with business sectors, and the use of technologies for the development of platforms and spaces for the transfer of knowledge and good practices at the regional and international levels, such as the case of the Circular Economy Coalition of Latin America and the Caribbean, created in 2021 with the aim of developing instruments in the region that trigger collaborative circular economy approaches between sectors, from a joint vision of circularity. The first report was published in 2022, which demonstrated the relevance of opening spaces for international and regional cooperation for future alliances, since support was documented for 49 initiatives with innovation laboratories. The project articulates 18 countries and currently has allies such as the IDB, UNEP, and the Ellen MacArthur Foundation, among others (UNEP, 2022).

It is important to explore how multi-sector partnerships contribute to the transition toward circular models, based on collaborative structural reforms, where it is possible to overcome the systemic barriers that typically limit the comprehensive implementation of circularity. The circular economy proposes economic growth without promoting the excessive exploitation of available resources. It is important to establish synergies between different sectors, including public–private governance models, as is the case in Latin American countries. Brazil is an example, where good results have been achieved; however, collaborative frameworks aligned with regulatory reforms with shared goals, innovative co-financing systems, and transparent and ongoing monitoring platforms are required (BID, 2022; UNEP, 2022). In the context of international coalitions and technology transfer, the European Union has promoted the Circular Economy Resource Center (CERC) to adapt collaborative systems and policies to diverse contexts by establishing international alliances for the transfer of knowledge and experiences. Some of the barriers identified for the aforementioned are related to barriers or limitations surrounding local and regional technical capacities and financing systems. Significant gaps exist, as, according to the World Bank, only 12% of climate fund financing is allocated to these initiatives (Banco Mundial, 2021). It is also important to envision tools that enable the integration of different sectors, including informal ones, which will have a significant impact on value chains, shifting from informality to inclusion and social justice programs, especially in developing countries. Some of the trends in multi-sectoral collaboration toward the circular economy are geared toward initiatives that were traditionally biased toward isolated sectoral approaches. However, strategic alliances have recently been promoted that, in addition to strengthening SDG 11 and SDG 17, are also impacting public policies through the establishment of regional coalitions. This has shifted to global approaches with the development of instruments for the implementation of the circular economy from diverse perspectives. These platforms also have a significant impact on collaborative governance, which allows for contexts of public–private collaboration and the design and innovation of appropriate circular public policies. These collaborative frameworks, in turn, guide regulatory spaces and contexts focused on co-responsibility mechanisms among stakeholders, such as Chile's Extended Responsibility Law. Along with the above, it is feasible to develop industrial symbiosis projects and value chains based on circularity, which allows for a circular categorization system that fosters technological innovation initiatives such as blockchain and artificial intelligence for material traceability, as mentioned in the 2025 Annual Trends Report (WasteCero, 2025; Elias Bibri et al., 2023; Lucanto, 2023; UIA World Congress of Architects, 2024).

12.2.1 Circular Economy Technologies for Multi-sector and Multi-scale Urban Planning

In the implementation of new technologies, platforms have been developed that optimize financial mechanisms and standardize criteria to foster multi-sector, multi-level co-financing and international cooperation opportunities. Some of the enabling platforms for their application in the circular economy include Internet of Things (IoT) for real-time monitoring of material flows, artificial intelligence for automated waste classification, as in the case of some recycling plants, blockchain for material traceability, and digital twins for simulating urban scenarios, which are also trending in the design of regenerative environments, as in the case of Zaragoza and Medellín (Narayan et al., 2025). To further this analysis, 100 initiatives have been documented, standing out from 2019 to 2025, as triggers of innovation in multi-sector ecosystems at the global and regional levels, where technology generates spaces for transformation.

12.3 The Rationale for Multi-sector Collaboration

12.3.1 Complex, Systemic Nature of Circular Economy Transitions

Circular economy transitions are inherently complex and systemic, involving changes across production processes, consumption patterns, material flows, and end-of-life management. Achieving these transformations requires coordinated action that goes beyond the capabilities of any single organization or sector (Laurenti et al., 2018). Multi-sector collaboration allows stakeholders to address interlinked environmental, economic, and social challenges simultaneously, fostering holistic solutions that consider the full lifecycle of products and resources (Silva et al., 2020).

12.3.2 *Interdependencies Across Value Chains*

Value chains in a circular economy are highly interdependent, as the efficiency of one actor often depends on the practices of upstream suppliers, downstream distributors, recyclers, and consumers (Narayana et al., 2024). For example, the effectiveness of recycling programs relies on product design, manufacturing decisions, logistics, and consumer participation. Collaboration across sectors ensures that these interdependencies are recognized and managed, enabling integrated solutions that optimize resource use, reduce waste, and create shared value throughout the system (Doğan et al., 2015).

12.3.3 *Benefits of Collaborative Approaches Over Isolated Efforts*

Collaborative approaches offer significant advantages compared with isolated initiatives. By pooling knowledge, resources, and capabilities, stakeholders can achieve economies of scale, reduce duplication, and accelerate innovation (Lian et al., 2023). Partnerships facilitate risk-sharing, enhance access to infrastructure, and create market incentives for circular products and services. Moreover, joint initiatives often gain greater legitimacy, attract public and private investment, and achieve broader societal impact than single-actor efforts, making collaboration a more effective strategy for systemic circular economy adoption (Aldieri et al., 2021).

12.3.4 *Barriers to Collaboration*

Despite the benefits, multi-sector collaboration faces several barriers. Institutional barriers include misaligned regulations, fragmented governance structures, and competition for resources (Ullah et al., 2021). Cultural barriers arise from differences in organizational priorities, risk tolerance, and decision-making styles across sectors. Informational barriers involve limited data sharing, lack of standardized metrics, and insufficient knowledge about best practices (Francis et al., 2024). Overcoming these challenges requires deliberate mechanisms for

Table 12.1 Roles and contributions of different sectors

Sector	*Key roles*	*Impact on circular economy*
Government	Policy, regulation, incentives, standard setting	Provides enabling environment, reduces uncertainty
Private sector	Technological innovation, investment, supply chain redesign	Implements practical solutions, drives economic value
Civil society/ NGOs	Advocacy, community engagement, monitoring	Ensures inclusivity, accountability, and social acceptance
Academia/ research	Knowledge creation, assessment tools, capacity-building	Provides evidence-based solutions, supports decision-making
Consumers	Behavioral change, demand creation	Drives market demand, supports adoption of circular solutions

Source: Compiled by the Authors

trust-building, transparent communication, joint planning, and shared accountability, ensuring that diverse stakeholders can effectively work together toward common circular economy goals (Ijaz et al., 2022) (Table 12.1).

12.4 Frameworks and Models of Collective Action

12.4.1 Overview of Collaborative Governance Models

Collaborative governance models provide structured approaches for multiple stakeholders, governments, private companies, civil society, and academia to work together in achieving shared circular economy objectives (Cappellaro et al., 2020). These models emphasize participatory decision-making, transparency, and accountability, fostering trust and long-term commitment. By creating mechanisms for dialogue, joint planning, and coordinated implementation, collaborative governance helps reconcile diverse interests, align incentives, and overcome barriers such as regulatory fragmentation, resource constraints, or power imbalances (Zhu, 2024). It serves as the overarching framework that guides various forms of multi-sector collaboration.

12.4.2 *Public–Private Partnerships (PPPs)*

Public–private partnerships are a widely used model for pooling resources, expertise, and risk to implement circular initiatives at scale. Governments provide regulatory support, policy incentives, or funding, while private companies contribute technological solutions, investment, and operational capabilities (Ugwu et al., 2024). PPPs are particularly effective in building shared infrastructure such as recycling facilities, remanufacturing hubs, or logistics networks that would be challenging for a single actor to establish (Saadany, 2021). By clearly defining roles, responsibilities, and benefits, PPPs foster long-term collaboration, accelerate innovation, and reduce financial and operational barriers for circular projects.

12.4.3 *Industrial Symbiosis and Eco-industrial Parks*

Industrial symbiosis and eco-industrial parks exemplify circular collaboration at the operational level, where companies co-locate or network to optimize resource flows. By sharing energy, water, materials, and by-products, firms reduce waste, lower costs, and enhance sustainability (Rosano & Schianetz, 2014). Eco-industrial parks often include joint facilities for recycling, energy generation, and logistics, creating localized circular systems that benefit all participants (Ya et al., 2023). These models demonstrate how systemic collaboration can convert waste streams into valuable inputs, foster innovation, and generate both environmental and economic value at a community or regional scale (Rajpal et al., 2024).

12.4.4 *Multi-stakeholder Platforms and Alliances*

Multi-stakeholder platforms and alliances provide structured forums for dialogue, knowledge exchange, and coordinated action across sectors (Mutmainnah et al., 2022). They bring together companies, regulators, NGOs, academia, and sometimes consumer groups to co-create solutions, share best practices, and align strategies. These platforms often focus on specific sectors, materials, or sustainability challenges, such as plastics, electronics, or food waste (Agmapisarn et al., 2024). By facilitating transparency, trust, and joint problem-solving, multi-stakeholder alliances help overcome coordination barriers, accelerate policy adoption, and scale circular initiatives beyond individual organizations or industries.

Table 12.2 Frameworks and models of collective action

Model/framework	*Key stakeholders involved*	*Primary benefits*
Collaborative governance	Government, private sector, NGOs, academia	Trust-building, coordinated strategy, long-term alignment
Public–private partnerships (PPPs)	Government, companies	Risk-sharing, resource pooling, scalability
Industrial symbiosis/ eco-industrial parks	Companies, local government	Waste reduction, cost efficiency, innovation
Multi-stakeholder platforms/alliances	Companies, NGOs, academia, regulators	Policy alignment, joint problem-solving
Innovation ecosystems/ clusters	Startups, universities, companies	Accelerated R&D, technology transfer, shared learning

Source: Compiled by the Authors

12.4.5 Innovation Ecosystems and Cluster-Based Collaboration

Innovation ecosystems and cluster-based collaboration emphasize regional or sectoral networks that foster experimentation, knowledge sharing, and co-development of circular solutions (Daniels & Gebhardt, 2021). Startups, established companies, universities, and research institutions collaborate within clusters to pilot technologies, test business models, and disseminate innovations (Estensen et al., 2016). These ecosystems leverage shared infrastructure, expertise, and funding mechanisms to accelerate R&D, facilitate commercialization, and enhance competitiveness. By connecting diverse actors around a common vision, innovation ecosystems create dynamic, resilient environments that support continuous circular innovation and long-term systemic transformation (Alatawi, 2025) (Table 12.2).

12.5 Roles and Contributions of Different Sectors

12.5.1 Government: Policy, Regulation, Incentives, Standard Setting

Governments play a pivotal role in enabling circular economy transitions by creating the regulatory and institutional frameworks that guide stakeholder behavior (Vrieze et al., 2022). This includes establishing laws and standards for waste management, product design, extended producer responsibility, and resource efficiency. Policymakers can also use economic instruments such as subsidies, tax incentives, or green public procurement to encourage investment in circular practices (Choudhury et al., 2025a, b). Beyond enforcement, governments facilitate multi-stakeholder coordination, provide funding for infrastructure or research, and act as conveners to align national, regional, and local circular economy initiatives (Odoyi & Riekkinen, 2022). Their leadership helps set long-term priorities and reduces uncertainty for private actors, accelerating the adoption of sustainable practices.

12.5.2 Private Sector: Technological Innovation, Investment, Supply-Chain Redesign

The private sector drives much of the practical implementation of circular solutions through innovation, capital deployment, and operational changes. Companies can redesign products for durability, reparability, and recyclability; optimize supply chains for resource efficiency; and develop new business models such as product-as-a-service or sharing platforms (Chourasiya & Malviya, 2025). Businesses also invest in recycling and remanufacturing infrastructure, research and development, and collaborations with startups or research institutions to accelerate technological breakthroughs (Groenewald, 2024). By aligning economic incentives with circular strategies, the private sector transforms ideas into scalable solutions that create both financial and environmental value.

12.5.3 Civil Society and NGOs: Advocacy, Community Engagement, Monitoring

Civil society organizations and NGOs play a critical role in advocating for circular policies, raising public awareness, and holding governments and companies accountable (Hoque et al., 2016). They engage communities, facilitate participatory planning, and ensure that social and environmental considerations are integrated into circular initiatives. NGOs often act as intermediaries between stakeholders, providing expertise, mobilizing volunteers, and monitoring the impacts of circular projects. Their advocacy helps shape regulatory frameworks, promotes ethical practices, and ensures that circular economy efforts are inclusive, transparent, and socially responsible (Reynolds, 2024).

12.5.4 *Academia and Research Institutions: Knowledge Creation, Assessment Tools, Capacity-Building*

Academic and research institutions provide the scientific and technical foundation necessary for circular economy implementation. They develop assessment tools, modeling frameworks, and indicators to evaluate material flows, environmental impacts, and economic feasibility (Finamore & Oltean-Dumbrava, 2024). Universities and research centers conduct applied research, pilot innovative technologies, and contribute to capacity-building programs that train professionals in circular practices (Guterman, 2024). Their collaboration with industry and government ensures that policies and business models are informed by robust evidence and emerging best practices, bridging the gap between theory and practical application.

12.5.5 *Consumers: Behavioral Change and Demand Creation*

Consumers influence circular economy outcomes by shaping demand for sustainable products, services, and practices. Through conscious purchasing, participation in recycling and take-back programs, and adoption of sharing or leasing models, consumers create market incentives for companies to implement circular strategies (Bucur, 2023). Education and awareness campaigns further empower individuals to reduce waste, extend product lifecycles, and embrace resource-efficient lifestyles. Consumer behavior thus not only drives adoption of circular offerings but also signals societal support for systemic transitions toward sustainability (Yang et al., 2023) (Table 12.3).

12.6 Enablers of Effective Collaboration

12.6.1 *Shared Vision and Common Goals*

A clearly articulated shared vision and common goals are the foundation of effective multi-sector collaboration in the circular economy. When stakeholders from government agencies and businesses to civil society and research institutions understand and commit to the same overarching objectives, it aligns actions, reduces conflicts, and enhances coordination (Akofio-Sowah et al., 2024). Co-creating goals with input from all sectors ensures that diverse perspectives are incorporated, increases buy-in, and

Table 12.3 Enablers and challenges of multi-sector collaboration

Category	*Key factors*	*Impact/significance*
Enablers	Shared vision & common goals	Aligns priorities, fosters commitment
	Data sharing & digital infrastructure	Supports evidence-based decisions, reduces duplication
	Financial mechanisms	Encourages investment, reduces risk
	Trust & transparent governance	Sustains long-term collaboration
	Skills development & cross-sector learning	Builds capacity, promotes innovation
Challenges	Power imbalances	Can reduce equity, hinder participation
	Misaligned incentives & time horizons	Slows collaboration, reduces commitment
	Resource constraints & coordination costs	Delays implementation, limits scalability
	Regulatory fragmentation	Creates barriers to multi-region initiatives
	Risk of greenwashing	Reduces credibility, undermines trust

Source: Compiled by the Authors

motivates participants to actively contribute to joint initiatives. Beyond goal alignment, a shared vision helps establish a long-term roadmap, setting measurable milestones and providing a sense of purpose that sustains engagement through challenges and uncertainties (Morrison et al., 2024).

12.6.2 Data Sharing and Digital Infrastructure

Access to accurate, timely, and interoperable data is critical for collaborative decision-making and system-level optimization. Digital infrastructure including cloud platforms, IoT-enabled monitoring, blockchain for traceability, and lifecycle assessment tools facilitates seamless collection, analysis, and transparent sharing of information among stakeholders (Farina et al., 2025). This visibility into material flows, resource efficiency, and environmental impacts enables partners to identify circular opportunities, benchmark progress, and make evidence-based decisions. Moreover, robust data systems reduce duplication of effort, support predictive analytics for supply chain optimization, and create a foundation for innovative business models that rely on real-time material and product information (Ripanti & Tjahjono, 2019).

12.6.3 Financial Mechanisms that Support Joint Initiatives

Collaborative circular economy projects often require significant upfront investment and shared risk. Financial mechanisms such as co-funding arrangements, public–private partnerships, innovation grants, blended finance, and performance-based incentives encourage stakeholders to pool resources and invest jointly in shared infrastructure, R&D, or pilot projects (Yurieva et al., 2021). These mechanisms not only reduce individual risk but also ensure equitable distribution of costs and benefits, making ambitious circular initiatives feasible for actors of varying sizes and capacities. Additionally, linking financial support to sustainability performance encourages long-term commitment and reinforces the systemic impact of the collaboration (Lakasse et al., 2024).

12.6.4 Trust Building and Transparent Governance

Trust is essential for sustaining multi-stakeholder collaboration, especially where interests, resources, and capacities differ. Transparent governance structures with clearly defined roles, decision-making processes, accountability mechanisms, and regular communication help build confidence, reduce conflicts, and ensure equitable participation (Adeusi et al., 2024). Practices such as joint planning sessions, independent monitoring, and conflict resolution protocols further reinforce trust. When stakeholders feel confident that commitments will be honored and information is shared openly, they are more likely to engage in risk-sharing, co-innovation, and experimental initiatives that push the boundaries of circular practice (Anwar et al., 2022).

12.6.5 Skills Development and Cross-Sector Learning

The success of collaborative circular initiatives depends on stakeholders having the technical expertise, systems thinking, and collaborative skills necessary to engage effectively (Hensley, 2024). Training programs, workshops, peer-to-peer learning, secondments, and cross-sector knowledge platforms facilitate continuous capacity-building (Mulder & Magni, 2022). By promoting mutual understanding, fostering innovation capabilities, and bridging knowledge gaps between industry, government, academia, and civil society, cross-sector learning strengthens the collective ability to design, implement, and scale circular solutions (El-Jardali et al.,

2018). Furthermore, cultivating soft skills such as negotiation, facilitation, and adaptive problem-solving enhances resilience and flexibility in navigating the complexity of multi-stakeholder systems (Pelletier et al., 2017).

12.6.6 Standardized Metrics and Impact Measurement

Standardized metrics and impact measurement are essential for ensuring accountability, comparability, and continuous improvement in circular economy collaborations (Lande et al., 2024). By establishing common indicators for resource efficiency, material recovery, waste reduction, carbon footprint, and social outcomes, stakeholders can objectively assess the effectiveness of initiatives across sectors and regions (Mallick et al., 2022). Standardization enables benchmarking, facilitates data-driven decision-making, and highlights best practices that can be replicated or scaled. Moreover, consistent measurement frameworks build trust among partners and investors, as progress can be verified transparently and fairly (Mai et al., 2024). Digital tools, lifecycle assessment methodologies, and reporting protocols further enhance accuracy, allowing organizations to track both environmental and socio-economic impacts over time (Morganti et al., 2023). Ultimately, robust metrics and systematic impact assessment transform circular initiatives from isolated projects into measurable, scalable, and evidence-based pathways toward a regenerative economy.

12.7 Case Studies: Global and Regional Circular Models

One hundred success stories, projects, and initiatives were selected at the global and regional levels to identify opportunities, challenges, and potential limitations based on shared knowledge, experiences, and lessons learned. These include areas such as strategic alliances, co-participation mechanisms and adaptive governance, innovation in regulatory frameworks, public policy design, citizen participation and co-creation models, technological barriers, and digital environments, among others. Each case is recognized for its relevance and innovation at the global and regional levels. A brief description is included, including strategies and actions that illustrate the practice and relevant data to validate the success of each case, as well as the projections and participating sectors (Table 12.4).

Table 12.4 Case studies of the world

Initiative	*Sectorial participants*	*Description*	*Success stories*
Circular City Coalition	ONU-Hábitat, Google, Ellen MacArthur Found. Gob	Global data and technology platform for circular cities. Public–private collaboration, AI for material flows	Amsterdam: waste mapping and optimized recycling
Ai4cities	Ciudades europeas, startups, empresas	Public procurement of AI solutions for emissions reduction. Co-creation, pilots with the private sector	Helsinki: AI for efficient traffic and energy
Urban Releaf	EIT Urban Mobility, universidades, ONGs, gobiernos	AI and sensors for green space regeneration. Citizen science, academic partnerships	Lisbon: mapping urban biodiversity
Circular Glasgow	Cámara Comercio, Zero Waste Scotland, empresas	Circular business ecosystem. Mentoring, workshops, collaborative pilots	Reuse of materials in construction
Smart Waste Seoul	Gobierno Seúl, Samsung, universidades	Smart waste management with AI. Government-business, big data, incentives	+30% recycling, fewer landfills
Regenerative Cities Network	ICLEI, WWF, gobiernos, sector privado	Global network of urban regenerative practices. Replicable pilots, blended financing	Quito: restoration of urban watersheds
Circular Buildings Coalition	Arup, Google, RICS, sector construcción	Circular buildings with AI in design and materials. Industry-technology, standards, pilots	London: smartly monitored buildings
Ai Urban Mobility Lab	MIT, Toyota, ciudades piloto	Optimizing urban mobility with AI: academia-industry, urban pilots	Boston: congestion reduction
Green Digital Finance Alliance	ONU Medio Ambiente, AntFinancial, bancos	AI and blockchain to finance sustainable urban projects. Fintech alliance, digital incentives	Shenzhen: green microcredits for recycling

(*continued*)

Table 12.4 (continued)

Initiative	*Sectorial participants*	*Description*	*Success stories*
Urbantech Challenge LATAM	BID, startups, gobiernos, aceleradoras	AI solutions for circularity and regeneration. Mentoring, pilots, regional scaling	Medellín: AI for waste tracking
Redciclach	Municipios chile, supermercados, startups	Digital platform and blockchain for recycling. Public–private partnership, traceability	Implementation in Santiago and supermarkets
City Lab Biobío	MIT, Gobierno de Biobío, universidades	AI for urban data analysis and well-being. Academia-government, multi-dimensional analysis	Improving the quality of life in Concepción
Smart Water Lisboa	Siemens, EDP, Câmara Municipal	Technological alliance, real-time monitoring	20% savings in water consumption
Barcelona SuperblockIa	Ayuntamiento, MIT	Urban redesign AI green spaces. Predictive simulations, co-design	Emissions reduction in Gràcia
Plastic Bank	IBM, ONGs, empresas de retail	Blockchain + AI for plastic waste. Digital incentives, traceability	+1000 tons/ month recovered in the Philippines
Economía Circular CDMX	Gobierno de CDMX, empresas	Digitalization and AI optimize resources. Public policies, tax incentives	−15% industrial waste
ZenRobotics Waste Sorting	Terex, municipios europeos	Automated waste sorting with AI. Robots, computer vision	95% accuracy in plastics
Copenhagen Carbon Neutral 2025	Gobierno, Algoritmia	Smart grid with AI for renewables. Wind and solar integration	−70% emissions (2024)
Bosque Urbano Bio Pappel	Bio Pappel, comunidades mexicanas	Production of 100% recycled paper. Circular chain, local partnerships	795 tons/year of recycled paper
IA Agricultura Regenerativa	FAO, Microsoft, UNAM	Monitoring urban gardens with AI. Open data platforms	+40% productivity in Rome
Blockchain Para Residuos Plásticos	Plastic Bank, IBM	Waste monetization with digital tokens. Incentives, blockchain traceability	Philippines: 1000 tons/month recovered

(*continued*)

Table 12.4 (continued)

Initiative	*Sectorial participants*	*Description*	*Success stories*
Red De Ciudades Resilientes	C40, ONU-Hábitat	Exchange of practices in urban resilience. Global network, climate finance	50 cities with adaptation plans
Algoritmos Verdes C40-Deepmind	C40, DeepMind	Energy optimization in data centers. AI for energy efficiency	30% savings in London
Fab City Barcelona	IAAC, Ayuntamiento	Hyperlocal manufacturing AI and 3D printing makerspaces, urban network	45% less transportation of materials
Simbiosis Industrial València	Empresas, agricultores	Industrial by-product exchange. Digital platform, matchmaking	10,000 tons of CO_2 avoided/year
Ia Generativa En Singapur	Gobierno, startups	+100 AI solutions for urban services. Open innovation competitions	Multilingual chatbots, optimized transportation
Red Ciudades Creativas Unesco	UNESCO, 246 ciudades	Culture as a cornerstone of urban sustainability. Multi-stakeholder cooperation	Medellín: districts of cultural innovation
Circulab París	Renault, Veolia	Digital twins for the circular economy. Predictive models, simulation	90% reuse of EV batteries
Gestión Inteligente Playas	The Ocean Cleanup, Chile	AI drones for microplastics. Computer vision, coastal mapping	200 km of clean beaches
Urban Mining Bruselas	Umicore, Google Cloud	Metal recovery with AI. Urban mining, blockchain	85% recycled metals
Red Mundial Ciudades Del Aprendizaje	UNESCO, gobiernos	Continuing education for sustainability. Citizen training	Seoul: bridging the digital divide
Ai4city Onu-Habitat	ONU-Habitat, Alibaba	Global AI standards for cities. Multi-stakeholder dialogue, technical guides	15 megacities with AI guides

(*continued*)

Table 12.4 (continued)

Initiative	*Sectorial participants*	*Description*	*Success stories*
Netzero C40	C40, empresas tecnológicas	Carbon neutrality in urban infrastructure. Green bonds, predictive AI	25 cities with zero emissions
Smart Mobility Curitiba	Gobierno, Volvo, universidades	Smart and sustainable public transport: AI for routes, modal integration	Model for Latin America
Redciclach Chile	Municipios, supermercados	Blockchain for recycling and traceability. Public–private partnership	Santiago supermarkets
City Lab Biobío	MIT, Gobierno Biobío	AI for urban well-being: multidimensional analysis	Improves quality of life in Concepción
Smart Water Lisboa	Siemens, EDP	Water management with AI and sensors. Real-time monitoring	20% water savings
Superblock Barcelona	Ayuntamiento, MIT	Urban redesign AI for green spaces. Simulation, co-design	Reducing emissions in Gràcia
Plastic Bank	IBM, ONGs	Blockchain para residuos plásticos. Incentivos digitales	1000 ton/mes recuperadas Filipinas
Economía Circular CDMX	Gobierno, empresas	AI to optimize industrial resources. Public policies	−15% industrial waste
ZenRobotics Waste Sorting	Terex, municipios	Waste sorting with AI. Robots, computer vision	95% precision plastics
Copenhagen Carbon Neutral	Gobierno, Algoritmia	AI electricity grid for renewables. Wind, solar, AI	−70% emissions
Bio Pappel Urbano	Bio Pappel, comunidades	Circular recycled paper. Circular chain	795 ton/year recycled
Ia Agricultura Regenerativa	FAO, Microsoft	AI in urban gardens. Open data	+40% productivity in Rome
Blockchain Residuos Plásticos	Plastic Bank, IBM	Monetizing waste tokens. Blockchain, AI	1000 tons/month in the Philippines
Ciudades Resilientes	C40, ONU-Hábitat	Urban resilience practices. Global network	50 adapted cities

(*continued*)

Table 12.4 (continued)

Initiative	*Sectorial participants*	*Description*	*Success stories*
Algoritmos Verdes C40	C40, DeepMind	Data center optimization. AI energy efficiency	30% savings London
Fab City Barcelona	IAAC, Ayuntamiento	Hyperlocal manufacturing AI. Makerspaces	45% less transportation
Simbiosis Industrial	Empresas, agricultores	Industrial by-products. Digital platform. VALENCIA	10,000 tons of CO_2 avoided
Ia Generativa Singapur	Gobierno, startups	+100 Urban AI solutions. Open innovation	Chatbots, optimized transportation
Bienestar Humano Territorial	CIT-UAI, gobiernos	AI for comprehensive urban diagnosis. Data science, well-being	Conception: inclusive planning
Red De Movilidad Compartida	BlaBlaCar, gobiernos	AI to optimize ride sharing. Digital platform, incentives	Reduction in traffic and emissions
Green Building Challenge	GBCI, sector construcción	Circular building certification: AI for monitoring and design	Green buildings in 30 cities
Urban Solar Grid	Tesla, gobiernos locales	Smart urban solar energy. AI for grid balancing	20% renewable energy on the grid
Smart Lighting Buenos Aires	Gobierno, Philips	Smart urban lighting. Sensors, AI, efficiency	−40% energy consumption
AiForClean Air	UNICEF, Intel	Hyperlocal pollution monitoring. Low-cost sensors, ML	−25% PM2.5 New Delhi
Circular Construction Hub	Holcim, Autodesk	Reversible modular design with AI. BIM circular, materials	Demountable buildings in Zurich
Food Rescue Alliance	Walmart, Too Good To Go	Food redistribution AI. Demand prediction	1 million tons of food saved
Urban Mining Platform	Umicore, Google Cloud	Metal recovery AI. Urban mining, blockchain	85% metal recycling in Brussels
Smart Parks	WWF, gobiernos locales	AI for biodiversity monitoring. Sensors, drones, AI	Urban park restoration

(*continued*)

Table 12.4 (continued)

Initiative	*Sectorial participants*	*Description*	*Success stories*
Green Roofs NYC	NYC, universidades	Green roofs AI for urban climate. Predictive models	−3 °C local temperature
Circular Electronics Pact	Dell, HP, gobiernos	Electronic collection and reuse. Blockchain traceability	500,000 devices recovered/year
Urbantech Challenge LATAM	BID, startups	AI for circularity and regeneration. Mentoring, pilots	Medellín: AI for waste
Ai4waste Singapur	NEA, startups	Automated waste sorting. Autonomous robots, ML	99% precision PET plastics
Regenera CDMX	Sedema, UNAM	Soil restoration with AI and drones. Urban agriculture	50 hectares rehabilitated in Xochimilco
Green Algorithms	C40, DeepMind	Carbon footprint reduction AI. Energy optimization	−30% energy for servers in London
Fab City Global	IAAC, gobiernos	Distributed digital manufacturing network. Hyperlocal manufacturing	45% less material transportation
Circular Economy Accelerator	Fundación Ellen MacArthur, empresas	Circular scaling startups. Mentoring, investment	+200 projects accelerated
Smart Waste Management	Gobierno, startups Chile	AI for collection routes. Sensors, big data	−20% operating costs
Urban Green Corridors	WWF, municipios	Green corridors AI for biodiversity. Ecological planning	+10% urban green areas
Digital Twin Amsterdam	Municipio, universidades	Digital twin for urban management. AI, simulation, open data	Traffic and energy optimization
AiForWaterMexico	CEMEX, universidades	AI for water monitoring. Sensors, big data	−15% leaks in Monterrey
Smart Mobility Madrid	EMT, startups	AI for public transport. Route optimization, data	−20% mobility efficiency
Urban Circularity Index	Fundación Ellen MacArthur, ciudades	Urban circularity index. AI for flow analysis	Benchmarking in 50 cities

(*continued*)

Table 12.4 (continued)

Initiative	*Sectorial participants*	*Description*	*Success stories*
City Data Hub Seoul	Gobierno, Samsung	Urban data platform. AI. Open data, predictive analytics	Improvement of public services
Smart Grid Medellín	EPM, universidades	Smart grid. Renewable energy balancing	–25% clean energy
Urban Forests Initiative	FAO, municipios	AI for urban forest planning. Drones, sensors	–100,000 trees planted
Green Digital Finance Alliance	ONU Medio Ambiente, AntFinancial	AI and blockchain for green projects. Fintech, digital incentives	Green microcredits in Shenzhen
Smart Cities LATAM	CAF, gobiernos	Urban AI solutions platform. Regional network, pilots	–30 cities implementing AI
Circular Economy Observatory	CEPS, universidades	Circular policy observatory. AI for impact analysis	Reports for the EU and LATAM
Ecosmart Mobility	Renault, gobiernos	Electric fleets. AI for mobility. Route optimization, data	–40% emissions from urban fleets
City AiConnect	City AI, gobiernos	Platform for cities. Generative AI. Sharing best practices	Cases in Dallas, Buenos Aires, Singapore
Urban Health Analytics	OMS, universidades	AI for urban health and air quality. Predictive models	Early warnings in 10 cities
Smart Waste Bogotá	Gobierno, startups	AI for routes and recycling. Sensors, big data	–25% municipal recycling
Urban Energy Storage	Tesla, municipios	Smart storage. AI. Demand prediction	–20% energy efficiency
Circular Fashion Lab	H&M, startups	AI for textile recycling. Blockchain, traceability	–10 million garments recovered
Smart Irrigation Dubai	DEWA, universidades	AI for efficient irrigation. Sensors, big data	–30% water consumption in parks
Sustainable Mobility Paris	RATP, startups	AI for low-carbon transport. Route optimization	–25% CO_2 in public transport

(*continued*)

Table 12.4 (continued)

Initiative	*Sectorial participants*	*Description*	*Success stories*
City Biodiversity Index	ICLEI, municipios	Urban biodiversity index. AI for monitoring	Benchmarking in 40 cities
Urban Circular Labs	BID, universidades	Circular innovation labs. Pilots, mentoring	–15 participating LATAM cities
AiFor Urban Planning	MIT, gobiernos	AI for regenerative urban design. Simulation, open data I	Projects in Boston and Quito
Smart City Challenge India	Gobierno, empresas	Urban AI solutions competition. Pilots, scaling	–100 cities with AI projects
Urban Mobility Platform	Siemens, municipios	AI platform for urban mobility. Open data, modal integration	–50 European cities
Circular Procurement Net	ICLEI, gobiernos	Circular public procurement network. AI for supplier analysis	–30 cities green purchasing
Sustainable Urban Logistics	DHL, startups	AI for low-carbon urban logistics. Route optimization	–20% emissions from deliveries
Green City Index	Siemens, Economist	Urban sustainability index. AI for comparative analysis	Ranking of 120 cities
Urban Food Systems Lab	FAO, universidades	AI for urban food systems. Predictive models	–10 cities with AI urban gardens
Smart Waste Platform	Veolia, municipios	AI platform for waste management. Sensors, big data	–30 cities implementing
Urban Circular Innovation Hub	Fundación Ellen MacArthur, empresas	Urban circular innovation hub. Mentoring, investment	+100 accelerated startups
AiFor Urban Resilience	ONU-Hábitat, IBM	AI for urban climate resilience. Predictive models, open data	Pilots in 10 vulnerable cities

Source: Compiled By the Authors

As can be seen, there is a wide range of initiatives that, while mostly focused on waste management, many of them are significantly transforming the industry and the ways in which we collaborate.

12.8 Materials and Methods, Hello from Road to Integrated Systems

There are methodologies such as the Circular Transition Indicator Framework (WBCSD) from which 12 measurement dimensions are proposed, and in some cases the so-called digital material passports have been issued, as in Finland, which are product lifecycle records through blockchain. In the case of Latin America, projects such as Circular Movement trained 580,000 people in regenerative design, achieving that 41% of SMEs adopted circular principles in their processes (King Research, 2024; Movimientocircular, 2025; Suau Ibáñez, 2019; Quesada, 2022; Elias Bibri et al., 2023). Initially, a methodology was developed to analyze 100 cases of good practices and emerging initiatives in various countries. This was carried out in a systemic analysis based on criteria such as innovation, temporality, impact factors, and cultural diversity. This analysis was conducted in international databases and official websites, focusing primarily on those in the global South that have been globally recognized for their creativity and innovation. The procedure for this research was configured in four stages: the first was based on the state of the art, which allowed for the analysis of theoretical and conceptual approaches; the second was developed from the analysis of 100 success stories or initiatives at a global level; the third stage identified 50 models or instruments for multi-sector alliances implemented at the global and regional levels; and the fourth stage involves a methodological proposal called the "Multi-Sector Circular Economy Implementation Model: A Comprehensive Strategic Framework," which synthesizes the main findings resulting from the study.

12.8.1 Integrated circular systems, methodology, and public policy design

Following the case analysis, 50 official and recent models, instruments, and methodological tools were identified that have guided practices with a global and regional scope. The most innovative ones with a global focus are shown in Image 12.1, and some with a regional focus are included in Image 12.2.

As mentioned above, an integrated model is suggested that can serve as a reference for the design of strategies, plans, and programs as instruments for implementing circular cities and economies from a multi-sectoral integrated approach. To this end, a five-phase model has been proposed,

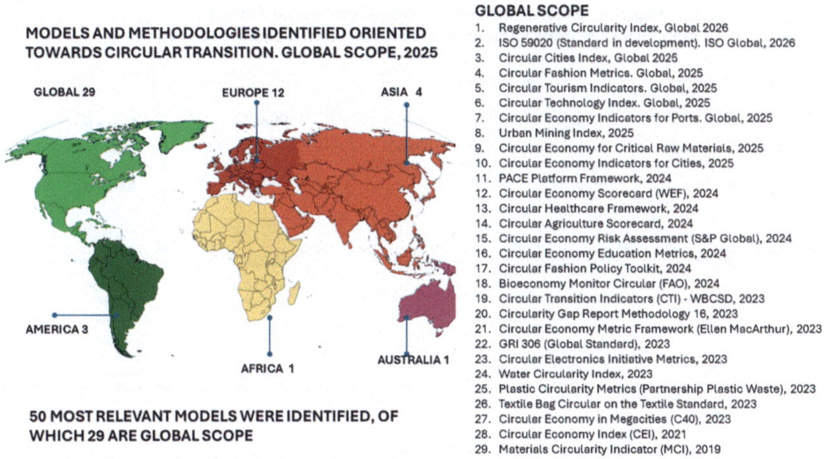

Image 12.1 Models and methodologies identified oriented toward circular transition by global scope, 2025. By Authors

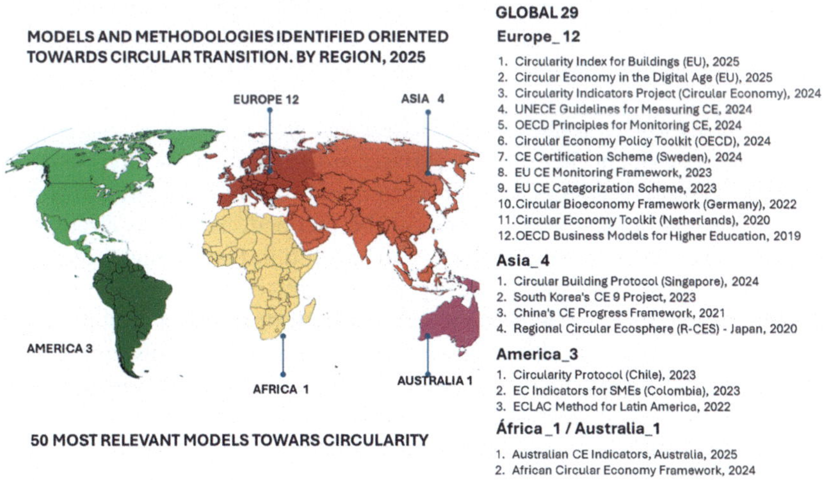

Image 12.2 Models and methodologies identified oriented toward circular transition by region, 2025. By Authors

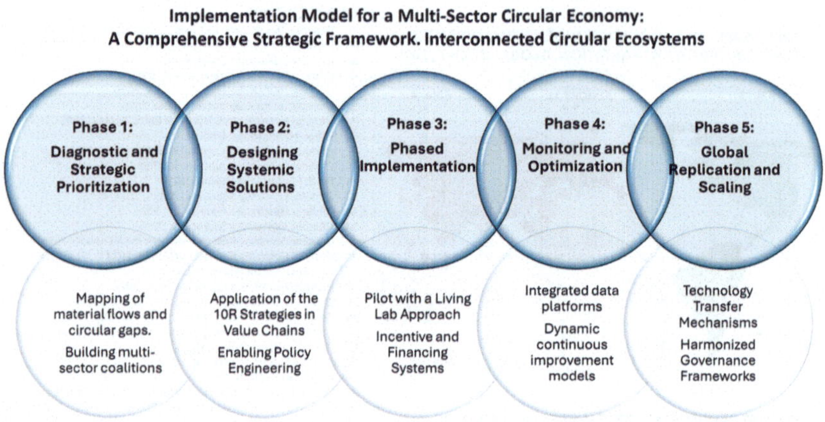

Image 12.3 Implementation model for a multi-sector circular economy: a comprehensive strategic framework. Interconnected Circular Ecosystems, 2025. By Authors

which, from a systemic perspective, guides processes that facilitate the implementation of the circular model. The conceptualization of the Implementation model and its phases are shown in Image 12.3.

Phase 1: Strategic Diagnosis and Prioritization of Needs
From the logic of an integrated approach and adaptability to diverse contexts, the first phase suggests the development of an initial diagnosis where the flow of materials and processes can be mapped, as well as circularity gaps. This refers to a reengineering of urban and industrial metabolic systems that can also be supported by the metrics of the aforementioned global systems, such as the Sectoral Circularity Index, with a digital monitoring system from the different sectors in the regions. This entails the design and development of a collaboration platform for the construction and consolidation of alliances, as shown by the European Union Plastics Agreement, which has registered the participation of 298 companies and 45 governments, innovation centers and startups, managers and financiers, as well as the community in 2025. In the case of Malaysia, there is evidence that circular industrial clusters increased their productivity by 28% based on new multi-sector participatory governance protocols (Ellen Macarthur, 2024).

Phase 2: Systemic Design—Possible Solutions

In the second phase, once the initial diagnosis visualizing multi-sector capabilities and limitations has been structured, successful models are identified that operate under three conditions that enhance the success of implementation strategies. These are the operational core, which can be supported by multi-sector consortia between academia, industry, innovation hubs, and R&D for the development of new technologies. The support network, with a financing platform, and the enabling environment that integrates harmonized legal and regulatory frameworks with roadmaps and instruments appropriate for the regions are also identified. To this end, appropriate technical instruments must be considered for their validation, replicability, and scalability, which will lead to specific proposals for process reengineering and productive transition mechanisms. This will involve, within the framework of circularity, the implementation of initiatives such as the 10R+ value chain strategies, prioritizing sectors such as construction, the fashion industry, mining, products and services, and manufacturing. This will have implications for existing policies at the innovation level within the regulatory and political framework for circularity, such as the European Action Plan toward 2030, which integrates circular design standards under the BS 8001:2017 standard, mixed financing schemes, and circularity fees for producers and service providers (European Commission, 2023b).

Phase 3: Management and Implementation Plan

Given the need to strengthen local practices, it is suggested that a management and implementation plan be developed that can be guided by pilot projects, with a living lab approach, as is the case with initiatives such as the LBC, which certifies buildings and living communities with a regenerative focus (LBCH, 2022). Phased implementation allows for the validation of initiatives and the establishment of areas for continuous improvement, where urban labs play an important role as hubs for experimentation and data collection. One example is Amsterdam's "Circular Innovation District" program, which, in its three stages (conceptualization, modular scaling + value chains, and replicability), has positively impacted 34 cities (European Commission, 2023a). Likewise, permanent co-financing mechanisms should be considered during this phase, through incentive systems and circular impact bonds, or so-called material passports, among others.

Phase 4: Process Monitoring and Optimization
In Phase 4, the design and development of integrated data platforms is proposed, where technologies such as blockchain and AI allow for the establishment of robust databases for data-driven decision-making and more efficient monitoring and evaluation processes, where multi-sector information systems converge in real time, which is vital. Some of the most successful examples are the Digital Product Passport, which guides the traceability of materials based on their lifecycle, and Singapore's National Circularity Dashboard, which crosses 15 databases in real time (Circularise, 2023; EPA, 2025).

Phase 5: Replicability of the Pilot Model and Scalability
In this phase, once the best practices and success stories have been validated, the tools implemented at the implementation model level will guide their application to new cases in different cities, based on technology transfer mechanisms, as mentioned with the Circular Economy Resource Center (CERC) in the European Union and international cooperation networks with the global South. This, in turn, involves developing new governance spaces with collaborative approaches to circularity.

12.9 Challenges and Risks

12.9.1 Power Imbalances Among Stakeholders

Power imbalances among stakeholders represent a significant challenge in multi-sector circular economy initiatives, as dominant actors—often large corporations or well-resourced governments—can disproportionately influence decision-making, priorities, and the allocation of benefits (Eikelenboom & Long, 2022). Smaller businesses, local communities, and marginalized groups may lack the resources, technical expertise, or political leverage to effectively participate, leaving their needs underrepresented and potentially exacerbating social and economic inequalities (Mezzina et al., 2022). These imbalances can lead to conflicts of interest, slower consensus-building, and decisions that prioritize short-term gains over systemic, long-term circular outcomes. Addressing this challenge requires deliberate mechanisms to level the playing field, such as inclusive governance structures, equitable benefit-sharing arrangements, transparent communication channels, and capacity-building programs for underrepresented stakeholders (Santarlacci et al., 2024). Without careful management, power asymmetries risk undermining trust, reducing collaboration effectiveness, and limiting the scalability and fairness of circular economy transitions (Saif et al., 2024).

12.9.2 Misaligned Incentives and Time Horizons

Misaligned incentives and differing time horizons pose a major risk to multi-sector circular economy initiatives, as stakeholders often operate under contrasting priorities and planning cycles (Blomberg et al., 2023). For instance, businesses may focus on short-term profitability, investors on quarterly returns, and governments on election cycles, while the benefits of circular projects such as resource efficiency, ecosystem restoration, or infrastructure development typically materialize over longer periods. This misalignment can discourage investment in collaborative initiatives, delay decision-making, and reduce commitment to shared goals (Schulders, 2023). Additionally, partners may prioritize actions that advance their own interests rather than systemic circular outcomes, creating friction and inefficiencies. Addressing this challenge requires designing incentive structures that align economic, environmental, and social objectives, such as performance-based contracts, long-term public–private partnerships, and shared-value models. Transparent communication, joint planning, and adaptive governance mechanisms can further help synchronize expectations and sustain engagement across diverse stakeholders.

12.9.3 Resource Constraints and Coordination Costs

Implementing circular economy initiatives often requires significant financial, technical, and human resources. Smaller organizations and local governments may struggle to contribute effectively, while larger actors may face challenges justifying high upfront investments (Kaswan et al., 2023). Additionally, coordinating multiple stakeholders across sectors involves substantial time, effort, and administrative costs, including meetings, data sharing, and alignment of strategies. Without efficient coordination mechanisms, projects can suffer delays, duplication of efforts, or inefficient allocation of resources. Overcoming these challenges requires streamlined governance structures, pooled funding mechanisms, and shared platforms for collaboration that reduce transaction costs and maximize collective impact (Vaman, 2024).

12.9.4 Regulatory Fragmentation

Circular economy efforts frequently span industries and jurisdictions, but inconsistent regulations, standards, and policies can create barriers to collaboration. For example, differing waste management rules, product standards, or extended producer responsibility requirements across regions can impede material exchanges, scaling of initiatives, and cross-border partnerships (Solomon et al., 2024). Fragmented regulatory environments increase complexity, create legal uncertainties, and discourage investment in circular solutions. Harmonizing standards, developing cross-sector policy frameworks, and fostering regional or national coordination are essential to reduce fragmentation and enable more seamless circular collaboration (Skondras et al., 2024).

12.9.5 Risk of Greenwashing in Cross-Sector Partnerships

Multi-sector collaborations may sometimes prioritize marketing or reputational gains over genuine circular outcomes, creating a risk of greenwashing. Companies or organizations may exaggerate their environmental claims, overstate progress, or selectively highlight successes while ignoring failures, undermining trust and stakeholder engagement (Routray, 2024). This risk is amplified when accountability mechanisms are weak, reporting is inconsistent, or performance metrics are unclear. Mitigating greenwashing requires robust transparency, independent verification, standardized reporting, and clear communication of goals, achievements, and challenges to ensure that partnerships produce tangible environmental and social benefits rather than superficial claims (Nugroho & Permata, 2023).

12.10 Policy and Strategic Recommendations

12.10.1 Designing Collaborative Policy Frameworks

Designing collaborative policy frameworks is crucial for fostering the multi-sector coordination required to advance a circular economy. Effective frameworks create the enabling conditions for diverse actors, governments, businesses, civil society, and research institutions to work toward shared circular goals through clear regulations, aligned incentives, and supportive governance structures (Gupta, 2024). Policymakers can facilitate collaboration by establishing cross-sector task forces, developing harmonized standards for material use and product design, and creating mechanisms for joint planning and data sharing. Policies that encourage co-investment such as public–private partnerships, innovation grants, and shared infrastructure funding help reduce risks and mobilize resources for circular initiatives (Jiaqi, 2023). Additionally, embedding participatory processes ensures that stakeholders contribute to shaping regulations, enhancing legitimacy, and improving outcomes. By integrating coherence across environmental, industrial, and social policies, collaborative frameworks can accelerate systemic change and create a stable foundation for long-term circular transformation.

12.10.2 Incentivizing Co-investment and Shared Infrastructure

Incentivizing co-investment and shared infrastructure is a powerful strategy for accelerating circular economy initiatives that require substantial upfront resources and coordination across sectors. By offering financial incentives such as tax credits, matching funds, low-interest loans, or risk-sharing mechanisms governments can encourage businesses, municipalities, and community organizations to collaboratively invest in circular systems like advanced recycling facilities, repair hubs, remanufacturing centers, and material recovery networks (Scarpellini et al., 2021). Shared infrastructure reduces duplication of efforts, spreads costs across multiple stakeholders, and creates economies of scale that make circular solutions more economically viable. Public–private partnerships can further support joint investments by aligning long-term interests, pooling expertise, and ensuring continuity beyond political cycles. Additionally, clear guidelines for ownership, operations, and benefit-sharing help build trust among partners and promote sustained engagement (Huang et al., 2015). Through well-designed incentives and collaborative funding models, co-investment becomes a catalyst for systemic circular transformation at regional and national levels.

12.10.3 Strengthening Local and Regional Governance Systems

Strengthening local and regional governance systems is essential for embedding circular economy principles into everyday decision-making and ensuring that circular initiatives respond to the specific needs of communities (Nwankwo et al., 2024). Local and regional authorities are uniquely positioned to coordinate stakeholders, integrate circular strategies into urban planning, and manage key sectors such as waste, mobility, and public procurement. By enhancing institutional capacity, improving regulatory coherence, and fostering interdepartmental collaboration, these governments can create enabling environments that support innovation and long-term investment (Zhang, 2025). Regional platforms and councils can also facilitate dialogue among municipalities, businesses, and civil society, ensuring that policies are aligned across jurisdictions and that resources are allocated efficiently (Martinez & Köhler, 2016). Furthermore, empowering local actors with data, technical expertise, and participatory tools strengthens accountability and encourages community-led solutions. When governance systems are robust and well-coordinated, they become powerful drivers of inclusive, place-based circular transitions.

12.10.4 Cultivating Innovation Ecosystems

Cultivating innovation ecosystems is vital for accelerating the development and deployment of circular economy solutions by bringing together diverse actors, startups, established companies, researchers, investors, and public institutions into dynamic, collaborative environments (Wielopolski & Bulthuis, 2022). These ecosystems create fertile ground for experimentation by providing access to shared facilities such as innovation labs, maker spaces, pilot plants, and test beds where new circular technologies and business models can be developed and refined (Sant-Geronikolou et al., 2019). Supportive policies, incubators, and accelerators help entrepreneurs overcome early stage barriers, while networks and clusters facilitate knowledge exchange and cross-sector learning. By fostering partnerships between academia and industry, innovation ecosystems also ensure that scientific advances translate into practical applications, enhancing competitiveness and scalability. Moreover, access to funding, mentorship, and digital tools strengthens the capacity of innovators to address systemic challenges such as material efficiency, product longevity, and waste valorization (Chowdhury et al., 2024). When nurtured effectively, these ecosystems become engines of continuous innovation, driving the transition toward a resilient and regenerative circular economy.

12.10.5 Establishing Transparent Reporting and Accountability Mechanisms

Establishing transparent reporting and accountability mechanisms is critical for ensuring that circular economy initiatives deliver measurable results and maintain credibility among stakeholders. Clear, standardized metrics for material flows, resource efficiency, emissions reductions, and social impacts enable organizations to track progress, identify gaps, and make evidence-based decisions (Walden et al., 2021). Transparency also fosters trust in multi-sector collaborations by ensuring that all partners adhere to agreed-upon commitments and that benefits and responsibilities are fairly distributed. Digital tools such as open data platforms, lifecycle assessment software, and blockchain-based traceability systems can enhance the accuracy and accessibility of reporting, allowing stakeholders and the public to validate claims and monitor performance in real time (Movaffaghi & Yitmen, 2023). Independent audits, third-party certifications, and regulatory oversight further reinforce accountability and minimize risks of greenwashing. By institutionalizing robust reporting practices, governments and organizations create a culture of continuous improvement and ensure that circular initiatives contribute meaningfully to environmental, economic, and social goals.

12.11 Future Directions

12.11.1 Role of Digital Technologies in Enabling Circular Collaboration

Digital technologies are poised to play a transformative role in scaling circular collaboration by enabling seamless information flows, transparency, and coordinated action across entire value chains. Tools such as the Internet of Things (IoT) can track materials and products throughout their lifecycle, providing real-time data that allows partners to optimize resource use, predict maintenance, and design more durable systems (Khan et al., 2021). Artificial intelligence and advanced analytics facilitate the identification of circular opportunities such as potential industrial symbiosis matches or hotspot areas for waste reduction while automating complex decision-making processes. Blockchain and other distributed ledger technologies enhance trust by creating tamper-proof records of material provenance, certifications, and transactions, reducing barriers to cooperation and ensuring accountability in multi-party ecosystems (Shaikh et al., 2022). Together, these technologies create a digital backbone that enables stakeholders from diverse sectors to coordinate more effectively, share risks and benefits, and co-create innovative circular business models that would be difficult to achieve through traditional mechanisms.

12.11.2 Potential of Global Networks and Transnational Partnerships

Global networks and transnational partnerships hold significant potential to accelerate the circular economy by connecting stakeholders across borders, harmonizing standards, and facilitating the exchange of knowledge, technology, and best practices. As resource flows and supply chains are inherently global, collaboration at the international level enables countries and industries to jointly address systemic challenges such as plastic pollution, electronic waste, and critical material scarcity that cannot be solved within national boundaries alone (Olivieri et al., 2021). Multilateral initiatives, regional alliances, and international organizations help create shared policy frameworks, mobilize funding, and support capacity-building in emerging economies, fostering more equitable participation in the circular transition (Singh et al., 2024). These networks also provide platforms for experimentation and innovation, allowing pilot projects in one region to be scaled and adapted elsewhere. Ultimately, transnational cooperation strengthens global resilience, reduces fragmentation, and catalyzes a more coherent and inclusive shift toward circularity worldwide.

12.11.3 Emerging Models of Circular Entrepreneurship

Emerging models of circular entrepreneurship are redefining how value is created by placing resource efficiency, regeneration, and longevity at the core of business design. Entrepreneurs are increasingly adopting models such as product-as-a-service, where ownership shifts from consumers to providers, enabling companies to maximize product lifespans and recover materials after use (Guo, 2024). Repair, remanufacturing, and upcycling ventures are also gaining momentum, transforming waste into high-value products while fostering local economic development (Choudhury et al., 2025a, b). Digital platforms that facilitate sharing, leasing, and peer-to-peer exchange further expand access to goods without increasing material throughput (Wagner & Lange, 2021). Additionally, bio-based innovation and regenerative business models such as those that restore ecosystems or incorporate biological cycles into production are attracting investment as sustainability becomes a competitive advantage. Together, these entrepreneurial pathways demonstrate how creativity, experimentation, and systems thinking can drive scalable, market-ready solutions that support a resilient and fully circular economy.

12.11.4 Anticipating Social Impacts and Ensuring Inclusivity

Anticipating social impacts and ensuring inclusivity are essential for a circular economy transition that benefits all segments of society rather than reinforcing existing inequalities. As industries adopt new technologies and business models, it is crucial to address potential disruptions to labor markets by investing in reskilling, upskilling, and creating opportunities in repair, remanufacturing, recycling, and green innovation sectors (Morandini et al., 2023). Inclusive approaches also require engaging marginalized communities in decision-making processes and ensuring that policies and investments reflect diverse needs and do not disproportionately burden low-income groups. Equitable access to circular services such as affordable repair options, shared mobility, and community-based reuse initiatives helps broaden participation and distribute benefits more fairly (Hamzani et al., 2025). The circular transition must also consider cultural contexts, gender dynamics, and regional disparities to avoid one-size-fits-all solutions. By embedding social justice principles into planning and governance, circular economy efforts can foster not only environmental resilience but also social cohesion and long-term prosperity.

12.12 Conclusions

As observed, there is a broad scope for developing multi-sector alliances based on emerging markets, driven by the circular economy and cities. Creative capacity geared toward innovation will be a key factor in the sectors' willingness to reconfigure their value chains in favor of circularity. Among the findings, significant trends include the development of continuous monitoring and learning systems to strengthen capacities across various sectors, and the development of dynamic circularity indices, such as the MADI system and blockchain for REP in Chile (UNEP, 2022). Potential barriers can be identified, such as distrust in industries or institutions in the context of conventional sectoral competition, as well as digital gaps in capabilities, infrastructure, and financing, since investment costs for technology implementation are often high.

Furthermore, bureaucracy is also present when there is no culture of co-responsibility, and the regulatory framework is often fluid or weak. Likewise, resistance to change and a short-term vision without a future context are mentioned. To this end, any initiative should consider fostering Integrated Governance scenarios in which public–private mechanisms and NGOs can participate through open data platforms. This will not only provide greater transparency into information but also contribute to harmonized regulatory frameworks and consistent metrics across core indicators.

The study has shown that the circular economy goes beyond being a trend; it is a vision of the future in which people embrace new ways of living and reconfigure processes across industry, institutions, regulations, habitual consumption practices, and the economic development of cities. The circular economy requires actions based on education and a paradigm shift from a regenerative culture, in addition to triggering joint efforts to develop a space for systemic collaboration where multi-sectoral alliances are identified as catalysts in the processes of co-ownership, management, and implementation of actions that allow for the institutionalization of processes in interconnected, holistic models. The potential of multi-sectoral alliances lies in leveraging knowledge, will, and resources to drive development, innovation, and the emergence of new circular industries, reflecting a vision of a more sustainable, resilient, and regenerative future in neighborhoods, cities, and territories.

References

Adeusi, K., Jejeniwa, T., & Jejeniwa, T. (2024). Advancing financial transparency and ethical governance: Innovative cost management and accountability in higher education and industry. *International Journal of Management & Entrepreneurship Research, 6*(5), 1533–1546. https://doi.org/10.51594/ijmer.v6i5.1099

Agmapisarn, C., Seetha, P., & Fakfare, P. (2024). Optimizing surplus food redistribution: Leveraging digital platforms to advance sustainable consumption. *Sustainable Development, 33*(3), 3831–3846. https://doi.org/10.1002/sd.3322

Akofio-Sowah, L., Soraghan, M., Rajagopalan, S., Sakyi, K., Kwamie, A., Jennings, B., Dalglish, S., et al. (2024). Prioritising and including children in intersectoral policymaking: Uncovering immunity to change in Ghana. *BMJ Global Health, 9*(12), e013251. https://doi.org/10.1136/bmjgh-2023-013251

Alatawi, F. (2025). Knowledge-oriented leadership and innovation in digital governance. *International Journal of Electronic Government Research, 21*(1), 1–23. https://doi.org/10.4018/ijegr.384514

Aldieri, L., Brahmi, M., Bruno, B., & Vinci, C. (2021). Circular economy business models: The complementarities with sharing economy and eco-innovations investments. *Sustainability, 13*(22), 12438. https://doi.org/10.3390/su132212438

Anwar, M., Wong, W., & Tseng, M. (2022). Examining voluntary engagement barriers in knowledge sharing practices for supply chain innovation. *International Journal of Knowledge Management, 18*(1), 1–23. https://doi.org/10.4018/ijkm.291099

Banco Mundial. (2021). *Convertir las aspiraciones en realidad para que la acción climática tenga valor.*

BID. (2022). Accelerador de Políticas y Proyectos de Economía Circular en América Latina y el Caribe. *Proyectos Regionales.* Retrieved from https://www.iadb.org/es/proyecto/RG-T4079

Blomberg, A., Kujala, J., & Heikkinen, A. (2023). *Multi-stakeholder networks in a circular economy transition: A typology of stakeholder relationships*, 133–164. https://doi.org/10.1007/978-3-031-31937-2_5.

Bucur, L. (2023). Exploring the role of consumers in promoting a circular economy: Increasing awareness and engagement. *Proceedings of the International Conference on Business Excellence, 17*(1), 38–47. https://doi.org/10.2478/picbe-2023-0006

Cappellaro, F., Cutaia, L., Margareci, G., Scalbi, S., Paola, S., Segreto, M., Valpreda, E. et al. (2020). *The role of collaborative and integrated approach towards a smart sustainable district: the real case of Roveri industrial district*, 135–148. https://doi.org/10.1007/978-3-030-36660-5_9

Choudhury, M., Suhag, M., Jha, P., Harne, K. R., Samanta, P., Choudhury, M., Borgohain, R. B., Majumdar, S., Kravchenko, Y., & Choudhary, R. (2025a). Urban waste management and challenges in emerging smart city systems governance and circular economy perspectives. *Ecological Engineering & Environmental Technology, 26*(10), 243–266. https://doi.org/10.12912/27197050/211014

Choudhury, M., Cheshmehzangi, A., & Majumdar, S. (2025b). The circular economy paradigm: Core concepts and value. In S. Majumdar, M. Choudhury, & A. Cheshmehzangi (Eds.), *Sustainable urban future. Urban sustainability.* Springer. https://doi.org/10.1007/978-981-95-2976-6

Chourasiya, R., & Malviya, R. (2025). Factors influencing technology adoption for sustainability in manufacturing firms: An empirical analysis and literature review. *Sustainable Development, 33*(5), 6734–6757. https://doi.org/10.1002/sd.3482

Chowdhury, R., Barman, S., Choudhury, M., Kim, K. H., & Bhattacharya, S. S. (2024). Earthworm modifies microbial community and functional genes

for lignocellulosic waste valorization: Isolating plant-growth-promoting bacteria via next generation sequencing. *International Biodeterioration & Biodegradation, 193*, 105854. https://doi.org/10.1016/j.ibiod.2024.105854
Circularise. (2023). R-strategies for a circular economy. *Circularise.* Retrieved from https://www.circularise.com/blogs/r-strategies-for-a-circular-economy
Daniels, C., & Gebhardt, C. (2021). Special issue: Higher education, science and research systems for transformative change in Africa – What role for innovation policy? *Industry and Higher Education, 35*(5), 553–558. https://doi.org/10.1177/09504222211029901
Doğan, S., Çalğici, P., Arditi, D., & Günaydın, H. (2015). Critical success factors of partnering in the building design process. *METU Journal of the Faculty of Architecture, 32*(2), 61–78. https://doi.org/10.4305/metu.jfa.2015.2.4
Eikelenboom, M., & Long, T. (2022). Breaking the cycle of marginalization: How to involve local communities in multi-stakeholder initiatives? *Journal of Business Ethics, 186*(1), 31–62. https://doi.org/10.1007/s10551-022-05252-5
Elias Bibri, S., Ayyoob Sharifi, A., & Krogstie, J. (2023). Environmentally sustainable smart cities and their converging AI, IoT, and big data technologies and solutions: An integrated approach to an extensive literature review. *Energy Informatics, 6*, 9. https://doi.org/10.1186/s42162-023-00259-2
El-Jardali, F., Ataya, N., & Fadlallah, R. (2018). Changing roles of universities in the era of SDGs: Rising up to the global challenge through institutionalising partnerships with governments and communities. *Health Research Policy and Systems, 16*(1). https://doi.org/10.1186/s12961-018-0318-9
Ellen Macarthur. (2024). *Austin Materials Marketplace Foundation.* Retrieved from https://www.ellenmacarthurfoundation.org/circular-examples/developing-the-materials-marketplace-austin
EPA. (2025). *Circular economy implementation plan online platform.* United States Environmental Protection Agency. Retrieved from https://www.epa.gov/circulareconomy/circular-economy-implementation-plan-online-platform
Estensen, L., Bakken, T., & Ekambaram, A. (2016). *Technology transfer as driver for innovation and automation in SMEs – Experiences and results from different programs in Norway.* https://doi.org/10.2991/iwama-16.2016.49
European Commission. (2023a). *Methodology for the implementation of a circular economy at the local and regional scale.* Circular Cities and Regions Initiative (CCRI). Retrieved from https://circular-cities-and-regions.ec.europa.eu/support-materials/papers-and-reports/methodology-implementation-circular-economy-local-and-regional
European Commission. (2023b). *Circular City Actions Framework: Bringing the circular economy to every city.* Retrieved from https://circular-cities-and-regions.ec.europa.eu/support-materials/tools-and-methods/circular-city-actions-framework-bringing-circular-economy-every

Farina, G., Kocian, A., Brunori, G., Chessa, S., Lai, M., Nardi, D., Moreschi, L., et al. (2025). Interoperable traceability in agrifood supply chains: Enhancing transport systems through IoT sensor data, blockchain, and dataspace. *Sensors, 25*(11), 3419. https://doi.org/10.3390/s25113419

Finamore, M., & Oltean-Dumbrava, C. (2024). Circular economy in construction – Findings from a literature review. *Heliyon, 10*(15), e34647. https://doi.org/10.1016/j.heliyon.2024.e34647

Francis, C., Johnson, A., & Wilson, R. (2024). Supported decision-making interventions in mental healthcare: A systematic review of current evidence and implementation barriers. *Health Expectations, 27*(2). https://doi.org/10.1111/hex.14001

Goswami, S., Choudhury, M., & Agarwal, S. (Eds.). (2025). *Electronic waste: Impact on health, animals, and the environment* (1st ed.). CRC Press. https://doi.org/10.1201/978100358231

Groenewald, E. (2024). Circular economy strategies in supply chain management: Towards zero waste. *pst, 48*(1), 464–480. https://doi.org/10.52783/pst.291

Guo, S. (2024). Circular business models in sustainable fashion brands: Innovation, practice, and impact. *Journal of Humanities Arts and Social Science, 8*(1), 210–214. https://doi.org/10.26855/jhass.2024.01.035

Gupta, R. (2024). *A qualitative inquiry into the implementation challenges of circular economy practices in supply chains.* https://doi.org/10.21203/rs.3.rs-4284012/v1

Guterman, N. (2024). Still further challenges to advancing social work research: Response to Feldman. *Research on Social Work Practice, 35*(1), 21–23. https://doi.org/10.1177/10497315241264527

Hamzani, A., Soeharto, A., Taufiq, T., Wibowo, D., & Kunantiyorini, A. (2025). Balancing utilitarianism with access to environmental justice: An Indonesian case study. *Environmental Policy and Law, 55*(4–5), 152–165. https://doi.org/10.1177/18785395251374205

Hensley, N. (2024). Student-led discussions for sustainability education: An autoethnographic exploration. *International Journal of Sustainability in Higher Education, 26*(3), 505–517. https://doi.org/10.1108/ijshe-07-2023-0295

Hoque, A., Clarke, A., & Huang, L. (2016). Lack of stakeholder influence on pollution prevention. *Organization & Environment, 29*(3), 367–385. https://doi.org/10.1177/1086026615623057

Huang, T., Ferris, E., Crossley, R., Guillermin, M., Costa, S., & Cawley, J. (2015). A protocol for developing an evaluation framework for an academic and private-sector partnership to assess the impact of major food and beverage companies' investments in community health in the United States. *BMC Obesity, 2*(1). https://doi.org/10.1186/s40608-015-0066-0

Ijaz, F., Aslam, W., Alsanad, A., Aslam, Z., Ullah, I., & ullah, F. (2022). A critique on task allocation processes in distributed agile software development. *Scientific Programming, 2022*, 1–19. https://doi.org/10.1155/2022/8312253

Jiaqi, X. (2023). A conceptual research of public private partnership (ppp) modeling on the pandemic recovery in China. *Journal of Digitainability Realism & Mastery (Dream), 2*(01), 34–41. https://doi.org/10.56982/dream.v2i01.95

Kaswan, M., Sabale, D., & Rathi, R. (2023). Integrating circular economy aspects with manufacturing planning: An MSME perspective. *E3s Web of Conferences, 453*, 01007. https://doi.org/10.1051/e3sconf/202345301007

Khan, M., Imtiaz, S., Parvaiz, G., Hussain, A., & Bae, J. (2021). Integration of internet-of-things with blockchain technology to enhance humanitarian logistics performance. *IEEE Access, 9*, 25422–25436. https://doi.org/10.1109/access.2021.3054771

King Research. (2024). *Tamaño del mercado de la economía circular.* Retrieved from https://www.kingsresearch.com/es/circular-economy-market-1046

Kowsalya, P., Jennifer Flora, G., Kousalya, A., Toha, M. A., Khan, P. A., Choudhury, M., Parthasarathi, T., et al. (2025). Circular systems unveiled: Essential components and their impact on recycling, reuse, and resource recovery. In S. Majumdar, M. Choudhury, & A. Cheshmehzangi (Eds.), *Sustainable urban future. Urban sustainability* (pp. 55–71). Springer. https://doi.org/10.1007/978-981-95-2976-6_3

Lakasse, S., Amril, A., Syamsuri, H., & Jusman, I. (2024). The role of green finance in sustainable business strategies: Opportunities and challenges for business organizations. *Jurnal Penelitian Inovatif, 4*(2), 665–672. https://doi.org/10.54082/jupin.379

Lande, O., Layode, O., Naiho, H., Adeleke, G., Udeh, E., Labake, T., Johnson, E., et al. (2024). Circular economy and cybersecurity: Safeguarding information and resources in sustainable business models. *Finance & Accounting Research Journal, 6*(6), 953–977. https://doi.org/10.51594/farj.v6i6.1214

Laurenti, R., Singh, J., Frostell, B., Sinha, R., & Binder, C. (2018). The socio-economic embeddedness of the circular economy: An integrative framework. *Sustainability, 10*(7), 2129. https://doi.org/10.3390/su10072129

LBCH. (2022). *Living building challenge.* (I. L. Institute, Producer) Retrieved from https://living-future.org/

Lian, Y., Li, Y., & Cao, H. (2023). How does corporate ESG performance affect sustainable development: A green innovation perspective. *Frontiers in Environmental Science, 11.* https://doi.org/10.3389/fenvs.2023.1170582

Lucanto, D. (2023). Advanced circular design, a life cycle approach. Methods and tools for an eco-innovative life cycle approach for buildings energy and resource optimization. In F. Calabrò, L. Della Spina, & M. Piñeira Mantiñán (Eds.), *New metropolitan perspectives. NMP 2022. Lecture notes in networks and systems* (Vol. 482). Springer. https://doi.org/10.1007/978-3-031-06825-6_180

Mai, W., Ambashe, M., & Ohueri, C. (2024). Artificial intelligence ethics best practices model for financial decision-making in Chinese financial institutions. *International Journal of Information Technologies and Systems Approach, 17*(1), 1–18. https://doi.org/10.4018/ijitsa.337388

Majumdar, S., Choudhury, M., & Cheshmehzangi, A. (Eds.). (2025). *Sustainable urban future: Embracing circular economy in waste management*. Springer Singapore. https://doi.org/10.1007/978-981-95-2976-6

Mallick, P., Pigosso, D., Salling, K., & McAloone, T. (2022). Designing configurator for take-back for a circular economy – A conceptual framework. *Proceedings of the Design Society, 2*, 1041–1050. https://doi.org/10.1017/pds.2022.106

Martinez, M., & Köhler, J. (2016). Civil society participation in the health system: The case of Brazil's health councils. *Globalization and Health, 12*(1). https://doi.org/10.1186/s12992-016-0197-1

Mezzina, R., Gopikumar, V., Jenkins, J., Saraceno, B., & Sashidharan, S. (2022). Social vulnerability and mental health inequalities in the "syndemic": Call for action. *Frontiers in Psychiatry, 13*. https://doi.org/10.3389/fpsyt.2022.894370

Morandini, S., Fraboni, F., Angelis, M., Puzzo, G., Giusino, D., & Pietrantoni, L. (2023). The impact of artificial intelligence on workers' skills: Upskilling and reskilling in organisations. *Informing Science the International Journal of an Emerging Transdiscipline, 26*, 039–068. https://doi.org/10.28945/5078

Morganti, L., Demutti, M., Fotoglou, I., Coscia, E., Perillo, P., & Pracucci, A. (2023). Integrated platform-based tool to improve life cycle management and circularity of building envelope components. *Buildings, 13*(10), 2630. https://doi.org/10.3390/buildings13102630

Morrison, E., Adu, D., Guo, Y., Kimani, D., & Saa, V. (2024). Assessing the impact of board sustainability committees on greenhouse gas performance: Evidence from industrialised European countries. *Business Strategy and the Environment, 34*(2), 2529–2555. https://doi.org/10.1002/bse.4073

Movaffaghi, H., & Yitmen, İ. (2023). Framework for dynamic circular economy in the building industry: Integration of blockchain technology and multi-criteria decision-making approach. *Sustainability, 15*(22), 15914. https://doi.org/10.3390/su152215914

Movimientocircular. (2025). https://www.kingsresearch.com/es/circular-economy-market-1046. Retrieved from https://movimentocircular.io/es/economia-circular

Mulder, I., & Magni, A. (2022). A collaborative learning infrastructure to build capacity for urban transformations. *Interaction Design and Architecture(s), 52*, 119–140. https://doi.org/10.55612/s-5002-052-007

Mutmainnah, M., Alfian, A., & Djafar, F. (2022). A multi-stakeholder forum as an approach to address partnership and land tenure conflicts in north konawe district, Southeast Sulawesi. https://doi.org/10.4108/eai.14-9-2021.2317181

Narayan, S., Yadav, D., Jha, S., Khatun, B., Choudhary, R., Verma, U. K. Choudhury, M. & Kumar, D. (2025). From past to present: The evolution of waste management and the rise of circular economy principles. In *Sustainable urban future: Embracing circular economy in waste management* (pp. 23–53). Springer Nature. https://doi.org/10.1007/978-981-95-2976-6_2

Narayana, A., Verma, U., Vijayvargy, L., Sahoo, A., & Shastri, R. (2024). Network building capabilities for a sustainable and circular economy. *International Journal of Mathematical Engineering and Management Sciences, 9*(2), 305–322. https://doi.org/10.33889/ijmems.2024.9.2.016

Nugroho, H., & Permata, I. (2023). Sustainable auditing: Practices and perceptions in the accounting industry. *Golden Ratio of Auditing Research, 3*(2), 57–69. https://doi.org/10.52970/grar.v3i2.377

Nwankwo, T., Ejairu, E., Awonuga, K., Usman, F., & Nwankwo, E. (2024). Conceptualizing sustainable supply chain resilience: Critical materials manufacturing in Africa as a catalyst for change. *International Journal of Science and Research Archive, 11*(1), 2427–2437. https://doi.org/10.30574/ijsra.2024.11.1.0314

Odoyi, E., & Riekkinen, K. (2022). Housing policy: An analysis of public housing policy strategies for low-income earners in Nigeria. *Sustainability, 14*(4), 2258. https://doi.org/10.3390/su14042258

Olivieri, D., Yu, Z., Tabin, G., Thapa, R., & Greenberg, P. (2021). Characterising transnational ophthalmic surgical partnerships by engagement and training. *Clinical and Experimental Ophthalmology, 49*(4), 347–356. https://doi.org/10.1111/ceo.13920

Pelletier, D., Gervais, S., Hafeez-ur-Rehman, H., Sanou, D., & Tumwine, J. (2017). Boundary-spanning actors in complex adaptive governance systems: The case of multisectoral nutrition. *The International Journal of Health Planning and Management, 33*(1). https://doi.org/10.1002/hpm.2468

Qudrat-Ullah, H. (2025). *Navigating complexity. AI and systems thinking for smarter decisions.* Springer. https://doi.org/10.1007/978-3-031-82742-6

Quesada, G. U. (2022). Culturas regenerativas: enfoques alternativos al desarrollo sostenible basados en personas y entornos económicos emergentes. *Regeneratio, 1*(1), 75–88. https://doi.org/10.55924/ucireg.v1i1.3

Rajpal, A., Choudhury, M., Goswami, S., Chakravorty, A., & Raghavan, V. (Eds.). (2024). Waste management and treatment: Advances and innovations (1st ed.). CRC Press. https://doi.org/10.1201/9781003258377.

Reynolds, S. (2024). *Stakeholder engagement and its impact on supply chain sustainability in the context of renewable energy.* https://doi.org/10.20944/preprints202406.0080.v1

Ripanti, E., & Tjahjono, B. (2019). Unveiling the potentials of circular economy values in logistics and supply chain management. *The International Journal of Logistics Management, 30*(3), 723–742. https://doi.org/10.1108/ijlm-04-2018-0109

Rosano, M., & Schianetz, K. (2014). Measuring sustainability performance in industrial parks: A case study of the Kwinana industrial area. *International Journal of Sustainable Development, 17*(3), 261. https://doi.org/10.1504/ijsd.2014.064181

Routray, B. (2024). The paradox of lopsided ESG branding: Critiquing the over-emphasis on environmental aspects at the cost of marginalizing social and governance dimensions in corporate sustainability communications. https://doi.org/10.36227/techrxiv.171387686.66534143/v1

Saadany, A. (2021). Inventory management in reverse logistics with imperfect production, learning, lost sales, subassemblies, and price/quality considerations. https://doi.org/10.32920/ryerson.14643948.v1

Saif, M., Blay-Roger, R., Zeeshan, M., Bobadilla, L., Reina, T., Nawaz, M., Odriozola, J., et al. (2024). Navigating the legislative interventions, challenges, and opportunities in revolutionizing textile upcycling/recycling processes for a circular economy. *ACS Sustainable Resource Management, 1*(11), 2338–2349. https://doi.org/10.1021/acssusresmgt.4c00242

Santarlacci, A., Ângelo, H., Souza, Ã., Lima, M., Joaquim, M., Miguel, É., Carneiro, J., et al. (2024). Benefit sharing governance framework: Pathways for financial benefit sharing in traditional communities. *Sustainability, 16*(7), 2650. https://doi.org/10.3390/su16072650

Sant-Geronikolou, S., Martínez-Ávila, D., & Koulouris, A. (2019). Academic libraries on the creative industries track: The perception of Spanish and Brazilian professionals. *Education for Information, 35*(4), 377–398. https://doi.org/10.3233/efi-180203

Scarpellini, S., Gimeno, J., Tarragona, M., & Llera-Sastresa, E. (2021). Financial resources for the investments in renewable self-consumption in a circular economy framework. *Sustainability, 13*(12), 6838. https://doi.org/10.3390/su13126838

Schulders, M. (2023). Use of public-private partnerships for revitalization initiatives in Poland. *International Journal of Public Administration Management and Economic Development, 8*(2). https://doi.org/10.60026/ijpamed.v8i2.115

Shaikh, Z., Khan, A., Байтенова, Л., Zambinova, G., Yegina, N., Ivolgina, N., Barykin, S., et al. (2022). Blockchain hyperledger with non-linear machine learning: A novel and secure educational accreditation registration and distributed ledger preservation architecture. *Applied Sciences, 12*(5), 2534. https://doi.org/10.3390/app12052534+

Silva, H., Lefebvre, A., Oliveira, R., & Lehoux, P. (2020). Fostering responsible innovation in health: An evidence informed assessment tool for innovation stakeholders. *International Journal of Health Policy and Management*. https://doi.org/10.34172/ijhpm.2020.34

Singh, A., Pandey, A., Santhosh, D., Ganavi, N., Sarma, A., Deori, C., D, S., et al. (2024). A comprehensive review on greenhouse gas emissions in agriculture and evolving agricultural practices for climate resilience. *International Journal of Environment and Climate Change, 14*(5), 455–464. https://doi.org/10.9734/ijecc/2024/v14i54206

Skondras, A., Nastis, S., Skalidi, I., Theofilou, A., Bakousi, A., Mone, T., Stylianidis, E., et al. (2024). Governance strategies for sustainable circular bioeconomy development in Europe: Insights and typologies. *Sustainability, 16*(12), 5140. https://doi.org/10.3390/su16125140

Solomon, N., Simpa, P., Adenekan, O., & Obasi, S. (2024). Circular economy principles and their integration into global supply chain strategies. *Finance & Accounting Research Journal, 6*(5), 747–762. https://doi.org/10.51594/farj.v6i5.1133

Suau Ibáñez, C. (2019). Circular design in everyday urbanism: Towards regenerative and restorative dynamic spaces in cities. *Wellbeing in daily built environments* (Special Issue). https://doi.org/10.13135/2384-8677/3390

Ugwu, M., Adewusi, A., & Nwokolo, N. (2024). The role of public-private partnerships in building clean energy infrastructure in the United States and Nigeria. *International Journal of Management & Entrepreneurship Research, 6*(4), 1049–1068. https://doi.org/10.51594/ijmer.v6i4.984

UIA World Congress of Architects. (2024). Design for Climate Adaptation. In B. Faircloth, M. Pedersen Zari, M. Ramsgaard Thomsen, & M. Tamke (Eds.), *Proceedings of the UIA World Congress of Architects Copenhagen*. Springer. https://doi.org/10.1007/978-3-031-36320-7

Ullah, S., Ahmad, N., Khan, F., Bădulescu, A., & Bădulescu, D. (2021). Mapping interactions among green innovations barriers in manufacturing industry using hybrid methodology: Insights from a developing country. *International Journal of Environmental Research and Public Health, 18*(15), 7885. https://doi.org/10.3390/ijerph18157885

UNEP. (2022). *La coalición de América Latina y el Caribe comparte una visión de la economía circular para la región*. Programa para el MEdio Ambiente UNEP ONU. Retrieved from https://www.unep.org/es/noticias-y-reportajes/comunicado-de-prensa/la-coalicion-de-america-latina-y-el-caribe-comparte-una

Vaman, R. (2024). Research on emerging infectious diseases: Pros and cons of centralised government controls. *Indian Journal of Medical Ethics, 9*(2), 169–170. https://doi.org/10.20529/ijme.2024.016

Vrieze, A., Koppelmäki, K., Morrow, O., & Wagenberg, C. (2022). *Circularity by design: Governance landscape of the metropolitan region Amsterdam*. https://doi.org/10.18174/567976

Wagner, J., & Lange, S. (2021). Towards digital growth-independent societies. *Ökologisches Wirtschaften - Fachzeitschrift, 36*(O1), 44–48. https://doi.org/10.14512/oewo360144

Walden, J., Steinbrecher, A., & Marinkovic, M. (2021). Digital product passports as enabler of the circular economy. *Chemie Ingenieur Technik, 93*(11), 1717–1727. https://doi.org/10.1002/cite.202100121

WasteCero. (2025). *Tendencias de Sostenibilidad 2025.* WasteCero. Retrieved from https://www.wastecero.com/tendencias-de-sostenibilidad-2025/

Wielopolski, M., & Bulthuis, W. (2022). The better building initiative – A collaborative ecosystem involving all stakeholders as catalyst to accelerate the adoption of circular economy innovations in the construction sector. *Circular Economy and Sustainability, 3*(2), 719–733. https://doi.org/10.1007/s43615-022-00205-6

Ya, C., Masukujjaman, M., Sobhani, F., Hamayun, M., & Alam, S. (2023). Green logistics, green human capital, and circular economy: The mediating role of sustainable production. *Sustainability, 15*(2), 1045. https://doi.org/10.3390/su15021045

Yang, J., Gao, L., Liu, Q., & Guo, Z. (2023). Assessing environmental impacts of response strategies for sustainable food system transformation. *Sustainable Development, 32*(3), 2435–2453. https://doi.org/10.1002/sd.2795

Yurieva, T., Voropaeva, L., Beliakova, M., & Adamchuk, N. (2021). Infrastructure investment projects: Financing and management mechanisms. *Journal of Modelling in Management, 17*(4), 1623–1638. https://doi.org/10.1108/jm2-12-2020-0323

Zhang, J. (2025). Does digital government promote collaborative innovation? Evidence from the e-government pilot policy in China. *PLoS One, 20*(10), e0334131. https://doi.org/10.1371/journal.pone.0334131

Zhu, J. (2024). Efforts made by international organizations in facing challenges to international governance – regulating armed conflict by the United Nations and regional organizations. *Lecture Notes in Education Psychology and Public Media, 44*(1), 75–81. https://doi.org/10.54254/2753-7048/44/20230081

PART IV

Future Horizons: Overcoming Challenges in Urban Transformation

CHAPTER 13

Technological Barriers: Identifying and Overcoming Implementation Gaps

Alcardo Alex Barakabitze, David Mihigo, and Neema Simon Sumari

13.1 Introduction

13.1.1 Background

As cities continue to grow and urban populations expand, governments worldwide are increasingly adopting smart technologies to enhance sustainability, resilience, and livability. Smart city initiatives leveraged digital tools, data analytics, and connected infrastructure to improve the efficiency of energy, transport, waste, and water systems (Choudhury et al., 2025a, b). When integrated with circular economy principles, which emphasize resource reuse, recycling, and waste minimization, these technologies offer transformative potential for building sustainable and

A. A. Barakabitze • N. Simon Sumari
Department of Informatics and Information Technology, College of Applied and Natural Sciences, Sokoine University of Agriculture, Morogoro, Tanzania
e-mail: alcardo.alex@sua.ac.tz

D. Mihigo (✉)
Zhongfa Aviation Institute of Beihang University, Hangzhou, China

S. Majumdar, M. Choudhury (eds.), *City of the Future*,
https://doi.org/10.1007/978-3-032-18026-1_13

inclusive urban environments. Despite this potential, many smart and circular city initiatives struggle to scale or sustain beyond pilot stages. Their implementation is often constrained by fragmented governance, limited interoperability of digital systems, insufficient technical capacity, and weak policy alignment (Majumdar et al., 2025). The technological implementation gaps, such as inadequate digital infrastructure, poor data governance, and a lack of financing, undermine the long-term effectiveness of such initiatives. Consequently, cities face challenges in translating smart and circular concepts into practical, system-wide transformations. Examples of promising technologies illustrate both the opportunities and limitations in this transition. Smart energy systems, such as IoT-enabled grids and renewable integration platforms, enhance energy efficiency, while AI-based waste management systems optimize recycling and reduce reliance on landfills (Kowsalya et al., 2025). Smart mobility innovations, including electric vehicles and intelligent transport systems, improve accessibility and reduce emissions. Similarly, IoT-enabled water systems and digital twin buildings promote resource efficiency and sustainable urban management (Choudhury et al., 2025a, b). When effectively coordinated, these technologies can serve as catalysts for circularity and low-carbon urban development. However, achieving this vision requires more than technological deployment; it demands systemic reform (Narayan et al., 2025). Cities must strengthen policy and governance frameworks, promote education and capacity building, create innovative financing mechanisms, and encourage community-driven participation to ensure sustainability and equity (Sharma et al., 2023). This study, therefore, seeks to identify and categorize the key barriers to the adoption of smart technologies and circular economy practices and to propose actionable strategies and innovations to bridge implementation gaps. By doing so, it contributes to shaping future cities that are smart, circular, and resilient in their pursuit of sustainable urban transformation.

13.1.2 *Defining the Problem: What Are Technological Implementation Gaps?*

Technological implementation gaps refer to the disconnects that emerge when promising urban innovations, designed to improve efficiency, sustainability, and quality of life, fail to translate into lasting, scaled solutions. In the context of smart and circular cities, technological barriers include insufficient digital infrastructure, weak system interoperability, limited technical expertise, and regulatory or institutional misalignment (Agyemang et al., 2024; Mutambik, 2024). These barriers are not purely technical; they often stem from gaps in governance, funding, and societal readiness, which together hinder the full deployment and integration of smart technologies. There are several distinct types of implementation gaps commonly observed:

- Design–Deployment Gap: Technologies that are well conceived in concept or in pilot/trial phases often falter upon deployment because they are not adapted to local urban conditions, whether that be power/infrastructure constraints, cultural or social norms, or maintenance and operational realities (Ajaj & Buniya, 2024; Poshaia & Intauno, 2025).
- Innovation–Adoption Gap: Even when technologies have been deployed, adoption by users, institutions, or the broader system can lag due to issues such as distrust, lack of awareness, digital literacy, or misaligned incentives (Nusir et al., 2023; Ntanda & Carolissen, 2025).
- Policy–Technology Mismatch: This occurs when regulatory, legal, or institutional frameworks lag behind technological developments, or when policies are uncoordinated, ambiguous, or do not account for emergent technologies. In such cases, technologies may be constrained by outdated laws, insufficient data protection frameworks, or unclear standards (Ajaj & Buniya, 2024; Mutambik, 2024).

These gaps matter profoundly for smart and circular cities. Without closing the loop, investments in smart grids, intelligent transport, waste recycling systems, or urban sensing often remain confined to pilot projects rather than scaling into systemic urban solutions. Gaps translate into wasted resources, risk of public mistrust, inequalities in who benefits from innovations, and missed opportunities for sustainability. For cities pursuing circularity, where resource loops, longevity, reuse, and efficiency are central, these implementation barriers hinder progress toward reduced waste, optimized resource flows, and resilient infrastructures. Only by understanding and tackling design–deployment, innovation–adoption, and policy–technology mismatches can cities hope to move from isolated experiments to integrated, sustainable, and inclusive urban systems (Table 13.1).

Table 13.1 Classification of implementation gaps

Gap type	*Description*	*Example*	*Required action*
Design–Deployment	System not adapted to local needs	Smart bins with unreliable sensors	Conduct pre-deployment trials
Innovation–Adoption	Low uptake despite deployment	Smart mobility apps unused by citizens	Awareness campaigns and incentives
Policy–Technology	Misaligned laws and systems	Data-sharing regulations lag	Establish adaptive governance

Source: Compiled by Authors

13.1.3 Purposes

The primary objective of this study is to examine the technological implementation gaps that hinder the successful adoption, scaling, and sustainability of smart and circular city initiatives. The main contributions to urban innovation practice:

The study advances the conceptual understanding of "technological implementation gaps" by developing a multidimensional analytical framework that integrates perspectives from innovation studies, urban governance, and the circular economy. This framework captures the interdependencies between technological design, adoption, and policy mechanisms that shape urban transformation processes. Drawing on recent case studies and empirical evidence from diverse urban contexts, including Africa, Asia, and Europe, the paper identifies recurring patterns and context-specific drivers of implementation failure. It highlights how infrastructural deficiencies, fragmented governance, and weak regulatory ecosystems collectively limit the transition from pilot projects to fully realized smart and circular systems. The study provides evidence-based recommendations for policymakers and city authorities to foster cross-sectoral collaboration, promote technological interoperability, and establish adaptive governance mechanisms for circular innovation. It also proposes financing and capacity-building strategies to overcome the digital divide and institutional inertia that hinder smart city development. This research positions itself at the forefront of the discourse on sustainable urban innovation. It contributes to ongoing policy dialogues on how cities can leverage AI, IoT, big data analytics, and blockchain technologies to close resource loops and enhance urban resilience amid rapid digitalization and climate challenges.

13.2 Case Studies and Associated Methods

13.2.1 Case Studies: Barriers in Practice

The study presents case studies on urban innovation technologies, including smart waste management systems, circular housing retrofits, mobility platforms, and e-waste management solutions (Table 13.2). Emphasis was placed on identifying both technical limitations and systemic barriers affecting the deployment and sustainability of these technologies. Four illustrative case studies were selected to examine real-world barriers and highlight the practical dimensions of technological implementation gaps in India, Africa, America, and Europe.

13.2.1.1 Smart Waste Systems in Urban India

Indian cities have implemented advanced digital waste management systems, including IoT-enabled bins and AI-driven route optimization platforms (Fig. 13.1). Despite technological sophistication, many initiatives fail due to mismatches between infrastructure readiness and user behavior, insufficient maintenance capacity, and low citizen engagement. These challenges highlight gaps in design, deployment, and the

Table 13.2 Global case studies on technological implementation gaps in smart and circular cities

Case study	*Technology focus*	*Region*	*Key barriers*	*Lessons learned*
Smart Waste System	IoT-enabled bins, AI-based route optimization	India	Poor maintenance, low engagement	User education and local co-design improve sustainability
Circular Housing Retrofits	Energy efficiency and modular reuse	Europe	High costs, weak coordination	Public–private financing and shared standards help scale adoption
Mobility-as-a-Service (MaaS)	Integrated transport via apps	North America	Interoperability issues, legal delays	Need for shared data policies and unified mobility standards
E-waste Management	IoT & blockchain tracking	Africa	Informal sector exclusion, weak policy	Integrate informal recyclers and develop PPP frameworks

Source: Compiled by Authors

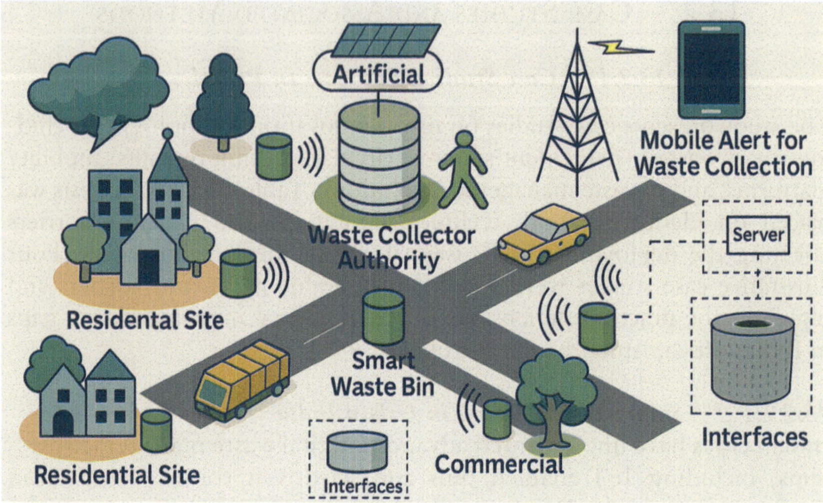

Fig. 13.1 Smart waste collection system (Modified from Dholera's Smart Waste Collection System in India (see reference OO))

adoption of innovation, providing lessons for other emerging economies where socio-technical alignment is critical (Choudhury et al., 2025a, b).

The Smart Waste Collection System exemplifies an integrated, technology-driven approach to urban waste management by combining smart infrastructure, IoT connectivity, and data intelligence. The system's foundation lies in smart bins equipped with sensors that monitor waste levels in real time, ensuring timely collection and effective segregation at the source. Through IoT integration, these bins communicate seamlessly with GPS-enabled collection trucks and waste management centers, enabling optimized routes and efficient resource use. A centralized monitoring system consolidates data from all devices, providing administrators with real-time insights for decision-making and operational planning. Furthermore, data analytics and machine learning enhance predictive capabilities by analyzing waste-generation trends, optimizing collection efficiency, and facilitating proactive interventions, making Dholera's system a model for sustainable, smart urban waste management (Fig. 13.2).

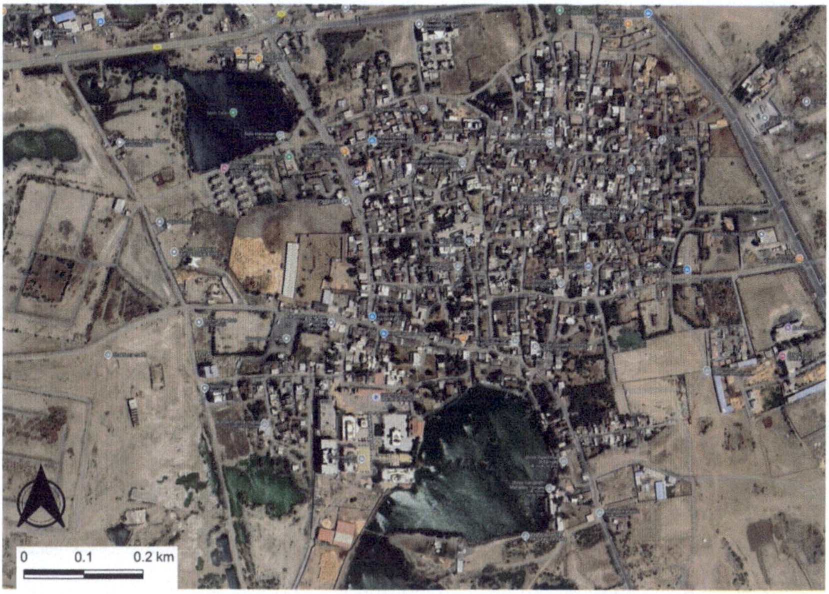

Fig. 13.2 Evidence of the current layouts of Dholera city, Gujarata, India Source: Google Earth Engine, 2025

13.2.1.2 Circular Housing Retrofits in Western Europe

European cities have increasingly pursued circular construction initiatives, such as energy-efficient building retrofits and modular material reuse programs, as shown in Fig. 13.3. Institutional complexity, high costs, and fragmented stakeholder responsibilities often hamper implementation. These barriers demonstrate how policy–technology mismatches and financing gaps can slow adoption, even in regions with advanced infrastructure and regulatory frameworks (Bharathy et al., 2025).

13.2.1.3 Mobility-as-a-Service (MaaS) in North America

Mobility-as-a-Service (MaaS) in North America represents a transformative shift in urban transportation, integrating various modes of transport, such as public transit, ride-hailing, car-sharing, bike-sharing, and micromobility, into a single, digital platform that offers seamless journey planning, booking, and payment (see Fig. 13.4). Driven by advancements in mobile technology, data analytics, and smart infrastructure, MaaS aims to

Fig. 13.3 A circular approach to retrofitting (Modified from Constructive Voices, see (reference PP) Source: Author, 2025

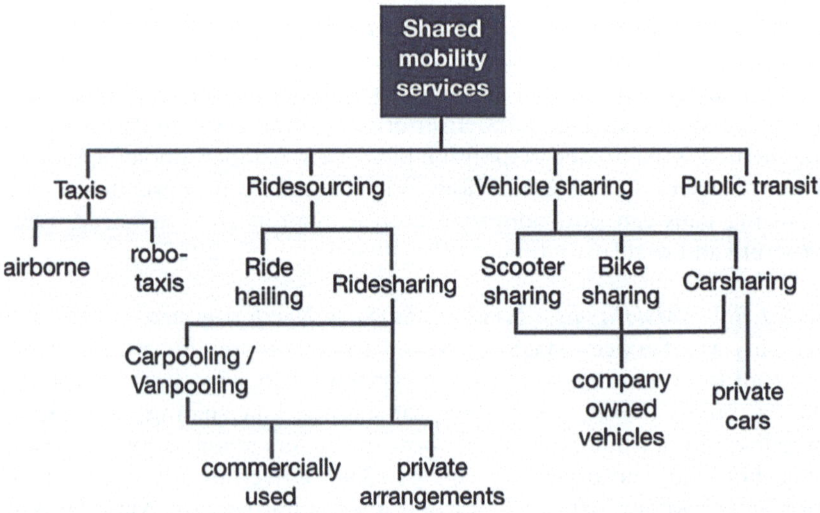

Fig. 13.4 Mobility-as-a-Service (MaaS) in North America (Shaheen et al., 2016)

reduce private car dependency, improve accessibility, and enhance sustainability across cities. In major urban centers like New York, Toronto, and San Francisco, MaaS initiatives are being implemented through partnerships between public transit authorities and private mobility providers, enabling users to plan multimodal trips and pay for them through unified mobile apps. These systems leverage real-time data to optimize routes and reduce congestion, while promoting greener transport options such as electric vehicles and shared mobility. Despite challenges related to data integration, regulation, and equitable access, MaaS continues to gain traction in North America as a key enabler of smart, efficient, and user-centered urban mobility. However, the adoption is frequently slowed by interoperability issues across transport operators, data-sharing limitations, and regulatory delays, highlighting both technological and governance barriers. This case underscores the importance of aligning digital systems, institutional mandates, and legal frameworks to enable scalable urban mobility innovations.

The **Electric Mobility as a Service (eMaaS) system architecture** proposed by Reyes García et al. (2020) extends the traditional MaaS concept by integrating electric mobility infrastructure, data analytics, and multi-provider ecosystems into a unified framework. The architecture is structured around three main functional blocks: Shared e-Mobility, Data and Analytics Extension, and Other Mobility Providers and Third-Party Systems, supported by a vertical data layer that connects all components. This design ensures interoperability, scalability, and real-time intelligence across electric vehicle fleets, charging infrastructure, and user interfaces, enabling a seamless, sustainable, and user-centric mobility experience (Turoń, 2024).

The Shared e-Mobility block forms the core of the architecture, encompassing shared electric mobility services such as e-car, e-scooter, and e-bike sharing, electric ride-hailing, and peer-to-peer electric vehicle sharing. It integrates critical subsystems for fleet management, telematics, vehicle tracking, user authentication, payment processing, and booking coordination (Luo et al., 2023). The inclusion of public and private charging stations, charge-point management systems, and virtual fleets enables dynamic aggregation of assets across vendors and operators. This integration enables flexible access to the charging infrastructure. It supports advanced features such as trip planning, vehicle matching, and user preference learning, which are essential for optimizing resource utilization and enhancing user satisfaction in electric mobility ecosystems (Schnieder & West, 2019).

The Data and Analytics Extension acts as the intelligence backbone of the eMaaS system, enabling data-driven decision-making, predictive analytics, and system optimization. It incorporates a smart data broker to connect heterogeneous data sources, vehicles, chargers, IoT devices, and smart city platforms, along with analytics and optimization engines for demand forecasting, route optimization, and energy management (Yang & Han, 2021). Meanwhile, the Other Mobility Providers and Third-Party Systems block ensures interoperability with external systems, including public transport, taxis, insurance, and energy utilities. Collectively, the architecture emphasizes flexibility, interoperability, and ethical data use, providing a vendor-agnostic, future-proof model for integrating electric and shared mobility within a single digital ecosystem (Alanazi & Alenezi, 2024).

13.2.1.4 E-Waste Management in African Smart Cities

African urban centers face unique challenges in e-waste management due to rapidly growing technology consumption and informal recycling sectors. Key barriers include the exclusion of informal actors, the lack of integrated policies, and limited enforcement mechanisms, which impede sustainable resource recovery and the integration of the circular economy (Khanal et al., 2024). This case illustrates the critical need for inclusive governance models and tailored policy frameworks in emerging smart cities (Fig. 13.5).

E-waste management has emerged as a critical challenge and opportunity in the development of African smart cities. With the rapid adoption of digital technologies, smartphones, and Internet of Things (IoT) devices, electronic waste generation has surged across the continent, posing significant environmental, health, and socio-economic risks (Huang et al., 2024). African smart cities such as Kigali, Nairobi, and Cape Town are increasingly integrating smart and circular approaches to manage e-waste sustainably. These include the use of digital tracking systems, IoT-enabled collection points, and blockchain-based traceability platforms to monitor electronic products throughout their life cycle and ensure responsible recycling (Ahmad et al., 2021). Moreover, public-private partnerships

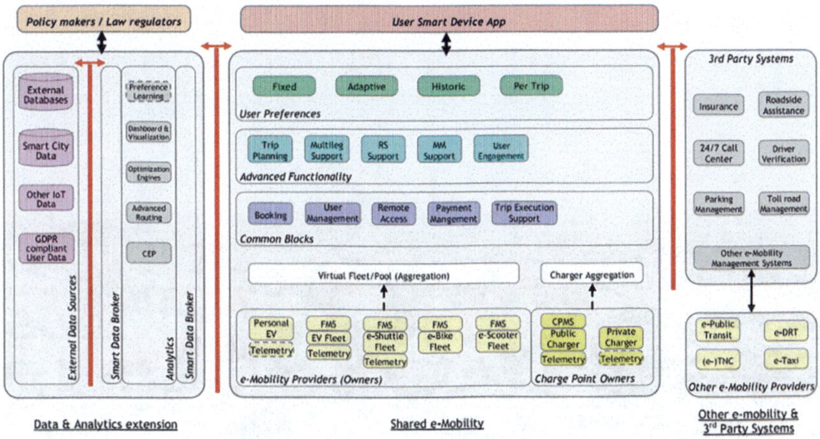

Fig. 13.5 The eMaaS system architecture

(PPPs) and innovation hubs are promoting e-waste recycling startups and supporting youth-led green enterprises that extract valuable materials from discarded electronics (Goswami et al., 2025). However, challenges persist, including limited regulatory enforcement, inadequate recycling infrastructure, and informal sector dominance in e-waste handling (Gogoi et al., 2025). To overcome these barriers, African governments are adopting data-driven policies, harmonizing regional e-waste frameworks, and investing in capacity-building programs for safe recycling practices. Integrating e-waste management into the broader smart city agenda not only supports environmental sustainability but also stimulates the circular economy by recovering valuable resources, creating green jobs, and advancing Africa's transition toward a digitally inclusive and resource-efficient future (Fig. 13.6).

13.2.2 An Insight and Associated Techniques

13.2.2.1 Understanding Recent Trends on This Research Topic

This chapter adopts a comparative case study approach to investigate technological implementation gaps in smart and circular cities across diverse geographic and socio-economic contexts. The approach allows for the in-depth exploration of systemic, technical, and institutional barriers while

Fig. 13.6 E-waste management in South Africa (see on reference SS). Source: Author, 2025

providing opportunities to generalize insights across different urban environments. The study focuses on cities in Asia, Europe, North America, and Africa, representing varied stages of smart city development, governance structures, and circular economy adoption. This selection enables the identification of common patterns and context-specific challenges in implementing innovative urban technologies. Data for this study were collected through secondary research, including peer-reviewed journals, technical reports, government publications, and urban policy documents from 2023 to 2025. Comparative analysis techniques were employed to identify recurring patterns and contextual differences across case studies. Barriers were categorized according to a conceptual framework encompassing design deployment gaps, innovation–adoption gaps, and policy–technology mismatches, enabling systematic analysis of technological implementation challenges. The study also synthesized insights from expert commentaries and urban innovation assessments to derive actionable recommendations for bridging implementation gaps.

13.2.2.2 Understanding Latest Trend on This Topic and Analysis
To understand the trend in this subject title, this study focuses on publications from 2023 to 2025, including recent publications, urban policy

documents, technical reports, and case studies. Since this domain is rapidly advancing, driven by new technologies, governance reforms, and sustainability frameworks, drawing on recent publications from 2023 to 2025 ensures that the analysis reflects current realities, policy directions, and implementation challenges. Sources were selected to provide a comprehensive representation of technological implementation experiences across diverse geographic contexts, including Asia, Europe, North America, and Africa. For each chosen case study, information was systematically gathered on the smart or circular city technology deployed, the socio-technical environment, institutional frameworks, stakeholder participation, and observed barriers to implementation. In addition, the study reviewed expert commentaries, governmental assessments, and urban innovation evaluations to triangulate findings and ensure reliability. This approach enabled the collection of rich, context-rich data on technological gaps, adoption challenges, and policy–technology mismatches.

13.2.2.3 Analytical Approach

The analysis employed a qualitative comparative framework to identify patterns, similarities, and differences in implementation barriers across the selected case studies. Collected data were organized into three primary categories of technological implementation gaps: design-deployment gaps, innovation-adoption gaps, and policy–technology mismatches. Each case study was analyzed to determine the specific factors contributing to these gaps, including infrastructural limitations, stakeholder engagement issues, regulatory misalignment, and financing challenges. The study also applied a cross-contextual synthesis method, enabling the identification of recurring challenges and best practices applicable across different urban environments. The findings were interpreted in the context of smart and circular city theory, allowing for the development of evidence-based recommendations to bridge technological gaps, enhance adoption, and support sustainable urban transformation.

13.3 Barriers

13.3.1 Technological Barriers: Identifying and Overcoming Implementation Gaps

Technological implementation gaps in smart and circular cities often stem from foundational infrastructure and systemic challenges (see Table 13.3).

Table 13.3 Categorization of technological barriers

S/N	*Category*	*Strategy*	*Expected impact*
1	Infrastructure	Upgrade broadband & IoT systems	Enables scalable smart networks
2	Governance	Create inter-ministerial STI councils	Improves policy coordination
3	Capacity Building	Develop digital literacy & training hubs	Builds local technical expertise
4	Financing	Introduce blockchain-based PPPs	Enhances transparency and attracts investors
5	Social Innovation	Run co-design hackathons	Increases citizen trust and inclusion

Source: Compiled by the Author

These barriers hinder the seamless integration and scalability of urban innovations, affecting their effectiveness and sustainability. Key obstacles include legacy infrastructure, interoperability issues, connectivity deficits, and specific challenges related to the circular economy.

13.3.1.1 Infrastructure and Systemic Challenges

13.3.1.1.1 Legacy Infrastructure

Many cities continue to rely on outdated infrastructure that was not designed to support modern smart technologies. This legacy infrastructure often lacks the capacity to integrate advanced systems such as IoT sensors, smart grids, and data analytics platforms. The mismatch between old and new systems can lead to inefficiencies and increased costs, impeding the transition to smart urban environments (Gupta, 2023a, b).

13.3.1.1.2 Lack of Interoperability

Interoperability remains a significant challenge in smart cities, as diverse technologies and platforms often operate in silos. The absence of standardized protocols and data formats complicates the integration of various systems, leading to fragmented services and reduced efficiency. Recent studies highlight the need for common standards and frameworks to enhance system compatibility and data sharing across different urban services (Biswas et al., 2023; Hardi et al., 2025).

13.3.1.1.3 Connectivity Gaps

Reliable and widespread connectivity is crucial for the functioning of smart city technologies. However, many urban areas, especially in developing regions, face significant connectivity gaps. These include limited broadband access, inadequate network infrastructure, and digital divides that exclude marginalized communities from the benefits of smart initiatives. Addressing these gaps is essential to ensure equitable access to smart city services (Caragliu & Del Bo, 2023; Kumar, 2025).

13.3.1.1.4 Circular Economy-Specific Challenges

Integrating circular economy principles into urban systems presents unique challenges. These include the need for specialized infrastructure for resource recovery, waste management, and material tracking. Additionally, there is often a lack of policies and regulations that support circular practices, as well as insufficient public awareness and participation. Overcoming these barriers requires coordinated efforts across various sectors and levels of government to create enabling environments for circular economy initiatives (Anttiroiko, 2023a, b; Liu, 2024).

13.3.1.2 Governance, Policy, and Institutional Barriers

In addition to technological and infrastructure-related challenges, governance, policy, and institutional barriers play a critical role in shaping the success or failure of smart and circular city initiatives. Effective implementation requires coordinated institutional frameworks, streamlined regulatory processes, and long-term political commitment; however, many cities face structural and policy-related impediments that hinder innovation adoption (Sharma et al., 2012).

13.3.1.2.1 Fragmented Urban Governance

Many urban areas suffer from fragmented governance structures, where multiple agencies, departments, and stakeholders are responsible for overlapping or disconnected aspects of urban management. This fragmentation can lead to poor coordination, duplication of efforts, and unclear accountability, which impedes the implementation of integrated smart city technologies or circular economy initiatives (Mutambik, 2024; Ajaj & Buniya, 2024). The lack of a centralized or harmonized urban governance framework often results in pilot projects remaining siloed, rather than being scaled across municipal systems.

13.3.1.2.2 Rigid Procurement and Bureaucracy

Rigid procurement policies and bureaucratic procedures present another major barrier to innovation. Smart city technologies often require flexible contracting, rapid procurement, and adaptive funding mechanisms, yet existing bureaucratic processes are slow, complex, and risk-averse. These limitations delay technology deployment, hinder experimentation, and discourage private sector participation, particularly in public–private partnership arrangements (Nusir et al., 2023; Poshaia & Intauno, 2025).

13.3.1.2.3 Lack of Standards and Regulation

The absence of coherent standards, regulatory frameworks, and interoperability protocols further compounds implementation challenges. Without clear guidelines for data management, privacy protection, cybersecurity, or technical interoperability, cities struggle to integrate new technologies across departments or sectors. This regulatory gap can undermine public trust, reduce investment incentives, and create barriers to scaling innovations beyond pilot phases (Biswas et al., 2023; Hardi et al., 2025).

13.3.1.2.4 Short-Termism in Political Cycles

Urban development initiatives are often subject to short-term political cycles, where priorities and funding commitments change with successive administrations. This short-termism discourages investment in long-horizon smart city or circular economy projects that require multi-year planning, sustained maintenance, and policy consistency. As a result, projects may be abandoned, underfunded, or redirected before delivering measurable outcomes, exacerbating implementation gaps and limiting the transformative potential of technological innovations (Caragliu & Del Bo, 2023; Gupta, 2023a, b). Addressing these governance, policy, and institutional barriers requires comprehensive urban planning, streamlined regulatory processes, multi-stakeholder collaboration, and political commitment that extends beyond election cycles. By fostering integrated governance and regulatory frameworks, cities can create an enabling environment that supports the deployment, scaling, and sustainability of smart and circular urban technologies.

13.3.1.3 Human Capital and Skills Deficits

The success of smart and circular cities depends fundamentally on the quality and readiness of human capital. However, across many developing and developed contexts alike, a skills and capacity deficit persist, slowing the effective implementation of digital and circular innovations. These gaps span the technical workforce, urban communities, and institutional actors, affecting the sustainability and scalability of smart and circular city solutions.

13.3.1.3.1 Workforce Readiness

Many cities face a shortage of skilled professionals capable of developing, deploying, and maintaining advanced technologies such as the IoT, AI, blockchain, and big data analytics (Wang et al., 2024). The rapid evolution of these technologies has outpaced the education and training systems that prepare the workforce. This gap results in underutilization of urban data infrastructure, inefficient system management, and a reliance on external expertise. For example, AI-driven predictive maintenance systems in waste management or transport require expertise in data analytics and algorithmic modeling, skills that are still limited within local municipal teams (El Houssaini & Alami, 2024). Smart workforce development platforms using AI-based e-learning systems, digital twins, and virtual reality (VR) training can help upskill urban technicians and planners for these emerging roles.

13.3.1.3.2 Lack of Community Digital Literacy

Beyond technical professionals, communities themselves often lack the digital literacy needed to engage effectively with smart city services and circular economy programs. Limited awareness and low trust in digital technologies can lead to low adoption of smart waste segregation systems, mobile recycling apps, or energy monitoring tools (Alam & Rahman, 2023). For instance, citizen participation in IoT-based recycling platforms or blockchain-enabled waste traceability systems depends on user understanding and behavioral change. Implementing AI-powered community education platforms, gamified digital inclusion programs, and mobile learning applications in local languages can bridge this divide and increase citizen engagement in sustainable digital practices (Gaur & Sharma, 2025).

13.3.1.3.3 Capacity in the Circular Economy

Building expertise in circular economy (CE) practices, such as waste valorization, product life extension, and industrial symbiosis, requires cross-disciplinary knowledge that integrates environmental science, materials engineering, and digital innovation. However, few institutions offer training tailored to integrating digital technologies into CE models (Baldassarre et al., 2024). For example, technologies like 3D printing, digital product passports (DPPs), smart material tracking, and AI-powered lifecycle assessment tools are essential for operationalizing circularity in cities. However, limited local expertise and weak institutional collaboration hinder their large-scale implementation. Digital training ecosystems, massive open online courses (MOOCs) on circular innovation, and university–industry knowledge hubs can enhance professional capacities to bridge this gap. It is worth noting that the human capital and skills deficits require a holistic digital capacity-building framework that aligns education, vocational training, and community engagement with the technological requirements of smart and circular cities. Governments can foster a digitally competent workforce and a digitally literate citizenry, and ensure that technological innovation translates into inclusive, sustainable urban transformation.

13.3.1.4 Financial and Economic Constraints

As cities strive to transition toward smart and circular models, financial and economic barriers remain among the most significant challenges limiting their successful implementation. The high capital costs, inadequate long-term funding models, and poor return-on-investment (ROI) visibility often hinder municipalities, especially in developing regions, from scaling up smart and sustainable initiatives (OECD, 2020; UN-Habitat, 2023).

13.3.1.4.1 High Capital Costs of Smart Technologies

The initial investment needed to implement smart city technologies is often prohibitively high. These costs encompass infrastructure modernization, acquisition of advanced technologies, and system integration (Albino et al., 2015). Municipalities in developing countries frequently face limited fiscal space and competing budgetary priorities, restricting their ability to invest in digital transformation. To address this, Public-Private Partnerships (PPPs) can be leveraged to share financial risk and attract private sector expertise. Blockchain technology enhances these partnerships by ensuring transparency and accountability in contractual agreements (Kshetri, 2021). Moreover, Smart Financing Platforms, which aggregate capital from crowdfunding, venture capital, and green bonds, can utilize data analytics and AI to match projects with investors, increasing access to diverse funding streams (World Bank, 2022).

13.3.1.4.2 Inadequate Long-Term Funding Models

Smart city projects often suffer from a reliance on short-term or project-based funding mechanisms that do not support their long-term sustainability (European Commission, 2021). As a result, many initiatives fail to progress beyond pilot stages. To overcome this, Performance-Based Financing Solutions can be adopted, supported by technologies such as IoT sensors and real-time data monitoring systems that demonstrate the tangible outcomes of smart projects, such as improved energy efficiency or waste reduction, thereby justifying continued funding (Yigitcanlar et al., 2019). Additionally, automated grant management systems can streamline the process of applying for and managing government incentives or green subsidies through digital platforms, enhancing efficiency and accountability (UNDP, 2022).

13.3.1.4.3 Poor ROI Visibility

A major deterrent for investors and city administrators is the lack of clear visibility regarding the ROI of smart city initiatives. Many benefits, such as improved livability or environmental quality, are intangible or realized over the long term (Nam & Pardo, 2011). To improve ROI visibility, data analytics and visualization tools can model the financial and social impacts of various smart city scenarios. Predictive analytics can quantify potential cost savings and environmental benefits, supporting evidence-based decision-making (Hashem et al., 2016). Similarly, Digital Twin technologies, virtual replicas of physical urban systems, enable cities to simulate interventions and analyze their impacts in real time. This allows stakeholders to evaluate the performance, maintenance costs, and long-term benefits of different technologies before committing large investments (Batty, 2018).

13.3.1.5 Socio-Cultural and Behavioral Barriers

Socio-cultural and behavioral factors are critical non-technical barriers to the successful deployment of smart and circular city solutions. These barriers shape how citizens perceive, accept, and interact with new technologies; they also influence policymaker choices and the legitimacy of urban digital programs.

13.3.1.5.1 Resistance to Change

Resistance to change in the urban context arises from multiple sources: entrenched administrative routines, professional cultures in municipal services, vested interests tied to legacy systems, and simple citizen inertia when services, workflows, or daily habits are altered. Empirical and review studies from 2023–2024 emphasize that resistance is not only cognitive (fear of the unknown) but institutional short political cycles, fragmented governance, and weak participatory channels mean that many smart pilots never scale because stakeholders (public servants, local businesses, communities) are insufficiently involved or see little benefit for their roles. Addressing resistance, therefore, requires co-design, staged pilots that demonstrably reduce burdens on existing actors, and institutional incentives (e.g., capacity building, career recognition for digital innovation) that realign incentives across municipal departments.

13.3.1.5.2 Technophobia and Trust Deficits

Technophobia (low digital confidence, anxiety, or active avoidance of technology) and broader trust deficits (mistrust of data collection, algorithmic decisions, or private tech providers) are converging barriers in the 2023–2025 literature. Health-sector and smart-city studies from 2024 show that technophobia and perceived poor usability reduce the uptake of digital services, while a lack of transparent governance and the explainability of algorithms drive public distrust. Systematic reviews and surveys from 2024–2025 highlight that trust in AI and automated systems depends strongly on explainability, accountability mechanisms, and demonstrable safeguards for privacy and fairness; without these, citizens either disengage or demand restrictive policies that can stifle innovation. Effective countermeasures include human-centered design to reduce anxiety and improve usability, transparent communication and explainable algorithmic decision-making, community pilots with independent evaluation, and institutional trust-building (audits, oversight bodies, public dashboards) to demonstrate concrete protections and benefits.

13.3.1.5.3 Equity and Inclusion Issues

Equity and inclusion are central socio-cultural constraints: smart city technologies often advantage the tech savvy, well resourced, and better educated, while excluding older adults, low-income households, migrants, and people with disabilities. Recent reviews (2024) frame inclusion as multidimensional material access (devices, connectivity), skills and digital literacy, motivational/attitudinal access, and meaningful participation in decision-making and show that failure on any of these dimensions produces uneven outcomes and legitimacy deficits. Policy responses supported by the literature include targeted subsidy programs for devices/connectivity, community digital literacy and peer-mentor schemes, co-design practices that engage marginalized groups from project inception, and metrics to measure distributional outcomes (who benefits, who pays) rather than only system performance. Inclusive procurement rules and accessible design standards (e.g., meeting WCAG for web services) are practical levers that reduce exclusion while improving adoption and long-term sustainability.

13.3.1.5.4 Short Synthesis and Policy Implications

Between 2023 and 2025, the literature emphasizes that socio-cultural barriers are not peripheral; they are structural determinants of whether smart city investments deliver public value. Successful strategies combine: (1) participatory governance and co-design to reduce institutional and citizen resistance; (2) human-centered design, explainability, and accountability frameworks to reduce technophobia and build trust; and (3) targeted inclusion policies (connectivity subsidies, skills programs, accessibility standards) to close the digital divide and ensure equitable benefits. Implemented together, these measures help shift smart city programs from vendor-driven technology deployments to socially legitimate, inclusive urban transformations.

13.4 Smart and Circular City Initiatives

The findings of this study reveal that while smart and circular city initiatives hold significant promise for advancing urban sustainability, their implementation remains constrained by a web of socio-technical, institutional, and behavioral barriers (Majumdar et al., 2024; González et al., 2023). The literature indicates that technological deployment alone cannot guarantee transformation; systemic coordination, inclusive governance, and cultural adaptation are equally critical. Urban innovations such as AI-driven waste management, IoT-based mobility systems, and digital circular economy platforms frequently stall due to weak interoperability, fragmented governance, and skill deficits. These issues are exacerbated by short-term policy horizons and inadequate financing models that prevent pilot initiatives from scaling into citywide programs. Recent evidence suggests that cities that align digital infrastructure investments with robust policy frameworks and participatory governance structures demonstrate greater resilience and greater adoption success (Mutambik, 2024; Hardi et al., 2025; Liu, 2024). Thus, bridging the technological implementation gap requires cross-sectoral collaboration, harmonized regulatory standards, and adaptive governance mechanisms that sustain innovation beyond political cycles.

Beyond institutional reforms, human and behavioral factors emerge as decisive in shaping the effectiveness and inclusivity of smart urban transitions. Resistance to change, technophobia, and digital trust deficits limit both public participation and policy momentum. Studies by Alquran et al. indicate that citizens' hesitation toward automation, coupled with limited digital literacy, often leads to underutilization of smart systems such as e-governance portals and waste-tracking applications. Moreover, inequities in access to digital infrastructure, education, and decision-making reinforce exclusion among vulnerable populations. To address these behavioral and cultural divides, human-centered design, participatory urban planning, and transparent algorithmic governance are vital. Building institutional trust through community engagement, co-design, and visible accountability frameworks ensures that citizens view technology as a means of empowerment rather than control.

Finally, achieving sustainable, smart, and circular cities depends on integrating technological, social, and economic innovations within an equitable policy ecosystem (Majumdar et al., 2024). Financially, cities must explore adaptive funding mechanisms, such as blockchain-based public–private partnerships and performance-based financing, to ensure long-term viability. Equally, investments in human capital, digital literacy, and skills development for the circular economy are foundational to scaling innovation (Baldassarre et al., 2024; Wang et al., 2024). Governments and development partners must therefore move beyond isolated technology pilots toward systemic reforms that link data-driven decision-making, inclusive governance, and education for sustainability. By addressing infrastructural, governance, and socio-cultural barriers holistically, cities can transform smart and circular initiatives from experimental projects into sustainable, equitable, and resilient urban futures.

13.5 Overcoming the Gaps: Strategies and Recommendations

The successful implementation of smart and circular cities requires a multi-pronged approach that addresses existing infrastructure gaps, governance inefficiencies, financial limitations, and socio-cultural challenges. The following strategies and recommendations offer practical pathways to accelerate urban sustainability, inclusivity, and digital transformation.

13.5.1 Strategies

13.5.1.1 Infrastructure and System Solutions

Cities must prioritize the modernization of digital and physical infrastructure to enable efficient data management, energy utilization, and resource recycling. Integrating IIoT, AI, and digital twins can optimize energy grids, waste management, and transportation systems (Cheng et al., 2024). Public-private partnerships (PPPs) should also be promoted to finance resilient urban systems and ensure interoperability across sectors.

13.5.1.2 Policy and Governance Reform

Effective governance requires adaptive, cross-sectoral, and participatory frameworks that support innovation. Governments should establish national digital and circular economy strategies aligned with sustainability goals, while promoting open data policies and standards for interoperability (Mekki et al., 2024). Integrating citizen participation platforms can enhance accountability and trust, ensuring policies reflect community priorities.

13.5.1.3 Capacity Building and Education

Developing a digitally skilled workforce and fostering community literacy are central to the smart city transition. Initiatives should focus on technical training in AI, data analytics, and renewable energy systems while embedding sustainability concepts into formal and informal education systems (Olawale et al., 2023). Partnerships between academia, industry, and government can establish innovation hubs for continuous learning and experimentation.

13.5.1.4 *Financing Innovation*

Cities must adopt blended finance mechanisms that combine public funding, green bonds, and venture capital to support innovation in smart technologies. Emerging models like performance-based financing and outcome-oriented funding can ensure accountability and long-term project sustainability (Li & Chen, 2025). Digital platforms leveraging blockchain can also enhance transparency and attract investors.

13.5.1.5 *Community-Led Innovation*

Encouraging grassroots innovation is crucial for ensuring inclusivity and social acceptance of smart city initiatives. Local innovation labs, co-design workshops, and hackathons can empower citizens to co-create solutions tailored to local contexts. Such participatory models enhance social resilience and ensure that technology adoption aligns with cultural and behavioral realities (Kumar et al., 2024).

13.5.2 *Recommendations*

13.5.2.1 *Infrastructure and System Solutions*

Governments should invest in resilient and interoperable smart infrastructure systems, emphasizing renewable energy integration, data-driven transportation, and circular waste management. Collaboration between ICT providers, municipalities, and utilities can enhance the scalability of digital infrastructure projects.

13.5.2.2 *Policy and Governance Reform*

Develop regulatory frameworks that foster innovation while ensuring data privacy, cybersecurity, and ethical use of technology. Establishing inter-ministerial science, technology, and innovation (STI) platforms can streamline coordination and eliminate policy fragmentation across sectors.

13.5.2.3 *Capacity Building and Education*

Implement nationwide digital literacy programs and targeted upskilling initiatives to prepare the workforce for Industry 4.0 and green economy transitions. Encourage universities and vocational institutions to integrate interdisciplinary curricula on sustainability, circular economy, and emerging technologies.

13.5.2.4 Financing Innovation

Promote fiscal incentives and risk-sharing mechanisms to attract private investment in smart city initiatives. Develop smart financing tools, such as AI-driven investment analytics, to assess project feasibility and return on investment (ROI), reducing uncertainty and encouraging innovation.

13.5.2.5 Community-Led Innovation

Adopt participatory governance frameworks that institutionalize citizen engagement in decision-making. Provide microgrants and incubation support for community-driven innovations addressing local sustainability challenges, thereby enhancing ownership and long-term impact.

13.6 Toward a Resilient and Inclusive City of the Future

13.6.1 Integrating Circularity and Technology Holistically

Creating a truly smart and sustainable city demands an understanding that technology and circularity are not parallel ambitions but interconnected pathways. Smart technologies can only fulfill their promise when embedded within systems that regenerate resources, reduce waste, and promote long-term efficiency. Likewise, circular practices gain new momentum when supported by digital innovations that track material flows, enable predictive maintenance, and foster transparent decision-making. By merging these two dimensions, cities can evolve into dynamic ecosystems where efficiency, equity, and environmental stewardship reinforce one another.

13.6.2 Reframing Barriers as Design Challenges

Obstacles to technological implementation are often treated as limitations; yet, they can become opportunities for creative redesign. Viewing barriers through a design lens shifts attention from what is missing to what can be reimagined. Technical incompatibilities may inspire new interoperability standards; gaps in data governance may lead to stronger ethical frameworks; and resistance to change can encourage more participatory planning. This mindset turns disruption into a learning process where each constraint becomes a prompt for more adaptive, context-aware, and human-centered solutions.

13.6.3 Call to Action

The future city will not emerge from technology alone, but from the collective will to use innovation responsibly and inclusively. Policymakers, technologists, and citizens share equal responsibility in shaping the systems that define urban life. It is time to move beyond isolated experiments toward coordinated, transparent, and scalable models of urban transformation. Building resilience and inclusivity requires courage—to challenge outdated infrastructures, rethink governance, and place people at the core of progress. The call is clear: the smart city of tomorrow begins with the thoughtful, circular, and equitable choices made today.

13.7 Conclusion

As cities transition toward smarter, more sustainable futures, technological barriers remain pivotal challenges that determine the pace and success of the transformation. The journey toward circular and data-driven urban systems is not merely about adopting advanced tools; it is about cultivating the conditions that allow technology to thrive within social, economic, and institutional contexts. Overcoming implementation gaps requires more than technical fixes; it calls for adaptive governance, inclusive collaboration, and a clear vision that aligns innovation with human and environmental well-being. By recognizing the interplay between infrastructure readiness, policy frameworks, and digital literacy, urban leaders can move beyond fragmented experimentation toward systemic progress. The "city of the future" will emerge not through the technology itself, but through the collective capacity to integrate, manage, and evolve these tools responsibly. Addressing today's technological barriers, therefore, is not just an operational necessity; it is an ethical and strategic foundation for shaping resilient, equitable, and regenerative urban systems.

References

Agyemang, E., Anderson, B., Patiño, J., & Tremolieres, M. (2024). Toward achieving smart cities in Africa: Challenges to data use and the way forward. *Data & Policy, 6*, E13. Cambridge University Press & Assessment.

Ahmad, R., Salah, K., Jayaraman, R., Yaqoob, I., & Omar, M. (2021). Blockchain for waste management in smart cities: A survey. *IEEE Access, 9*, 131520–131541. https://doi.org/10.1109/access.2021.3113380

Ajaj, R., & Buniya, M. K. (2024). A case study on the barriers towards achieving sustainable Smart City for Abu Dhabi. *Iet Smart Cities, 6*(2), 112–128. Iet Research Journal.

Alam, S., & Rahman, H. (2023). Digital literacy and citizen participation in smart urban ecosystems: A developing-country perspective. *Sustainability, 15*(14), 11245. https://doi.org/10.3390/Su151411245

Alanazi, F., & Alenezi, M. (2024). Interoperability for intelligent traffic management systems in smart cities. *International Journal of Electrical and Computer Engineering (Ijece), 14*(2), 1864. https://doi.org/10.11591/ijece.v14i2.pp1864-1874

Albino, V., Berardi, U., & Dangelico, R. M. (2015). Smart cities: Definitions, dimensions, performance, and initiatives. *Journal of Urban Technology, 22*(1), 3–21. 10.1080/10630732.2014.942092.

Anttiroiko, A. V. (2023a). Governance of smart-green transition in urban regions. *Sustainability, 15*(2), 1234. https://doi.org/10.3390/Su15021234

Anttiroiko, A.-V. (2023b). Smart circular cities: Governing the relationality, spatiality, and digitality in the promotion of circular economy in an urban region. *Sustainability, 15*(17), 12680. https://doi.org/10.3390/Su151712680

Baldassarre, B., Bocken, N. M. P., & Ritala, P. (2024). Digital innovation and the circular economy: Skills and education gaps in practice. *Journal of Cleaner Production, 433*, 139585. https://doi.org/10.1016/J.Jclepro.2024.139585

Batty, M. (2018). Digital twins and smart cities. *Environment and Planning B: Urban Analytics and City Science, 45*(5), 817–820. 10.1177/2399808318796416.

Bharathy, N., Kowsalya, P., David, E. M., Choudhury, M., Khan, P. A., Aidoo, M. K., Parthasarathi, T. et al. (2025). Breaking barriers: Innovative approaches and technologies for climate resilience: A case study from European countries. In *Sustainable synergy: Harnessing ecosystems for climate resilience* (pp. 177–190). Springer Nature. https://doi.org/10.1007/978-3-031-77957-2_12

Biswas, S., Yao, Z., Yan, L., Alqhatani, A., Bairagi, A. K., Asiri, F., & Masud, M. (2023). Interoperability benefits and challenges in Smart City services: Blockchain as a solution. *Electronics, 12*(4), 1036. https://doi.org/10.3390/Electronics12041036

Caragliu, A., & Del Bo, C. (2023). Smart cities and the urban digital divide. *Nature Sustainability, 6*(7), 732–740. https://doi.org/10.1038/S42949-023-00117-W

Cheng, L., Zhang, X., & Huang, B. (2024). Digital twins for smart and sustainable urban infrastructure: A systematic review. *Journal of Cleaner Production, 438*, 138982. https://doi.org/10.1016/J.Jclepro.2023.138982

Choudhury, M., Majumdar, S., Goswami, S., & Sillanpää, M. (Eds.). (2025a). *Smart wastewater systems and climate change: Innovations through spatial intelligence*. Royal Society of Chemistry. https://doi.org/10.1039/9781837676880

Choudhury, M., Suhag, M., Jha, P., Harne, K. R., Samanta, P., Choudhury, M., Borgohain, R. B., Majumdar, S., Kravchenko, Y., & Choudhary, R. (2025b). Urban waste management and challenges in emerging smart city systems governance and circular economy perspectives. *Ecological Engineering & Environmental Technology, 26*(10), 243–266. https://doi.org/10.12912/27197050/211014

El Houssaini, A., & Alami, A. (2024). Ai-driven training and workforce readiness for Smart City development. *Smart Cities, 7*(3), 412–428. https://doi.org/10.3390/Smartcities7030024

European Commission. (2021). *Financing the transition to smart and sustainable cities*. European Commission. Retrieved from https://Ec.Europa.Eu/Info/Publications/Financing-Smart-Sustainable-Cities_En

Gaur, R., & Sharma, D. (2025). Gamification and digital inclusion in urban sustainability education. *International Journal of Urban Management and Technology, 9*(2), 77–93. https://doi.org/10.1016/Ijumt.2025.02.006

Gogoi, N., Borah, D., & Choudhury, M. (2025). Electronic wastes: The man-made crisis. In S. Goswami, M. Choudhury, & S. Agarwal (Eds.), *Electronic waste: Impact on health, animals, and the environment* (pp. 16–24). CRC Press. https://doi.org/10.1201/9781003582311-2

González, E., Kandpal, V., Machado, M., Martens, M., & Majumdar, S. (2023). A bibliometric analysis of circular economies through sustainable smart cities. *Sustainability, 15*(22), 15892. https://doi.org/10.3390/su152215892

Goswami, S., Choudhury, M., & Agarwal, S. (Eds.). (2025). *Electronic waste: Impact on health, animals, and the environment* (1st ed.). CRC Press. https://doi.org/10.1201/978100358231

Gupta, A. (2023a). Developing capabilities in Smart City ecosystems: A multi-level approach. *Urban Studies, 60*(9), 1875–1892. https://doi.org/10.1177/01708406231164114

Gupta, N. (2023b). Editorial: Smart cities challenges, technologies and trends. *Frontiers in Big Data, 6*, 1258051. https://doi.org/10.3389/Fdata.2023.1258051

Hardi, R., Nurmandi, A., Purwaningsih, T., & Manaf, H. A. (2025). Smart City governance and interoperability: Enhancing human security in Yogyakarta and Makassar, Indonesia. *Frontiers in Political Science, 7*, 1553177. https://doi.org/10.3389/Fpos.2025.1553177

Hashem, I. A. T., Chang, V., Anuar, N. B., Adewole, K., Yaqoob, I., Gani, A., Chiroma, H., et al. (2016). The role of big data in Smart City development. *International Journal of Information Management, 36*(5), 748–758. 10.1016/J.Ijinfomgt.2016.05.002.

Huang, C., Chen, Q., Lin, Z., Su, W., Zhang, Z., & Liu, S. (2024). The economic impact and application challenges of iot technology in smart home and infrastructure. *Journal of Innovation and development, 6*(1), 35–39. https://doi.org/10.54097/y54jkp60

Khanal, A., Sherpa, P., Chataut, P., Khanal, A., & Giri, S. (2024). Transforming e-waste management: Challenges and opportunities. *International Research Journal of Multidisciplinary Technovation*, 108–115. https://doi.org/10.54392/irjmt2429

Kowsalya, P., Jennifer Flora, G., Kousalya, A., Toha, M. A., Khan, P. A., Choudhury, M., Parthasarathi, T., et al. (2025). Circular systems unveiled: Essential components and their impact on recycling, reuse, and resource recovery. In S. Majumdar, M. Choudhury, & A. Cheshmehzangi (Eds.), *Sustainable urban future. Urban sustainability* (pp. 55–71). Springer. https://doi.org/10.1007/978-981-95-2976-6_3

Kshetri, N. (2021). Blockchain and the economics of smart cities. *Technological Forecasting and Social Change, 168*, 120776. https://doi.org/10.1016/J.Techfore.2021.120776

Kumar, H. (2025). Developing smart cities: Challenges and opportunities. *Urban Studies, 62*(3), 567–589. https://doi.org/10.1177/00420980231012345

Kumar, R., Singh, P., & Patel, A. (2024). Community-led innovation and social resilience in circular smart cities. *Cities, 152*, 104984. https://doi.org/10.1016/J.Cities.2024.104984

Li, J., & Chen, Y. (2025). Financing sustainable smart cities: The role of innovation ecosystems and digital finance. *Technological Forecasting and Social Change, 209*, 122564. https://doi.org/10.1016/J.Techfore.2025.122564

Liu, Y. (2024). Circular economy in urban systems: How to measure the impact? *Waste Management & Research, 39*(2), 197–198. https://doi.org/10.1177/0734242x231160520

Luo, Q., Ye, Z., & Jia, H. (2023). A charging planning method for shared electric vehicles with the collaboration of mobile and fixed facilities. *Sustainability, 15*(22), 16107. https://doi.org/10.3390/su152216107

Majumdar, S., Kandpal, V., & Anthopoulos, L. G. (2024). *Smart cities: Innovations, challenges and future perspectives.* Springer Nature. https://doi.org/10.1007/978-3-031-59846-3

Majumdar, S., Choudhury, M., & Cheshmehzangi, A. (2025). *Sustainable urban future: Embracing circular economy in waste management.* Springer Singapore. https://doi.org/10.1007/978-981-95-2976-6

Mekki, M., Alalouch, C., & Al-Salti, Z. (2024). Smart City governance and interoperability: Pathways to sustainable policy frameworks. *Sustainability, 16*(2), 743. https://doi.org/10.3390/Su16020743

Mutambik, I. (2024). Unlocking the potential of sustainable smart cities: Barriers and strategies. *Sustainability, 16*(12), 5061. Mdpi.

Nam, T., & Pardo, T. A. (2011). Conceptualizing Smart City with dimensions of technology, people, and institutions. In *Proceedings of the 12th Annual International Digital Government Research Conference on Digital Government Innovation in Challenging Times*, 282–291. https://doi.org/10.1145/2037556.2037602.

Narayan, S., Yadav, D., Jha, S., Khatun, B., Choudhary, R., Verma, U. K. Choudhury, M., & Kumar, D. (2025). From past to present: The evolution of waste management and the rise of circular economy principles. In *Sustainable urban future: Embracing circular economy in waste management* (pp. 23–53). Springer Nature Singapore. https://doi.org/10.1007/978-981-95-2976-6_2

Ntanda, A., & Carolissen, R. (2025). Technology's dual role in smart cities and social equality: A systematic literature. *Journal of Local Government Research and Innovation, 6*, A238. Jolgri.Org.

Nusir, M., Alshirah, M., & Alghsoon, R. (2023). Investigating Smart City adoption from the citizen's insights: Empirical evidence from the Jordan context. *Peerj Computer Science, 9*, E1289. Pmc

OECD. (2020). *Smart cities and inclusive growth: Building on the outcomes of the OECD programme on smart cities.* OECD Publishing. https://doi.org/10.1787/Eef43182-En

Olawale, O., Mensah, P., & Adebayo, T. (2023). Education for digital transformation and green economy in Africa: A policy perspective. *Applied Economics Letters, 31*(5), 450–456. https://doi.org/10.1080/13504851.2023.2219702

Poshaia, L., &Intauno, K. (2025). Building blocks for smart cities adoption and associated implementation drawbacks in Sub-Saharan Africa. *International Journal of Urban Sciences* (Ahead of Print). Taylor & Francis Online.

Reyes García, J. R., et al. (2020). State of the art of mobility as a service (MaaS) ecosystems and architectures an overview of, and a definition, ecosystem and system architecture for electric mobility as a service (eMaaS). *World Electric Vehicle Journal, 11*(1), 7. https://doi.org/10.3390/wevj11010007

Schnieder, M., & West, A. (2019). *Evaluation of alternative battery charging schemes for one-way electric vehicle smart mobility sharing systems based on real urban trip data*, 296–301. https://doi.org/10.1109/rtsi.2019.8895556

Shaheen, S. et al. (2016). *Shared mobility current practices and guiding principles.* U.S. Department of Transportation, Federal Highway Administration. Available at: https://ops.fhwa.dot.gov/publications/fhwahop16022/fhwahop16022.pdf

Sharma, A., Cook, S., Tjandraatmadja, G., & Gregory, A. (2012). Impediments and constraints in the uptake of water sensitive urban design measures in greenfield and infill developments. *Water Science & Technology, 65*(2), 340–352. https://doi.org/10.2166/wst.2012.858

Sharma, A., Choudhury, M., Agarwal, S., Sharma, R., & Sharma, R. (2023). Corporate social responsibility and roles of developers for sustainability in companies. In *The route towards global sustainability: Challenges and management practices* (pp. 313–332). Springer International Publishing. https://doi.org/10.1007/978-3-031-10437-4_16

Turoń, K. (2024). Personalization of the car-sharing fleet selected for commuting to work or for educational purposes – An opportunity to increase the attractiveness of systems in smart cities. *Smart Cities, 7*(4), 1670–1705. https://doi.org/10.3390/smartcities7040066

UNDP. (2022). *Digital finance for sustainable development.* United Nations Development Programme. Retrieved from https://www.undp.org/publications/digital-finance-sustainable-development

Un-Habitat. (2023). *State of the world's cities 2023: Financing the urban future.* Un-Habitat. Retrieved from https://unhabitat.org/state-of-the-worlds-cities-2023

Wang, L., Zhang, Q., & Chen, X. (2024). Bridging the Ai skills gap for smart infrastructure management in emerging economies. *Technological Forecasting and Social Change, 197*, 122895. https://doi.org/10.1016/J.Techfore.2024.122895

World Bank. (2022). *Innovative financing for smart and resilient cities.* The World Bank. Retrieved from https://www.worldbank.org/en/topic/urbandevelopment/publication/innovative-financing-smart-cities

Yang, T., & Han, J. (2021). Integrated management strategy with feasible smartness over heterogeneous iot environments. *Electronics, 10*(2), 149. https://doi.org/10.3390/electronics10020149

Yigitcanlar, T., Kamruzzaman, M., Buys, L., & Teimouri, R. (2019). Smart cities and sustainability: A review and implications. *Sustainable Cities and Society, 45*, 640–654. https://doi.org/10.1016/J.Scs.2018.11.033

CHAPTER 14

Tech-Driven Transformation: Leveraging AI and Blockchain for Circular Economies

Ravi Rajak, Swarnava Dey, and Prachi Wasnik

14.1 Introduction

14.1.1 The Digital Revolution in Circular Economics

The move from traditional "take–make–dispose" production systems to circular, regenerative models is one of the most important economic shifts of the twenty-first century. The extraction of virgin resources, short product lifecycles, and large amounts of trash have been linked to economic expansion for decades; this strategy is no longer sustainable with growing environmental constraints, resource scarcity, and climate commitments (Corvellec & Stål, 2019). As a result, governments, corporations, and urban areas are moving towards circular economic systems that prolong

R. Rajak (✉)
Department of Economics, University of Bristol, Bristol, UK
e-mail: ravi.india.2019@bristol.ac.uk

S. Dey
School of Environmental Studies, Jadavpur University, Kolkata, India

P. Wasnik
Dyson School of Design Engineering, Imperial College London, London, UK

S. Majumdar, M. Choudhury (eds.), *City of the Future*,
https://doi.org/10.1007/978-3-032-18026-1_14

the useful life of materials, restore natural systems, and separate value creation from resource depletion, but this shift is much more complicated than simply increasing recycling or reducing waste (Agócsová & Chodasová, 2021). Deep supply chain visibility, accurate product lifecycle knowledge, and real-time coordination between manufacturers, customers, recyclers, and regulators are all necessary for a circular economy (Garay-Rondero et al., 2019).

This change is being made possible by digital technology, particularly blockchain and artificial intelligence (AI). AI offers the intelligence needed to extend product lifetimes, estimate demand, manage resource flows, and automate recovery procedures (Khinvasara et al., 2024). This is enhanced by blockchain, which provides secure, immutable traceability of resources, product histories, and environmental impacts. When combined, these technologies provide the digital foundation required for large-scale, reliable, and profitable circular systems (Giorgi, 2024).

14.1.2 Emergence of Smart Technologies in Economic and Industrial Systems

Smart technologies have reshaped modern economic and industrial systems by introducing data-driven, automated, and highly interconnected modes of production and resource management. This emergence is part of a larger shift towards digitally enhanced industrial ecosystems, where processes rely on continuous sensing, real-time optimisation, and transparent flows of information across supply chains. This shift is often described as the core of the fourth industrial revolution, in which the convergence of AI, blockchain, IoT, cyber-physical systems, and advanced analytics has pushed industries towards more connected, adaptive environments (Kagermann et al., 2013; Lasi et al., 2014). AI has become a foundational driver of this shift. Its predictive capabilities help industries optimise energy use, reduce downtime, and improve material efficiency. For instance, predictive maintenance supported by machine learning models has been shown to reduce machine failures by up to 30% and extend equipment life by 20–40% (Lee et al., 2015). In waste and recycling systems, AI-powered image recognition and robotics now outperform manual sorting. AMP Robotics' deployments in the US demonstrate real-world gains, with AI systems achieving sorting accuracy above 98% for specific material streams and increasing throughput significantly. Blockchain strengthens this ecosystem by ensuring the reliability, traceability, and

Segment	2023 Value	2030 Projection	CAGR
Total Circular Economy	$553B	$1.32T	13.19%
Digital Component	$2.34B	$10.60B	24.40%
AI/ML Segment	-	-	36.54%

Fig. 14.1 Linear vs. circular economy transformation model. Source: DataM Intelligence (2025), Precedence Research (2025)

interoperability of data that circulates within digital industrial networks. Studies of blockchain-enabled supply chains show measurable improvements in provenance tracking and fraud reduction, particularly in materials-intensive sectors such as electronics and minerals (Kshetri, 2017; Saberi et al., 2019). Regulatory momentum has reinforced the integration of these technologies into circular economic strategies. The European Union's upcoming digital product passport explicitly identifies blockchain and interoperable digital identifiers as core enablers of product traceability, repair data management, and material recovery (European Commission, 2023). Together, these developments show that smart technologies are not simply improving industrial efficiency; they are reshaping the structural logic of economic systems. Transparency and continuous optimisation replace linear extraction and opaque supply chains. As AI and blockchain mature, they enable new circular pathways: from digital marketplaces to automated component recovery and lifecycle monitoring, they offer a practical and scalable foundation for regenerative industrial systems (Fig. 14.1).

It is increasingly recognised that achieving both operational scalability and economic sustainability in circular systems requires AI-driven supply chains, material recovery, and product lifecycle optimisation (Majumdar et al., 2025). When taken as a whole, these estimates show a dual dynamic: digital technologies, especially artificial intelligence (AI), are emerging as the key driver of the circular economy's growth and efficacy, while the circular economy itself is becoming a core economic subject.

14.2 Digital Technologies in Circular Economies

14.2.1 Evolution of Circular Economy Concepts

The concept of the circular economy has evolved significantly over recent decades, moving from a simple waste-minimisation approach to a comprehensive regenerative system. Early foundational models proposed by Boulding and Stahel emphasised closing material loops and designing out waste, which has since expanded into a holistic approach encompassing entire product lifecycles (Ellen MacArthur Foundation, 2019). Contemporary circular economy frameworks prioritise reuse, remanufacturing, and sustainable business models that create economic value while minimising environmental impact (Peyravi et al., 2024). This evolution reflects a fundamental shift from linear "take-make-dispose" models towards regenerative systems that maintain resource value within the economy for extended periods.

However, despite growing awareness and academic discourse, recent evidence indicates concerning trends. The global circularity rate has declined from 9.1% in 2018 to 7.2% in 2023, representing a 21% decrease over five years, even as discussions about the circular economy have nearly tripled in the same period (Circle Economy Foundation, 2024). This paradox highlights the gap between theoretical understanding and practical implementation of the circular economy principle.

14.2.2 Role of AI and Blockchain in Supply Chains and Sustainability

Artificial intelligence and blockchain technologies have emerged as critical enablers for advancing circular economy practices, particularly within complex supply chain networks (Hong & Xiao, 2024). AI applications in circular systems include predictive analytics for equipment maintenance, resource-optimisation algorithms, and automated waste-sorting systems that achieve 85–95% accuracy, compared to 60–70% in traditional systems (McKinsey and Company, 2024a, b). These technologies enable real-time monitoring of product lifecycles and facilitate data dissemination among stakeholders, creating more responsive and efficient circular systems. Table 14.1 shows various AI techniques for circular economy processes.

Blockchain technology addresses fundamental trust and transparency challenges by providing immutable, verifiable records of product lifecycles

Table 14.1 AI techniques in circular economy processes

Circular economy stage	*AI techniques*	*Key functions*
Product design	Machine learning, generative design	Optimise material usage, predict lifecycle impact
Manufacturing & production	Predictive analytics, robotics	Reduce waste, optimise resource consumption
Supply chain & logistics	Demand forecasting, route optimisation	Minimise inventory waste, optimise transportation
Waste management & recycling	Image recognition, anomaly detection	Identify recyclable materials, sort waste efficiently

and material flows. Major automotive manufacturers, including Ford and Volvo Cars, are implementing blockchain solutions to trace cobalt in electric vehicle batteries, capturing origin details, chain-of-custody information, and compliance verification. The integration of blockchain with circular economy initiatives enables secure material traceability, automated smart contracts for circular transactions, and decentralised marketplaces for recovered materials.

14.2.3 Current Research on Digital Circular Economy Tools

The digital circular economy market has experienced substantial growth, reaching $2.34 billion in 2024 and projected to grow to $25.4 billion by 2034 at a compound annual growth rate (CAGR) of 24.40% (DataM Intelligence, 2025; Precedence Research, 2025). Research indicates that the artificial intelligence and machine learning segment within this market is expected to be the fastest-growing category, with a CAGR of 36.54% from 2024 to 2029. Academic research has increasingly focused on digital tools supporting circular economy goals, including Internet of Things (IoT), big data analytics, and digital product passports (van Capelleveen et al., 2023). Digital product passports have gained regulatory prominence through the European Union's Circular Economy Action Plan and the Ecodesign for Sustainable Products Regulation, which came into force in July 2024 (European Commission, 2023). These tools enable comprehensive product traceability, facilitate material recovery, and support data-driven decision-making for sustainability initiatives.

Current implementations demonstrate significant potential: AI-powered waste management systems can process 300–500 items per minute compared to 100 items per minute in traditional systems, while achieving material recovery rates of 70–85% versus 45–55% in conventional approaches.

14.2.4 Gaps in Implementation and Theoretical Integration

Despite promising technological capabilities, significant barriers constrain the widespread adoption of digital circular economy solutions. High initial investment costs, ranging from $10,000 to over $100,000 per company for digital product passport implementation alone, represent substantial barriers for many organisations. Technical complexity, regulatory fragmentation, and lack of standardised frameworks further impede implementation.

A systematic literature review by Jhon et al. (2025) identified policy-related issues, lack of predictability, psychological barriers, and information vulnerability as critical impediments to digitalisation-led circular economy transitions. The research reveals that while over 70% of global organisations recognise circular economy opportunities, most have yet to implement comprehensive circular economy initiatives.

Furthermore, theoretical integration remains limited across different sectors and regions. Most research and practical applications have concentrated on the manufacturing and electronics sectors, with insufficient attention to agriculture, healthcare, and service industries (Fatima et al., 2024). The lack of unified frameworks for understanding how digital technologies interact systemically with circular economy principles represents a significant knowledge gap.

14.3 An Insight on AI and Blockchain Applications in Circular Economy Systems

This chapter addresses identified gaps by providing a comprehensive analysis of AI and blockchain applications in circular economy systems, synthesising recent market data, case studies, and policy developments.

The chapter contributes to existing literature by: (1) providing updated market analysis with verified statistics from authoritative sources; (2) demonstrating real-world applications through detailed case studies from electronics and automotive sectors; (3) addressing implementation barriers with specific cost-benefit analyses; and (4) offering policy

Table 14.2 Blockchain use cases in circular economies

Blockchain features	*Function in circular economy*	*Benefits*
Smart contracts	Automate transactions and compliance	Transparency, efficiency, trust
Traceability & provenance	Track materials and products	Reduced fraud, improved sustainability reporting
Tokenisation & incentives	Reward sustainable behaviour	Encourages participation, monetises sustainability
Decentralised data management	Securely share data across stakeholders	Interoperability, data integrity

Source: Compiled by the authors

recommendations based on current regulatory developments including the European Union's digital product passport requirements. Table 14.2 shows various uses of blockchain technologies for circular economies.

Unlike previous studies that focus on individual technologies or specific sectors, this chapter adopts a systems-based approach that examines the synergistic effects of AI–blockchain integration across multiple circular economy applications. This integrated perspective provides novel insights into how digital technologies can collectively accelerate the transition from linear to circular economic models while addressing scalability and implementation challenges (Nguyen et al., 2022), from the initial phases of design and resource extraction to production, optimisation, supply-chain coordination, and end-of-life recovery.

14.3.1 Optimising Resource Flows and Supply Chains

Circular economies are changing because of artificial intelligence. Its ability to transform massive, dispersed, and rapidly evolving data into understandable, valuable insights is its greatest asset. AI enables the operation of circular systems with a degree of accuracy that conventional techniques cannot match. AI facilitates real-time optimisation of circular supply networks at the operational level (Isensee et al., 2021). Machine learning tools can track materials, components, and products throughout their lifecycles, identify inefficiencies, and quickly modify system processes. This ongoing feedback loop helps businesses predict material shortages, plan repairs before malfunctions, and more precisely coordinate recovery or recycling operations. Materials circulate longer, waste is reduced, and system-wide resource efficiency increases (Pal et al., 2019).

AI also makes it easier for stakeholders to share information, which is crucial for circularity (Kowsalya et al., 2025). Producers, recyclers, logistics partners, and regulators can access the same verified data through tools such as digital product passports, sensor-linked databases, and lifecycle analytics platforms. Throughout the value chain, this shared visibility facilitates quicker, more assured decision-making and lessens information asymmetry.

14.4 Impacts

14.4.1 *Economic Impact*

AI accelerates the development of new business models, such as automated recovery systems, product-as-a-service offerings, and predictive maintenance services, thereby generating new revenue streams in the manufacturing, logistics, and recycling industries. When taken as a whole, these elements position AI not only as an efficiency tool but also as a long-term competitive advantage in a world where growth is becoming increasingly resource-constrained (Tsarouhas & Papaevangelou, 2024) (Table 14.3).

14.4.2 *Predictive Analytics for Lifecycle Management*

Predictive analytics is one of the most potent AI applications in circular economy systems since it is essential to prolonging the usable life of goods

Table 14.3 AI applications in circular supply chain optimisation

Application area	*Technology*	*Efficiency gain (%)*	*Economic impact*
Demand forecasting	Machine learning	20–30 reduction in waste	€100–200B annually
Route optimisation	AI algorithms	15–25 cost reduction	€50–100B annually
Inventory management	Predictive analytics	30–40 improvement	€75–150B annually
Quality control	Computer vision	40–60 defect reduction	€25–50B annually

Source: Compiled by the authors from Successive Digital (2024), Averroes AI (2025), Omdena (2025), McKinsey and Company (2024a, b), Analytics Insight (2025)

and machinery. AI models can detect minor signs of degradation long before they manifest as problems by assessing continuous streams of operational data, such as vibration patterns, temperature changes, loading conditions, and energy use (Özsoy, 2023). Maintenance becomes a deliberate, data-driven operation instead of a reactive, resource-intensive one thanks to this early detection. By reducing unforeseen malfunctions and unscheduled downtime, this focused strategy not only prolongs asset life but also stabilises production cycles. This results in more efficient workflows, fewer interruptions in circular supply chains, and more dependable component recirculation for reuse or refurbishment across industries (Lodhi et al., 2024). Additionally, predictive lifecycle management provides a useful design feedback loop. Engineers can choose materials that offer a longer service life or easier end-of-life recovery, increase durability in future models, and identify recurring weak points using insights gathered during a product's usage phase. Over time, this reduces reliance on virgin materials and helps create more robust circular product designs (Nyström et al., 2021). It has been demonstrated that predictive maintenance solutions increase machinery operating life by 20–30%, significantly reducing the number of components discarded too early (Taşcı et al., 2023). Businesses reduce material consumption and waste generation by preventing equipment failures and extending repair intervals, thereby improving the environment and lowering operating expenses. By ensuring resources remain in productive use for as long as possible and supporting clever, effective lifecycle management, these enhancements closely align with the goals of the circular economy (Rajpal et al., 2024).

14.4.3 Intelligent Waste Management Systems

Robotics, sophisticated sensor technologies, and AI-powered vision systems have revolutionised the identification, sorting, and processing of waste. These technologies can differentiate between plastics, metals, textiles, glass, organics, and even complex composite materials that have historically presented significant issues for recyclers by evaluating optical, spectral, and material fingerprints in real time. AI continuously learns from new trash streams, enhancing its detection accuracy and adapting to new materials and packaging types, unlike manual or mechanical sorting (Choudhury et al., 2025a, b, c, d) (Table 14.4).

Table 14.4 Performance metrics for AI-powered waste management systems

Metric	*Traditional systems*	*AI-powered systems*	*Improvement*
Sorting accuracy	60–70%	85–95%	+25–35%
Processing speed	100 items/min	300–500 items/min	+200–400%
Material recovery rate	45–55%	70–85%	+25–30%
Operating costs	Baseline	−30-50%	30–50% reduction

Source: 2025 ScienceDirect (2024), McKinsey & Company (2024)

14.4.4 *Automated Remanufacturing and Recycling*

AI is revolutionising recycling and remanufacturing by enabling levels of accuracy, speed, and consistency that are not possible with manual operations. Large-scale remanufacturing is not economically viable in traditional systems because product returns are often erratic and difficult to evaluate (Gry et al., 2023). AI alters this by analysing returned goods in real time, assessing their condition, and automatically directing them to the best recovery route (Sculley et al., 2015).

AI systems assess each returned product and categorise components according to wear, performance history, and material integrity, as shown in the process diagram. Robotics and AI-guided vision systems then support the disassembly process, recognising fasteners, sequencing removal stages, and limiting damage to reusable parts. AI is revolutionising remanufacturing and recycling. This guarantees the preservation and reintroduction of components with residual life value into the production cycle instead of their disposal. These capabilities significantly increase the economic viability of remanufacturing. AI lowers processing costs, increases recovery accuracy, and reduces labour effort by automating complex sorting and disassembly processes (Choudhury et al., 2025a, b, c, d). By capturing parts and materials that would have otherwise been lost due to incorrect classification or manual-handling restrictions, it also increases material efficiency. Higher recovery rates, a more reliable supply of remanufactured parts, and a more seamless introduction of secondary materials into the market are the outcomes.

14.5 Blockchain Technology: Trust and Transparency in Circular Systems

Blockchain helps bridge the gaps in openness and trust that often impede circular economy systems. It provides verifiable information on product lifecycles, material transfers, and sustainability performance (Saberi et al., 2019) by documenting each transaction in a secure, tamper-resistant ledger. This is particularly helpful in circular value chains, where goods pass through several players, and precise data determines whether materials can be recycled, mended, or reused. Blockchain enables stakeholders to confidently view a product's composition, ownership history, and environmental impact through digital product identities and verifiable records. By automating procedures such as leasing agreements, repair obligations, and payment for material returns, smart contracts further simplify circular activities (Christidis & Devetsikiotis, 2016).

14.5.1 Material Traceability and Product Passports

In practice, blockchain supports digital product passports, certification of recycled content, and authentication of remanufactured goods helping reduce fraud and prevent greenwashing. As global regulations increasingly demand end-to-end traceability, blockchain provides the secure digital backbone needed to scale high-integrity circular systems (Choudhary et al., 2024) (Table 14.5).

Table 14.5 Digital product passport structure and economic benefits

Component	*Information tracked*	*Market value (2024)*	*Implementation cost (per company)*
Material origin	Source, extraction method	Included in DPP market	\$10,000–\$100,000+ (varies widely)
Manufacturing	Process, energy use	\$184 M–\$214 M (global, 2024)	\$10,000–\$100,000+ (varies widely)
Transportation	Route, emissions	\$1.7–\$1.8B (global, 2030)	\$10,000–\$100,000+ (varies widely)
End-of-life	Disposal, recovery		

Source: MarketsandMarkets (2025), Renoon (2025), inriver (2025), GS1 Hong Kong (2025)

14.5.2 Smart Contracts for Circular Transactions

Smart contracts are essential to improving the scalability and efficiency of circular economy models. Once predetermined conditions are met, these self-executing digital agreements automatically initiate actions, such as payments, deposits, repair obligations, or ownership transfers (Khan et al., 2021). Smart contracts minimise delays, enhance compliance, and provide a more reliable foundation for circular trades by eliminating the need for manual verification or third-party mediation (Sahoo et al., 2024). Transactions in circular systems frequently involve several parties, including manufacturers, repair facilities, logistics companies, recyclers, and customers, all of whom rely on precise information and prompt coordination (Henriques et al., 2023). By ensuring obligations and rewards are applied promptly and transparently, smart contracts simplify these exchanges. This makes it possible for business models that would otherwise be too complicated or expensive to run manually, like product-as-a-service, automated deposit-return schemes, pay-per-use leases, and verified take-back programmes.

Smart contracts can reduce transaction and coordination costs significantly by automating verification, payments, and material monitoring, making previously economically marginal circular activities financially feasible. In addition to improving operational effectiveness, these cost savings enhance stakeholder trust and foster broader participation in circular value chains (Zhang, 2024). In general, smart contracts facilitate the transition of circular systems from partially manual, trust-based procedures to interconnected, autonomous digital ecosystems that operate more quickly, accurately, and profitably.

14.5.3 Decentralised Circular Marketplaces

Blockchain enables customers and suppliers to trade circular materials, reconditioned parts, and recovery services directly on decentralised markets. These platforms cut transaction costs, lessen reliance on intermediaries, and improve price discovery for secondary commodities by moving verification and coordination to a secure distributed ledger (Dang et al., 2019). The suppliers list products or recovered materials with blockchain-verified origin, quality, and sustainability information, as depicted in the architecture. Because records cannot be changed or selectively published, buyers can examine this data with confidence (Cao et al., 2023). Then, to limit delays and disputes, smart contracts automate important processes, including delivery confirmation, return conditions, and payment release. In the absence of a central governing body, the network's validators verify listings and transactions. This framework expands access to circular value chains by facilitating participation from smaller recyclers, repair facilities, and remanufacturers. Overall, decentralised circular marketplaces provide a more effective, transparent, and scalable environment for circular commerce while also unlocking the value of secondary materials (Wang et al., 2019; Sahoo et al., 2024).

14.6 The Economic Impact of AI–Blockchain Integration

A significant technical frontier that is changing how firms manage resources, coordinate value chains, and verify sustainability outcomes is the integration of AI and blockchain into circular economy systems. Investment in AI-driven optimisation and blockchain-enabled transparency has increased across manufacturing, shipping, waste management, and consumer goods as businesses become more aware of the financial benefits of predictive intelligence and tamper-proof data. This change is part of a larger shift towards digital infrastructures that can enable large-scale, resource-efficient, high-integrity circular models (Nham, 2022).

14.6.1 Market Size and Growth Projections

The combined application of blockchain and AI is anticipated to open new industry opportunities as adoption grows, ranging from data-verified product-as-a-service models to automated take-back programmes and decentralised material marketplaces (Table 14.6).

Table 14.6 Regional market distribution for digital circular economy

Region	*2024 market share (%)*	*2030 projection (%)*	*Growth rate (%)*
North America	35	32	12.5 CAGR
Europe	30	28	11.8 CAGR
Asia-Pacific	25	30	18.2 CAGR
Rest of world	10	10	13.0 CAGR

Source: Precedence Research (2025), Next Move Strategy Consulting (2025)

14.6.2 Cost Reduction and Efficiency Gains

By simplifying procedures that were previously laborious, tedious, and error-prone, the integration of AI and blockchain in circular systems significantly lowers operating expenses. These technologies improve efficiency in waste processing operations, supply chains, quality assurance procedures, and regulatory compliance—areas that frequently determine whether circular methods are economically feasible at scale (Dursun et al., 2022) (Table 14.7).

14.6.3 New Business Models and Revenue Streams

Circular business models that were previously too complicated, expensive, or dangerous to adopt are now possible because of the convergence of blockchain technology and artificial intelligence. Organisations may create revenue models that reward durability, resource efficiency, and continuous material rotation by fusing real-time intelligence with reliable, unchangeable data (Saberi et al., 2019).

14.6.3.1 Product-as-a-Service (PaaS)

Manufacturers may retain ownership of their products and offer them as services, such as machinery leasing or lighting-as-a-service, thanks to AI-powered monitoring and predictive maintenance (Tukker, 2004; Lee et al., 2015). PaaS models are financially appealing to both providers and customers because real-time performance data reduces maintenance costs, extends product lifecycles, and ensures predictable service delivery.

Table 14.7 Cost reduction impact of AI–blockchain integration

Area	*Traditional approach*	*AI-blockchain approach*	*Cost reduction (%)*
Supply chain management	Manual tracking	Automated transparency	30–50
Quality assurance	Sampling-based	Real-time monitoring	40–60
Waste management	Linear processing	Circular optimisation	25–40
Compliance reporting	Manual documentation	Automated verification	50–70

Source: Institut Montaigne (2024), Cambridge Econometrics (2014)

14.6.3.2 Material Banking
Materials can be treated as long-term assets thanks to blockchain-verifiable records of their composition, origin, and quality. Recovered resources can be confidently retained, exchanged, or reintegrated into production after verification. This enables the creation of new financial instruments based on certified secondary resources, such as material credits or collateral-backed reserves (Ellen MacArthur Foundation & Google, 2019).

14.6.3.3 Circular Marketplaces
Compared to traditional exchanges, AI algorithms are significantly more accurate at matching supply and demand for secondary materials, refurbished components, and repair services (Wang et al., 2019). Decentralised marketplaces become more effective, lower search costs, and generate new revenue streams for recyclers, remanufacturers, and repair services by integrating this intelligence with blockchain-verified listings.

14.6.3.4 Performance-Based Contracts
Agreements that are contingent on circular criteria, such as product longevity, maintenance frequency, material recovery rates, or emissions reduction, are enabled by smart contracts. Performance-based circular contracts are easier to enforce and more appealing to industry and government customers thanks to automated verification on the blockchain, which guarantees transparency and trust (Vatiero, 2024).

Together, these models strengthen the economic foundation of circular systems by balancing long-term profitability with sustainability and resource efficiency, while also diversifying revenue streams (Saberi et al., 2019).

14.6.3.5 Case Studies: Real-World Applications
Industrial use across waste management, manufacturing, and global supply chains shows how smart technologies are enabling more circular economic practices. In the recycling sector, artificial intelligence has become a practical tool for improving material recovery. Cities such as Amsterdam and Singapore deploy AI-driven smart waste-sorting and collection systems that optimise routes, predict waste volumes, and improve recycling rates. These smart city initiatives combine IoT sensors, AI analytics, and data platforms to reduce landfill use and resource extraction (Choudhary et al., 2025). Traceability challenges in the plastics and chemicals industries have prompted companies to experiment with blockchain to improve

data integrity across supply chains. Circularise, working with partners such as Covestro and Mitsubishi Chemical, has piloted decentralised traceability platforms that record material origins, sustainability attributes, and mass-balance claims. Companies like Qodequay utilise AI-driven lifecycle intelligence to design products with modularity and disassembly in mind, facilitating repair, reuse, and recycling. Automotive manufacturing provides another strong example. Renault's XCEED platform uses a blockchain ledger to track compliance data for thousands of components across multiple suppliers (Renault Group, 2020). Large-scale logistics show both the progress and the limits of digital transformation. IBM and Maersk's TradeLens platform attempted to create an integrated blockchain network connecting carriers, ports, and customs organisations. Although the programme ended in 2022, it demonstrated clear improvements in documentation speed and shipment visibility (IBM & Maersk, 2022). Across these cases, a typical pattern emerges: AI helps generate high-quality, actionable data from messy physical processes. At the same time, blockchain and similar digital tools secure data and enable its sharing across supply chains. Together, they enable industrial systems to manage materials more intelligently and support transparent, efficient circular models.

14.7 Challenges and Implementation Barriers

14.7.1 Technical and Scalability Issues

14.7.1.1 Blockchain Performance Constraints

Despite strong potential, several technical constraints limit large-scale deployment of AI–blockchain systems in circular economy applications. Blockchain throughput, latency, and consensus overhead remain major bottlenecks, especially when high-frequency lifecycle data must be recorded across supply chains. Even permissioned ledgers face trade-offs between speed, security, and decentralisation, which complicates continuous material-level tracking in industrial settings. Despite significant potential, several technical challenges constrain widespread adoption. Solutions such as layer-2 scaling, permissioned ledgers, and alternative consensus methods help, but the trade-offs between speeds, decentralisation, and security are still unresolved for many industrial use cases (Khan et al., 2021).

14.7.1.2 Data and Model Limitations in AI Systems

AI-driven sorting, prediction, and optimisation depend on large, clean, and consistently labelled datasets. In real operations, waste streams fluctuate, sensors drift, and data gaps appear, lowering model reliability. These conditions demand frequent retraining, domain adaptation, and human oversight, which increases operational complexity and limits automation at scale (Antony Jose et al., 2024).

14.7.1.3 Interoperability Challenges Across IoT, Cyber Physical Systems (CPS), and Ledgers

Industrial systems rely on heterogeneous sensors, protocols, and data formats, making it difficult to synchronise IoT and CPS data with blockchain records or digital product passports (Alam, 2022). Without common data standards, unified identity schemes, and robust middleware, integration becomes costly, fragile, and difficult to scale across multiple suppliers and jurisdictions.

14.7.1.4 Systems-Level and Regulatory Barriers

Broader concerns around data governance, security, and compliance with emerging regulations amplify technical issues. Protecting commercially sensitive information while enabling transparency remains a significant challenge. Together, these factors are creating a technical landscape where scaling remains possible but requires improvements in ledger efficiency, data-centric AI engineering, and cross-platform interoperability (Ferreira et al., 2024).

14.7.2 Economic and Financial Barriers

14.7.2.1 Investment Requirements

Circular systems provided by blockchain and AI demand a substantial initial investment despite their long-term efficiency gains. Sensors, data infrastructure, digital product passports, automated sorting technologies, integration software, and employee training are all necessary investments for businesses (Jones, 2024). These expenses are challenging for many companies, particularly SMEs, to absorb in the absence of outside incentives or assured returns. The requirement to modify current IT systems and production lines increases the financial risk, making the shift both capital-intensive and disruptive to the company.

14.7.2.2 Economic Viability Threshold

The economic underpinnings have not yet changed sufficiently to make high-integrity circular models globally feasible, despite growing interest in circular practices worldwide. This decrease indicates that circular solutions remain unable to compete with lower-cost linear methods, particularly when raw materials remain cheap and environmental externalities are undervalued. The commercial case for circularity is weakened across many areas by unstable secondary-material markets, inadequate infrastructure for recovery, and uneven regulatory incentives (Boorsma et al., 2022).

14.7.3 Regulatory and Policy Challenges

One of the biggest challenges to scaling circular systems provided by blockchain and AI is still regulatory fragmentation. Rules governing data exchange, digital identities, trash classification, product stewardship, and recycling vary across nations and within regions within the same nation. These discrepancies impede market expansion and system interoperability, making it challenging for businesses to develop solutions that function flawlessly across national boundaries (Henninger & Mashatan, 2021). Divergent regulations on digital ledgers, data privacy, and the legal acceptance of smart contracts all contribute to the complexity of blockchain. There is confusion about the design and governance of circular data platforms, as companies may be subject to stringent data protection regulations in one country while facing little oversight in another.

A practical policy framework requires economic methods that make circular behaviours financially appealing. Extended producer responsibility promotes products made for recycling and repair, tax incentives reduce the cost of implementing circular technologies, green procurement generates steady demand for verified circular goods, and carbon pricing reduces the competitiveness of resource-intensive linear production. By coordinating economic incentives with the objectives of the circular economy, these tools facilitate the broader use of circular systems enabled by blockchain and AI (Wulandari, 2022).

14.8 Policy Framework and Regulatory Considerations

Regulations about the circular economy are likewise inconsistent. While some areas lack clear frameworks completely, others have well-defined extended producer responsibility programmes, right-to-repair laws, or requirements for digital product passports. This patchwork discourages innovation, hinders investment, and keeps businesses from realising economies of scale (Jensen et al., 2024).

Scaling the integration of blockchain and AI in circular economy systems requires a favourable policy environment. Adoption, interoperability, and long-term investment are hampered by regulatory systems that have not yet kept pace with rapid technological advancements. Therefore, for companies to confidently implement digital circular solutions, effective policy frameworks must offer clarity, incentives, and safeguards (Yerram, 2022).

14.8.1 Government Initiatives and Support

By providing funds for research, supporting demonstration projects, and establishing incentives for business engagement, governments play a key role in driving the adoption of circular systems enabled by blockchain and artificial intelligence. To test digital traceability tools, create scalable models for material recovery and reuse, and develop new circular technologies, the programme brings together academics, business partners, and legislators.

14.8.2 International Cooperation and Standards

Stronger international cooperation is needed to reduce legislative and regulatory fragmentation. Circular value chains frequently span several nations, but the regulations governing product design, waste classification, material traceability, data sharing, and digital verification vary by jurisdiction. Businesses face high compliance costs and limited cross-border scalability for digital circular solutions in the absence of universal standards (Sharma et al., 2023).

A more cohesive global environment would result from the development of common frameworks for material composition reporting, smart contract governance, digital product passports, and environmental effect assessment (Choudhury et al., 2025a, b, c, d). Recognising that transparent, interoperable data is essential to circular systems, international organisations like the Global Battery Alliance, UNEP, and ISO are already working towards harmonised methodologies.

Standardised guidelines for lifecycle data, digital product passports, and innovative contract governance would lower ambiguity and encourage broader adoption. Targeted incentives, such as tax breaks for remanufacturing, procurement requirements for certified secondary materials, and subsidies for circular digital infrastructure, can help bridge the economic divide between linear and circular processes.

To preserve confidence in digital traceability systems, policymakers must also provide robust safeguards for data privacy, cybersecurity, and the ethical use of AI. To enable AI and blockchain to accelerate the shift to a circular economy, a balanced regulatory strategy that incorporates clear standards, financial incentives, and robust protections is necessary (Luo, 2023).

14.8.3 Incentive Mechanisms and Economic Instruments

Economic strategies that make circular practices financially attractive need to be supported by a strong policy framework. Tax incentives can reduce the upfront cost of investing in circular technologies. Extended producer responsibility schemes encourage companies to design products that are easier to repair, reuse, and recycle. Green public procurement ensures a stable demand for certified circular products. Carbon pricing makes emissions-intensive linear production less competitive and strengthens the business case for circular approaches. When these measures work together, they align economic incentives with environmental goals and support wider adoption of circular systems enabled by AI and blockchain (Lăzăroiu et al., 2023).

14.9 Future Outlook and Emerging Trends

14.9.1 Technological Convergence and Innovation

By the end of 2025, artificial intelligence is expected to be incorporated into around 30% of smart city applications, signalling a significant shift in the management of urban systems (Herath & Mittal, 2022). By enabling real-time optimisation of energy use, transportation flows, waste management, and public services, this rapid adoption of AI will transform the sustainability landscape. AI will become increasingly crucial for managing cyclical resource flows and supporting data-driven environmental planning as cities install more networked sensors and digital infrastructure (Narayan et al., 2025).

Several significant trends in technological convergence will influence the next ten years. Real-time circular system optimisation will be enabled by the marriage of AI and IoT, with linked sensors continuously feeding data into intelligent decision-making models. Complete lifecycle monitoring will be enabled by blockchain interactions with digital twins, which will produce verified, immutable records of how goods are made, used, repaired, and returned. Many circular operations, including sorting, disassembly, and remanufacturing, will become more autonomous as machine learning and sophisticated robotics combine. Ultimately, large-scale circular transitions will be accelerated by the ability to coordinate extremely complex resource flows between cities and industries through the combination of quantum computing and optimisation algorithms (Choudhury et al., 2025a, b, c, d).

14.9.2 Market Evolution and Competitive Landscape

The market for the digital circular economy is growing quickly as industries recognise the economic gains of automation and data-driven optimisation. Global analyses suggest that digital technologies will play a significant role in scaling circular business models, as the broader circular tech market is projected to grow steadily throughout the next decade. This momentum is evident in sectors with complex material flows (electronics, automotive, and packaging), where companies are increasingly adopting AI-driven and blockchain-based systems. AI-driven resource processing is already reshaping competition. Studies show that machine vision and robotic sorting systems deliver higher accuracy, throughput, and material purity than manual operations, giving early adopters a clear operational edge (Fang et al., 2023; Lubongo et al., 2024). Blockchain technologies are also influencing market dynamics by strengthening lifecycle traceability and credibility of sustainability claims. Research demonstrates that distributed ledgers reduce information asymmetry and support secure tracking of materials across supply chains (Kouhizadeh & Sarkis, 2018; Saberi et al., 2019). Policy developments are intensifying this competitive landscape. China's intelligent manufacturing roadmap emphasises real-time analytics, IoT integration, and smart factory infrastructure as significant resource catalysts for efficiency (MIIT, 2022).

Overall, the competitive landscape suggests that market leaders will be those able to combine high-fidelity data, automated resource processing, and trusted lifecycle traceability into integrated circular business models. Companies that lag in digitalisation risk higher operational costs, weaker regulatory compliance, and limited access to emerging circular value chains.

14.9.3 Societal and Environmental Impact Projections

The shift to a more circular economy carries strong job-creation potential. By 2030, circular business models—such as product-as-a-service, remanufacturing, repair, and recycling—are expected to create around 700,000 new jobs in the European Union. These roles will range from high-tech positions in AI-enabled waste management, digital product passports, and automated sorting, to hands-on jobs in repair services, refurbishing centres, and regional material recovery facilities (De Pascale et al., 2023). Growth is also expected in logistics, quality assessment, circular product design, and digital sustainability services. This expansion reflects both the labour-intensive nature of circular activities and the rising need for new skills in data analytics, predictive maintenance, and circular system management. Altogether, it signals a significant transformation in Europe's employment landscape.

14.9.3.1 Environmental Benefits

Greenhouse gas emissions can fall substantially as circular systems mature; this shift is significant given that nearly 45% of global emissions are linked to how products are made and consumed. Circularity also improves resource efficiency, with the potential to raise productivity by roughly 3% each year. Better material recovery systems can deliver 60–80% reductions in waste, easing pressure on landfills while conserving natural resources (Mondal et al., 2023; Alkan & Pauliuk, 2025).

14.9.3.2 Social Impact

Circular practices contribute to cleaner cities by improving air and water quality, especially in densely populated areas. They enhance resource security by reducing dependence on imported raw materials. They also strengthen economic resilience, allowing communities to withstand supply disruptions while supporting long-term sustainability.

14.10 Future Recommendations

Circular systems enabled by blockchain and AI are expected to become much more autonomous, interoperable, and commercially significant over the course of the next ten years. Waste sorting, remanufacturing, and predictive maintenance will all benefit from AI-driven automation, and digital product passports will offer more comprehensive, real-time lifespan data for optimisation. It is anticipated that blockchain networks will develop into reliable infrastructures for global material traceability, facilitating more stable secondary material markets and lowering fraud in streams of recycling and reuse. Product-as-a-service, material banking, and performance-based contracts, backed by smart contracts and ongoing monitoring, will become more prevalent business models. Major areas' policies, particularly those of the US, UK, China, and EU, are also likely to converge on common standards for digital verification, carbon accounting, and traceability. When taken as a whole, these developments hint at a future in which circular systems are more predictable, transparent, and economically feasible, establishing AI–blockchain integration as a key facilitator of global resource resilience.

14.11 Conclusion

A unique chance to balance long-term environmental resilience with economic growth is presented by the convergence of blockchain and artificial intelligence in circular economy systems. Industry, governments, and technology suppliers must work together to realise this promise. However, the magnitude of the potential benefits—increased productivity, reduced reliance on materials, and improved climate outcomes—makes the investment both strategically necessary and worthwhile. By 2030, businesses and economies that adopt circular models enabled by blockchain and artificial intelligence early on will have a distinct competitive advantage. They will demonstrate improved environmental performance, operate with more resource efficiency, open new revenue streams, and confidently satisfy growing regulatory and customer demands.

References

Agócsová, Á., & Chodasová, Z. (2021). Innovative methods in sustainable urban development and water management. *Tem Journal*, 757–765. https://doi.org/10.18421/tem102-33

Alam, T. (2022). Blockchain-based internet of things: Review, current trends, applications, and future challenges. *Computers, 12*(1), 6. https://doi.org/10.3390/computers12010006

Alkan, A., & Pauliuk, S. (2025). Reducing greenhouse gas emissions and material use for passenger vehicles in Türkiye: A scenario-based analysis with material efficiency strategies. *Resources, Conservation and Recycling, 221*, 108385. https://doi.org/10.1016/j.resconrec.2025.108385

Analytics Insight. (2025). *AI trends shaping global supply chains in 2025*. https://www.analyticsinsight.net

Antony Jose, S., Cook, C. A. D., Palacios, J., Seo, H., Torres Ramirez, C. E., Wu, J., & Menezes, P. L. (2024). Recent advancements in artificial intelligence in battery recycling. *Batteries, 10*(12), 12. https://doi.org/10.3390/batteries10120123

Averroes AI. (2025). *Stats & impact of AI in manufacturing: 2025 insights*. https://averroes.ai.

Boorsma, N., Polat, E., Bakker, C., Peck, D., & Balkenende, R. (2022). Development of the circular product readiness method in circular design. *Sustainability, 14*(15), 9288. https://doi.org/10.3390/su14159288

Cao, P., Duan, G., Tu, J., Jiang, Q., Yang, X., & Li, C. (2023). Blockchain-based process quality data sharing platform for aviation suppliers. *IEEE Access, 11*, 19007–19023. https://doi.org/10.1109/access.2023.3246984

Choudhary, D., Sangwan, K., & Singh, A. (2024). Blockchain-enabled architecture for lead acid battery circularity. *Scientific Reports, 14*(1). https://doi.org/10.1038/s41598-024-67404-z

Choudhary, R., Kumar, A., Narayan, S., Mishchenko, I., Choudhury, M., Bhardwaj, L. K., & Boiko, O. (2025). Integrating fuzzy logic into life cycle assessment for sustainable municipal solid waste management in metro cities. *Ecological Engineering & Environmental Technology, 26*(7), 316–331. https://doi.org/10.12912/27197050/205500

Choudhury, M., Cheshmehzangi, A., & Majumdar, S. (2025a). The circular economy paradigm: Core concepts and value. In S. Majumdar, M. Choudhury, & A. Cheshmehzangi (Eds.), *Sustainable urban future. Urban sustainability*. Springer. https://doi.org/10.1007/978-981-95-2976-6

Choudhury, M., Dixit, G., & Majumdar, S. (Eds.). (2025b). *Sustainable synergy: Harnessing ecosystems for climate resilience*. Springer Nature. https://doi.org/10.1007/978-3-031-77957-2

Choudhury, M., Majumdar, S., Goswami, S., & Sillanpää, M. (Eds.). (2025c). *Smart wastewater systems and climate change: Innovations through spatial intelligence*. Royal Society of Chemistry. https://doi.org/10.1039/9781837676880

Choudhury, M., Suhag, M., Jha, P., Harne, K. R., Samanta, P., Choudhury, M., Borgohain, R. B., Majumdar, S., Kravchenko, Y., & Choudhary, R. (2025d). Urban waste management and challenges in emerging smart city systems governance and circular economy perspectives. *Ecological Engineering & Environmental Technology, 26*(10), 243–266. https://doi.org/10.12912/27197050/211014

Christidis, K., & Devetsikiotis, M. (2016). Blockchains and smart contracts for the internet of things. *IEEE Access, 4*, 2292–2303.

Circle Economy Foundation. (2024). *The circularity gap report 2024*. Circle Economy Foundation.

Corvellec, H., & Stål, H. (2019). Qualification as corporate activism: How Swedish apparel retailers attach circular fashion qualities to take-back systems. *Scandinavian Journal of Management, 35*(3), 101046. https://doi.org/10.1016/j.scaman.2019.03.002

Dang, C., Zhang, J., Kwong, C., & Li, L. (2019). Demand side load management for big industrial energy users under blockchain-based peer-to-peer electricity market. *IEEE Transactions on Smart Grid, 10*(6), 6426–6435. https://doi.org/10.1109/tsg.2019.2904629

DataM Intelligence. (2025). *Digital circular economy market report 2025–2032*. DataM Intelligence.

De Pascale, A., Di Vita, G., Giannetto, C., Ioppolo, G., Lanfranchi, M., Limosani, M., & Szopik-Depczyńska, K. (2023). The circular economy implementation at the European Union level. Past, present and future. *Journal of Cleaner Production, 423*, 138658. https://doi.org/10.1016/j.jclepro.2023.138658

Dursun, E., Ülker, Y., & Günalay, Y. (2022). Blockchain's potential for waste management in textile industry. *Management of Environmental Quality an International Journal, 34*(4), 1174–1197. https://doi.org/10.1108/meq-03-2022-0085

Ellen MacArthur Foundation, & Google. (2019). *Artificial intelligence and the circular economy: AI as a tool to accelerate the transition*. Ellen MacArthur Foundation. https://ellenmacarthurfoundation.org/articles/artificial-intelligence-and-the-circular-economy

European Commission. (2023). *Digital product passport proposal under the ecodesign for sustainable products regulation (ESPR)*.

Fang, B., Yu, J., Chen, Z., Osman, A. I., Farghali, M., Ihara, I., & Yap, P. S. (2023). Artificial intelligence for waste management in smart cities: A review. *Environmental Chemistry Letters, 21*(4), 1959–1989. https://doi.org/10.1007/s10311-023-01561-0

Fatima, S., Grover, P., Phutela, N., & Kandpal, V. (2024). Exploring the paradigm shift towards sustainability: A systematic literature review on circular economy and eco-innovation. *AIMS Environmental Science, 11*(6), 847–882. https://doi.org/10.3934/environsci.2024047

Ferreira, J., Elvas, L., Correia, R., & Mascarenhas, M. (2024). Enhancing ehr interoperability and security through distributed ledger technology: A review. *Healthcare, 12*(19), 1967. https://doi.org/10.3390/healthcare12191967

Garay-Rondero, C., Flores, J., Smith, N., Morales, S., & Aldrette-Malacara, A. (2019). Digital supply chain model in industry 4.0. *Journal of Manufacturing Technology Management, 31*(5), 887–933. https://doi.org/10.1108/jmtm-08-2018-0280

Giorgi, S. (2024). Assessment methods and digital tools to support circular building strategies. *Iop Conference Series Earth and environmental Science, 1402*(1), 012047. https://doi.org/10.1088/1755-1315/1402/1/012047

Gry, S., Niederlaender, M., Lodi, A. N., Mutz, M., & Werth, D. (2023). Advances in AI-based garment returns prediction and processing: A conceptual approach for an AI-based recommender system. In *ICSBT* (pp. 15–25).

Henninger, A., & Mashatan, A. (2021). Distributed interoperable records: The key to better supply chain management. *Computers, 10*(7), 89. https://doi.org/10.3390/computers10070089

Henriques, R., Figueiredo, F., & Nunes, J. (2023). Product-services for a resource-efficient and circular economy: An updated review. *Sustainability, 15*(15), 12077.

Herath, H. M. K. K. M. B., & Mittal, M. (2022). Adoption of artificial intelligence in smart cities: A comprehensive review. *International Journal of Information Management Data Insights, 2*(1), 100076. https://doi.org/10.1016/j.jjimei.2022.100076

Hong, Z., & Xiao, K. (2024). Digital economy structuring for sustainable development: The role of blockchain and artificial intelligence in improving supply chain and reducing negative environmental impacts. *Scientific Reports*. https://doi.org/10.1038/s41598-024-53760-3

IBM & Maersk. (2022, November 29). *TradeLens to be discontinued: Statement on the closure of the blockchain-enabled global trade platform* [Official announcement]. IBM Blockchain; A.P. Moller–Maersk. https://www.maersk.com/news

Isensee, C., Griese, K., & Teuteberg, F. (2021). Sustainable artificial intelligence: A corporate culture perspective. *Nachhaltigkeitsmanagementforum | Sustainability Management Forum, 29*(3–4), 217–230. https://doi.org/10.1007/s00550-021-00524-6

Jensen, S., Kristensen, J., Christensen, A., & Wæhrens, B. (2024). An ecosystem orchestration framework for the design of digital product passports in a circular economy. *Business Strategy and the Environment, 33*(7), 7100–7117. https://doi.org/10.1002/bse.3868

Jones, K. (2024). The future of fx trading: Exploring the intersection of ai, open innovation, and industry evolution. *Technology and Investment, 15*(02), 117–133. https://doi.org/10.4236/ti.2024.152007

Kagermann, H., Wahlster, W., & Helbig, J. (2013). *Securing the future of German manufacturing industry: Recommendations for implementing the strategic initiative Industrie 4.0.* Acatech – National Academy of Science and Engineering.

Khan, D., Jung, L. T., & Hashmani, M. A. (2021). Systematic literature review of challenges in blockchain scalability. *Applied Sciences, 11*(20), 9372. https://doi.org/10.3390/app11209372

Khinvasara, T., Ness, S., & Shankar, A. (2024). Leveraging ai for enhanced quality assurance in medical device manufacturing. *Asian Journal of Research in Computer Science, 17*(6), 13–35. https://doi.org/10.9734/ajrcos/2024/v17i6454

Kouhizadeh, M., & Sarkis, J. (2018). Blockchain practices, potentials, and perspectives in greening supply chains. *Sustainability, 10*(10), 3652. https://doi.org/10.3390/su10103652

Kowsalya, P., Jennifer Flora, G., Kousalya, A., Toha, M. A., Khan, P. A., Choudhury, M., Parthasarathi, T., et al. (2025). Circular systems unveiled: Essential components and their impact on recycling, reuse, and resource recovery. In S. Majumdar, M. Choudhury, & A. Cheshmehzangi (Eds.), *Sustainable urban future. Urban sustainability* (pp. 55–71). Springer. https://doi.org/10.1007/978-981-95-2976-6_3

Kshetri, N. (2017). Blockchain's roles in strengthening cybersecurity and protecting privacy. *Telecommunications Policy, 41*(10), 1027–1038. https://doi.org/10.1016/j.telpol.2017.09.003

Lasi, H., Fettke, P., Kemper, H. G., Feld, T., & Hoffmann, M. (2014). Industry 4.0. Business & Information. *Systems Engineering, 6*(4), 239–242. https://doi.org/10.1007/s12599-014-0334-4

Lăzăroiu, G., Bogdan, M., Geamănu, M., Hurloiu, L., Luminița, L., & Ștefănescu, R. (2023). Artificial intelligence algorithms and cloud computing technologies in blockchain-based fintech management. *Oeconomia Copernicana, 14*(3), 707–730. https://doi.org/10.24136/oc.2023.021

Lee, J., Bagheri, B., & Kao, H. A. (2015). A cyber-physical systems architecture for industry 4.0-based manufacturing systems. *Manufacturing Letters, 3*, 18–23. https://doi.org/10.1016/j.mfglet.2014.12.001

Lodhi, S., Hussain, H., & Hussain, I. (2024). Using ai to increase heat exchanger efficiency: An extensive analysis of innovations and uses. *International Journal of Multidisciplinary Sciences and Arts, 3*(4), 1–14. https://doi.org/10.47709/ijmdsa.v3i4.4617

Lubongo, C., Bin Daej, M. A. A., & Alexandridis, P. (2024). Recent developments in technology for sorting plastic for recycling: The emergence of artificial

intelligence and the rise of the robots. *Recycling, 9*(4), 59. https://doi.org/10.3390/recycling9040059
Luo, X. (2023). Artificial intelligence and corporate innovation: Intelligent transformation and development trends under technological empowerment. *Advances in Economics Management and Political Sciences, 45*(1), 201–207. https://doi.org/10.54254/2754-1169/45/20230285
Majumdar, S., Choudhury, M., & Cheshmehzangi, A. (2025). *Sustainable urban future: Embracing circular economy in waste management*. Springer Singapore. https://doi.org/10.1007/978-981-95-2976-6
McKinsey & Company. (2024a). *State of AI in 2024: Adoption, impact, and emerging risks*. https://www.mckinsey.com
McKinsey & Company. (2024b). *The circular economy: Moving from theory to practice*. McKinsey Global Institute.
MIIT. (2022). *China's intelligent manufacturing development plan*. Ministry of Industry and Information Technology.
Mondal, T., Choudhury, M., Kundu, D., Dutta, D., & Samanta, P. (2023). Landfill: An eclectic review on structure, reactions and remediation approach. *Waste Management, 164*, 127–142. https://doi.org/10.1016/j.wasman.2023.03.034
Narayan, S., Yadav, D., Jha, S., Khatun, B., Choudhary, R., Verma, U. K., Choudhury, M., & Kumar, D. (2025). From past to present: The evolution of waste management and the rise of circular economy principles. In *Sustainable urban future: Embracing circular economy in waste management* (pp. 23–53). Springer Nature Singapore. https://doi.org/10.1007/978-981-95-2976-6_2
Nguyen, D., Tuong, N., & Pham, H. (2022). A blockchain-based framework for developing traceability applications towards sustainable agriculture in Vietnam. *Security and Communication Networks, 2022*, 1–10. https://doi.org/10.1155/2022/1834873
Nham, N. T. H. (2022). Making the circular economy digital or the digital economy circular? Empirical evidence from the European region. *Technology in Society, 70*, 102023.
Nyström, T., Whalen, K., Diener, D., Hollander, M., & Boyer, R. (2021). Managing circular business model uncertainties with future adaptive design. *Sustainability, 13*(18), 10361. https://doi.org/10.3390/su131810361
Omdena. (2025). *Top AI companies transforming supply chains in 2025*. https://omdena.com
Özsoy, T. (2023). The role of artificial intelligence in facilitating the transition to a circular economy. *Nişantaşı Üniversitesi Sosyal Bilimler Dergisi, 11*(2), 369–389. https://doi.org/10.52122/nisantasisbd.1345828
Pal, R., Shen, B., & Sandberg, E. (2019). Circular fashion supply chain management: Exploring impediments and prescribing future research agenda. *Journal*

of Fashion Marketing and Management, 23(3), 298–307. https://doi.org/10.1108/jfmm-07-2019-166

Peyravi, B., Peleckis, K., Peleckis, K., Limba, T., & Peleckienė, V. (2024). The circular economy practices in the European Union: Eco-innovation and sustainable development. *Sustainability, 16*(13), 5473. https://doi.org/10.3390/su16135473

Precedence Research. (2025). *Digital circular economy market size to hit USD 25.40 Bn by 2034.*

Rajpal, A., Choudhury, M., Goswami, S., Chakravorty, A., & Raghavan, V. (Eds.). (2024). Waste management and treatment: Advances and innovations (1st ed.). CRC Press. https://doi.org/10.1201/9781003258377

Renault Group. (2020, September 10). Renault group launches XCEED: A blockchain solution to enhance vehicle traceability [Press release]. *Groupe Renault.* https://media.renaultgroup.com

Saberi, S., Kouhizadeh, M., Sarkis, J., & Shen, L. (2019). Blockchain technology and its relationships to sustainable supply chain management. *International Journal of Production Research, 57*(7), 2117–2135. https://doi.org/10.1080/00207543.2018.1533261

Sahoo, S., Kumar, S., Sivarajah, U., Lim, W. M., Westland, J. C., & Kumar, A. (2024). Blockchain for sustainable supply chain management: Trends and ways forward. *Electronic Commerce Research, 24*(3), 1563–1618.

Sculley, D., Holt, G., Golovin, D., Davydov, E., Phillips, T., Ebner, D., Dennison, D. et al. (2015). Hidden technical debt in machine learning systems. *Advances in Neural Information Processing Systems, 28.*

Sharma, A., Choudhury, M., Agarwal, S., Sharma, R., & Sharma, R. (2023). Corporate social responsibility and roles of developers for sustainability in companies. In *The route towards global sustainability: Challenges and management practices* (pp. 313–332). Springer International Publishing. https://doi.org/10.1007/978-3-031-10437-4_16

Successive Digital. (2024). Modernisation in manufacturing: Digital strategies for operational efficiency. *Successive Technologies.* https://successive.tech/blog/modernisation-in-manufacturing.

Taşcı, B., Omar, A., & Ayvaz, S. (2023). Remaining useful lifetime prediction for predictive maintenance in manufacturing. *Computers & Industrial Engineering, 184*, 109566. https://doi.org/10.1016/j.cie.2023.109566

Tsarouhas, P., & Papaevangelou, N. (2024). Critical steps and conditions to be included in a business model in logistics, seeking competitive advantage from the perspective of the modern digital age and industry 4.0. *Applied Sciences, 14*(7), 2701. https://doi.org/10.3390/app14072701

Tukker, A. (2004). Eight types of product–service system: Eight ways to sustainability? Experiences from SusProNet. *Business Strategy and the Environment, 13*(4), 246–260.

van Capelleveen, G., Vegter, D., Olthaar, M., & van Hillegersberg, J. (2023). The anatomy of a passport for the circular economy: A conceptual definition, vision and structured literature review. *Resources, Conservation & Recycling Advances, 17*, 200131. https://doi.org/10.1016/j.rcradv.2023.200131

Vatiero, M. (2024). The "dark side" of smart contracts: A contract theory perspective. *European Management Review, 21*(3), 516–519. https://doi.org/10.1111/emre.12676

Wang, Y., Singgih, M., Wang, J., & Rit, M. (2019). Making sense of blockchain technology: How will it transform supply chains? *International Journal of Production Economics, 211*, 221–236.

Wulandari, H. (2022). Cryptocurrency price volatility analysis on bitcoin and altcoins before and during the covid-19 pandemic in Indonesia. *Journal of International Conference Proceedings, 5*(5). https://doi.org/10.32535/jicp.v5i5.2008

Yerram, S. (2022). Smart contracts for efficient supplier relationship management in the blockchain. *American Journal of Trade and policy, 9*(3), 119–130. https://doi.org/10.18034/ajtp.v9i3.700

Zhang, Z. (2024). The application of blockchain technology in enterprise value chain cost management. *Pacific International Journal, 7*(1), 143–149. https://doi.org/10.55014/pij.v7i1.541

CHAPTER 15

Governance for Sustainability: Designing Policies for Tomorrow's Urban Challenges

Saurav Narayan, Himanshu, Basona Khatun, Pallav Suman, Chandra Kant Raj, Deepak Kumar, and Moharana Choudhury

15.1 Introduction

The twenty-first century has been marked by rapid and complex urban transformation, characterised by intensified demographic pressures, widening socioeconomic divides, and escalating environmental uncertainties (Angel et al., 2011; United Nations, Department of Economic and Social Affairs, Population Division, 2019; Rajpal et al., 2024). Cities have increasingly become strategic spaces where global challenges intersect

S. Narayan • B. Khatun (✉)
Department of Law, School of Legal Studies, Central University of Punjab, Bathinda, Punjab, India

Himanshu
Department of English, University of Delhi, Delhi, India

P. Suman
Department of Chemistry and Biotechnology, Tallinn University of Technology, Tallinn, Estonia
e-mail: pallav.suman@taltech.ee

S. Majumdar, M. Choudhury (eds.), *City of the Future*,
https://doi.org/10.1007/978-3-032-18026-1_15

with local realities. They function simultaneously as engines of economic dynamism and as central points of vulnerability, making urban governance a decisive factor in shaping humanity's collective future. With projections indicating that nearly 70% of the world's population will reside in urban areas by 2050, the demands placed on infrastructure networks, public services, and governance institutions are set to intensify at an unprecedented scale (United Nations, Department of Economic and Social Affairs, Population Division, 2019). This demographic path underscores the urgency of transforming how cities are planned, governed, and made robust in the face of climate instability, resource stress, and accelerating social inequalities.

Within this evolving global landscape, sustainable urban governance has emerged not simply as an aspirational principle but as a structural necessity for balancing economic growth, ecological stewardship, and social wellbeing (Meuleman, 2019; United Nations Human Settlements Programme, 2015). The Sustainable Development Goals (SDGs) provide a comprehensive normative framework for guiding this transformation (Bhattacharjee & Narayan, 2024). SDG 9 advances the agenda for resilient infrastructure and innovation-driven development; SDG 11 provides for inclusive, safe, resilient, and sustainable cities; and SDG 16 emphasizes the imperative of accountable, transparent, and participatory institutions (United Nations, 2015a, b, c, d). Collectively, these goals offer a synergistic blueprint for reconfiguring urban governance systems to address complex, interlinked challenges.

Responding to these global imperatives, this chapter develops a multidisciplinary examination of the forces shaping contemporary urban environments and identifies pathways for designing policies aligned with sustainability and resilience objectives. The opening sections investigate

C. K. Raj
College of International Education, Jinzhou Medical University, Jinzhou, Liaoning Province, People's Republic of China

D. Kumar
Dnipro State Medical University, Dnipro, Ukraine

M. Choudhury
Environmental Research and Management Division, Voice of Environment (VoE), Guwahati, Assam, India

the structural underpinnings of urban complexity, including demographic expansion, spatial inequality, infrastructure deficits, and climate-related vulnerabilities, which illustrate why piecemeal or sector-specific solutions are insufficient (United Nations Human Settlements Programme, 2015; Satterthwaite et al., 2020). To complement this analysis, the chapter employs a bibliometric assessment using the Scopus database, mapping global research trends across fields such as urban planning, sustainability transitions, circular economy approaches, and policy innovation. This evidence provides insight into dominant scholarly themes and methodological shifts, demonstrating how academic discourse informs and shapes policy decision-making.

Building on this conceptual and empirical foundation, the chapter explores how urban governance can be strategically aligned with the SDGs. By examining SDGs 9, 11, and 16, it outlines an integrated governance model grounded in cross-sectoral coordination, community participation, digital innovation, and robust institutional frameworks (Meuleman, 2019; United Nations Development Programme, 2014). These theoretical discussions are illustrated through comparative case studies across Singapore, Curitiba, and Barcelona cities that exemplify diverse yet transferable models of resilient infrastructure development, inclusive mobility, and data-enabled participatory governance (Acuto et al., 2018). Their paths show that long-term vision, institutional coherence, and civic engagement are central to sustainable urban transformation.

The chapter also highlights persistent barriers to urban governance systems, including institutional fragmentation, political contestation, fiscal constraints, digital divides, and risks of exclusion (United Nations Human Settlements Programme, 2015; Satterthwaite et al., 2020). Understanding these constraints is crucial to designing grounded, resilient policy responses. The final sections propose actionable strategies to enhance institutional capacity, promote ethical and inclusive technological adoption, and strengthen adaptive monitoring and evaluation frameworks. Through this multidimensional analysis, the chapter positions sustainable urban governance as a guiding compass for cities navigating the pressures of the present while preparing for the uncertainties of the future (Fig. 15.1).

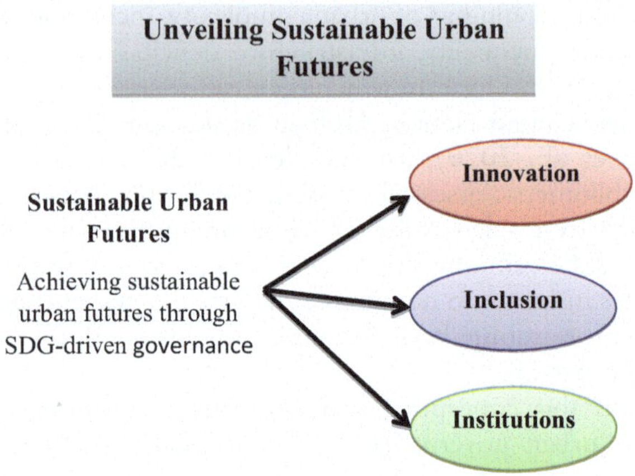

Fig. 15.1 Urban governance for sustainability

15.2 Bibliometric Insights into Urban Planning, Sustainability, and Circular Economy

To evaluate the current state of research on urban planning, designing, policies, sustainability, and the circular economy at the international level, a bibliometric analysis was conducted using the Scopus database. The search query employed was: TITLE-ABS-KEY (urban AND planning OR designing AND policies OR sustainability OR circular AND economy) AND PUBYEAR > 2014 AND PUBYEAR < 2025. The analysis focused on publications from 2015 to 2024 to capture recent trends and emerging themes in the field. Key findings obtained from the search results are graphically presented in Figs. 15.2, 15.3, and 15.4.

An analysis of publication output in the Scopus database was conducted to track research trends in urban planning, policy design, sustainability, and the circular economy. The data reveal a consistent growth in scholarly interest over the past decade. Beginning with 169 publications in 2015, the number rose steadily, peaking at 514 in 2024. Notable increases occurred after 2019, indicating a surge in global attention to the intersection of urban development and sustainability post-pandemic. The years 2022 to 2024 alone accounted for nearly 43% of all publications during the period studied, highlighting strong, accelerating momentum in the field (Fig. 15.2).

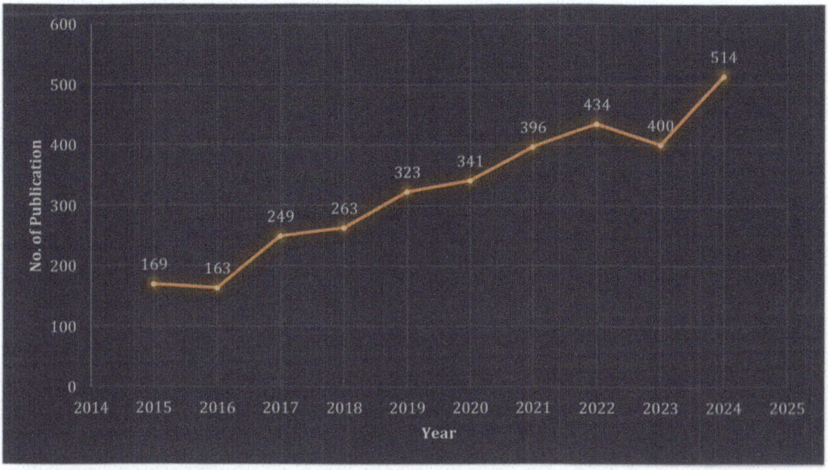

Fig. 15.2 Yearly publication trends (2015–2024)

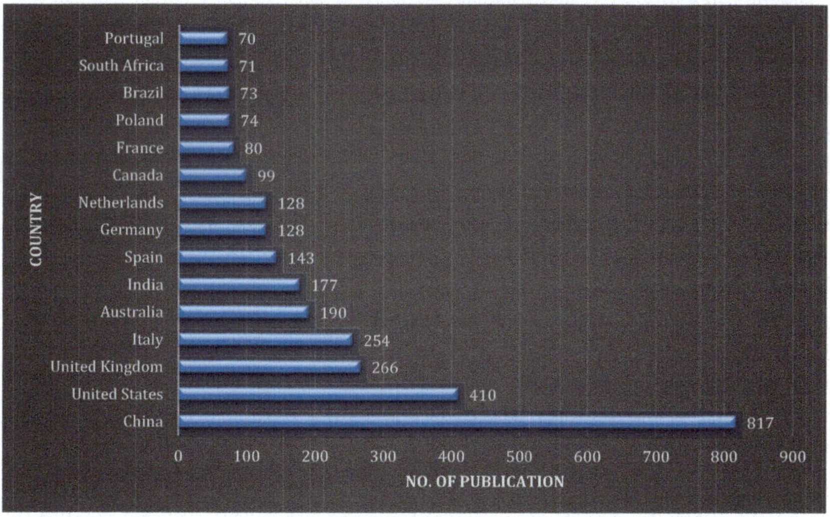

Fig. 15.3 Number of publications on *urban planning*, *designing policies*, *sustainability*, and *the circular economy* (2015–2024) by country

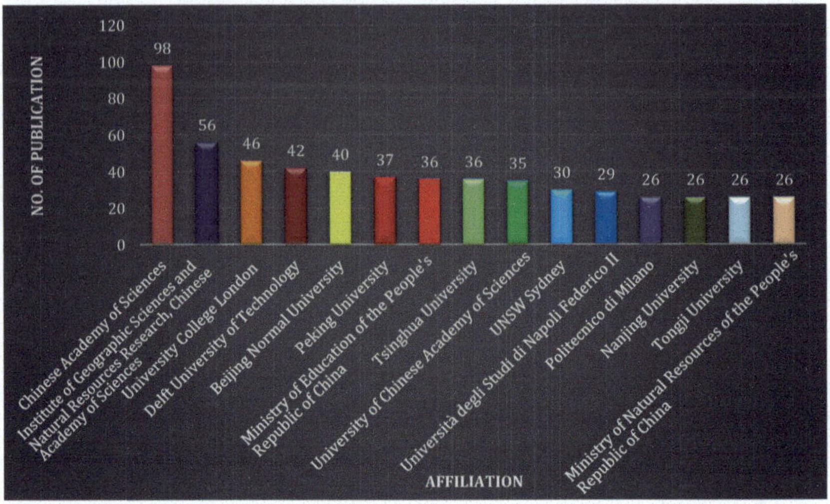

Fig. 15.4 Affiliation-wise research on *Urban Planning, Designing Policies, Sustainability, and Circular Economy*

The country-wise distribution of research on urban planning, policy design, sustainability, and the circular economy from 2015 to 2024 shows a clear dominance by China, which produced 817 research documents, making it the most prolific contributor in this domain. The United States ranks second with 410 publications, while the United Kingdom ranks third with 266 publications. Italy contributed 254 research papers, placing it closely behind the United Kingdom. Australia stands fifth with 190 publications, ahead of India, which produced 177 documents during the same period. Other notable contributors include Spain (143 documents), Germany (128 documents), and the Netherlands (also 128 documents). Canada made a significant contribution with 99 documents. France (80 documents), Poland (74), Brazil (73), South Africa (71), and Portugal (70) round out the top 15 countries. Collectively, these countries account for 2,980 research documents on the topic (Fig. 15.3).

From 2015 to 2024, an analysis of global research output highlights the leading institutions that have contributed significantly to scholarly work in the fields of Urban Planning, Policy Design, Sustainability, and the Circular Economy. The Chinese Academy of Sciences stands out as the top contributor, publishing 98 research documents during this period. It is followed by the Institute of Geographic Sciences and Natural Resources Research, Chinese Academy of Sciences, with 56 publications. University College London is also a major contributor, with 46 documents. Delft University of Technology leads with 42 research outputs, followed by Beijing Normal University with 40. Peking University recorded 37 publications, and both the Ministry of Education of the People's Republic of China and Tsinghua University contributed 36 each. The University of Chinese Academy of Sciences produced 35 documents, while UNSW Sydney published 30. Università degli Studi di Napoli Federico II followed with 29 documents. Politecnico di Milano, Nanjing University, Tongji University, and the Ministry of Natural Resources of the People's Republic of China each contributed 26 research publications, reflecting a shared position among notable contributors (Fig. 15.4).

This bibliometric analysis underscores the growing global engagement with urban planning, sustainability, and the circular economy over the past decade. The marked increase in publications, particularly from 2020 onward, reflects a heightened awareness and academic response to pressing environmental and socioeconomic challenges, especially in the post-pandemic context. China has emerged as the leading contributor both at the national and institutional levels, followed by strong performances from the United States, the United Kingdom, and key European and Asia-Pacific countries. The presence of a diverse set of high-performing institutions further highlights the international scope and interdisciplinary nature of this research area. As momentum continues, this trend signals an evolving, dynamic knowledge base poised to inform more sustainable urban development policies and practices worldwide.

15.3 The Urban Challenge: Context and Complexity

Urban areas today sit at the intersection of demographic, environmental, and socioeconomic pressures that shape their development paths. As cities expand, they become hubs of innovation and economic opportunity, yet also sites of deep inequality, ecological strain, and governance challenges. Addressing urban complexity requires understanding how population growth, climate risks, and structural inequities overlap and reinforce one another. These interlocking issues highlight the urgency for integrated, inclusive, and sustainable urban planning approaches.

15.3.1 Demographic Pressures and Infrastructure Gaps

Rapid urban population growth has strained urban systems in both developing and developed contexts (United Nations, Department of Economic and Social Affairs, Population Division, 2019; World Bank, 2021). As millions migrate to urban centres in search of better livelihoods, cities struggle to provide adequate housing, efficient transportation, clean water, and sanitation (World Bank, 2021). The pace of urbanization often exceeds municipal governments' capacity to expand infrastructure networks, leading to service delivery deficits and overburdened public utilities.

Often lacking basic services, informal settlements proliferate as a response to housing shortages, leading to spatial inequality and socio-environmental vulnerabilities (Angel et al., 2011; Satterthwaite et al., 2020; UN-Habitat, 2020). These settlements frequently emerge on peripheral or hazard-prone land where formal development is limited, reinforcing long-term cycles of marginalization. Additionally, rising land prices and inadequate regulatory frameworks often push low-income populations into substandard housing, further deepening urban inequality. The cumulative effect is a fragmented urban landscape where access to essential services becomes uneven and socially stratified.

15.3.2 Environmental and Climate Risks

Urban environments significantly contribute to climate change and are highly vulnerable to its effects (Intergovernmental Panel on Climate Change, 2023; United Nations Human Settlements Programme (UN-Habitat), 2015). Dense concentrations of infrastructure, vehicles, and industrial activity intensify environmental pressures, particularly in rapidly urbanizing regions. Urban heat islands, air and water pollution, and increased flood risks due to impervious surfaces threaten the sustainability of cities (Intergovernmental Panel on Climate Change, 2023). Replacing natural landscapes with built environments reduces ecological resilience and heightens exposure to extreme weather events.

Climate change intensifies these risks, disproportionately affecting low-income and marginalized communities who often reside in disaster-prone areas (Intergovernmental Panel on Climate Change, 2023; Satterthwaite, 2013; United Nations, 2015a, b, c, d). These populations typically lack the resources to adapt, relocate, or recover from climate-related shocks, reinforcing social and spatial vulnerabilities. Moreover, weak urban planning, outdated drainage systems, and insufficient green infrastructure increase the likelihood that climate hazards translate into large-scale disasters. The compounding nature of environmental risks underscores the necessity for climate-responsive and equitable urban development strategies.

15.3.3 Socioeconomic Inequities

Cities also emphasize economic opportunities and inequalities (Organisation for Economic Co-operation and Development [OECD], 2016; UN-Habitat, 2020). While urban areas often serve as engines of growth, their benefits are unevenly distributed, creating stark contrasts between affluent neighbourhoods and underserved communities. Marginalized populations frequently experience limited access to education, healthcare, and employment, reinforcing cycles of poverty (United Nations, 2015a, b, c, d; UN-Habitat, 2020). These barriers reduce social mobility and restrict individuals from fully participating in urban economic life.

Spatial segregation driven by housing markets, planning decisions, and historical marginalization intensifies these disparities, making inclusive urban development a critical imperative (Kabisch & Haase, 2014). In many cities, discriminatory zoning practices, inadequate affordable housing policies, and speculative real estate dynamics push vulnerable groups into isolated or deteriorating areas. This spatial divide not only affects economic prospects but also shapes exposure to environmental hazards, access to public services, and overall well-being. Addressing socioeconomic inequities, therefore, requires coordinated policies that integrate housing, social welfare, transportation, and labour market reforms to create more inclusive and resilient urban futures (Fig. 15.5).

15.3.4 The Governance–Practice Gap: Why Implementation Often Fails

Despite growing global consensus on the importance of sustainability-oriented urban governance, many cities continue to struggle to translate policy aspirations into tangible outcomes (Almulhim & Yigitcanlar, 2025). Urban governance systems often operate within a gap between what is designed and what is delivered, shaped by political cycles, resource asymmetries, and varying administrative capacities (Asadzadeh et al., 2023;

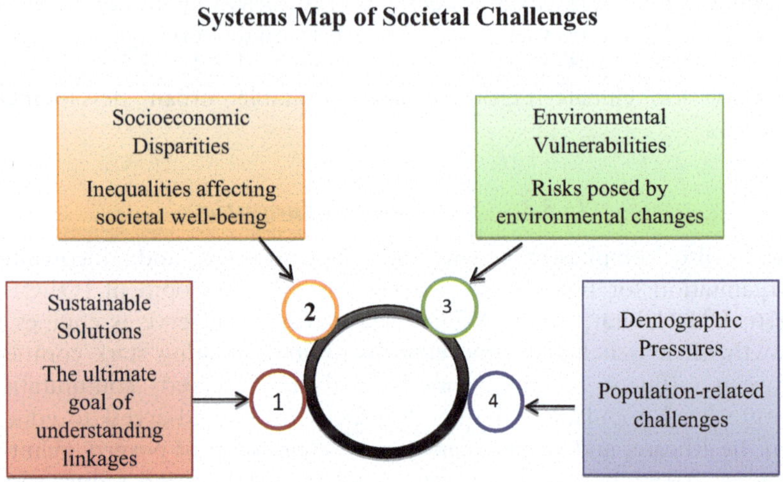

Fig. 15.5 Dimensions of the urban challenge

Secchi et al., 2024). Even when national frameworks articulate long-term sustainability commitments, local authorities may lack the technical expertise, institutional autonomy, or financial instruments needed to implement them (Kuzemko & Britton, 2020). As a result, well-intentioned strategies frequently remain confined to planning documents, unable to penetrate the operational routines of city departments or the socioeconomic contexts of neighbourhoods.

Another dimension of this governance–practice gap lies in the uneven distribution of power among urban stakeholders. Large infrastructure projects, for instance, are often shaped by private actors with substantial economic leverage, whereas community needs enter the conversation much later, if at all (Razzano & Bernardi, 2024). This imbalance impedes inclusive planning and can undermine public trust, particularly in contexts where marginalized groups already perceive decision-making processes as distant or opaque. Addressing this gap, therefore, requires not only institutional reforms but also cultural change: cities must cultivate governance approaches that value collaboration, transparency, and iterative learning. Without these shifts, even the most innovative policy frameworks may fail to achieve meaningful or equitable urban transformation (Eakin et al., 2022).

15.4 Aligning Urban Governance With the SDGs

Urban governance plays a central role in shaping sustainable, resilient, and equitable cities (Meyer & Auriacombe, 2019). The SDGs provide a holistic framework that guides local governments in addressing interconnected challenges such as infrastructure deficits, environmental stress, and social inequalities (Guha & Chakrabarti, 2019). Aligning urban governance with SDG principles enables policymakers to integrate innovation, inclusivity, and environmental stewardship into city planning and management (Reddy, 2016). By linking institutional reform with infrastructure development and community participation, cities can build systems that respond effectively to the complexities of urbanization. This multi-dimensional alignment helps create urban environments that are both future-ready and socially just (Fig. 15.7).

15.4.1 SDG 9: Resilient Infrastructure and Innovation

SDG 9 emphasizes the development of quality, reliable, and sustainable infrastructure to foster inclusive industrialization and innovation (Abdel Aziz et al., 2024). This goal is critical in cities where infrastructure underpins economic activity and daily life. Investments in smart mobility, renewable energy grids, digital infrastructure, and low-carbon construction significantly enhance urban resilience by reducing environmental impacts and improving the efficiency of service delivery. These innovations not only mitigate emissions but also support job creation, stimulate local economies, and elevate urban quality of life.

For example, Kigali's investment in green industrial parks demonstrates the alignment of technology and industry with environmental and social goals (Negesa et al., 2023). Such projects illustrate how eco-friendly industrialization can reduce resource consumption, promote circular economy practices, and create greener employment pathways. They also highlight the importance of integrating sustainability principles into infrastructure development to prepare cities for rapid population growth and climate-related disruptions.

15.4.2 SDG 11: Sustainable Cities and Communities

SDG 11 targets safe, affordable housing, accessible transport, green public spaces, and disaster resilience (United Nations, 2015a, b, c, d). Achieving these objectives requires comprehensive, inclusive urban planning that recognizes the diverse needs of city residents. Participatory urban planning and ecological urbanism are key tools in this effort, enabling communities to contribute to decision-making processes and ensuring that development outcomes reflect shared priorities. Examples from different global contexts demonstrate the transformative potential of SDG 11-aligned initiatives. Community-driven housing in Thailand shows how resident-led processes can deliver secure, affordable homes while strengthening social networks. Urban farming in Detroit revitalizes vacant land, enhances food security, and promotes community resilience. Nature-based flood solutions in Rotterdam illustrate how integrating ecological design into urban landscapes can reduce flood risks and enhance environmental quality (UN-Habitat, 2020). These cases highlight how sustainable planning strengthens both environmental resilience and social inclusion.

15.4.3 SDG 16: Institutions, Justice, and Inclusion

SDG 16 calls for transparent, accountable, and inclusive governance (United Nations, 2015a, b, c, d). Strong institutions are essential in cities because they manage complex systems such as mobility, housing, utilities, and emergency services. Strengthening the rule of law, fighting corruption, and fostering citizen participation are fundamental for implementing SDG-aligned policies, as effective governance ensures that public resources are used responsibly and equitably.

Civic technology platforms, open data initiatives, and deliberative budgeting mechanisms, as seen in cities like Seoul and Barcelona, embody these principles (United Nations, 2022). These innovations allow citizens to participate in decision-making, hold institutions accountable, and influence public spending priorities. By promoting transparency and inclusiveness, such governance tools help build trust between governments and communities, which is essential for sustaining long-term urban development efforts aligned with the SDGs.

15.4.4 Interlinkages and Integration

Integrating SDGs 9, 11, and 16 reveals synergies that enhance policy coherence. Sustainable infrastructure development (SDG 9) requires transparent and equitable governance systems (SDG 16) to ensure projects benefit all residents, while also maintaining environmental

Fig. 15.6 A tree-lined road demonstrating sustainable, well-governed infrastructure integrated with natural landscapes. (Photo by: Moharana Choudhury)

sustainability and resilience (SDG 11) (Fei et al., 2021). Alignment across these goals strengthens cities' ability to deliver long-term, inclusive development outcomes.

Participatory governance fosters local ownership of urban infrastructure, ensuring that investments reflect community priorities and enhance sustainability (Fig. 15.6). Meanwhile, innovation—whether technological, institutional, or social—boosts efficiency in public administration and supports adaptive management (Sebunya & Gichuki, 2024). This systems-thinking approach enables cities to respond holistically to interconnected challenges such as climate change, inequality, and rapid urbanization. It reinforces the importance of integrated urban governance as a pathway towards achieving multiple SDGs simultaneously (Fig. 15.7).

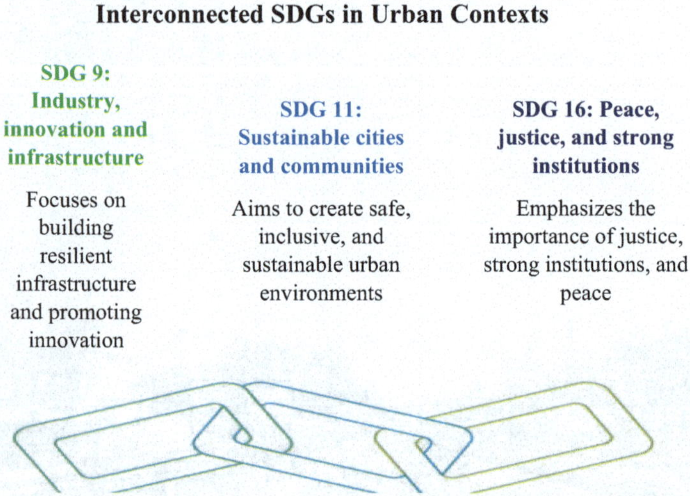

Fig. 15.7 SDG synergy wheel

15.4.5 Urban Governance in a Multipolar World: Emerging Geopolitical Pressures

Contemporary urban governance is also increasingly shaped by geopolitical dynamics that extend far beyond municipal boundaries (Essien & Jesse, 2025). Cities today operate in a multipolar global environment marked by shifting alliances, competing technological systems, and new forms of transnational influence (Da Cruz et al., 2019). As nations race to lead in digital infrastructure, renewable energy technologies, and climate adaptation solutions, urban areas often become the testing grounds for competing models of development. The diffusion of "*city diplomacy*," in which municipalities negotiate directly with international networks, donors, and private technology firms, further complicates the governance landscape (Muhammad, 2023). While such engagement can accelerate access to innovation and finance, it may also expose cities to geopolitical dependencies that are not always aligned with local priorities.

These pressures require urban leaders to navigate an increasingly complex field of actors, interests, and norms. For example, smart-city technologies sourced from different global suppliers carry with them not only technical specifications but also distinct governance philosophies, ranging from data-sovereignty models to surveillance-heavy administrative cultures (Anthony & Sarshar, 2025; Maier & Weinberger, 2024). Similarly, climate finance mechanisms, though essential for adaptation and mitigation, can impose reporting requirements or investment conditions that privilege specific development pathways over others (Wabwile & Ondabu, 2025; UNFCCC, 2025). In this context, sustainable urban governance must incorporate a geopolitical sensibility, ensuring that cities maintain autonomy in decision-making while engaging constructively with global partners (Medeiros & Van Der Zwet, 2020). Strengthening local policy intelligence, diversifying international collaborations, and fostering transparent procurement practices can help safeguard urban futures in an era where global politics increasingly intersects with local planning.

15.5 Principles of Effective Policy Design for Urban Sustainability

Effective urban sustainability policy requires an integrated approach that links legislation, governance, technology, and community participation. As cities face mounting pressures from climate change, population growth, and resource constraints, policy design must be both forward-looking and adaptive (Graham, 2025; Feng et al., 2025). Aligning legal frameworks with participatory processes and technological innovation enables urban systems to respond to complex, multidimensional challenges. These principles ensure that sustainability is embedded not only in planning documents but also in daily governance practices and long-term institutional strategies. Together, they create a foundation for cities to transition towards equitable, resilient, and environmentally responsible futures (Fig. 15.9).

15.5.1 Legislative Frameworks and Multidimensional Integration

Practical urban sustainability requires coherent legal structures integrating economic, environmental, and social priorities (OECD, 2017a, b). Legislative frameworks (Narayan & Singha, 2024) must be updated to include carbon-neutrality targets, enforce zoning regulations that promote density and mixed land use, and incentivize green infrastructure (Gogoi et al., 2024). These frameworks provide the stability and direction necessary for long-term planning, ensuring that urban development aligns with climate commitments and sustainability objectives.

For instance, municipal adaptation processes in Constance, Germany, exemplify the integration of climate information into urban planning (Harrs et al., 2024). Such cases demonstrate how embedding climate data into regulatory systems enables evidence-based decision-making and supports locally tailored adaptation strategies. Multidimensional policy alignment ensures that land-use planning, housing policy, and transport systems reinforce rather than conflict (OECD, 2017a, b). In practice, this integration prevents fragmented governance and promotes holistic urban development in which environmental safeguards, economic vitality, and social well-being mutually support one another.

15.5.2 Stakeholder Engagement and Participatory Governance

Robust policy design must be participatory and inclusive, involving citizens, civil society, academia, and the private sector in co-design. Engaging diverse stakeholders ensures that policies reflect community needs and incorporate expertise from multiple perspectives, thereby enhancing their legitimacy and effectiveness. Participatory budgeting initiatives in Porto Alegre and neighbourhood planning in London demonstrate the efficacy of citizen engagement in tailoring solutions to local needs (Fung, 2015). These models show how collaborative processes can allocate resources more equitably and strengthen local democratic practices.

Inclusive governance reduces the implementation gap by fostering trust, transparency, and shared responsibility. When residents have meaningful input, they are more likely to support, adopt, and sustain urban sustainability initiatives. Moreover, participatory governance helps identify context-specific challenges such as unequal access to services or environmental risks that top-down policies often overlook. Through iterative dialogue and co-production, cities can build policies that are not only technically sound but socially embedded (Routh et al., 2025).

15.5.3 Technology, Innovation, and Data-Driven Approaches (Fig. 15.8)

Technology serves as a critical enabler of urban policy innovation (Fig. 15.6). Smart grids, the Internet of Things (IoT), Geographic Information Systems, and Artificial Intelligence (AI) improve efficiency in

Fig. 15.8 Harmonious coexistence: Urban infrastructure blending seamlessly with natural greenery for sustainable development. (Photo by: Moharana Choudhury)

transport, energy, and waste management. In India, the Smart Cities Mission leverages integrated command centres to monitor urban functions in real time (Ministry of Housing & Urban Affairs, 2018). Moreover, open data platforms allow for greater transparency, accountability, and informed citizen participation (Fig. 15.9).

15.6 Learning from Global Experiences: Case Studies

Urban systems worldwide offer a rich repertoire of strategies for building resilience, sustainability, and citizen-centric governance. Learning from diverse contexts enhances policymakers' ability to design contextually appropriate responses to complex urban challenges. This section examines three international examples, Singapore, Curitiba, and Barcelona, which have each adopted distinctive yet mutually reinforcing approaches to

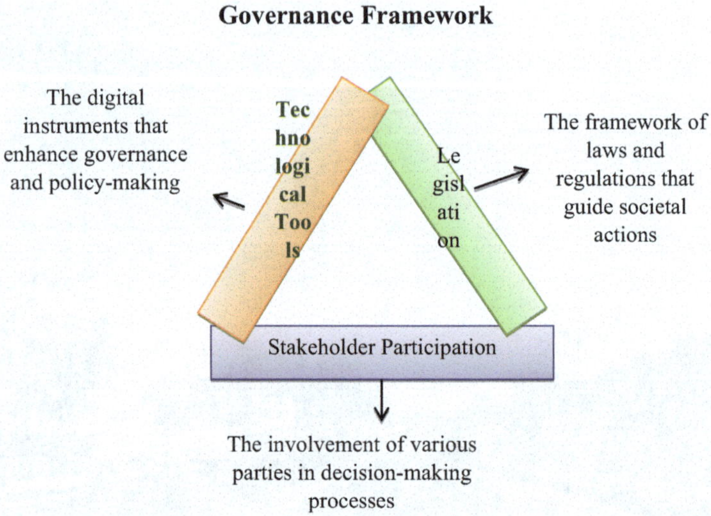

Fig. 15.9 Pillars of policy design for urban sustainability

infrastructure planning, mobility design, and digital governance. Together, these cases illustrate how long-term vision, institutional coherence, and adaptive capacity shape the trajectory of successful urban transformation. They also demonstrate that no single model is universally applicable; instead, cities must selectively adapt global experiences to their own socio-political and environmental realities.

15.6.1 Singapore's Resilient Infrastructure Model

Singapore presents a highly coordinated and forward-looking model of resilient infrastructure development, shaped by acute land scarcity and environmental vulnerability. The city-state has pursued extensive land reclamation and vertical expansion to secure space for housing, industry, and social amenities, thereby embedding compact, vertically layered urbanism into its long-term development trajectory (Chan et al., 2018). Complementing this are widespread green roofs, sky parks, and high-density spatial configurations that enhance energy efficiency and mitigate urban heat island effects (Tan et al., 2013).

A defining feature of Singapore's resilience strategy is its integration of hydrological planning and landscape design. The Public Utilities Board's Active, Beautiful, Clean Waters (ABC Waters) Programme transforms conventional drainage structures into multifunctional blue-green systems. Naturalised canals, bioswales, and detention ponds collectively serve ecological, aesthetic, and social purposes, reinforcing resilience while enhancing public spaces (Chan et al., 2018; Public Utilities Board, 2018). This synthesis of engineering, ecological design, and participatory urbanism demonstrates how integrated planning frameworks can deliver robust climate preparedness alongside high-quality urban environments.

15.6.2 *Curitiba's Sustainable Urban Planning*

Curitiba, Brazil, offers an enduring example of integrated, socially responsive, and cost-effective urban planning. The city's Bus Rapid Transit (BRT) system, introduced in the 1970s, established a paradigm for high-capacity surface transit that remains globally influential for its affordability, speed, and operational efficiency (Asimeng & Jauregui-Fung, 2025). Crucially, Curitiba's planners aligned land-use zoning with the BRT corridors, guiding urban density towards linear growth patterns and thereby reducing car dependence and travel times (Rabinovitch & Leitman, 2015).

Curitiba also developed pioneering waste management and recycling programmes that link ecological responsibility with social inclusion. Through community outreach, civic education, and incentive schemes such as exchanging recyclable materials for food or transport vouchers, the city achieved high public participation and strengthened environmental stewardship (Rabinovitch & Leitman, 2015). These innovations illustrate how governance systems rooted in accessibility, affordability, and civic engagement can generate long-term sustainability outcomes while reinforcing social equity.

15.6.3 *Barcelona's Smart City Initiatives*

Barcelona exemplifies a mature approach to digital urbanism that combines technological sophistication with strong democratic norms. The city employs digital twins, extensive sensor networks, and Internet of Things (IoT) applications to optimize urban services, including energy management, water distribution, waste collection, and mobility planning (Batty et al., 2012). These tools facilitate real-time analysis and predictive modelling, enabling urban managers to address environmental pressures and operational inefficiencies with greater precision.

What distinguishes Barcelona's smart city model is its commitment to participatory governance. The Decidim platform institutionalizes digital democracy by enabling residents to propose, deliberate, and vote on policy initiatives, thereby embedding civic oversight within technological innovation (Morozov & Bria, 2018; Barandiaran et al., 2024). This combination of advanced data-driven management and robust participatory mechanisms represents a deliberate shift away from technocratic smart city paradigms towards more inclusive, rights-based digital governance.

15.6.4 Synthesis and Lessons Learned

The experiences of Singapore, Curitiba, and Barcelona demonstrate that multiple pathways exist for achieving sustainable and resilient urban development. Singapore emphasizes climate-resilient infrastructure, water-sensitive urbanism, and strong institutional coordination (Kumaraswamy et al., 2025). Curitiba foregrounds integrated mobility planning and socially oriented environmental programmes that strengthen inclusivity and reduce spatial inequalities. Barcelona illustrates how technological innovation, when paired with participatory governance, can enhance transparency, service delivery, and civic trust (Barandiaran et al., 2024; Morozov & Bria, 2018).

Despite their varied contexts, these cases converge on the central principles of integration, long-term vision, institutional coherence, and stakeholder participation. They also reveal that successful models emerge from iterative experimentation, adaptive regulation, and the alignment of infrastructure, technology, and community needs. For policymakers, the key lesson is not replication but contextual adaptation: global exemplars must be interpreted through local environmental conditions, regulatory capacities, and socio-political dynamics (Batty et al., 2012).

15.7 Overcoming Barriers to Implementation

Efforts to transition towards sustainable and resilient urban systems inevitably confront a range of structural, institutional, financial, and social barriers. These challenges are often deeply embedded in cities' governance architecture and are shaped by historical trajectories, political economies, and administrative norms. Understanding these constraints is essential to designing reform pathways that are both feasible and durable. This section outlines three major categories of barriers: institutional and political constraints, financial and resource limitations, and the challenge of ensuring equity, drawing on global examples to demonstrate how cities navigate and negotiate these obstacles.

15.7.1 Institutional and Political Challenges

Institutional fragmentation remains one of the most pervasive obstacles to effective urban governance. Many cities operate within governance systems characterized by overlapping mandates, unclear jurisdictional boundaries, and weak horizontal and vertical coordination, all of which impede coherent planning and policy execution (United Nations Human Settlements Programme [UN-Habitat], 2020). Bureaucratic silos and entrenched administrative hierarchies frequently obstruct cross-sectoral collaboration, while political contestations between local and national authorities create further uncertainty. In several metropolitan regions, reforms designed to accelerate sustainability transitions are slowed or derailed by bureaucratic inertia, patronage networks, and vested interests within urban political economies (World Bank, 2015).

Examples from the Global South vividly illustrate these challenges. In Nairobi, fragmented responsibilities between county and national agencies have diminished regulatory coherence, particularly in transport and land management (Swapan & Khan, 2021). Similarly, governance in Dhaka is hindered by overlapping institutional jurisdictions, leading to fragmented service delivery and inadequate metropolitan planning. These patterns echo earlier observations in African and Asian cities where unclear administrative mandates and political interference compromise long-term urban development strategies (Watson, 2009). Addressing these institutional deficiencies requires structural reforms, enhanced inter-agency coordination, and depoliticized decision-making.

15.7.2 Financial and Resource Limitations

Sustainable urban transformation demands substantial financial investment, yet many municipalities face chronic fiscal constraints that limit their ability to implement large-scale infrastructure or environmental projects. Restricted local tax bases, statutory borrowing limits, and heavy dependence on national government transfers reduce municipalities' fiscal autonomy and planning horizon (United Nations Human Settlements Programme [UN-Habitat], 2020). Urban authorities often struggle to secure stable revenue streams for climate resilience, mobility improvements, or basic service upgrades, making them reliant on donor financing or ad hoc project-based funding (Suzuki et al., 2013).

In response to these constraints, cities have increasingly explored innovative financing mechanisms. Municipal green bonds, climate finance instruments, and various forms of public–private partnerships represent alternative avenues for mobilizing capital while diversifying funding sources. Johannesburg's green bond, for example, demonstrates how cities can leverage capital markets to raise funds for low-carbon transport, renewable energy, and energy-efficiency initiatives (Ngwenya & Simatele, 2020). Such instruments not only expand financial capacity but also enhance municipal visibility in global climate finance architectures. Nevertheless, achieving equitable and accountable financial innovation requires strong regulatory oversight and robust public governance.

15.7.3 *Ensuring Equity in Policy Outcomes*

Equity constitutes a central dimension of sustainable urban development, yet it is often the most difficult to achieve in practice. Without explicit redistributive safeguards, sustainability initiatives can inadvertently exacerbate socio-spatial inequality, trigger gentrification, or reinforce exclusionary urban patterns (Anguelovski et al., 2018). Digitalization efforts risk deepening the digital divide, while green neighbourhood upgrades may displace long-standing low-income residents due to rising land values (Shirazi & Keivani, 2017). These dynamics erode public trust in urban governance and undermine the legitimacy of sustainability agendas.

Policy instruments such as inclusive zoning regulations, targeted subsidies for vulnerable groups, and community benefits agreements offer tangible pathways to align sustainability goals with equity considerations (United Nations Human Settlements Programme [UN-Habitat], 2020). Medellín, Colombia, provides a compelling example: through integrated urban transformation initiatives grounded in social equity, such as cable-car transit, library parks, and participatory planning, the city reduced historical marginalization and improved accessibility for low-income communities (Dávila, 2013). Such experiences underscore the need for

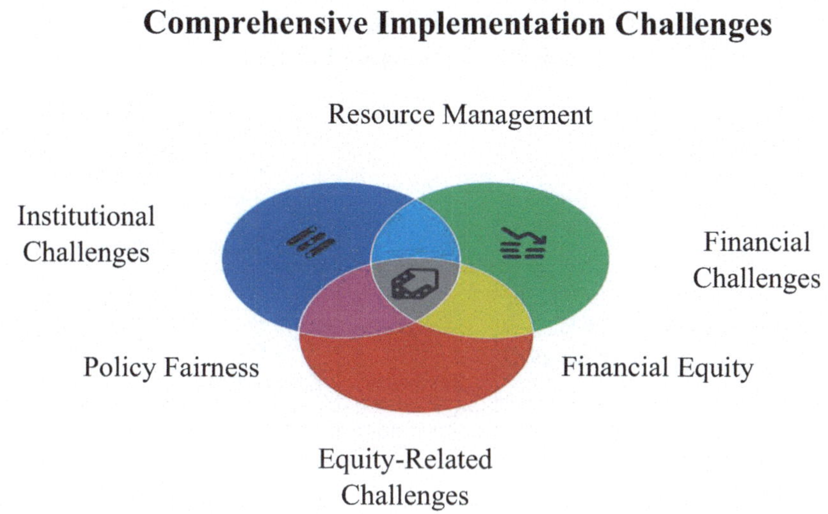

Fig. 15.10 Barriers to implementation

equity-centred governance that recognizes sustainability not merely as a technical challenge but as a social and ethical imperative (Fig. 15.10).

15.8 Advancing Sustainable Urban Futures: Through Adaptive and Inclusive Governance

Contemporary cities face rising uncertainty driven by climate change, technological disruption, and social inequality, demanding governance models that move beyond traditional planning. The following sections explore how anticipatory governance, community co-production, ethical use of technology, data rights, climate justice, and sustainable finance can collectively strengthen urban resilience. Together, these approaches provide a roadmap for cities seeking to build sustainable, inclusive, and future-ready urban systems (Fig. 15.11).

15.8.1 Urban Foresight and Anticipatory Governance

Urban governance is increasingly confronted with forms of uncertainty that cannot be addressed through conventional, linear planning models

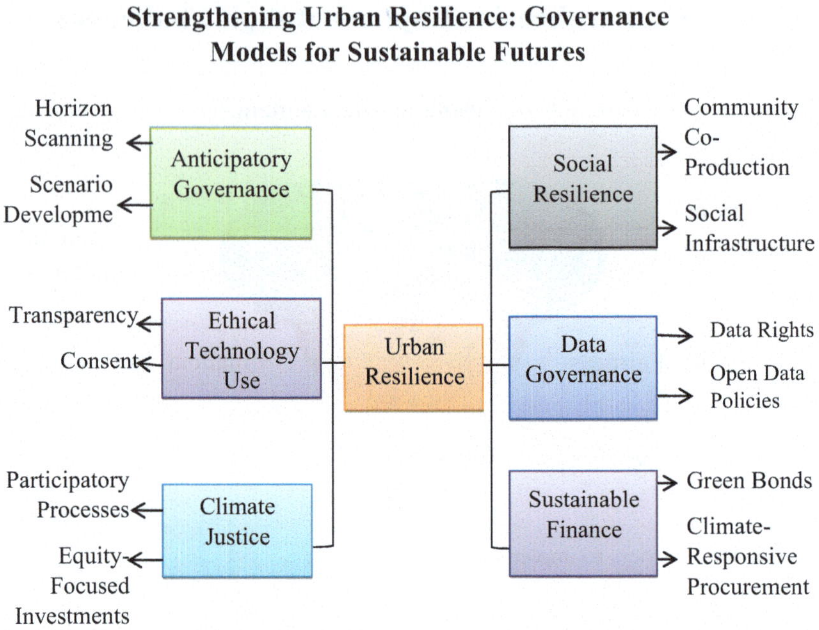

Fig. 15.11 Governance models linking anticipatory governance, ethical technology use, climate justice, social resilience, data governance, and sustainable finance to strengthen urban resilience

(Asadzadeh et al., 2023; Abujder Ochoa et al., 2024). Climate instability, technological disruption, demographic shifts, and geopolitical volatility demand governance approaches that extend beyond short-term risk management. Urban foresight and anticipatory governance offer a strategic response to this complexity by equipping city administrations with the tools to explore multiple future trajectories, assess emerging risks, and design policies that remain robust across diverse conditions (Fuerth, 2009; United Nations Department of Economic and Social Affairs, 2021).

Foresight methods such as horizon scanning, scenario development, and systems modelling help planners identify weak signals of change and prepare for challenges that might otherwise be overlooked (National Academies of Sciences, Engineering, and Medicine, 2020). These approaches foster a culture of institutional reflexivity, encouraging policymakers to question entrenched assumptions and consider alternative pathways for urban development. Cities that have integrated anticipatory practices into their governance frameworks often exhibit greater agility in responding to shocks, whether environmental, economic, or societal. This orientation is particularly critical for rapidly developing urban regions, where infrastructure decisions made today will shape environmental outcomes and social equity for decades (Akavarapu, 2025; Edmondson et al., 2025). Anticipatory governance shifts the focus from reactive problem-solving to proactive stewardship, enabling cities to navigate uncertainty with greater coherence and long-term vision. By embedding foresight into regulatory cycles, budget planning, and policy evaluation, urban systems strengthen their overall resilience and capacity for transformation (Muñoz-Erickson et al., 2021).

15.8.2 Social Resilience and Community Co-production

While institutional reforms and technological innovation play a central role in advancing sustainable urban governance, the resilience of cities is equally shaped by the social capacities embedded within communities (Noaime et al., 2025). Social resilience, the ability of individuals and neighbourhoods to absorb, adapt, and reorganize in the face of disruption, forms a crucial yet often under-recognized pillar of urban sustainability (Zeng et al., 2022; Cariolet et al., 2019). During crises such as pandemics, floods, or economic downturns, community networks frequently act as first responders, providing support systems that complement or even substitute formal institutional responses (Fortnam et al., 2024).

Community co-production has emerged as a practical framework for integrating these social capacities into urban governance (Sorrentino et al., 2018). Unlike traditional participatory models, which often limit residents to consultative roles, co-production involves communities directly in the design, delivery, and evaluation of public services (Khine et al., 2021). This approach deepens civic ownership, builds trust in local institutions, and generates locally grounded knowledge essential for tailoring interventions to diverse neighbourhood conditions. Examples such as community-led environmental monitoring, neighbourhood disaster preparedness committees, and collaborative housing initiatives demonstrate the transformative potential of co-production for both resilience and social equity (Simon et al., 2020). Strengthening social infrastructure libraries, community centres, shared public spaces, and digital commons further enhances a city's capacity to withstand and recover from shocks (Hocking et al., 2025). Research consistently shows that communities with strong social bonds and accessible services recover more quickly and more equitably. Integrating social resilience into urban policy frameworks, therefore, shifts sustainability discourse beyond physical infrastructure to include the relationships, networks, and collective capacities that make cities adaptive and inclusive. By recognising social resilience as a strategic asset, policymakers can build governance systems that are both people-centred and future-ready, while cultivating environments that are equitable, resilient, and future-ready (Hageer, 2025; Abdillah et al., 2025).

15.8.3 Ethical Dimensions of Urban Technology

As digital technologies have become embedded in urban governance, from predictive policing algorithms to integrated command centres, the ethical dimensions of their deployment demand careful consideration (Trigyn Technologies, 2023; Sustainability Directory, 2025). While technological innovation is often celebrated for improving efficiency, such systems also shape social behaviour, access to services, and the distribution of power within cities. Without deliberate safeguards, data-driven decisions may reflect systemic biases, reproduce patterns of exclusion, or erode personal privacy (Zhang et al., 2025). Ethical lapses in the use of surveillance infrastructure or algorithmic tools can undermine public trust, particularly in contexts where citizens already perceive institutions as opaque or unaccountable (Sanchez et al., 2025).

An ethical governance framework requires transparency, explicit consent, and precise mechanisms for redress when rights are violated. This also includes ongoing audits of technological systems to assess fairness, accuracy, and unintended consequences. Cities adopting smart technologies must therefore pair innovation with robust deliberation, ensuring that digital tools enhance freedom and well-being rather than constrain them. As global debates on digital rights gain prominence, urban authorities have a critical opportunity to establish norms that balance innovation with dignity, inclusion, and democratic oversight.

15.8.4 Data Governance and Urban Rights

The rapid expansion of data-intensive urban systems from IoT-enabled mobility networks to real-time environmental monitoring has elevated data governance to a central concern of urban sustainability (Omrany et al., 2024; Okonta & Vukovic, 2024). Data, once viewed as a neutral resource, now functions as a strategic asset that shapes political power, economic opportunity, and public accountability (Purtova & Van Maanen, 2024). Effective data governance frameworks must therefore address questions of ownership, access, security, and interoperability, ensuring that data-driven decision-making aligns with broader social values and human rights principles (Bernardo et al., 2024; Acev et al., 2025).

In many cities, data ecosystems remain fragmented, controlled by multiple public agencies and private technology providers. This fragmentation creates challenges for transparency and limits the potential for integrated urban planning. Comprehensible data governance architecture can mitigate these challenges by establishing standards for data sharing, open-data policies, and citizen-centric data rights (Ogunkan & Ogunkan, 2025; Lnenicka et al., 2022). Such frameworks empower residents to understand how their data is used, consent to its collection, and challenge misuse when necessary. By embedding data justice into urban governance, cities can balance the benefits of digital innovation with the imperative to protect civic freedoms and promote equitable access to technological advancements.

15.8.5 Climate Justice and Inclusive Urban Transitions

Although climate adaptation and mitigation strategies are central to sustainable urbanism, their implementation often raises profound questions of justice (Lobo et al., 2023). Climate risks such as heat stress, flooding, and air pollution disproportionately affect low-income and marginalized communities (Pellerey et al., 2025). However, these same groups frequently have the least influence over planning decisions and the fewest resources to recover from environmental shocks. Climate justice frameworks seek to correct this imbalance by ensuring that sustainability efforts do not reproduce historical inequalities but instead promote equitable distribution of climate benefits and burdens (Levy & Patz, 2015; Cañizares et al., 2024).

Integrating climate justice into urban governance requires a commitment to participatory processes, targeted investments in vulnerable areas, and inclusive policy design (IIED, 2024). Tools such as social vulnerability mapping, equity-focused impact assessments, and community-driven resilience planning help planners identify which populations are most at risk and tailor interventions accordingly (Abrash Walton et al., 2021; Ran et al., 2020). Moreover, climate justice extends beyond adaptation to include fair access to green jobs, clean energy, and sustainable mobility. By foregrounding justice, cities can cultivate climate transitions that are not only environmentally sound but socially transformative.

15.8.6 Sustainable Urban Finance and Investment Architectures

Achieving long-term sustainability requires substantial and predictable financial resources, yet many cities face structural financial constraints and limited access to credit markets. As global climate finance expands, urban authorities must navigate an increasingly complex landscape of green bonds, blended finance, multilateral climate funds, and public–private partnerships (Thakre, 2024). Each instrument carries distinct risks and opportunities, making financial governance a critical but often under-examined dimension of sustainable urban development (United Nations Human Settlements Programme, 2015; Hansson et al., 2019).

Sustainable finance frameworks encourage cities to align investment decisions with long-term ecological and social goals. This involves embedding sustainability criteria into municipal budgets, establishing climate-responsive procurement policies, and developing pipelines of bankable green infrastructure projects (C40 Cities Climate Leadership Group, 2025). Cities such as Copenhagen and Vancouver have demonstrated the value of integrating sustainability metrics into financial planning, strengthening investor confidence while advancing low-carbon development pathways (Jiang et al., 2025). At the same time, safeguards are essential to ensure that private-sector participation does not compromise equity, accountability, or public ownership of critical assets.

By adopting transparent financing strategies and leveraging diverse funding mechanisms, cities can build resilient financial ecosystems that support transformative infrastructure, inclusive mobility, and climate adaptation (Meng et al., 2024; Meraj & Hashimoto, 2025). Urban finance, therefore, becomes not only a technical tool but a strategic instrument for shaping sustainable and equitable urban futures.

15.9 Pathways Forward: Recommendations for Policymakers

As cities confront intensifying pressures from climate vulnerability and demographic change, governance fragmentation and technological disruption, policymakers must adopt strategic frameworks that are both integrative and future-oriented. Contemporary scholarship emphasizes the necessity of evidence-based governance, grounded in reliable data, participatory processes, and multi-level institutional coordination (United Nations, 2019). Drawing from theoretical insights and demonstrated global practices (UN-Habitat, 2020), the following recommendations advance actionable pathways for strengthening urban sustainability in alignment with SDGs 9 (industry, innovation, and infrastructure), 11 (sustainable cities and communities), and 16 (peace, justice, and strong institutions) (United Nations, 2015a, b, c, d). These recommendations recognize that sustainable development requires not only technical solutions but also institutional legitimacy, social inclusion, and ethical governance.

15.9.1 Strengthening Institutional Capacity

Enhancing institutional capacity is foundational to achieving effective and resilient urban governance. Urban administrations must invest in continuous human resource development to equip municipal staff with advanced competencies in sustainability planning, systems thinking, spatial analysis, and data-driven decision-making. Structured training programmes, embedded within municipal academies or national urban institutes, can support long-term capacity accumulation and reduce reliance on external consultants (UN-Habitat, 2020).

In addition to personnel development, procedural reforms are essential for improving administrative efficiency and public trust. Establishing independent anti-corruption authorities and embedding transparency mandates within municipal legal frameworks can mitigate political interference and enhance accountability. Cities must also streamline regulatory procedures such as permitting, procurement, and compliance monitoring to reduce bureaucratic delays that often hinder sustainability initiatives.

Strengthening intergovernmental coordination is particularly critical in metropolitan regions where responsibilities are distributed across levels of government. Coordination platforms such as metropolitan councils, interdepartmental working groups, and national-local policy alignment committees can harmonize regulations, promote resource sharing, and ensure coherence between national development strategies and local implementation priorities (OECD, 2017a, b). Such mechanisms help mitigate fragmentation, foster collaborative problem-solving, and enable long-term policy stability.

15.9.2 *Leveraging Technology Responsibly*

Technological innovation holds immense potential for enhancing urban governance, yet its deployment must be approached with caution, ethical reflection, and a commitment to inclusion. Rights-based technology governance requires that digital transformations not exacerbate inequality or undermine civil liberties. Digital literacy campaigns can help bridge socio-digital divides, enabling all residents, including marginalized groups, to participate meaningfully in e-governance platforms and digital public services.

Governments must also adopt comprehensive ethical frameworks to regulate data governance, algorithmic decision-making, and urban surveillance. Without clear safeguards, AI-driven tools and data-intensive systems can reproduce biases, erode privacy, or restrict democratic freedoms (Morozov & Bria, 2018). Transparent guidelines on data ownership, consent, interoperability, and algorithmic accountability should therefore be institutionalized within urban governance structures.

Investment in robust urban data infrastructures can significantly improve municipal responsiveness and crisis preparedness. Real-time dashboards, spatial analytics, and integrated command centres support dynamic decision-making across functions such as mobility, health, and emergency management. The experiences of Seoul and Helsinki during the COVID-19 pandemic exemplify how data-based systems combined with public communication and behavioural transparency enabled rapid containment measures and efficient service delivery (United Nations, 2022). These examples demonstrate that technology, when leveraged responsibly, can strengthen urban resilience and institutional trust.

15.9.3 Adaptive Monitoring and Evaluation

Monitoring and evaluation (M&E) systems form the backbone of effective policy implementation, enabling cities to assess progress, identify bottlenecks, and recalibrate strategies. Moving beyond traditional audit-based models, contemporary M&E emphasizes continuous feedback, transparency, and adaptive learning. Real-time monitoring platforms that integrate citizen feedback, performance metrics, and geospatial data enable agile policymaking and mid-course corrections, particularly in fast-changing urban environments.

Participatory evaluation approaches further enhance the legitimacy and contextual relevance of policy assessments. Community mapping, citizen science initiatives, and urban observatories empower residents to contribute data, articulate local priorities, and co-produce knowledge (Fung, 2015). These tools not only strengthen trust but also ensure that evaluation systems reflect lived experiences and diverse urban realities.

Cities such as Amsterdam offer illustrative examples of adaptive policy cycles, using iterative learning processes, scenario testing, and experimental neighbourhood-level interventions to respond to emerging challenges (Rauws & De Roo, 2016). Such adaptive governance models underscore the importance of flexibility, institutional reflexivity, and continuous learning in navigating complex urban transitions.

Together, these pathways emphasize the necessity of anticipatory governance that is transparent, inclusive, and capable of evolving with new challenges. By committing to ethical technological deployment, institutional strengthening, and adaptive evaluation, cities can position themselves as laboratories of innovation and frontline actors in the global pursuit of sustainable development. Through sustained collaboration between governments, communities, and knowledge institutions, urban centres can accelerate progress towards the SDGs.

15.10 Conclusion

Urbanization continues to accelerate, redesigning social, economic, and ecological landscapes at a pace that demands active, inclusive, and forward-looking governance. This chapter offers a comprehensive analysis of the multidimensional forces shaping contemporary urban challenges, highlighting the interwoven dynamics of demographic expansion, environmental pressures, socioeconomic inequality, and institutional constraints.

By integrating these issues within the framework of the Sustainable Development Goals, particularly SDGs 9, 11, and 16, this underscores the need for governance systems that are both analytically rigorous and normatively grounded in equity, resilience, and transparency. Through the combined use of bibliometric analysis, theoretical perspectives, and global case studies, the chapter has demonstrated that sustainable urban governance must be context-sensitive yet globally informed. From Singapore's climate-resilient infrastructure planning to Curitiba's integrated transit-land use model and Barcelona's participatory digital governance, the examples discussed reveal that effective urban transformation hinges on institutional coherence, long-term planning, and meaningful civic engagement. These diverse urban experiences underscore the importance of adapting global innovations to local political, social, and ecological realities.

The chapter has also identified persistent and structural barriers ranging from fragmented administrative mandates and limited fiscal autonomy to entrenched inequalities and uneven technological access that continue to obstruct sustainable urban transitions. Recognizing these challenges is essential for designing policies that are both feasible and enduring. As argued throughout the text, transformative urban governance requires strengthened institutional capacities, cross-sectoral integration, and governance models that embed ethics, justice, and public accountability.

Looking ahead, the chapter's promoters for governance approaches are anticipatory, evidence-based, and people-centred. This includes adopting forward-looking regulatory frameworks, responsible technological innovation, inclusive stakeholder engagement, and adaptive monitoring systems. Such approaches enable urban administrations to respond dynamically to evolving conditions while ensuring that policy actions remain aligned with broader societal goals. Ultimately, cities that embrace integrative, equitable, and adaptive governance models can position themselves as laboratories of sustainable development and as catalysts for achieving national and global aspirations.

By bridging theory, empirical insights, and actionable policy recommendations, this chapter presents a vision of urban governance capable of steering cities through the uncertainties of the twenty-first century. The imperative is clear to cultivate governance systems that empower communities, protect ecological resources, foster innovation, and build urban futures that are resilient, just, and inclusive for generations to come.

References

Abdel Aziz, K. M., Daoud, A. O., Singh, A. K., & Alhusban, M. (2024). Integrating digital mapping technologies in urban development: Advancing sustainable and resilient infrastructure for SDG 9 achievement – A systematic review. *Alexandria Engineering Journal.* Advance online publication. https://doi.org/10.1016/j.aej.2024.12.078

Abdillah, A., Widianingsih, I., Buchari, R. A., & Nurasa, H. (2025). Resilience governance: A concepts analysis and research synthesis. *Discover Environment, 3*(1), 50. https://doi.org/10.1007/s44274-025-00229-3

Abrash Walton, A., Marr, J., Cahillane, M. J., & Bush, K. (2021). Building community resilience to disasters: A review of interventions to improve and measure public health outcomes in the northeastern United States. *Sustainability, 13*(21), 11699. https://doi.org/10.3390/su132111699

Abujder Ochoa, W. A., Iarozinski Neto, A., Vitorio Junior, P. C., Calabokis, O. P., & Ballesteros-Ballesteros, V. (2024). The theory of complexity and sustainable urban development: A systematic literature review. *Sustainability, 17*(1), 3. https://doi.org/10.3390/su17010003

Acev, D., Biyani, S., Rieder, F., Aldenhoff, T. T., Blazevic, M., Riehle, D. M., & Wimmer, M. A. (2025). Systematic analysis of data governance frameworks and their relevance to data trusts. *Management Review Quarterly.* https://doi.org/10.1007/s11301-025-00545-1

Acuto, M., Parnell, S., & Seto, K. C. (2018). Building a global urban science. *Nature Sustainability, 1*(1), 2–4. https://doi.org/10.1038/s41893-017-0013-9

Akavarapu, S. V. (2025). A systematic literature review furthering the participatory futures and governance debate to capacities. *Futures, 173,* 103670. https://doi.org/10.1016/j.futures.2025.103670

Almulhim, A. I., & Yigitcanlar, T. (2025). Understanding smart governance of sustainable cities: A review and multidimensional framework. *Smart Cities, 8*(4), 113. https://doi.org/10.3390/smartcities8040113

Angel, S., Parent, J., Civco, D. L., Blei, A. M., & Potere, D. (2011). The dimensions of global urban expansion: Estimates and projections for all countries, 2000–2050. *Progress in Planning, 75*(2), 53–107. https://doi.org/10.1016/j.progress.2011.04.001

Anguelovski, I., Connolly, J. J. T., Masip, L., & Pearsall, H. (2018). Assessing green gentrification in historically disenfranchised neighborhoods: A longitudinal and spatial analysis of Barcelona. *Urban Geography, 39*(3), 458–491. https://doi.org/10.1080/02723638.2017.1349987

Anthony, B., & Sarshar, S. (2025). Enhancing data sovereignty to improve intelligent mobility services in smart cities. *Urban Governance, 5*(1), 20–31. https://doi.org/10.1016/j.ugj.2025.02.002

Asadzadeh, A., Fekete, A., Khazai, B., Moghadas, M., Zebardast, E., Basirat, M., & Kötter, T. (2023). Capacitating urban governance and planning systems to drive transformative resilience. *Sustainable Cities and Society, 96*, 104637. https://doi.org/10.1016/j.scs.2023.104637

Asimeng, E. T., & Jauregui-Fung, F. (2025). Tailoring bus rapid transit to the complex realities of African cities: Critical issues and public policy planning approaches. *Transportation Research Interdisciplinary Perspectives, 30*, 101351. https://doi.org/10.1016/j.trip.2025.101351

Barandiaran, X. E., Calleja-López, A., Monterde, A., & Romero, C. (2024). Decidim, a technopolitical network for participatory democracy: Philosophy, practice and autonomy of a collective platform in the age of digital intelligence. Spring https://doi.org/10.1007/978-3-031-50784-7

Batty, M., Axhausen, K. W., Giannotti, F., Pozdnoukhov, A., Bazzani, A., Wachowicz, M., Ouzounis, G., & Portugali, Y. (2012). Smart cities of the future. *The European Physical Journal Special Topics, 214*, 481–518. https://doi.org/10.1140/epjst/e2012-01703-3

Bernardo, B. M. V., Mamede, H. S., Barroso, J. M. P., & Dos Santos, V. M. P. D. (2024). Data governance & quality management—Innovation and breakthroughs across different fields. *Journal of Innovation & Knowledge, 9*(4), 100598. https://doi.org/10.1016/j.jik.2024.100598

Bhattacharjee, N., & Narayan, S. (2024). From shelter to the streets: The feminine face of homelessness in contemporary democracies. *The International Journal of Human Rights, 28*(5), 852–882. https://doi.org/10.1080/13642987.2024.2327374

C40 Cities Climate Leadership Group. (2025, February 24). *Sustainable finance action and advocacy: A roadmap for Global South cities*. https://www.c40.org/news/sustainable-finance-roadmap/

Cañizares, J. C., Copeland, S., & Doorn, N. (2024). Embedding justice considerations in climate resilience. *Ethics, Policy & Environment, 27*(1), 63–88. https://doi.org/10.1080/21550085.2023.2197824

Cariolet, J.-M., Vuillet, M., & Diab, Y. (2019). Mapping urban resilience to disasters – A review. *Sustainable Cities and Society, 51*, 101746. https://doi.org/10.1016/j.scs.2019.101746

Da Cruz, N. F., Rode, P., & McQuarrie, M. (2019). New urban governance: A review of current themes and future priorities. *Journal of Urban Affairs, 41*(1), 1–19. https://doi.org/10.1080/07352166.2018.1499416

Dávila, J. D. (Ed.). (2013). *Urban mobility and poverty: Lessons from Medellín and Soacha, Colombia*. Development Planning Unit, UCL and Universidad Nacional de Colombia. https://www.ucl.ac.uk/bartlett/publications/2013/jun/urban-mobility-and-poverty-lessons-medellin-and-soacha-colombia-spanish-version

Eakin, H., Keele, S., & Lueck, V. (2022). Uncomfortable knowledge: Mechanisms of urban development in adaptation governance. *World Development, 159*, 106056. https://doi.org/10.1016/j.worlddev.2022.106056

Edmondson, D., Engström, R., Pilli-Sihvola, K., Rinne, J., Perjo, L., Udovyk, O., & Puerari, E. (2025). *Anticipatory governance in city climate planning and policy: How four European Mission cities are integrating forward-looking governance into their climate-neutrality strategies* [Report]. NetZeroCities. https://netzerocities.app/resource-4632

Essien, E., & Jesse, E. E. (2025). Urban governance and political influence in contemporary urban settings of midsized cities in Nigeria. *Urban Governance, 5*(2), 169–181. https://doi.org/10.1016/j.ugj.2025.05.003

Fei, W., Opoku, A., Agyekum, K., Oppon, J. A., Ahmed, V., Chen, C., & Lok, K. L. (2021). The critical role of the construction industry in achieving the sustainable development goals (SDGs): Delivering projects for the common good. *Sustainability, 13*(16), 9112. https://doi.org/10.3390/su13169112

Feng, W., Yang, L. E., Zeng, F., Jia, J., Zhang, H., & Wu, Z. (2025). Enabling urban climate resilience through integrated optimization of urban design. *Frontiers in Sustainable Cities, 7*, 1657008. https://doi.org/10.3389/frsc.2025.1657008

Fortnam, M., Hailey, P., Witter, S., & Balfour, N. (2024). Resilience in interconnected community and formal health (and connected) systems. *SSM - Health Systems, 3*, 100027. https://doi.org/10.1016/j.ssmhs.2024.100027

Fuerth, L. S. (2009). Foresight and anticipatory governance. *Foresight, 11*(4), 14–32. https://doi.org/10.1108/14636680910982412

Fung, A. (2015). Putting the public back into governance: The challenges of citizen participation and its future. *Public Administration Review, 75*(4), 513–522. https://doi.org/10.1111/puar.12361

Gogoi, N., Choudhury, M., Ul Hasan, M. S., Changmai, B., Baruah, D., & Samanta, P. (2024). Soil organic carbon pool in diverse land utilization patterns in North-East India: An implication for carbon sequestration. *Environment, Development and Sustainability*, 1–27. https://doi.org/10.1007/s10668-024-05037-y

Graham, C. C. (2025). Localizing climate action in cities: Applying historical ecology to urban climate governance. *Global Environmental Change Advances, 5*, 100024. https://doi.org/10.1016/j.gecadv.2025.100024

Guha, J., & Chakrabarti, B. (2019). Achieving the sustainable development goals (SDGs) through decentralisation and the role of local governments: A systematic review. *Commonwealth Journal of Local Governance, 22*, 1–21. https://doi.org/10.3316/INFORMIT.896493850496702

Hageer, Y. (2025). Bridging equity and resilience: A systematic review of social sustainability in climate change mitigation and adaptation. *Environmental*

Science & Policy, 173, 104243. https://doi.org/10.1016/j.envsci.2025.104243

Hansson, S., Arfvidsson, H., & Simon, D. (2019). Governance for sustainable urban development: The double function of SDG indicators. *Area Development and Policy, 4*(3), 217–235. https://doi.org/10.1080/23792949.2019.1585192

Harrs, J.-A., Reinhart, V., Vögt, V., Scheib, J. P. P., Tewes, T., Pohl, T., & Rechid, D. (2024). Integration of climate information into urban climate change adaptation: A case study of municipal processes in Constance. *Climate Services, 35*, 100494. https://doi.org/10.1016/j.cliser.2024.100494

Hocking, L., Opdyke, A., Siguan, A., & Aldrich, D. P. (2025). Linking social infrastructure and social capital: Barangay halls and basketball courts as cornerstones of community resilience. *International Journal of Disaster Risk Reduction, 130*, 105868. https://doi.org/10.1016/j.ijdrr.2025.105868

Intergovernmental Panel on Climate Change. (2023). Climate change 2022: Impacts, adaptation and vulnerability. *Contribution of Working Group II to the Sixth Assessment Report of the Intergovernmental Panel on Climate Change* (H.-O. Pörtner, D. C. Roberts, M. Tignor, et al., Eds.). Cambridge University Press. https://doi.org/10.1017/9781009325844

International Institute for Environment and Development. (2024, June 7). *Climate action for equitable cities: Working with informal communities to develop low-carbon, resilient futures.* https://www.iied.org/climate-action-for-equitable-cities-working-informal-communities-develop-low-carbon-resilient-futures

Jiang, H., Lu, J., Zhang, R., & Xiao, X. (2025). Investigation of diverse urban carbon emission reduction pathways in China: Based on the technology–organization–environment framework for promoting socio-environmental sustainability. *Land, 14*(2), 260. https://doi.org/10.3390/land14020260

Kabisch, N., & Haase, D. (2014). Green justice or just green? Provision of urban green spaces in Berlin, Germany. *Landscape and Urban Planning, 122*, 129–139. https://doi.org/10.1016/j.landurbplan.2013.11.016

Khine, P. K., Mi, J., & Shahid, R. (2021). A comparative analysis of co-production in public services. *Sustainability, 13*(12), 6730. https://doi.org/10.3390/su13126730

Kumaraswamy, T. R., Choudhury, M., Ponappa, S., & Majumdar, S. (2025). A significance of climate resilience: Strategies for local, regional, and national levels. In M. Choudhury, G. Dixit, & S. Majumdar (Eds.), *Sustainable synergy: Harnessing ecosystems for climate resilience. Climate change management.* Springer. https://doi.org/10.1007/978-3-031-77957-2_13

Kuzemko, C., & Britton, J. (2020). Policy, politics and materiality across scales: A framework for understanding local government sustainable energy capacity applied in England. *Energy Research & Social Science, 62*, 101367. https://doi.org/10.1016/j.erss.2019.101367

Levy, B. S., & Patz, J. A. (2015). Climate change, human rights, and social justice. *Annals of Global Health, 81*(3), 310. https://doi.org/10.1016/j.aogh.2015.08.008

Lnenicka, M., Nikiforova, A., Luterek, M., Azeroual, O., Ukpabi, D., Valtenbergs, V., & Machova, R. (2022). Transparency of open data ecosystems in smart cities: Definition and assessment of the maturity of transparency in 22 smart cities. *Sustainable Cities and Society, 82*, 103906. https://doi.org/10.1016/j.scs.2022.103906

Lobo, J., Aggarwal, R. M., Alberti, M., Allen-Dumas, M., Bettencourt, L. M. A., Boone, C., Brelsford, C., Castán Broto, V., Eakin, H., Bagchi-Sen, S., Meerow, S., D'Cruz, C., Revi, A., Roberts, D. C., Smith, M. E., York, A., Lin, T., Bai, X., Solecki, W., Pataki, D., Bojórquez Tapia, L., Rockman, M., Wolfram, M., Schlosser, P., & Gauthier, N. (2023). Integration of urban science and urban climate adaptation research: Opportunities to advance climate action. *Npj Urban Sustainability, 3*, 32. https://doi.org/10.1038/s42949-023-00113-0

Maier, F., & Weinberger, M. (2024). Metaverse meets smart cities—Applications, benefits, and challenges. *Future Internet, 16*(4), 126. https://doi.org/10.3390/fi16040126

Medeiros, E., & Van Der Zwet, A. (2020). Sustainable and integrated urban planning and governance in metropolitan and medium-sized cities. *Sustainability, 12*(15), 5976. https://doi.org/10.3390/su12155976

Meng, J., Ye, Z., & Wang, Y. (2024). Financing and investing in sustainable infrastructure: A review and research agenda. *Sustainable Futures, 8*, 100312. https://doi.org/10.1016/j.sftr.2024.100312

Meraj, G., & Hashimoto, S. (2025). Bridging the adaptation finance gap: The role of nature-based solutions for climate resilience. *Sustainability Science, 20*(3), 1093–1107. https://doi.org/10.1007/s11625-025-01655-1

Meuleman, L. (2019). *Metagovernance for sustainability: A framework for implementing the sustainable development goals*. Routledge.

Meyer, N., & Auriacombe, C. (2019). Good urban governance and city resilience: An afrocentric approach to sustainable development. *Sustainability, 11*(19), 5514. https://doi.org/10.3390/su11195514

Ministry of Housing & Urban Affairs. (2018). *Integrated Command and Control Center Maturity Assessment Framework and Toolkit* (Draft Version 1.0). Government of India. https://smartcities.gov.in/sites/default/files/2023-07/MoHUA%20ICCC%20Maturity%20Assessment%20Framework%20IMAF%20toolkit_vf211218.pdf

Morozov, E., & Bria, F. (2018). *Rethinking the smart city: Democratizing urban technology*. Rosa Luxemburg Stiftung. https://rosalux.nyc/wp-content/uploads/2021/02/RLS-NYC_smart_cities_EN.pdf

Muhammad, A. (2023). *The geopolitical implications of shifting alliances in a multipolar world.* https://doi.org/10.5281/ZENODO.8382207

Muñoz-Erickson, T. A., Selkirk, K., Hobbins, R., Miller, C., Feagan, M., Iwaniec, D. M., Miller, T. R., & Cook, E. M. (2021). Anticipatory resilience bringing back the future into urban planning and knowledge systems. In Z. A. Hamstead, D. M. Iwaniec, T. McPhearson, M. Berbés-Blázquez, E. M. Cook, & T. A. Muñoz-Erickson (Eds.), *Resilient urban futures* (pp. 159–172). Springer International Publishing. https://doi.org/10.1007/978-3-030-63131-4_11

Narayan, S., & Singha, S. (2024). Legal framework for forest rights in India: A critical analysis. *Indian Journal of Law and Justice, 15*(2), 187–211. http://42.104.73.51:91/nbujournal/law/v15_no2.pdf

National Academies of Sciences, Engineering, and Medicine. (2020). *Horizon scanning and foresight methods.* National Academies Press. https://doi.org/10.17226/25753

Negesa, D., Geme, T., Nijman-Ross, E., & Ntawuhiganayo, E. B. (2023). Eco-industrial transformation in Uganda: Performance evaluation of industrial parks using the international EIP framework. *Frontiers in Sustainability, 4.* https://doi.org/10.3389/frsus.2023.1286611

Ngwenya, N., & Simatele, M. D. (2020). The emergence of green bonds as an integral component of climate finance in South Africa. *South African Journal of Science, 116*(1–2). https://doi.org/10.17159/sajs.2020/6522

Noaime, E., Alalouch, C., Mesloub, A., Hamdoun, H., Gnaba, H., & Alnaim, M. M. (2025). Urban centrality as a catalyst for city resilience and sustainable development. *Land, 14*(7), 1333. https://doi.org/10.3390/land14071333

OECD. (2016). *Making cities work for all: Data and actions for inclusive growth.* OECD Publishing. https://doi.org/10.1787/9789264263260-en

Ogunkan, D. V., & Ogunkan, S. K. (2025). Exploring big data applications in sustainable urban infrastructure: A review. *Urban Governance, 5*(1), 54–68. https://doi.org/10.1016/j.ugj.2025.02.003

Okonta, D. E., & Vukovic, V. (2024). Smart cities software applications for sustainability and resilience. *Heliyon, 10*(12), e32654. https://doi.org/10.1016/j.heliyon.2024.e32654

Omrany, H., Al-Obaidi, K. M., Hossain, M., Alduais, N. A. M., Al-Duais, H. S., & Ghaffarianhoseini, A. (2024). IoT-enabled smart cities: A hybrid systematic analysis of key research areas, challenges, and recommendations for future direction. *Discover Cities, 1*(1), 2. https://doi.org/10.1007/s44327-024-00002-w

Organisation for Economic Co-operation and Development. (2017a). *Land-use planning systems in the OECD: Country fact sheets.* OECD Publishing. https://www.oecd.org/en/publications/land-use-planning-systems-in-the-oecd_9789264268579-en.html

Organisation for Economic Co-operation and Development. (2017b). *Multi-level governance reforms: Overview of OECD country experiences*. OECD Publishing. https://doi.org/10.1787/9789264272866-en

Pellerey, V., Torabi Moghadam, S., & Lombardi, P. (2025). A systematic review of justice integration to climate resilience: Current trends and future directions. *Urban Climate, 59*, 102250. https://doi.org/10.1016/j.uclim.2024.102250

PhD Finance Candidate, Department of Accounting and Finance at the School of Business, KCA University, Wabwile, W. A. E., Ondabu, I. T., & Senior Lecturer, Department of Accounting and Finance at the School of Business, KCA University. (2025). Funding climate action: A systematic review of climate finance efficiency and impact. *Journal of Economics, Finance and Management Studies, 08*(04). https://doi.org/10.47191/jefms/v8-i4-49

Public Utilities Board. (2018). *Active, beautiful, clean waters (ABC Waters) programme – Celebrating 10 years of transforming our waterbodies*. https://www.pub.gov.sg/Resources/Publications/ABC-Waters

Purtova, N., & Van Maanen, G. (2024). Data as an economic good, data as a commons, and data governance. *Law, Innovation and Technology, 16*(1), 1–42. https://doi.org/10.1080/17579961.2023.2265270

Rabinovitch, J., & Leitman, J. (2015). Urban planning in Curitiba. In R. T. LeGates & F. Stout (Eds.), *The city reader* (6th ed., pp. 526–535). Routledge. https://doi.org/10.4324/9781315748504

Rajpal, A., Choudhury, M., Goswami, S., Chakravorty, A., & Raghavan, V. (Eds.). (2024). Waste management and treatment: Advances and innovations (1st ed.). CRC Press. https://doi.org/10.1201/9781003258377

Ran, J., MacGillivray, B. H., Gong, Y., & Hales, T. C. (2020). The application of frameworks for measuring social vulnerability and resilience to geophysical hazards within developing countries: A systematic review and narrative synthesis. *Science of the Total Environment, 711*, 134486. https://doi.org/10.1016/j.scitotenv.2019.134486

Rauws, W., & De Roo, G. (2016). Adaptive planning: Generating conditions for urban adaptability. Lessons from Dutch organic development strategies. *Environment and Planning B: Planning and Design, 43*(6), 1052–1074. https://doi.org/10.1177/0265813516658886

Razzano, C. C., & Bernardi, M. (2024). Empowering urban resilience: Grassroots volunteering for a renewed shock-response governance in post-pandemic Milan. *Urban Governance, 4*(4), 374–387. https://doi.org/10.1016/j.ugj.2024.12.006

Reddy, P. S. (2016). Localising the sustainable development goals (SDGs): The role of local government in context. *African Journal of Public Affairs, 9*(2), 1–15. https://hdl.handle.net/2263/58190

Routh, R., Bhavsar, D., & Pandya, P. (2025). Exploring participatory governance in Indian cultural heritage: Perceptions, practices, and challenges. *Built Heritage, 9*(1), 55. https://doi.org/10.1186/s43238-025-00224-w

Sanchez, T. W., Brenman, M., & Ye, X. (2025). The ethical concerns of artificial intelligence in urban planning. *Journal of the American Planning Association, 91*(2), 294–307. https://doi.org/10.1080/01944363.2024.2355305

Satterthwaite, D. (2013). The political underpinnings of cities' accumulated resilience to climate change. *Environment and Urbanization, 25*(2), 381–391. https://doi.org/10.1177/0956247813500902

Satterthwaite, D., Archer, D., Colenbrander, S., Dodman, D., Hardoy, J., Mitlin, D., & Parnell, S. (2020). Building resilience to climate change in informal settlements. *One Earth, 2*(2), 143–156. https://doi.org/10.1016/j.oneear.2020.02.002

Sebunya, J., & Gichuki, A. (2024). The impact of participatory planning on sustainable development: A literature review. *Journal of Strategic Management, 4*(4), 1–9. https://doi.org/10.70619/vol4iss4pp1-9

Secchi, M., Tomassoni, F., & Allegretti, G. (2024). Local best practices: Urban governance and the ongoing platformization process. In S. Mezzadra, N. Cuppini, M. Frapporti, & M. Pirone (Eds.), *Capitalism in the platform age* (pp. 305–331). Springer International Publishing. https://doi.org/10.1007/978-3-031-49147-4_18

Shirazi, M. R., & Keivani, R. (2017). Critical reflections on the theory and practice of social sustainability in the built environment – A meta-analysis. *Local Environment, 22*(12), 1526–1545. https://doi.org/10.1080/13549839.2017.1379476

Simon, K., Diprose, G., & Thomas, A. C. (2020). Community-led initiatives for climate adaptation and mitigation. *Kōtuitui: New Zealand Journal of Social Sciences Online, 15*(1), 93–105. https://doi.org/10.1080/1177083X.2019.1652659

Sorrentino, M., Sicilia, M., & Howlett, M. (2018). Understanding co-production as a new public governance tool. *Policy and Society, 37*(3), 277–293. https://doi.org/10.1080/14494035.2018.1521676

Suzuki, H., Cervero, R., & Iuchi, K. (2013). *Transforming cities with transit: Transit and land-use integration for sustainable urban development*. World Bank. https://doi.org/10.1596/978-0-8213-9745-9

Swapan, M. S. H., & Khan, S. (2021). Urban informality and parallel governance systems: Shaping citizens' engagements in urban planning processes in Bangladesh. *Local Economy, 36*(1), 1–17. https://doi.org/10.1080/13563475.2021.1899902

Tan, P. Y., Wang, J., & Sia, A. (2013). Perspectives on five decades of the urban greening of Singapore. *Cities, 32*, 24–32. https://doi.org/10.1016/j.cities.2013.02.001

Thakre, M. (2024, May 22). *Enhancing access to urban climate finance: Learnings from four cities in India* (ORF Issue Brief No. 711). Observer Research Foundation. https://www.orfonline.org/research/enhancing-access-to-urban-climate-finance-learnings-from-four-cities-in-india

Trigyn Technologies. (2023). *Legal & ethical considerations for smart cities and IoT.* https://www.trigyn.com/insights/legal-and-ethical-considerations-smart-cities-and-iot

UNFCCC. *Introduction to climate finance.* Retrieved November 19, 2025., from https://unfccc.int/topics/introduction-to-climate-finance

UN-Habitat. (2020). *World cities report 2020: The value of sustainable urbanization.* https://unhabitat.org/sites/default/files/2020/10/wcr_2020_report.pdf

United Nations. (2015a). *Goal 11: Make cities and human settlements inclusive, safe, resilient and sustainable.* https://sdgs.un.org/goals/goal11

United Nations. (2015b). *Goal 16: Promote just, peaceful and inclusive societies.* https://sdgs.un.org/goals/goal16

United Nations. (2015c). *Sustainable development goals.* https://sdgs.un.org/goals

United Nations. (2015d). *Transforming our world: The 2030 agenda for sustainable development.* A/RES/70/1. https://sdgs.un.org/2030agenda

United Nations. (2022). *United Nations E-government survey 2022: The future of digital government.* https://publicadministration.un.org/egovkb/en-us/Reports/UN-E-Government-Survey-2022

United Nations Department of Economic and Social Affairs. (2021). *Strategy guidance note on strategic planning and foresight.* United Nations. https://unpan.un.org/sites/default/files/Events/Events%202022/Strategy%20note%20%20strategic%20foresight%20Mar%202021_1.pdf

United Nations, Department of Economic and Social Affairs, Population Division. (2019). *World urbanization prospects: The 2018 revision* (ST/ESA/SER.A/420). https://population.un.org/wup/assets/WUP2018-Report.pdf

United Nations Development Programme. (2014). *Governance for sustainable development: Integrating governance in the post-2015 development framework* (Discussion Paper). https://www.undp.org/publications/discussion-paper-governance-sustainable-development

United Nations Human Settlements Programme. (2015). *Financing sustainable urban development.* https://www.un.org/esa/ffd/wp-content/uploads/sites/2/2015/03/Financing-Urban-Development_UN-Habitat.pdf

Watson, V. (2009). The planned city sweeps the poor away...: Urban planning and 21st century urbanisation. *Progress in Planning, 72*(3), 151–193. https://doi.org/10.1016/j.progress.2009.06.002

World Bank. (2015). *Competitive cities for jobs and growth: What, who, and how* (Report No. 101673). World Bank Group. https://documents.

worldbank.org/en/publication/documents-reports/documentdetail/902411467990995484/competitive-cities-for-jobs-and-growth-what-who-and-how

World Bank. (2021). *Urban development*. https://www.worldbank.org/en/topic/urbandevelopment/overview

Zeng, X., Yu, Y., Yang, S., Lv, Y., & Sarker, M. N. I. (2022). Urban resilience for urban sustainability: Concepts, dimensions, and perspectives. *Sustainability, 14*(5), 2481. https://doi.org/10.3390/su14052481

Zhang, Y., Lin, Y., Zheng, G., Liu, Y., Sukiennik, N., Xu, F., Xu, Y., Lu, F., Wang, Q., Lai, Y., Tian, L., Li, N., Fang, D., Wang, F., Zhou, T., Li, Y., Zheng, Y., Wu, Z., & Guo, H. (2025). MetaCity: Data-driven sustainable development of complex cities. *The Innovation, 6*(2), 100775. https://doi.org/10.1016/j.xinn.2024.100775

CHAPTER 16

Citizen-Centric Design: Envisioning Inclusive and Sustainable Urban Environments

Suiven Joun Paul Tume, Donizete Beck, Shinjita Das, and Sushobhan Majumdar

16.1 Introduction

Citizen-centric urban planning design prioritises the needs, preferences, behaviours, and lived experiences of residents as the primary drivers of urban development decisions (Lux, 2025). Contrasting with traditional

S. J. P. Tume
Department of Geography and Planning, The University of Bamenda, Bamenda, Cameroon

D. Beck
Escola de Economia e Negócios, Pontifícia Universidade Católica de Campinas, Campinas, Brazil

S. Das
Design and Social Context, School of GUSS, RMIT University, Melbourne, VIC, Australia

S. Majumdar (✉)
Department of Geography, Pandit Raghunath Murmu Smriti Mahavidyalaya, Bankura, West Bengal, India

S. Majumdar, M. Choudhury (eds.), *City of the Future*,
https://doi.org/10.1007/978-3-032-18026-1_16

top-down planning, which often focuses on infrastructure efficiency, economic growth, or expert-led visions, citizen-centric design flips the script by placing people at the centre of the planning process, using participatory methods such as co-design workshops, digital surveys, community forums, and data from citizen feedback platforms, and iteratively testing and refining urban interventions based on real-world user feedback (Perillo, 2025; Ravi & Brück, 2025). It builds on people-centred design principles and adapts them to the scale of cities.

Citizen-centric design represents a transformative approach in contemporary urban planning, emphasising the active involvement of citizens as co-creators rather than passive recipients of the built environment (Preston et al., 2020). Its significance lies in redefining how cities are imagined, developed, and governed, placing human experience, local knowledge, and community values at the core of decision-making. By prioritising inclusivity, accessibility, and sustainability, citizen-centric design bridges the gap between policy intentions and lived realities, ensuring that urban interventions genuinely respond to diverse social needs. This approach not only enhances transparency and trust in planning processes but also strengthens civic identity and resilience, enabling communities to adapt collaboratively to social and environmental change (Majumdar et al., 2025). Ultimately, citizen-centric design reframes urban planning as a participatory, iterative, and empathetic practice, aiming to create cities that are not merely functional but also equitable, meaningful, and reflective of the people who inhabit them (Foth, 2018).

Inclusivity and sustainability are foundational principles of citizen-centric design, shaping how urban environments are planned and experienced. Inclusivity ensures that cities reflect the diversity of their inhabitants by actively considering the needs, perspectives, and aspirations of all social groups, particularly those historically marginalised in urban decision-making (Itair et al., 2023). It promotes equitable access to public spaces, services, and opportunities, fostering a sense of belonging and social cohesion. Sustainability, on the other hand, extends this vision by integrating ecological responsibility and long-term resilience into the urban fabric (Majumdar, 2019). It emphasises the balance between human development and environmental stewardship, encouraging design strategies that reduce resource consumption, mitigate climate impacts, and enhance urban biodiversity (Dixit et al., 2025; Kumaraswamy et al., 2025). When pursued together, inclusivity and sustainability reinforce one another: inclusive processes lead to more contextually appropriate and enduring solutions. At the same time, sustainable systems create healthier and more just environments for all citizens (Choudhury et al., 2025; Majumdar, 2020). Together, they form the ethical and practical foundation of citizen-centric urban design, ensuring that cities evolve not only efficiently, but equitably and responsibly (Malhotra et al., 2020).

The role of citizens in shaping urban environments has evolved from passive beneficiaries of planning decisions to active participants and co-creators in the design and governance of cities (Müller et al., 2023). Traditionally, urban planning was dominated by top-down models driven by expert knowledge and institutional authority. However, growing awareness of social diversity, technological innovation, and environmental challenges has prompted a paradigm shift towards more participatory and collaborative approaches (Pichler-Milanovič & Foški, 2015). Citizens are now increasingly engaged through community consultations, participatory budgeting, digital mapping, and co-design processes that democratize urban decision-making. This evolution reflects a broader understanding that cities function as dynamic ecosystems shaped by everyday human experiences and collective action. By recognising citizens as vital agents of change, urban planning becomes more adaptive, inclusive, and context-sensitive, responding not only to technical and economic imperatives but also to social values, cultural identities, and environmental priorities (Ives & Baker, 2024). Ultimately, the evolving role of citizens signifies a reimagining of urban governance where the city is co-produced through shared knowledge, dialogue, and stewardship. The key elements include empathy, inclusivity, iterative prototyping, and data-driven feedback (Table 16.1).

Table 16.1 Core elements of citizen-centric urban design

Element	*Description*
Empathy	Deep understanding of diverse user groups (elderly, children, low-income residents, immigrants) through ethnographic research
Inclusivity	Ensuring marginalised voices are amplified, not just the loudest or most privileged
Iterative prototyping	Testing small-scale interventions (pop-up parks) before city-wide rollout
Data-driven feedback	Using tools like mobile apps (FixMyStreet) or sensors to collect real-time citizen input

Source: Compiled by the authors

16.2 Historical Context and Theoretical Foundations

16.2.1 Traditional Urban Design Approaches Versus Citizen-Centric Paradigms

Citizen-centric urban design did not emerge in a vacuum. It is the culmination of over a century of critique against top-down, expert-driven planning and parallel movements in participatory democracy, social science, and design thinking (Nyongesa et al., 2025; Goncalves, 2025). An attempted chronological framework goes back to the late nineteenth and twentieth centuries (Table 16.2).

16.2.2 Key Theories

Citizen-centric urban design rests on interdisciplinary pillars that have undergone evolution. Some of these theoretical underpinnings include: human-centred design (HCD), participatory design (PD), social practice theory, deliberative democracy, actor-network theory (ANT), right to the city, and behavioural urbanism (Table 16.3).

16.2.3 Sustainability Frameworks Relevant to Urban Development

Two of the most relevant sustainability frameworks for urban development are the circular economy (CE) and green urbanism. The circular economy is focused on the efficient utilisation of resources and the minimisation of waste, with as little energy and new resource use as possible. Initially, the CE was based on "3-Rs": Reduction, Recycling, and Reuse (Yuan et al., 2006). Recycling plays a key role in the CE. The concept exceeds recycling over time from the well-known 3Rs principle to 10Rs, including vital principles such as: Refuse, Recover, Reduce, Reuse, Recycle, Repurpose, Remanufacture, Refurbish, Repair, Rethink (Pesta et al., 2023). R1: Refuse R2: Rethink R3: Reduce R4: Reuse R5: Repair R6: Refurbish R7: Remanufacture R8: Repurpose R9: Recycle R10: Recovery (Crump et al., 2019) (Fig. 16.1).

Refuse prompts us to challenge the necessity of new construction or superfluous components altogether (Pešta et al., 2023). Rethink encourages us to re-evaluate traditional building approaches. Reuse or Repurpose

Table 16.2 Chronological framework for citizen-centric urban design

Period	*Key events & critiques*	*Shift towards citizen-centricity*
Late nineteenth–early twentieth century	Industrial cities: overcrowding, slums; Ebenezer Howard's *Garden Cities* (1898) and Patrick Geddes' *regional survey* method emphasise lived experience over geometric plans	First seeds: Planners begin *observing* citizens, not just drawing lines
1930s–1950s: modernist era	Le Corbusier's *Ville Radieuse*, CIAM's Athens Charter (1933); Cities as "machines for living". Urban renewal bulldozes neighbourhoods (e.g., US public housing projects)	Backlash begins: Residents displaced; social fabric destroyed
1961: Jane Jacobs	*The Death and Life of Great American Cities* argues for "eyes on the street", mixed uses, and organic growth. First major intellectual assault on top-down planning	Paradigm shift: Citizens' *behavioural patterns* become design inputs
1969: Sherry Arnstein	Published "A Ladder of Citizen Participation" in *Journal of the American Institute of Planners.* Defines eight rungs from manipulation to citizen control	Framework born: Participation is now a measurable spectrum
1970s: Advocacy Planning	Paul Davidoff coined the term. Planners act as advocates for underrepresented groups Community Design Centres emerge (e.g., Urban Planning Aid in Boston)	Institutionalisation: Citizens are co-planners, not just consultees
1980s: Participatory budgeting	Originated in Porto Alegre, Brazil (1989) residents directly allocate municipal budgets Spreads globally	Power transfer: From consultation to decision-making authority
1990s: New urbanism & tactical urbanism	Andrés Duany and Elizabeth Plater-Zyberk push for walkable, mixed-use neighbourhoods. David Engwicht's "mental speed bumps" inspire pop-up interventions	Prototyping with people: Test ideas before permanent build
2000s: digital revolution	Web 2.0, smartphones, open data. Platforms like FixMyStreet (UK, 2007) and SeeClickFix enable real-time citizen reporting	Scalable feedback loops: From annual town halls to 24/7 input
2010s–present: smart cities + co-design	EU's *Human Smart Cities* and Singapore's *Smart Nation* integrate IoT with participatory governance. Tools like Pol.is, Decidim, and Commonplace are used in Barcelona, Taipei, and Reykjavik	Hybrid governance: Data + deliberation = adaptive cities

Source: Compiled by the authors

Table 16.3 Core theories of citizen-centred urban design

Theory	*Origin*	*Key message*	*Applications*
Human-centred design (HCD)	IDEO (1990s), Don Norman's *The Design of Everyday Things* (1988)	Design must solve problems for real users, not abstract ideals	Empathy mapping, journey maps, and prototyping adapted to public space (Project for Public Spaces "Place Game")
Participatory design (PD)	Scandinavian workplace democracy (1970s) workers co-design tools	Pelle Ehn's *Work-Oriented Design of Computer Artefacts* (1988)	Residents co-create parks, transit routes, housing layouts (co-design workshops in Medellín's *Metrocable*)
Social practice theory	Elizabeth Shove, Alan Warde, Theodore Schatzki	Everyday practices (walking, socialising, shopping) shape—and are shaped by—material environments	Urban design must support desired practices, not just aesthetics
Deliberative democracy	Jürgen Habermas, John Dryzek, Archon Fung	Legitimate decisions emerge from inclusive, reasoned public discourse	Citizens' juries, participatory budgeting, digital deliberation platforms
Actor-network theory (ANT)	Bruno Latour	Humans *and* non-humans (benches, apps, trees) co-produce urban experience	Sensors, street furniture, and apps become active participants in feedback loops
Right to the city	Henri Lefebvre (1968), expanded by David Harvey	Citizens have a right to shape the city that shapes them	Reclaim the Commons, Occupy, community land trusts
Behavioural urbanism	Twentieth and twenty-first centuries	Small "nudges" (painted crosswalks, timed lights) can shift citizen behaviour sustainably	Combines behavioural economics (Thaler, Sunstein) with urban design

Source: Compiled by the authors

echoes the need to build on existing materials as a fundamental aspect of circularity. By reusing structures and repurposing existing buildings, we extend the life of resources and reduce waste, curbing the use of virgin and non-renewable materials in new constructions. Reduction implies

Fig. 16.1 An Urban Park in Victoria, Australia demonstrating how citizen-centric design fosters inclusive public spaces that support social interaction, ecological balance, and community wellbeing

reducing the use or extraction of new resources from nature. At the same time, Recycling involves the Recycling of materials and energy that ended up as waste, reviving them and making them a leading resource for production. Reuse means using waste as an input in the circular economy (Kowsalya et al., 2025). Refurbish/Remanufacture emphasises the revitalisation of building components and systems. It involves upgrading existing elements to enhance performance and extend their lifespan. Besides the "10Rs", other principals include: durability, adaptability, and disassembly. Durability emphasises the need for buildings to last. Adaptability emphasises the need to design buildings that can be easily adapted to changing uses. This incorporates flexible layouts and dismountable components that can be reconfigured with ease.

16.3 Principles of Citizen-Centric Urban Design

Citizen-centric urban design goes beyond constructing buildings and paving roads, focusing on creating spaces that reflect the lives, needs, and hopes of the people who live there (Jakonen, 2023). This marks a shift from top-down planning to more inclusive and participatory methods that acknowledge the diverse and changing needs of urban residents (Gehl, 2013; Healey, 2020). Grounded in the principles of inclusivity, accessibility, transparency, participation, and flexibility, this approach envisions cities not merely as physical landscapes but as dynamic human systems responsive to diversity, resilience, and justice (Sandercock & Lyssiotis, 2003; Soja, 2013). It highlights fairness, a focus on people, and responsiveness, aiming to create urban environments that are not just functional but also inclusive and adaptable to shifting societal needs (Fig. 16.2).

Fig. 16.2 A sidewalk in Australia illustrating the role of citizen-centric design in creating inclusive, walkable, and socially connected urban spaces

16.3.1 Inclusivity: Addressing Diverse Needs

Historically, urban design has catered to a small demographic, often overlooking marginalised groups and reinforcing spatial injustices through one-size-fits-all planning models (Fincher & Iveson, 2017; Soja, 2013). With a focus on people-centred development, inclusive design acknowledges the socio-demographic diversity of urban populations (age, gender, ethnicity, ability, and socioeconomic differences). It promotes planning practices grounded in equity and justice. For instance, in patriarchal society, men are considered generic citizens or typical users of urban spaces, which has led women and other sexual gender minorities to feel excluded from the public sphere (Agarwal, 2022). This gap is addressed by gender inclusivity planning, which includes the voices of women, girls, sexual and gender minorities, and all individuals, to promote equitable access and safety for everyone (Chance & Trbovc, 2021). Similarly, age-inclusive design principles have led to safe, accessible, and age-friendly cities (Buffel et al., 2012; McGinley et al., 2022), incorporating features such as rest areas, walkable neighbourhoods, and intergenerational community spaces.

Designs should also reflect disability-inclusive planning guided by the principles of *Universal Design*, which advocate for environments usable by all people to the greatest extent possible, without the need for adaptation or specialised design (Imrie & Hall, 2001; Steinfeld & Maisel, 2012). Additionally, socioeconomic inclusivity goes beyond infrastructure to address spatial inequality and access to opportunity (Afacan & Afacan, 2011). Mechanisms such as inclusionary zoning, participatory budgeting, and mixed-income housing have been implemented to redistribute resources and reduce exclusion from urban cores (Fainstein, 2014; Marcuse & Madden, 2024). Emerging models such as community land trusts and cooperative housing support long-term affordability and local ownership, and will foster more stable and equitable neighbourhoods (Angotti, 2011). Although structural barriers persist, these inclusive designs should be the foundation for cities to ensure that all individuals can participate and have equal access to opportunities, resources, and services, regardless of their background or characteristics (Fig. 16.3).

Fig. 16.3 Re-designing public transport nodes, i.e., a bus stop in Victoria, Australia, highlights how citizen-informed planning can create more equitable and sustainable cities

16.3.2 Accessibility: Designing for Physical, Social, and Digital Inclusion

City design and planning should also extend beyond physical mobility and recognise the complex interplay among people, their physical environment, social participation, and access to digital infrastructure (Afacan & Afacan, 2011; Bertolini & Dijst, 2003). Accessibility is a fundamental prerequisite for the realisation of basic rights and freedoms (WHO, 2007). While physical features such as ramps, curb cuts, elevators, and way-finding aids remain essential, truly inclusive cities must also promote representational and relational accessibility, ensuring that all individuals feel seen, respected, and able to engage fully in public life.

For example, Barcelona exemplifies this approach through tactile paving, audio-visual crossing signals, and low-floor public transport (Bakıcı et al., 2013). Similarly, Tokyo's transit system features elevators, universal signage, and accessible restrooms, benefiting not only those with disabilities but also seniors, caregivers, and families (Imrie & Hall, 2001). These interventions embed accessibility directly into the city's spatial fabric from the outset.

Relational accessibility goes further by fostering inclusive social spaces that support dignity and participation without stigma. This includes culturally sensitive signage, inclusive programming, and multilingual services (Garza, 2015). For example, Melbourne's Federation Square and Toronto's Regent Park have been intentionally designed to host multicultural festivals, Indigenous events, and community programmes that promote social cohesion and reduce symbolic exclusion (Amin, 2002; Canzler & Kaufmann, 2016). Such environments are especially important for migrants, Indigenous populations, and other historically marginalised groups.

In the context of smart cities and e-governance, digital inclusion should be given equal priority (Kuzior et al., 2023). However, the digital divide, the gap between those with access to digital tools and those without, is a complex and multifaceted issue (Heeks, 2022). This reinforces and exacerbates existing social inequalities, such as in education, employment, and healthcare, primarily due to disparities in digital access and literacy (Tewathia et al., 2020). This can be tackled by implementing public Wi-Fi and intuitive user-friendly interfaces, making the Internet accessible to deprived communities, and fostering multilingual e-services (Hollands, 2020; Selwyn, 2004). For example, cities like Tallinn offer integrated e-governance and digital ID platforms, New York's LinkNYC provides free Wi-Fi and device charging in public spaces, and Seoul's Smart City dashboard demonstrates the potential of inclusive digital design for underserved populations. Therefore, all three dimensions of accessibility, physical, social, and digital, should be grounded in the principle of *Universal Design*, also known as "Design for All" (Aslaksen et al., 1997). This approach will integrate usability, simplicity, and perceptibility into metropolitan cities, fostering inclusive, equitable, and adaptive environments that enable all residents to live, connect, and thrive within the urban fabric (Fig. 16.4).

Fig. 16.4 A railway construction site in Victoria, Australia, illustrating the importance of citizen-centric design in shaping transport infrastructure that responds to community needs and environmental priorities

16.3.3 Transparency and Participation: Mechanisms to Engage Citizens Meaningfully

Under democratic urban government, legitimacy depends on citizens' trust that governments act in accordance with the public will (Tyler, 1998), which is sustained through a dynamic interplay of transparency, information access, accountability, and meaningful participation. However, governments often fall short of meeting citizens' expectations, prompting growing demands for transparency, widely regarded as a remedy for "the deficiencies and operations of government where government claims to be democratic but where it falls short of its rhetoric" (Harrison & Sayogo, 2014). These deficiencies highlight the need for meaningful citizen participation, as emphasised in Arnstein's (2007) "A Ladder of Citizen Participation", which distinguishes between tokenistic involvement (e.g., informing or consultation) and real empowerment (e.g., partnership or delegated control). Such participation must be supported by transparency, such as open, timely, and accessible sharing of information, processes, and decisions that enable citizens to understand and shape their urban environment (Narayan et al., 2025).

Many cities are now adopting planning practices that allow citizens to engage directly by scrutinising decisions and participating meaningfully in shaping their urban environments (Meijer et al., 2012). For example, the City of Helsinki maintains an open data portal that provides real-time updates on urban development projects, allowing residents to track decisions and provide informed feedback (Hämäläinen, 2021). Similarly, Barcelona's "Decidim" platform serves as a digital democracy tool that allows residents to propose, debate, and vote on policies with complete transparency (Hämäläinen, 2021).

In parallel, participatory planning tools have also expanded beyond traditional forums to include deliberative assemblies, co-design workshops, and digital participatory platforms. For instance, participatory GIS and open urban data platforms have proven effective in allowing communities to shape zoning changes, infrastructure proposals, and climate resilience planning (Malakar & Roy, 2024; Sieber, 2006). Digital technologies further expand participatory opportunities, particularly among youth, persons with disabilities, and those unable to attend in-person events (Raja, 2016; Rizk & Hillier, 2022). These tools reduce physical, temporal, and social barriers to participation by allowing citizens to engage from their own locations and at their own pace. Given these benefits for the public, transparency must be institutionalised by incorporating open government platforms, public performance dashboards, and real-time feedback loops to build trust and increase the accountability of planning authorities. However, these tools must be designed carefully to avoid power imbalances and ensure that more dominant groups do not drown out marginalised voices.

16.3.4 Flexibility and Adaptability: Design to Accommodate Evolving Citizen Needs

As cities are constantly evolving, the scale and speed of transformation are becoming rapid and unprecedented, driven by advancements in artificial intelligence, climate risks, and social and economic inequalities (Nel et al., 2018). In this context, urban design must shift from static, blueprint-driven models to flexible, adaptive approaches that can respond to emerging challenges and opportunities.

Planners are now focusing on spatial adaptability (Liu et al., 2023), which rethinks how physical environments can accommodate fluctuating demands in use, population, and technology. Examples include retractable street furniture, movable partitions in public buildings, and reprogrammable civic spaces that support events, markets, and recreation (Kelbaugh & McCullough, 2008). Such models are adopted in cities like Rotterdam, where industrial docks are transformed into floodable public plazas (Errigo, 2018), and Copenhagen, where climate-adaptive parks support both urban resilience and community life (Jørgensen et al., 2022).

Flexibility in urban design must also extend beyond spatial adaptability to encompass built environment reuse, ecological resilience, governance adaptability, and social inclusivity. Projects like New York City's *High Line* and Seoul's *Cheonggyecheon Stream* illustrate how adaptive reuse can revive cultural identity, foster community cohesion, and restore ecological functions (Lang & Rothenberg, 2017; Lee & Jung, 2016). Similarly, ecological resilience is supported through nature-based solutions, such as green roofs, urban wetlands, bioswales, and permeable pavements, which mitigate environmental risks while providing biodiversity, cooling, and aesthetic benefits (Lee & Jung, 2016). Singapore's *Bishan-Ang Mo Kio Park* epitomises this integration by converting a drainage canal into a multifunctional park (Beatley, 2016).

On the governance front, adaptive co-management frameworks should be adapted to support flexible planning by incorporating stakeholder feedback and adapting to shifting local needs (Fabricius & Currie, 2015). Smart city initiatives can strengthen this by prioritising citizen-responsive systems over technocratic control and by embedding public input meaningfully into decision-making structures. Finally, social adaptability should also ensure that urban services evolve with societal changes, including flexible schools, mobile health clinics, and inclusive shared spaces that support diverse family structures and cultural values (Sepe, 2025). Embedding such flexibility across spatial, ecological, governance, and social dimensions enables cities to adapt to change while promoting inclusivity, resilience, and human-centred growth.

16.4 Tools and Methods for Engaging Citizens

Citizen engagement in urban design is essential for co-creating inclusive, equitable, and sustainable cities. It involves residents in decision-making processes to ensure that developments reflect community needs, values, and priorities, reducing conflicts and enhancing trust between planners and the public. Effective engagement spans multiple levels, from informing the public to empowering them to make decisions and leveraging both traditional and digital approaches. This guide outlines key methods and tools, drawing from established practices in urban planning (Wu et al., 2025). Methods can be categorised by the level of involvement, from low (informative) to high (decision-making). These draw on participatory planning frameworks and are adaptable across project stages, including preparation, proposal, and approval (Table 16.4).

Engaging citizens in urban planning can generate effective ideas for reinventing cities. Designing easy-to-grasp and effective tools for

Table 16.4 Methods of citizen-centred urban design

Method category	*Description*	*Examples*	*Benefits*
Informative participation	Provides one-way communication to keep residents updated on plans, using accessible language and channels. Ideal for early awareness	Awareness campaigns, local news updates, and administrative websites	Builds transparency and public buy-in; mitigates potential opposition
Preparatory/ consultative participation	Gathers feedback and ideas from residents as collaborators, often in early stages, to inform alternatives	Community workshops, public meetings, home gatherings (Gjestebud in Norway), World Café discussions	Establishes two-way dialogue; reaches diverse demographics via online/offline mix
Decision-making/ collaborative participation	Empowers residents to propose, vote, and influence outcomes, such as resource allocation	Participatory budgeting, citizen assemblies, LOOP Co-Creation Games	Strengthens democracy; leads to actionable, community-owned solutions
Knowledge production methods	Involves residents in data collection and analysis to map assets and needs	Photo Voice (using photos to express concerns), citizen panels, asset-based community development (ABCD)	Uncovers lived experiences; promotes equity in research and planning
Mobilisation methods	Activates community action through incentives and networks	Urban regeneration initiatives, time credits (e.g., Timekred for volunteering), dream banks for collecting improvement ideas	Builds long-term community capacity; fosters ongoing involvement

Source: Compiled by the authors

co-creating meaningful urban spaces remains a significant challenge and an emerging need. Such tools that can intelligently involve the community are in strong demand. It is required to design, develop, and implement well-executed engagement tools that open the horizons of evaluating and responding to urban-related problems while involving city councils, architects, ICT developers, and urban planners. These tools address technology-driven civic engagement with diverse, multifaceted outcomes. The tools also bridge gaps among existing groups of urban planning software: physical planning tools, physical/civic engagement tools, civic scenario-planning tools, and data analysis methods. The tools highlight the potential of such participatory urban planning tools to generate socioeconomic impacts, fostering an inclusive environment for exchanging and implementing urban planning-related ideas and for integrated, coherent citizen engagement (Fig. 16.5).

Fig 16.5 A rooftop garden in Australia exemplifying citizen-centric approaches to sustainable urban living integrating green infrastructure, community participation, and climate resilience

16.4.1 Participatory Workshops and Co-design Sessions

Participatory workshops and co-design sessions are practical mechanisms for translating citizen-centric design into action. These forums bring together residents, planners, architects, and other stakeholders to collaboratively identify challenges, generate ideas, and prototype solutions for urban spaces (Başnou et al., 2020). By fostering dialogue, shared decision-making, and iterative feedback, such sessions empower citizens to influence both the vision and implementation of projects. Beyond enhancing inclusivity, participatory workshops cultivate a sense of ownership and accountability, ensuring that urban interventions are socially relevant, context-sensitive, and responsive to community needs. In this way, co-design practices serve as a bridge between technical expertise and lived experience, making urban development a collaborative and adaptive process (Ye et al., 2023).

16.4.2 Digital Platforms and Smart City Technologies for Feedback and Collaboration

Digital platforms and smart city technologies are increasingly pivotal in enabling citizen-centric urban design by facilitating feedback, collaboration, and real-time engagement (Ugochukwu et al., 2024; Majumdar et al., 2024). Mobile applications, online surveys, interactive maps, and social media channels allow residents to share insights, report issues, and participate in decision-making processes regardless of location or mobility constraints (Fanta & Soběhart, 2021). Similarly, smart sensors, data dashboards, and IoT-enabled infrastructure provide planners with continuous information about urban dynamics, from traffic flow to environmental quality, which can be interpreted alongside citizen input. By integrating these tools, cities can create more responsive, transparent, and adaptive planning systems that complement lived experiences with data-driven insights. This convergence of technology and participation not only streamlines collaboration but also strengthens accountability, ensuring that urban interventions are both efficient and aligned with community priorities (Romanelli, 2020).

16.4.3 Community Mapping and Ethnographic Research Techniques

Community mapping and ethnographic research techniques are essential tools in citizen-centric urban design, providing deep insights into the spatial, social, and cultural dynamics of neighbourhoods (Alhusban et al., 2019). Community mapping involves residents actively documenting local assets, challenges, and usage patterns, revealing lived experiences that are often overlooked in conventional planning data (Ismael et al., 2024). Ethnographic methods, such as interviews, participant observation, and storytelling, allow planners and designers to understand daily practices, social networks, and cultural values that shape how people interact with urban spaces (McVey et al., 2024). Together, these approaches enable a more nuanced and context-sensitive understanding of cities, ensuring that interventions are grounded in the realities and aspirations of the communities they serve. By prioritising local knowledge, community mapping, and ethnography, these practices help create inclusive, responsive, and socially meaningful urban environments (Omole et al., 2024).

16.4.4 Use of Data Analytics and GIS

Data analytics and Geographic Information Systems (GIS) have become powerful tools in citizen-centric urban planning, enabling planners to systematically understand citizen behaviors, movement patterns, and spatial needs (Dammag et al., 2024; Majumdar, 2023). GIS allows for the visualisation and analysis of demographic, environmental, and infrastructure data, revealing patterns of accessibility, congestion, and service gaps across neighbourhoods (Azzam & Robinson, 2013). Combined with data analytics, including mobile location data, social media inputs, and sensor-generated metrics, these technologies provide a detailed and dynamic understanding of how citizens interact with urban spaces. This evidence-based approach complements participatory methods by grounding decision-making in quantifiable insights, helping planners identify priorities, optimise resource allocation, and design interventions that respond to real-world usage patterns. By integrating analytical and spatial tools with citizen feedback, urban design becomes both more precise and more responsive to the community's lived experiences (Majumdar, 2021; Slingerland et al., 2022).

16.5 Case Study

The people-centred approach is transformative because it addresses long-standing failures in urban planning where grand projects often ignore how people actually use spaces. Its significance spans social, economic, environmental, and governance dimensions (Luo et al., 2025). Social equity and inclusion reduce alienation and amplify marginalised voices. Cities designed without input (Brasília's modernist plan) often feel sterile. Citizen-centric projects like Copenhagen's 'people-first' streets prioritise cyclists and pedestrians, boosting livability. In Medellín, Colombia, cable cars and escalators in slums were co-designed with residents, dramatically reducing inequality and violence. People-centred design has better outcomes and innovations. There are higher project success rates, reflecting that user needs are being better met. Singapore's HDB public housing system iteratively incorporates resident feedback, achieving satisfaction rates of 80%+ or more. Again, early citizen input prevents expensive retrofits. The High Line in New York evolved from a community-led vision, avoiding the fate of many failed urban renewals.

This urban design also builds environmental resilience through sustainable behaviour change and behavioural nudges. An example of sustainable behaviour change is Rotterdam's water squares, playgrounds that double as flood storage. They emerged from resident workshops, encouraging community stewardship of climate solutions. Citizen input reveals how people move, leading to designs that reduce car dependency (Barcelona's superblocks). Democratic legitimacy, built on trust and adaptivity, is one of the outcomes of people-centred urban design. In an era of populist backlash against 'elite' planning, participatory processes (such as Porto Alegre's participatory budgeting) empower citizens, reducing NIMBYism and political gridlock. Cities become living labs. Helsinki's "city-as-a-service" model uses citizen data to adjust services such as snow clearance and public transport dynamically. Some real-world examples can be cited from Copenhagen (Denmark), Curitiba (Brazil), and New York (USA) (Table 16.5).

Table 16.5 Real-world examples of citizen-centric models

City/project	*Citizen-centric method*	*Impact*
Copenhagen, Denmark	Bicycle user surveys + participatory budgeting	World's most bike-friendly city; 62% of residents cycle daily
Curitiba, Brazil	Jaime Lerner's "urban acupuncture" with resident input	Bus Rapid Transit (BRT) system co-designed with users; 70% public transport usage
New York, USA	Janette Sadikhan's tactical urbanism (pop-up bike lanes)	Permanent infrastructure after citizen testing; reduced pedestrian deaths

16.6 Challenges and Barriers

People-centred (human-centred, participatory, or community-engaged) design aims to place residents' lived experiences, needs, and aspirations at the core of planning processes. Despite its proven benefits, greater equity, reduced opposition, and more sustainable outcomes, numerous structural, procedural, and social barriers persistently undermine its effective implementation. Below is a systematic analysis of the most critical challenges, supported by evidence from planning scholarship, case studies, and practitioner reports (Table 16.6).

16.7 Strategies for Advancing Inclusive and Sustainable Urban Design

16.7.1 Policy Recommendations

Inclusive and sustainable urban design integrates social equity, environmental resilience, and economic viability into the built environment. It moves beyond participation as an add-on to embed co-creation, systems thinking, and regenerative principles from visioning to maintenance. Below is a strategic framework, organised by scale, actor, and phase, with evidence-based tactics, metrics, and exemplary cases. These strategies can be grouped into policy and governance, process innovations, design principles, technical strategies, financing and economic models, capacity building and cultural shift, monitoring, evaluation, and adaptive management. Strategic foundations are basically policy and governance strategies. These include inclusive urbanism frameworks, permanent co-design units, and legislative community benefit agreements (CBAs) (Table 16.7).

Table 16.6 Challenges and barriers of citizen-centric urban design

Challenge	*Description*	*Evidence & examples*
Institutional and Governance Challenges		
Top-down decision-making cultures	Many municipal agencies and developers operate within hierarchical systems that prioritise technical expertise and regulatory compliance over citizen input	In many US cities, zoning boards retain final authority even after extensive public consultation, rendering participatory outputs advisory only
Lack of legal mandates	Few jurisdictions *require* meaningful participation beyond minimum public-hearing notices, allowing tokenistic engagement	The EU's Aarhus Convention mandates access to information and justice in environmental matters, yet implementation varies widely; in Southern Europe, public hearings are often pro-forma
Siloed departmental structures	Urban design spans transport, housing, parks, and health, but departments rarely coordinate participatory processes	London's "Healthy Streets" initiative initially faced resistance because transport planners did not integrate community health data collected by public-health teams
Short electoral cycles	Politicians prioritise visible, quick-win projects over long-term participatory processes that may span multiple administrations	Participatory budgeting in Porto Alegre, Brazil, weakened after changes in municipal leadership
Resources and Capacity		
Insufficient funding & staffing	Facilitating inclusive workshops, translation services, childcare, or digital platforms requires dedicated budgets that are often cut first	A 2023 survey of 120 European municipalities found 68% cited budget limitations as the primary barrier to participatory planning
Planner skill gaps	Many urban designers lack training in facilitation, conflict mediation, trauma-informed engagement, or digital tool management	Only 12% of UK planning schools include mandatory community-engagement modules
Time intensity	Genuine co-design can add 6–18 months to project timelines, clashing with developer schedules and grant deadlines	The "Grand Paris Express" metro extension faced delays when community co-design for station plazas extended the planning phase by 14 months

(*continued*)

Table 16.6 (continued)

Challenge	*Description*	*Evidence & examples*
Equity and Inclusion Challenges		
Under-representation of marginalised groups	Low-income residents, renters, youth, seniors, non-native speakers, and people with disabilities are systematically excluded due to timing, location, language, or digital access barriers	In Toronto's 2021–2023 tower-renewal consultations, <5% of participants were tenants despite tenants comprising 85% of residents in target buildings
Digital divide	Over-reliance on online platforms excludes those without reliable Internet or digital literacy	During COVID-19 remote consultations in India, only 28% of slum residents could participate via Zoom or Google Forms
Tokenism & "consultation fatigue"	Repeated requests for input without visible outcomes erode trust, particularly in historically marginalised communities.	Detroit's Strategic Neighborhood Fund faced backlash when initial community plans were overridden by developer interests
Power imbalances	Professional jargon, expert-led presentations, and uneven access to data intimidate lay participants	Use of technical 3D models without plain-language summaries reduced meaningful input in Singapore's Kampong Glam redesign
Methodological and Technical Barriers		
Scaling participation	Tools effective for 50 people (e.g., World Café) fail at city-wide scales without robust aggregation mechanisms	Copenhagen's cloudburst adaptation plan struggled to synthesise >12,000 citizen inputs into coherent design directives
Data integration & traceability	Citizen-generated qualitative data (stories, photos, sketches) is hard to integrate with GIS, traffic models, or cost estimates	Maptionnaire surveys in Helsinki collected rich spatial feedback, but only 22% of comments were ultimately geo-referenced for engineering use
Measuring impact	Lack of standardised metrics makes it hard to prove that participation improved outcomes, weakening the case for future investment	The International Association for Public Participation (IAP2) spectrum is widely used but rarely linked to quantifiable urban indicators (reduced crime, increased walkability)

(*continued*)

Table 16.6 (continued)

Challenge	*Description*	*Evidence & examples*
Political and Economic Pressures		
Developer-driven timelines	Private-sector partners often control budgets and demand rapid approvals, sidelining extended engagement	In London's Old Oak Common redevelopment, the developer's 24-month timeline compressed community design charrettes into two weekend events.
NIMBYism vs. YIMBYism	Vocal local opposition can derail inclusive processes, while pro-development groups may dominate digital platforms	San Francisco's SB 827 housing bill faced organised resistance despite broad participatory planning for transit corridors
Gentrification fears	Residents in low-income areas withhold support for fear that "improvements" will raise rents and displace them	Barcelona's Superblock programme initially met resistance in Poblenou until anti-displacement guarantees were co-negotiated

Table 16.7 Strategic foundations: policy & governance

Strategy	*Key actions*	*Metrics & tools*	*Exemplary Case*
Adopt inclusive urbanism frameworks	Mandate 15-Minute City, Doughnut Economics, or Just Sustainabilities in master plans. Require Equity Impact Assessments (EIA) alongside EIAs	Percentage of policies with EIA; Social Progress Index integration	Paris (France): *Plan Local d'Urbanismebioclimatique* (2024) mandates 15-min access + equity scoring
Establish permanent co-design units	Create cross-departmental teams with community liaisons, funded at 2–5% of capital budgets	Number of co-design projects/ year; % budget ring-fenced	Medellín (Colombia): Secretaría de Participación Ciudadana runs 16 local co-design labs
Legislate community benefit agreements (CBAs)	Require developers to co-negotiate anti-displacement, local hiring, and green space provisions	Number of CBAs; % affordable units preserved	Los Angeles (USA): CBA for Staples Center secured $1B in community benefits (2001–2025)

Source: Compiled by the authors

Process innovations carry visions to implementation in three stages: early-stage co-creation; inclusive data and feedback loops; and implementation and stewardship (Table 16.8).

Table 16.8 Process innovations: from vision to implementation

Tactic	*Description*	*Tools*	*Impact*
Early-stage co-creation			
Future search conferences	Two to three-day multi-stakeholder visioning with residents, planners, youth, businesses	Appreciative Inquiry; Scenario Mapping	Freiburg (Germany): Vauban district vision (1990s) led to car-free, solar-powered neighbourhood
Youth-led design jams	Gamified charrettes with Minecraft, Roblox, or physical model kits	Block by Block (UN-Habitat); Tinkercad	Nairobi (Kenya): Youth redesigned Dandora public spaces; 40% cost savings via local materials
Inclusive data & feedback loops			
Community-led GIS (CLGIS)	Residents map assets, hazards, and desires using low-cost apps	Maptionnaire Community; Ushahidi; KoboToolbox	Cape Town (South Africa): Khayelitsha residents mapped 1,200+ informal services → upgraded infrastructure
AI-augmented synthesis	Natural language processing to cluster 10,000+ comments into themes	Remesh; Polis; Thematic	Helsinki (Finland): 30,000 inputs for 2050 plan synthesised in 48 hours with 92% theme accuracy
Traceable feedback loops	"You Said, We Did" dashboards showing how input shaped outcomes	Commonplace; CitizenLab	Ghent (Belgium): 87% trust increase after visible incorporation of 2019 mobility inputs
Implementation & Stewardship			
community land trusts (CLTs)	Resident-controlled land ownership to prevent displacement	Deed restrictions; shared equity models	London (UK): 280+ CLTs preserve 15,000 affordable homes (2024)
Time Banking for maintenance	Residents earn credits for park upkeep, bike repair, etc.	*hOurworld*; *TimeRepublik*	Lisbon (Portugal): 12 parks, maintained 60% by time-bank volunteers

Source: Compiled by the authors

16.7.2 *Integration of Sustainability Metrics in Design and Planning*

Design principles and technical strategies include regenerative infrastructure, universal design 2.0 and modular & adaptive built form (Table 16.9).

16.7.3 *Capacity Building and Education*

Financing and economic models encompass green and social bonds, land value capture (LVC), and crowd-resourcing platforms (Table 16.10).

Table 16.9 Design principles & technical strategies

Principle	*Tactics*	*Metrics*	*Example*
Regenerative infrastructure	Green-blue grids, urban forests, permeable surfaces, circular material flows	m^2 biodiversity habitat/city; % stormwater managed on-site	Singapore: ABC Waters 1,200+ bioswales treat 90% runoff
Universal design 2.0	Beyond accessibility: sensory calm zones, gender-safe lighting, intergenerational play	% public spaces with UD certification.	Vienna (Austria): Gender Mainstreaming since 1999 → 70% parks with women-friendly lighting
Modular & adaptive built form	Prefab, demountable structures; zoning for mixed-use conversion	% buildings designed for three+ future uses.	Amsterdam (Netherlands): De Ceuvel circular office park on repurposed houseboats

Source: Compiled by the authors

Table 16.10 Financing & economic models

Model	*Mechanism*	*Example*
Green & social bonds	Municipal bonds tied to inclusive KPIs (e.g., % low-income access to transit)	Gothenburg (Sweden): 2019 Green Bond funded 40 km cycling lanes + 1,200 social housing units
Land value capture (LVC)	Tax uplift from transit/plazas funds for community facilities	Hong Kong: MTR funds 60% of rail via LVC; surpluses subsidise fares
Crowd-resourcing platforms	Micro-donations + matched public funds for hyper-local projects	Spacehive (UK): £18M raised for 1,500+ civic projects (2024)

Source: Compiled by the authors

Table 16.11 Capacity building & cultural shift

Action	*Format*	*Outcome*
Planner certification in co-design	IAP2 + trauma-informed training; mandatory CE credits	Toronto: 2023 cohort reduced consultation fatigue by 40%
Community design academies	Free 6–12 week courses for residents on planning basics	Portland (USA): 1,200+ graduates; 25% now serve on advisory boards
Public design literacy campaigns	Interactive exhibits, AR apps, and school curricula	Seoul (South Korea): "Design Seoul 2030" exhibit reached 2.1 M visitors

Source: Compiled by the authors

Table 16.12 Monitoring, evaluation & adaptive management

Framework	*Indicators*	*Tools*
Inclusive sustainability index (ISI)	40 metrics: walkability, air quality, Gini coefficient, % women feeling safe at night	ArcGIS Dashboards; UN-Habitat CPI
Regenerative outcomes star	Six-point evaluation: ecological restoration, social cohesion, and economic circulation	Community self-scoring workshops
Annual "State of the City" co-reports	Residents + officials co-author progress reports	Bristol (UK): One City Plan co-reports since 2020

Source: Compiled by the authors

Another sustainable strategy is capacity building and cultural shift. These actions include planner certification in co-design, community design academies and public design literacy campaigns (Table 16.11).

At the end, monitoring, evaluation and adaptive management must be robust. This takes into account the inclusive sustainability index (ISI), the regenerative outcomes star and annual "state of the city" co-reports (Table 16.12).

16.8 Future Directions

Citizen-centric urban design is evolving from consultative participation towards distributed co-governance, co-creation, and planetary-scale stewardship. Powered by exponential technologies, climate urgency, and post-pandemic social contracts, the field is shifting from designing for people to designing with people as proto-city operators. Below is a forward-looking synthesis across six transformative vectors, each with emerging prototypes, enabling conditions, and 2030–2050 projections. Advancing inclusive and sustainable urban design requires systemic alignment, policy mandates, process innovation, regenerative techniques, diverse financing, and cultural transformation. Cities like Medellín, Vienna, and Singapore prove that when equity and ecology are non-negotiable design inputs, outcomes include lower carbon footprints, reduced inequality, and higher life satisfaction. The toolkit exists; success hinges on political courage, resource commitment, and relentless feedback loops that keep people at the centre. From participation to distributed urban governance focuses on neighbourhood decentralised autonomous organisations (DAOs), urban commons, and citizen assemblies; AI-augmented co-creation ecosystems (generative urbanism, digital twins as public goods, sentiment-to-space large language models—LLMs); biophilic and regenerative urbanism at scale (urban symbiosis platforms—citizens "farm" biodiversity credits via balcony forests, green roofs, mycelium bricks; bio-digital interfaces—sensors in trees/plants feed dashboards; citizens "adopt" organisms via apps); education and cultural shift: from citizen to urban systems steward (urban design in every curriculum, lifelong urban stewards programme, narrative co-design). By 2050, it is expected that the average urban resident will spend three–five hours/week in active city stewardship, voting in virtual reality assemblies, tuning digital twins, earning biodiversity credits, and co-authoring urban narratives. The planner's role evolves from gatekeeper to orchestrator of human-AI-nature systems. Citizen-centric urban design is no longer a method; it is the operating system of the resilient, equitable, and regenerative city. Some milestones are anticipated (Table 16.13).

Table 16.13 Future directions for citizen-centric cities

Year	*Milestone*
2026	Global Charter for Citizen-Centric Cities (UN-Habitat): Mandates digital twins + DAO pilots
2028	AI Co-Design Act: EU requires generative urbanism in all >€50 M projects
2032	1,000 Cities run sortition + VR assemblies; 50% budget via participatory mechanisms
2040	Net-Zero + Equity Cities: Twin goals; failure = loss of international climate finance
2050	Planetary Urban Commons: Cities as nodes in an earth-wide regenerative system; citizens govern via Earth Assembly

16.9 Conclusion

Contrasting with traditional top-down planning, which often focuses on infrastructure efficiency, economic growth, or expert-led visions, citizen-centric design flips the script by placing people at the centre of the planning process; using participatory methods such as co-design workshops, digital surveys, community forums, and data from citizen feedback platforms, and iteratively testing and refining urban interventions based on real-world user feedback. It builds on people-centred design principles and adapts them to the scale of cities. The key elements include empathy, inclusivity, iterative prototyping, and data-driven feedback. Citizen-centric urban design is a culmination of over a century of critique against top-down, expert-driven planning and parallel movements in participatory democracy, social science, and design thinking. Citizen-centric urban planning design prioritises the needs, preferences, behaviours, and lived experiences of residents as the primary drivers of urban development decisions. Two of the most relevant sustainability frameworks for urban development are the circular economy and green urbanism. This marks a shift from top-down planning to more inclusive and participatory methods that acknowledge the diverse and changing needs of urban residents. This approach envisions cities not merely as physical landscapes but as dynamic human systems responsive to diversity, resilience, and justice. Citizen engagement in urban design is essential for co-creating inclusive, equitable, and sustainable cities, involving residents in decision-making processes to ensure that developments reflect community needs, values, and priorities, reducing conflicts and enhancing trust between planners and the public. Effective engagement spans multiple levels, from informing the public to empowering them in decision-making and leveraging both traditional and digital approaches. This guide outlines key methods and tools, drawing from established practices in urban planning. Methods can be categorised by the level of involvement, from low (informative) to high (decision-making). These draw on participatory planning frameworks and are adaptable across project stages, including preparation, proposal, and approval. People-centred urban design faces critical challenges, as evidenced by planning scholarship. These challenges can be overcome through inclusive, sustainable urban design that integrates social equity, environmental resilience, and economic viability into the built environment. It moves beyond participation as an add-on to embed co-creation, systems thinking, and regenerative principles from visioning to maintenance. By 2050, it is expected that the average urbanite will be more active in city stewardship, voting in virtual reality assemblies, tuning digital twins, earning biodiversity credits, and co-authoring urban narratives.

References

Afacan, Y., & Afacan, S. O. (2011). Rethinking social inclusivity: Design strategies for cities. *Proceedings of the Institution of Civil Engineers - Urban Design and Planning, 164*(2), 93–105. https://doi.org/10.1680/udap.2011.164.2.93

Agarwal, A. (2022). *Planning strategies for enhancing gender inclusivity in a City: A case of South Delhi*. Indian Institute of Technology Kharagpur, 10.

Alhusban, S., Alhusban, A., & Al-Betawi, Y. (2019). Suggesting theoretical urban neighborhood design concept by adopting the changing discourse of social capital. *Journal of Enterprising Communities People and Places in the Global Economy, 13*(3), 391–411. https://doi.org/10.1108/jec-09-2018-0064

Amin, A. (2002). Ethnicity and the multicultural city: Living with diversity. *Environment and Planning A, 34*(6), 959–980. https://doi.org/10.1068/a3537

Angotti, T. (2011). *New York for sale: Community planning confronts global real estate*. MIT Press.

Arnstein, S. R. (2007). A ladder of citizen participation. *Journal of the American Institute of Planners, 35*(4), 216–224. https://doi.org/10.1080/01944366908977225

Aslaksen, F., Bergh, S., Bringa, O. R., & Heggem, E. K. (1997). *Universal design: Planning and design for all.*

Azzam, T., & Robinson, D. (2013). Gis in evaluation. *American Journal of Evaluation, 34*(2), 207–224. https://doi.org/10.1177/1098214012461710

Bakıcı, T., Almirall, E., & Wareham, J. (2013). A smart city initiative: The case of Barcelona. *Journal of the Knowledge Economy, 4*, 135–148. https://doi.org/10.1007/s13132-012-0084-9

Başnou, C., Pino, J., Davies, C., Winkel, G., & Vreese, R. (2020). Co-design processes to address nature-based solutions and ecosystem services demands: The long and winding road towards inclusive urban planning. *Frontiers in Sustainable Cities, 2*. https://doi.org/10.3389/frsc.2020.572556

Beatley, T. (2016). Singapore City, Singapore: City in a garden. In *Handbook of biophilic city planning and design* (pp. 51–66). Springer. https://doi.org/10.5822/978-1-61091-621-9

Bertolini, L., & Dijst, M. (2003). Mobility environments and network cities. *Journal of Urban Design, 8*(1), 27–43.

Buffel, T., Phillipson, C., & Scharf, T. (2012). Ageing in urban environments: Developing 'age-friendly' cities. *Critical Social Policy, 32*(4), 597–617. https://doi.org/10.1177/0261018311430457

Canzler, W., & Kaufmann, V. (2016). *Tracing mobilities: Towards a cosmopolitan perspective*. Routledge.

Chance, S., & Trbovc, J. M. (2021). Integrating gender and inclusivity into research planning. In *Proceedings-SEFI 49th Annual Conference: Blended Learning in Engineering Education: Challenging, Enlightening, Lasting?*

Choudhury, M., Dixit, G., & Majumdar, S. (Eds.). (2025). *Sustainable synergy: Harnessing ecosystems for climate resilience*. Springer Nature Switzerland. https://doi.org/10.1007/978-3-031-77957-2

Crump, J., Jacob, K., King, P., Mangalagin, D., Zikgraf, C., Abiodun, B.J., Armiento, G., Bailey, R., Bowen, J., Dankelman, I., Djalante, R., Dutta, M., Hurley, F., Iraola, M. J., Kim, R. E., King, R., Kirilenko, A., dos Santos Lucon, O., Lyne, K., et al. (2019). *Systemic policy approaches for cross-cutting issues*. Global Environment Outlook, GEO 6: Healthy Planet, Healthy People. Cambridge University Press (pp. 425–447).

Dammag, A., Dai, J., & Dammag, B. (2024). Optimization strategies of sustainable urban land use planning using a hybridized technique based on swot-ahp, and gis in Ibb city, Yemen. https://doi.org/10.21203/rs.3.rs-3819775/v1

Dixit, G., Majumdar, S., & Choudhury, M. (2025). Introduction to ecosystem services for climate resilience: A theoretical framework. In M. Choudhury, G. Dixit, & S. Majumdar (Eds.), *Sustainable synergy: Harnessing ecosystems for climate resilience. Climate change management*. Springer. https://doi.org/10.1007/978-3-031-77957-2_1

Errigo, M. F. (2018). The Adapting City. Resilience through water design in Rotterdam. *TeMA-Journal of Land Use, Mobility and Environment, 11*(1), 51–64.

Fabricius, C., & Currie, B. (2015). Adaptive co-management. Adaptive management of social-ecological systems, 147–179, https://doi.org/10.1007/978-94-017-9682-8

Fainstein, S. S. (2014). The just city. *International Journal of Urban Sciences, 18*(1), 1–18. https://doi.org/10.1080/12265934.2013.834643

Fanta, M., & Soběhart, R. (2021). E-participatory and communication mobile applications: Case study of Prague, 139–147. https://doi.org/10.31410/eman.2021.139

Fincher, R., & Iveson, K. (2017). *Planning and diversity in the city: Redistribution, recognition and encounter*. Bloomsbury Publishing.

Foth, M. (2018). Participatory urban informatics: Towards citizen-ability. *Smart and Sustainable Built Environment, 7*(1), 4–19. https://doi.org/10.1108/sasbe-10-2017-0051

Garza, J. (2015). The role of language in the use of public spaces in Montréal.

Gehl, J. (2013). *Cities for people*. Island Press.

Goncalves, J. (2025). *Exploring the intersections of citizen engagement, urban planning and open science in climate action: Lessons from the TU Delft*. Citizen voice initiative. Department of Urbanism, Faculty of Architecture and the built environment, TU Delft, 1–7.

Hämäläinen, M. (2021). Urban development with dynamic digital twins in Helsinki city. *IET Smart Cities, 3*(4), 201–210. https://doi.org/10.1049/smc2.12015

Harrison, T. M., & Sayogo, D. S. (2014). Transparency, participation, and accountability practices in open government: A comparative study. *Government Information Quarterly, 31*(4), 513–525.

Healey, P. (2020). *Collaborative planning: Shaping places in fragmented societies*. Bloomsbury Publishing.

Heeks, R. (2022). Digital inequality beyond the digital divide: Conceptualising adverse digital incorporation in the global south. *Information Technology for Development, 28*(4), 688–704. https://doi.org/10.1080/02681102.2022.2068492

Hollands, R. G. (2020). Will the real smart city please stand up?: Intelligent, progressive or entrepreneurial? In *The Routledge companion to smart cities* (pp. 179–199). Routledge.

Imrie, R., & Hall, P. (2001). *Inclusive design: Designing and developing accessible environments*. Taylor & Francis.

Ismael, D., Hutton, N., Erten-Unal, M., Considine, C., Vandecar-Burdin, T., Davis, C., Chen, Y., et al. (2024). Community-centric approaches to coastal hazard assessment and management in Southside Norfolk, Virginia, USA. *Atmosphere, 15*(3), 372. https://doi.org/10.3390/atmos15030372

Itair, M., Shahrour, I., & Hijazi, I. (2023). The use of the smart technology for creating an inclusive urban public space. *Smart Cities, 6*(5), 2484–2498. https://doi.org/10.3390/smartcities6050112

Ives, C., & Baker, C. (2024). Engaging faith for a sustainable urban future. *Global Sustainability, 7*. https://doi.org/10.1017/sus.2024.32

Jakonen, O. (2023). People-centric and inclusive approach to planning of smart cities. *CSID Journal of Infrastructure Development, 6*(1), 9. https://doi.org/10.7454/jid.v6.i1.1087

Jørgensen, G., Fryd, O., Lund, A. A., Andersen, P. S., & Herslund, L. (2022). Nature-based climate adaptation projects, their governance and transitional potential-cases from Copenhagen. *Frontiers in Sustainable Cities, 4*, 906960. https://doi.org/10.3389/frsc.2022.906960

Kelbaugh, D., & McCullough, K. (2008). *Writing urbanism: A design reader*. Routledge.

Kowsalya, P., Jennifer Flora, G., Kousalya, A., Toha, M. A., Khan, P. A., Choudhury, M., Parthasarathi, T., et al. (2025). Circular systems unveiled: Essential components and their impact on recycling, reuse, and resource recovery. In

S. Majumdar, M. Choudhury, & A. Cheshmehzangi (Eds.), *Sustainable urban future. Urban sustainability* (pp. 55–71). Springer. https://doi.org/10.1007/978-981-95-2976-6_3

Kumaraswamy, T. R., Choudhury, M., Ponappa, S., & Majumdar, S. (2025). A significance of climate resilience: Strategies for local, regional, and national levels. In M. Choudhury, G. Dixit, & S. Majumdar (Eds.), *Sustainable synergy: Harnessing ecosystems for climate resilience. Climate change management*. Springer. https://doi.org/10.1007/978-3-031-77957-2_13

Kuzior, A., Pakhnenko, O., Tiutiunyk, I., & Lyeonov, S. (2023). E-governance in smart cities: Global trends and key enablers. *Smart Cities, 6*(4), 1663–1689. https://doi.org/10.3390/smartcities6040078

Lang, S., & Rothenberg, J. (2017). Neoliberal urbanism, public space, and the greening of the growth machine: New York City's high line park. *Environment and Planning A: Economy and Space, 49*(8), 1743–1761. https://doi.org/10.1177/0308518X16677969

Lee, M., & Jung, I. (2016). Assessment of an urban stream restoration project by cost-benefit analysis: The case of Cheonggyecheon stream in Seoul, South Korea. *KSCE Journal of Civil Engineering, 20*, 152–162. https://doi.org/10.1007/s12205-015-0633-4

Liu, Z., Fang, C., Liao, X., Fan, R., Sun, B., & Mu, X. (2023). Adaptation and adaptability: Deciphering urban resilience from the evolutionary perspective. *Environmental Impact Assessment Review, 103*, 107266. https://doi.org/10.1016/j.eiar.2023.107266

Luo, J., Liu, P., Kong, X., Shen, J., Wu, Q., & Xu, D. (2025). Urban digital twins for citizen-centric planning: A systematic review of built environment perception and public participation. *International Journal of Applied Earth Observation and Geoinformation, 143*(104746), 1–13. https://doi.org/10.1016/j.jag.2025.104746

Lux, M. S. (2025). Including citizens' perspective in advancing urban green infrastructure: A design-toolkit for private open spaces. *Sustainability, 17*, 6781. https://doi.org/10.3390/su17156781

Majumdar, S. (2019). Crisis for open space and loss of urban heritage around Kolkata city. *Research Journal of Humanities and Social Sciences, 10*(4), 969–972. https://doi.org/10.5958/2321-5828.2019.00158.X

Majumdar, S. (2020). *3. Sustainability in urban expansion of a metropolitan city: Impacts of urban growth towards the outer fringe of Kolkata* (pp. 59–72). https://doi.org/10.1515/9783110628593-003

Majumdar, S. (2021). *A Study of Dengue Disease Data by GIS in Kolkata City: An Approach to Healthcare Informatics.* In: Suresh, A., Paiva, S. (eds) Deep Learning and Edge Computing Solutions for High Performance Computing. EAI/Springer Innovations in Communication and Computing. Springer, Cham. https://doi.org/10.1007/978-3-030-60265-9_3

Majumdar, S. (2023). Uses of artificial intelligence and GIS technologies in healthcare services. In R. S. Tomar, R. K. Chauhan, & S. K. Nayak (Eds.), *Artificial intelligence and data science in contemporary healthcare*. CRC Press. https://doi.org/10.1201/9781003363606-18

Majumdar, S., Kandpal, V., & Anthopoulos, L. G. (2024). *Smart cities: Innovations, challenges and future perspectives*. Springer Nature Switzerland AG. https://doi.org/10.1007/978-3-031-59846-3

Majumdar, S., Choudhury, M., & Cheshmehzangi, A. (Eds.). (2025). *Sustainable urban future: Embracing circular economy in waste management*. Springer Singapore. https://doi.org/10.1007/978-981-95-2976-6

Malakar, K. D., & Roy, S. (2024). Community cartography and participatory GIS. In *Mapping geospatial citizenship: The power of participatory GIS* (pp. 61–72). Springer.

Malhotra, C., Anand, R., & Soni, V. (2020). Creating public services 4.0: Sustainable digital architecture for public services in India. *Indian Journal of Public Administration, 66*(3), 327–342. https://doi.org/10.1177/0019556120957421

Marcuse, P., & Madden, D. (2024). *In defence of housing: The politics of crisis*. Verso Books.

McGinley, C., Myerson, J., Briscoe, G., & Carroll, S. (2022). Towards an age-friendly design lens. *Journal of Population Ageing, 15*(2), 541–556. https://doi.org/10.1007/s12062-022-09367-5

McVey, L., Alvarado, N., Zaman, H., Healey, F., Todd, C., Issa, B., Randell, R., et al. (2024). Interactions that support older inpatients with cognitive impairments to engage with falls prevention in hospitals: An ethnographic study. *Journal of Clinical Nursing, 33*(5), 1884–1895. https://doi.org/10.1111/jocn.17006

Meijer, A. J., Curtin, D., & Hillebrandt, M. (2012). Open government: Connecting vision and voice. *International Review of Administrative Sciences, 78*(1), 10–29. https://doi.org/10.1177/0020852311429533

Müller, H., Rehn-Groenendijk, J., & Wasmer, A. (2023). Small-scale urban design interventions: A framework for deploying cities as resource for mental health and mental health literacy. *Frontiers in Psychology, 14*. https://doi.org/10.3389/fpsyg.2023.1112209

Narayan, S., Yadav, D., Jha, S., Khatun, B., Choudhary, R., Verma, U. K. Choudhury, M., & Kumar, D. (2025). From past to present: The evolution of waste management and the rise of circular economy principles. In *Sustainable urban future: Embracing circular economy in waste management* (pp. 23–53). Springer Nature Singapore. https://doi.org/10.1007/978-981-95-2976-6_2

Nel, D., du Plessis, C., & Landman, K. (2018). Planning for dynamic cities: Introducing a framework to understand urban change from a complex adaptive systems approach. *International Planning Studies, 23*(3), 250–263. https://doi.org/10.1080/13563475.2018.1439370

Nyongesa, W. J., Ondoro, C., & Ntonga, S. (2025). Does performance contracting drive citizen-centric service delivery: The case of Huduma Centres in Kenya? *The International Journal of Business & Management, 8*(10), 217–223.

Omole, F., Olajiga, O., & Olatunde, T. (2024). Sustainable urban design: A review of eco-friendly building practices and community impact. *Engineering Science & Technology Journal, 5*(3), 1020–1030. https://doi.org/10.51594/estj.v5i3.955

Perillo, M. (2025). Smart-inclusive Urbanery. Salute e integrazione digitale nella progettazioneb Urbana in ottica Lifecourse Smart-inclusive Urbanery/Health and Digital Integration in the Urban Design from a Lifecourse Perspective. *Strade Per La Gente/Streets For People*, 151–159.

Pešta, J., Vlasatá, B., Trubina, N., & Pavlů T., (2023). *Circular building designs. Strategies and tools. CirCon4Climate project.* European Climate Initiative (EUKI) of the German Federal Ministry for Economic Affairs and Climate Action (BMWK), 36.

Pichler-Milanovič, N., & Foški, M. (2015). Green infrastructure and urban resilience in central Europe: A solution for environmental and spatial challenges in the inner-city of Ljubljana, Slovenia. *Urbani Izziv, 26*(supplement). https://doi.org/10.5379/urbani-izziv-en-2015-26-supplement-004

Preston, S., Mazhar, M., & Bull, R. (2020). Citizen engagement for co-creating low carbon smart cities: Practical lessons from Nottingham city council in the UK. *Energies, 13*(24), 6615. https://doi.org/10.3390/en13246615

Raja, D. S. (2016). Bridging the disability divide through digital technologies. *Background paper for the World Development Report*, 1–35.

Ravi, K., & Brück, A. (2025). Citizen centred climate intelligence: Operationalising open tree data for urban cooling and eco-routing in Indian cities. *Citizen Climate Intelligence in Indian Cities*, 1–20. https://doi.org/10.48550/arXiv.2508.17648

Rizk, J., & Hillier, C. (2022). Digital technology and increasing engagement among students with disabilities: Interaction rituals and digital capital. *Computers and Education, 3*, 100099. https://doi.org/10.1016/j.caeo.2022.100099

Romanelli, M. (2020). Analysing the role of information technology towards sustainable cities living. *Kybernetes, 49*(7), 2037–2052. https://doi.org/10.1108/k-07-2019-0516

Sandercock, L., & Lyssiotis, P. (2003). *Cosmopolis II: Mongrel cities of the 21st century.* A&C Black.

Selwyn, N. (2004). Reconsidering political and popular understandings of the digital divide. *New Media & Society, 6*(3), 341–362. https://doi.org/10.1177/1461444804042519

Sepe, M. (2025). Contemporary approaches to healthy and livable public spaces: Proximity, flexibility, and diversification. *Urban Design International*, 1–15.

Sieber, R. (2006). Public participation geographic information systems: A literature review and framework. *Annals of the Association of American Geographers, 96*(3), 491–507. https://doi.org/10.1111/j.1467-8306.2006.00702.x

Slingerland, G., Murray, M., Lukosch, S., McCarthy, J., & Brazier, F. (2022). Participatory design going digital: Challenges and opportunities for distributed place-making. *Computer Supported Cooperative Work (CSCW), 31*(4), 669–700. https://doi.org/10.1007/s10606-022-09438-3

Soja, E. W. (2013). *Seeking spatial justice* (Vol. 16). U of Minnesota Press.

Steinfeld, E., & Maisel, J. (2012). *Universal design: Creating inclusive environments.* Wiley.

Tewathia, N., Kamath, A., & Ilavarasan, P. V. (2020). Social inequalities, fundamental inequities, and the recurrence of the digital divide: Insights from India. *Technology in Society, 61*, 101251. https://doi.org/10.1016/j.techsoc.2020.101251

Tyler, T. R. (1998). *Trust and democratic governance.* Trust and Governance/Russell Sage Foundation.

Ugochukwu, C., Ofodile, O., Okoye, C., & Akinrinola, O. (2024). Sustainable smart cities: The role of fintech in promoting environmental sustainability. *Engineering Science & Technology Journal, 5*(3), 821–835. https://doi.org/10.51594/estj.v5i3.906

WHO. (2007). *International classification of functioning, disability, and health: Children & Youth Version: ICF-CY.* World Health Organisation.

Wu, D., Liao, W., Yu, H., Majumdar, S., Xu, J., & Choudhury, M. (2025). Assessing traditional waste management practices in urban environments and recent trends. In S. Majumdar, M. Choudhury, & A. Cheshmehzangi (Eds.), *Sustainable urban future: Embracing circular economy in waste management.* Springer Nature. https://doi.org/10.1007/978-981-95-2976-6_5

Ye, X., Newman, G., Lee, C., Zandt, S., & Jourdan, D. (2023). Toward urban artificial intelligence for developing justice-oriented smart cities. *Journal of Planning Education and Research, 43*(1), 6–7. https://doi.org/10.1177/0739456x231154002

Yuan, Z., Bi, J., & Moriguichi, Y. (2006). The circular economy: A new development strategy in China. *Journal of Industrial Ecology, 10*(102), 4–8. https://doi.org/10.1162/108819806775545321

Correction to: City of the Future

Sushobhan Majumdar and Moharana Choudhury

Correction to:
S. Majumdar, M. Choudhury (eds.), *City of the Future*, https://doi.org/10.1007/978-3-032-18026-1

The original versions of Chapters 6 and 12 were inadvertently published with an incorrect spelling of the author's name. This has been corrected in the revised version. The author's name has been updated to **Sushobhan Majumdar** from **Sushobban Majumdar**.

The updated versions of these chapters can be found at
https://doi.org/10.1007/978-3-032-18026-1_6
https://doi.org/10.1007/978-3-032-18026-1_12

S. Majumdar, M. Choudhury (eds.), *City of the Future*,
https://doi.org/10.1007/978-3-032-18026-1_17